W0042124

Dyeing of Textile Substrates – II Fibres Other than Cotton and Blends

Dyeing of Textile Substrates – II
Fibres Other than Cotton and Blends

Mathews Kolanjikombil
Senior Textile Processing Technologist

Published by Woodhead Publishing India Pvt. Ltd.
Woodhead Publishing India Pvt. Ltd.,
303, Vardaan House, 7/28, Ansari Road,
Daryaganj, New Delhi - 110002, India
www.woodheadpublishingindia.com

First published 2020, Woodhead Publishing India Pvt. Ltd.
© Woodhead Publishing India Pvt. Ltd., 2020

This book contains information obtained from authentic and highly regarded sources. Reprinted material is quoted with permission. Reasonable efforts have been made to publish reliable data and information, but the authors and the publishers cannot assume responsibility for the validity of all materials. Neither the authors nor the publishers, nor anyone else associated with this publication, shall be liable for any loss, damage or liability directly or indirectly caused or alleged to be caused by this book.

Neither this book nor any part may be reproduced or transmitted in any form or by any means, electronic or mechanical, including photocopying, microfilming and recording, or by any information storage or retrieval system, without ermission in writing from Woodhead Publishing India Pvt. Ltd. The consent of Woodhead Publishing India Pvt. Ltd. does not extend to copying for general distribution, for promotion, for creating new works, or for resale. Specific permission must be obtained in writing from Woodhead Publishing India Pvt. Ltd. for such copying.

Trademark notice: Product or corporate names may be trademarks or registered trademarks, and are used only for identification and explanation, without intent to infringe.

Woodhead Publishing India Pvt. Ltd. ISBN: 978-93-88320-25-2
Woodhead Publishing India Pvt. Ltd. e-ISBN: 978-93-88320-26-9

Digitally Printed and bound by Replika Press Pvt. Ltd.

This Book is dedicated to all my colleagues who has been a part of my career without whom I could not have done what I have achieved in my career.

Preface

This is the fourth book in the 'Hand Book for Textile Processors' and Part II in Dyeing of Textile Substrates. This book includes the dyeing of all major fibres except Cotton – Natural, Regenerated, Synthetic and Blends.

As I have done in the previous book, I have tried to give the detailed dyeing procedures, with recipes, dyeing profile and precautions to achieve best results, etc., which processor can easily follow in his day to day work. Even though processes and procedures are given are practically proven but I will advice that one may have to make slight changes as per the conditions and type of machine, process water, etc., when adopting the recipe for your purpose. Some processes which are obsolete or not practiced widely by the processors now are given for academic interests and to get to understand how the new processes were evolved considering the economy and faster turnover which is the order of the day. When one decides to follow a particular process, the processor has to understand the machines on hand, the size of the dye lot, availability of chemicals, type of water, dyes, auxiliaries, etc. In most of the processes, I have tried to give the advantages and disadvantage which will enable you to select the process suitable for you. As far as dyeing of blends, I have tried to cover the dyeing methods for most common blends, even though one may come across other blends which are not given in the book. It may not be difficult for the dyer to design his own processes referring the dyeing methods of the individual fibre of the blend with a proper understanding of the processes.

I have not come across many books where the dyeing processes including preparation details of each blends are given for most of the blends available in the market. The blends dyeing in most cases are given in short, but for a better understanding the dyeing of individual fibres in this book (Volume II) and Volume I (Cotton) of the Dyeing of Textile Substrates can be referred.

Hope this book will be a handy reference book for the processor and also for textile students and will be accepted by the textile processing related personnel. The book is mainly based on my experience in the textile processing field.

Bangalore

Mathews Kolanjikombil

CONTENTS

Part III Dyeing of Synthetic fibres

Part I

Dyeing of Protein fibres

Chapter 1
Dyeing of wool

Wool is a general term applied to the various kinds of animal hairs which are capable of being spun into yarn. By far, the most important variety being the hair from sheep. Other than sheep wool there are other animal hairs for example, Mohair or Angora wool (comes from the Angora goat which is indigenous to Asia Minor), alpaca or paco (the hair of the alpaca, a native of the Chilean and Peruvian Andes),hairs of other animals such as the camel, horse and cow and the fur of hares and rabbits, etc. also comes in this category. The quality of wool or hair varies widely, not only according to the species of the animal but also according to the part of the animal from which it is taken. The length of the sheep wool can vary from 1"to 15" and diameter from 16 to 80 microns. The wool fibre is a horny cellular structure which is normally made up of three main parts i.e., cuticle, the cortical layer and the medulla. Chemically wool belongs to the group of proteins called Keratins. The wool Keratin consists of sulphur and a series of amino acids.

1.1   Preparation of wool
1.1.1   Processing of loose wool
1.1.2   Scouring of loose wool

Raw wool contains a great number of impurities which have to be removed before further processing. These consist of wool grease or wool fat and other external matters. Fleece wool may contain:

15–72%	Wool fibres
12–47%	Fat and suint (2–4% are water soluble potassium salts)
3–24%	Vegetable matter (burrs, straw, etc.) and dirt (excrement, sand and dried earth)
2–24%	Moisture

These impurities percentage may depend on the breed of sheep, the area in which the sheep are raised and husbandry methods. Wool grease is secreted from the sebaceous glands and is a greasy, often sticky substance, often yellow to brown in colour and is produced by the sheep to impart smoothness and flexibility to the wool. It is not a fat but a wax and it also protects the wool from external influences, it consists of about:

50–55% Fatty acid ester

40–45% Unsaponified matter

1–2% Free fatty acids

The wool fibre can contain following matter externally:

Suint or perspiration: Consists chiefly of potassium salts of various organic and inorganic acids.

Mineral matter: Mainly sand.

Vegetable matter: Such as burs, parts of twigs, etc. which are removed by carbonising.

Fecal matter and other substances: Such as sheep dips, ointments and brand marks.

These impurities which are external and internal have to be removed to obtain spinnable wool by various processes such as wool scouring, washing, carbonising, etc. The most important among them are wool scouring. The wool scouring should be an environmentally feasible process which should be effective in cleaning the contaminants from the wool, ensuring that the wool is in a physical and chemical condition for the further manufacturing processes like top making, spinning, weaving or knitting and to minimise entanglement, and retain the staple structure.

1.1.2.1   Scouring methods

The chief methods of scouring raw wool are:

1. Alkaline scouring.
2. Acid scouring.
3. Neutral scouring.
4. Extraction of wool grease with organic solvents.

Processes 1–3 are most commonly used.

1.1.2.2   Alkaline scouring (pH 8–10)

Anionic or nonionic detergent (with high emulsifying power, good capacity for dispersing and suspending oil, excellent protective colloid action, excellent resistance to salt, lime soap forming agents, low substantivity, no corrosive action, low foaming, etc. It also should permit straight forward recovery of wool grease if necessary–Lanoline is obtained from grease) and soda ash are used in alkaline scouring. A six bowl scouring range is ideal.

Scouring recipes are constructed with regard not only to the washing process used but also to the origin and fineness of the wool and thus to the composition of the wool suint. They also depend on the general degree of soiling so that, in the following recipes, a certain amount of leeway has been given as regards the amounts of the products to be added. The same applies to the amount of soda to be used and to the pH value of the liquors which naturally depends on the pH value of the water.

Recipes:

Bowl no.	Temperature	pH	Initial charge (g/l)	Feed rate (g/l/h)
1	55		Water	
2	50	9.5–10	1–3 Soda ash	0.5–1 Soda ash
3	45	9.5	0.5–2 Soda ash 0.2–5 Nonionicdetergent	Soda ash 0.05 Nonionic detergent
4	45	9	0.3 Nonionic detergent	0.05 Nonionic detergent
5	40		Water	0.02 Nonionic detergent
6	40		Rinsing water	

Alternate recipe

No.	Bath	Wetting agent(g/l)	Soda ash(g/l)	pH	Temp (°C)	Feeding rates kg/100 kg wool W. A.	Soda ash
1	Steeping bath		1.0–1.5	9–10	20–25		0.4–0.8
2	1st washing bath	0.4–0.6	0.8–1.2	8.5–9.5	55	0.2–0.4	0.3–0.6
3	2nd washing bath	0.2–0.5	0.4–0.8	8–9	50	0.2–0.3	0.3–0.5
4	3rd washing bath	0.2–0.4	0.3–0.5	8–9	50	0.2–0.3	0.2–0.4
5	Rinsing bath			Neutral	30–35		

1.1.2.3 Acid scouring (pH 4–6)

Acid scouring is done at a pH between 4 and 6,i.e., isoelectric zone, where the wool shows minimum swelling and maximum stability. An acid resistant detergent (in particular nonionic products) and an acid like formic, acetic or phosphoric acid are used.

Recipe

Bowl no.	Temperature	pH	Initial charge (g/l)		Feed rate (g/l/h)	
			Detergent	Formic acid	Detergent	Formic acid
1			Water		Water	
2	65–70	6.3–6.7	0.49	0.5		0.15–0.20
3	55–60	5.4–5.7	1.49	0.4–0.5	0.15–0.20	0.10–0.15
4	50	5.0–5.4	1.29	0.35	0.1	0.06–0.075
5	45	5.0–5.9	0.89	o.25	0.06	0.05
6 (Rinse)	40	6.3–6.5	Water		Water	

Notes:

1. Detergents generally used are aliphatic sulphonates.

2. Temperature of the scouring is mainly governed by two factors: (1) melting point of grease (45–50 °C) and(2) the maximum temperature at which the wool can be scoured before suffering damage from the alkali present. Max temperature used in presence of alkali 65 °C.

3. Soft water should be used for wool scouring.

4. Removal of tar tips: Branding the sheep with pitch bitumen, tar or red lead often causes heavy staining of the fibre tips that by normal treatment in the scouring is not sufficient to remove. The sorted contaminated wool fibres are soaked in a preparation of grease solvents and after few hours, thoroughly washed in a fresh bath preferably containing a synthetic detergent (0.5–1 g/l) and 0.5–1 ml/l ammonia and then rinsed.

5. The following products are all good solvents for oil, grease and tar pitch:

6. Hydrocarbons: benzene, toluene, etc.

7. Chlorinated hydrocarbons: trichloroethylene, carbon tetrachloride, dichloromethane

Hydroxy hydrocarbons: tetrahydronaphthalene

Hydroxyphenols: cyclohexanol.

But these solvents has a disadvantage that these degrease the wool too thoroughly that it imparts a harsh handle to the wool, rendering it brittle and often causing yellowing. On the other hand grease solvents of this kind sometimes in conjunction with soaps are successfully used for removing tar tips.

Experience has shown that the percentage of detergents consumed, calculated on the weight of the goods, is generally 1–2% wetting agent/detergent and 2–3% ammonia or 1–1.5% wetting agent/detergent and 3–4% soda ash. These amounts will, of course, vary according to conditions and depend also on the number of bowls used and the quality of the wool.

1.1.2.4   Machines used for washing/scouring of wool

The wool first goes through the dusting or opening machine to remove the loose external substances and mixing and then through a conveyor carries it to the scouring range. A common wool scouring machine consists of a set of bowls arranged in one line and containing the scouring liquor. In practice the raw wool is propelled gently through the scouring tank, then squeezed between rollers and then passed through another or more tanks in this way, until eventually it is rinsed in clean water. Such a multi-stage scouring environment is needed to remove dirt from the wool.

The bowls are usually fitted with a perforated false bottom or tray which allows the impurities, but not the wool, to settle at the actual bottom of the bowl. The impurities are periodically removed from here. A set consists of 4–7 bowls of which the first is known as the steeping or soaking bowl and the last is used a rinsing bowl. The bowls have a capacity of 2500–10,000 l, the first bowl being the largest which is mainly used for soaking. The wool is carried through the bowls either by swinging rakes or by rhythmically moving forks. The passage through the first bowl usually takes longer than through the subsequent bowls. The scouring liquor is passed through the bowls on the counter current principle. The production capacity may be 400–500 kg/h.

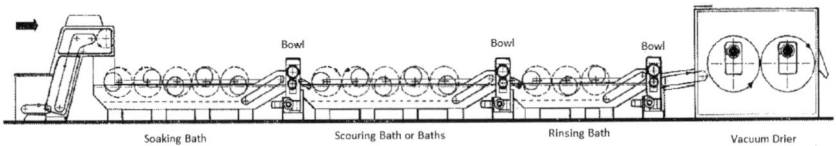

Wool scouring/washing range

The levels of wool wax and dirt (ash) remaining after conventional scouring are considerably higher on the top than base (see table below). The hard to remove wool wax is principally oxidised wool whose emulsion stability is considerably high.

Portion of fibre	Wool wax	Ash
Base	0.46	0.42
Top	1.18	4.73

1.1.2.5 Solvent scouring

Solvent scouring has many advantages over soap/soda scouring like –felting and entanglement associated with aqueous scouring are largely eliminated, wool grease and suint recovery is much more effective and aqueous effluent problems are avoided. This method removes almost 90% of wool wax. But it is not practiced very well due to the high cost, necessity for solvent recovery, possible toxicity and fire hazards to be set against it. The process uses a combination of non-polar (hexane, benzene, carbon tetrachloride) for wool grease removal and polar (isopropyl alcohol, white spirit, naphtha) solvents to suint removal. Some systems use 1,1,1-trichloroethane as solvent. Usual method is to apply solvent under high pressure jet on the material placed on a permeable conveyer and dislodges much of the dirt and suint apart from removal of wax from the fibre. The solvent content on the treated wool is reduced to about 45% by passing through the rollers and is further reduced to about 10% by centrifuging. Finally, the residual solvent is removed from the fibre by blowing hot air.

1.1.2.6 Drying

After the scouring the wool is squeezed through bowls and dried through a vacuum drier. The temperature should be kept below 100 °C. Drying is done to final moisture content between 19% and 15% (on dry weight basis). Generally, continuous driers are used with through flow air circulation worked in counter flow to maximise the thermal efficiency and drying capacity. Many types of driers are available which are classified according to the mode of transport of the wool.

Suction driers will have a chamber with perforated drums arranged inside it. The loose wool is fed over a belt conveyor to the perforated drums. The flat bed of the wool is sucked on to the drum due to the vacuum inside and passed over and under the drums alternately and moves forward till the end of the chamber. Fixed baffles located inside the drums restrict the air flow to the upper or lower half of each drums alternately. The wool gets dried when the hot air is sucked through them. The air can be heated by steam, or thermic oil or direct gas firing. Fresh air enters at the delivery end of the drier and flows in counter direction of the wool movement and partly exhausted at the entry portion of the drier where the air will have the maximum moisture.

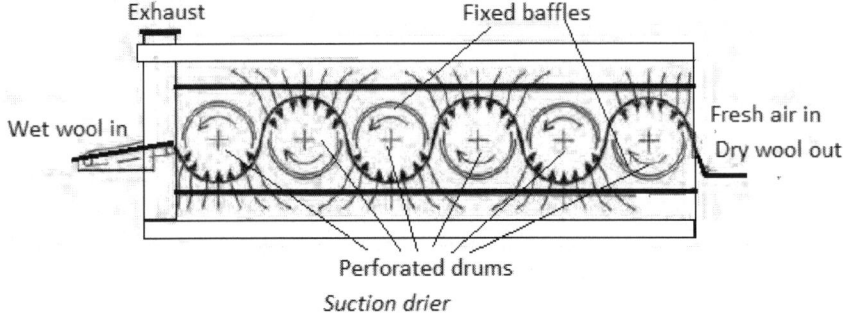

Suction drier

Another type commonly used driers are conveyor driers which works like a relax drier used in knit processing. The wool is fed on perforated endless slat conveyor which transports it through the hot chambers till the end of the driers. While the wool is conveyed the hot air is blown from above through nozzles. The air can be heated by steam or thermic oil radiators or

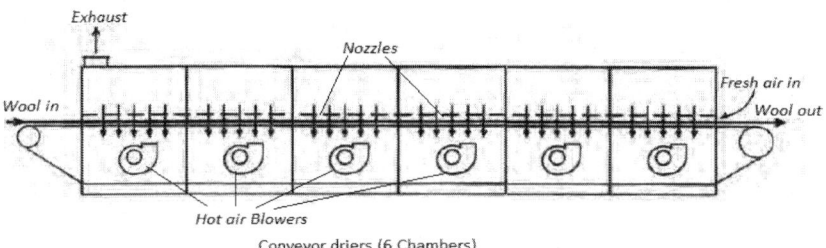

Conveyor driers (6 Chambers)

heated by gas firing. Air enters at the delivery end and exhaust is positioned at the feeding end as in the case of suction driers. There are some disadvantages for these driers like it occupies more space and drying by the blowing of the air over the loose wool bed is slow compared to suction driers and capital costs are also higher. Hence, the response for these types of driers from customers is not that good.

A new type of drier was developed incorporating both suction and conveyor driers, which was called unidrier. The main disadvantage of slow drying in conveyor drier was due to the fact that hot air was not passing through the wool bed as the air was only blown to the top. Blowing through the bottom was not practical as the wool bed may lifted off the conveyor. In the new type of conveyor two endless conveyors made out of polyester mesh was positioned in between alternate top and bottom blowers where the wool bed was passed in between these conveyors. The wool is carried through the dryer between these two conveyors while being subjected to reversing airflows in alternate sections of the dryer. The original design used air temperatures up to 150 °C, and relatively heavy bed weights of wool. Lower operating temperatures were used in later designs mainly due to industry concerns about wool yellowing. Because the wool mat is totally contained, through flow air velocities may be adjusted without the wool drop-off

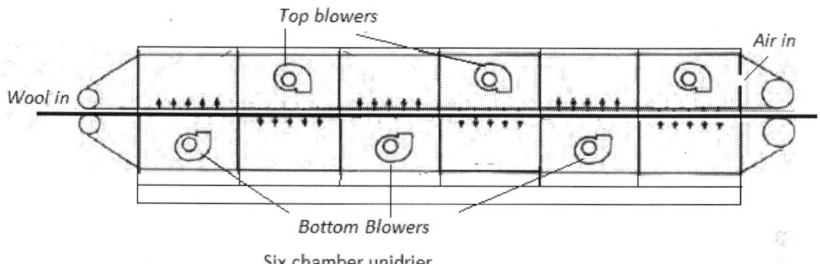

Six chamber unidrier

problems that might occur in a suction drum dryer. This means that control of air velocity with possible energy savings is a viable means of regain control. Cleaning requirements should be less frequent because it is less likely that fugitive wool will accumulate in the dryer.

1.1.2.7   Carbonising

It is necessary to remove the vegetable matter, etc. from wool, since these may adhere to wool even after carding. The carbonisation is done in loose wool form and also as piece goods.

Carbonising of loose wool: the scoured wool is steeped in:
Sulphuric acid:3.4–4.5 °Tw and 1 g/l wetting agent.

The wool is then squeezed and hydro extracted and baked for 1–3 h at 90–110 °C. The water evaporates and the concentrated sulphuric acid chars the adhering vegetable matter. After carbonising the wool is washed neutralised with soda solution 3 °Tw and then rinsed.

Notes:

1.   A wetting agent is a must because it effects a rapid and even penetration of the acid and ensures satisfactory carbonising. If wetting agent is not used, the strength of sulphuric acid has to be increased by 6–7° Tw but the results are not as good as the case where wetting agents are used.

2. Hydrochloric acid also can be used in place of sulphuric acid or acid liberating salts like aluminums chloride can also be used.

1.1.3   Processing of woolen yarn

Generally woolen yarns are manufactured by worsted, semi-worsted or woolen system (since it is a spinning technology we are not dealing the process in detail)

1.1.3.1   Worsted system

Virgin wool of the fibre length between 40 and 100 mm is used for worsted spinning. Worsted system mainly differs from other spinning in combing. Different systems are used for worsted spinning. A general flow chart is given below. For each process different systems/machineries available and manufacturer uses a combination which gives the best performance taking into consideration of the wool available for spinning.

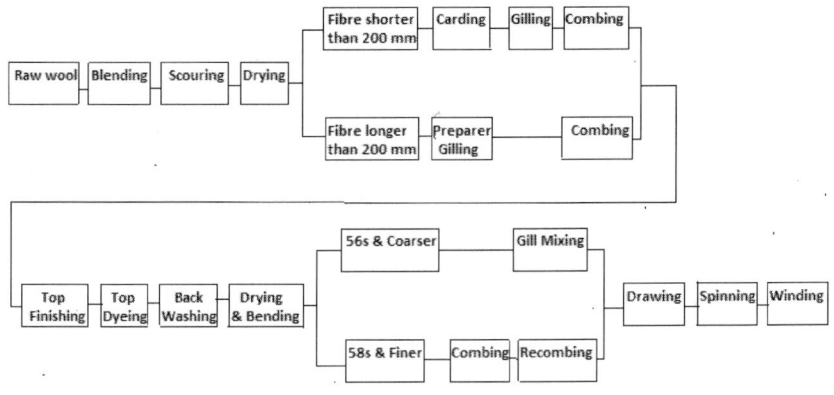

Flow chart of worsted system

1.1.3.2   Semi-worsted system

The semi-worsted system, mainly for synthetics and blends, essentially consists of carding, gilling and spinning. For medium and fine yarns a roving process is added before spinning. Raw wools are scoured and dried, opened, blended and lubricated prior to carding. The machinery used is very similar to that employed in worsted processing. This system has production and economic advantages over both the woollen and worsted system, but generally can not produce yarns of the same fineness, character or quality. The quality of semi-worsted yarns is somewhere between that of the worsted and woolen yarns, being bulkier, weaker and less regular than worsted yarns. A general flow chart is given below:

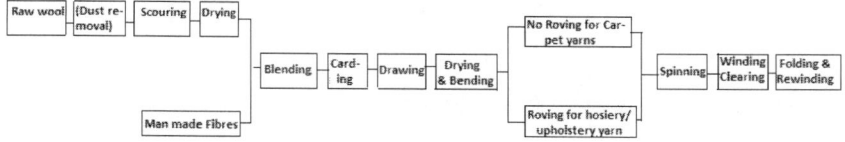

Flow chart of semi-worsted system

It should be noted that the combing is omitted in this system hence economical but it is mainly used for hand knitting yarns and carpet yarns which comes in the range of 50–500tex.

1.1.3.3   Woolen system

Generally used for 100% wool coarse yarns (in the range of 30–2000 tex), this process is the shortest of all the systems. Normal route is blending, opening, lubrication, carding and spinning. Due to short process the yarns produced by this system will not have the fibres well aligned, bulky, low density, hairiness, less soft and low twist. The method is used for yarns for particular end uses. A general flow chart is shown below:

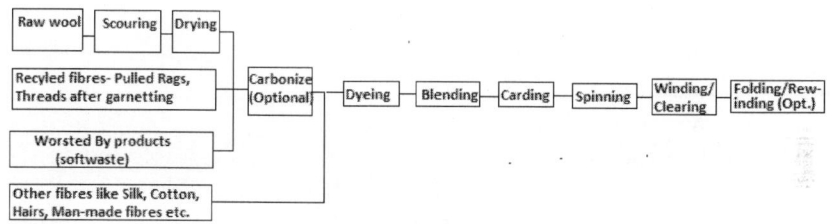

Flow-chart of woolen system

1.1.4   Raw wool processing

1.1.4.1   Bleaching

Raw wool is graded as per the natural colour. But it is usual to give some bleach to make it more attractive to the customer even some times florescent whitened to compete with cotton and other man made fibres. Usually reductive bleach or peroxide bleach is preferred. Bleaching agents used for this purpose include, hydros, sodium metabisulphite, hydrogen peroxide. Care should be taken to avoid yellowing after bleaching.

Peroxide gives a better whiteness which is also economical, is used more commonly. The generally followed procedure is as follows:

A bath is set at room temperature with

Quantity	Unit	Bath Addition
20	ml/l	Hydrogen peroxide 35%
0.4	g/l	Sodium pyrophosphate (Stabiliser)

pH is adjusted to around 8.5. Raw wool is introduced into the bath and slowly raised the temperature to 65 °C and maintained at this temperature for 3 h.

It is also usual to incorporate a bleaching treatment along with the scouring bath. In this case bath concentration will be less and acidic peroxide bleach or a reductive bleach is given after scouring. In practice, in the last compartment after scouring and washing 0.5–1.1% hydrogen peroxide at pH 4–6 is taken at around 55 °C (alternatively, a 0.4% sodium hydrosulphite solution at pH 5.7–6.7 and temperature 55 °C is taken). The wool bed is passed through the bath and squeezed with a small amount of peroxide or hydros on the material which will further act on the wool during the drying process also. The amount of the chemical carried forward is controlled such a way as to finish the reactions especially of hydrogen peroxide is completed and peroxide is fully consumed before bailing process to avoid any detrimental action in the bale form.

As mentioned earlier, wool has the tendency of yellowing after the bleaching or dyeing operation on storage. This is called hydrothermal yellowing, the chemical reaction of which is not very well clarified. It is believed that the yellowing is due to the α-keto acids present in the wool. It has been found that inclusion of hydroxylamine (HA) salts at 0.2% concentration in a conventional mildly acidic dyebath reduces hydrothermal yellowing by at least 50–70%. Higher temperature processing also may be a reason for yellowing. Hence it is better to avoid any processes including dyeing above 80 °C. Thus avoiding higher temperature processing and the presence of hyroxylamine can completely prevent yellowing of wool.

1.1.4.2   Moth proofing

Wool is prone to be attacked by insects, moths and beetles. Hence it is important to proof the material against these when wool or wool containing products are stored (or in usage) in raw wool form or manufactured yarn, fabric, carpets, etc. The main species attacking woolen materials are *Tineola bissiella* (moth), *Hofmannophila pseudosprettella* (brown house moth), *Anthrenus flavipes* (carpet beetle). Proofing woolen materials is very demanding as it has to comply with environmental regulation and to be effective through the lifetime of the wool textile or carpet. In addition in many cases it has to withstand hydrolysis in boiling dye baths, and possess adequate stability to sunlight and laundering or cleaning and shampoo-washing in case of carpets.

Environmental regulations have banned products like DDT, dieldrin, etc., and the effort to reduce organo chlorine residues in the environment has led steadily to replacement of the older products. Permethrin has a lower chlorine content and low mammalian toxicity but is toxic to fish and aquatic invertebrates vital to the nutritional food-chain of fresh-water fish. This has put pressure on development of application methods that generate little waste chemical discharge. Mitin AL combines permethrin with a hexahydropyrimidine derivative that has good properties of beetle protection. It is advisable to add the IR agent to dyebath formulations. In the case of wool–nylon blend yarns commonly used in carpets, the IR agent may be taken up predominantly by the polyamide and this portion is ineffective in terms of protection of wool. Some IR agents favour the wool component and are clearly to be preferred for such products.

Following are the main products in the market which are effective IR:

Chemical Name	Commercial Name	Manufacturer
Chlorophenylid	Molantin P	Chemapol
Cyfluthrin	Eulan SP	Bayer
Cyhalothrin	Cirrasol MPW	ICI (Aus)
Permethrin	Antitarma NTC	Dalton
Permethrin	Mitin BC	Ciba
Permethrin	SMA-V	Vickers
Permethrin	Perigen	Glaxo/Wellcome
Permethrin	Eulan SP	Bayer
Permethrin Hexahydropyrimidine derivative	Mitin AL	Ciba
Sulcoferon	Mitin FF	Ciba

- Products which are not confirming to new regulations are removed from the market

In some cases like in carpet production a relatively small quantity of wool is treated with higher amount of the IR products and mixed with higher quantities of untreated products to avoid limiting the effluent problems.

1.1.5 Pretreatment of yarn and fabrics

Even though the loose wool is scoured, carbonised the wool wax and other impurities would not have completely removed which can pose problems in dyeing and other processes. Other than this there can be lubricating oils (mineral oils, oleins and other oils), coning oil, etc. added on to the yarn during spinning. These have to be removed before taken for dyeing.

Normally, woolen and woolen blends fabrics/yarn is allowed to set before taking for scouring, dyeing, etc. So let us see the details of setting processes.

Setting of yarn is carried out by stretching the hanks to about 30–40% on a suitable frame and the stretched hanks are immersed in boiling water for 30 min–2 h. The hanks are allowed to cool under tension before removing from the frame. When released most of the extension is retained. When the free fibres are again boiled for an hour, some contraction occurs and the residual extension is defined as 'permanent set'.

1.1.5.1   Setting of woolen fabrics – crabbing

To avoid the distortions during following processes, the latent strain introduced into the yarn and fabric during the spinning and weaving operations has to be removed. This is done by the process called crabbing. It is basically a relaxation process, in the batch process, the cloth is passed in open width under guide rollers through boiling water in the first trough and is then tightly wound on the lower roller of the two large rollers in such a way that half of the rotating roller of cloth is immersed in the hot water. On rotation of the rolls the fabric is alternately treated hot and cold

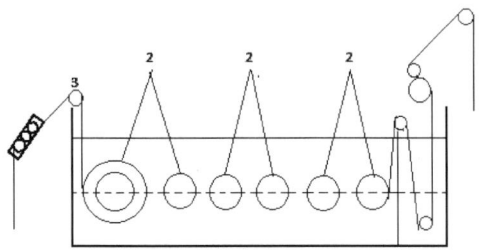

Flow diagram of a crabbing operation

and next the cloth goes to the next trough. This treatment takes about 5 min. The process is repeated by passing the cloth to the second unit. In this way the outer layer of the cloth in the first treatment becomes the inner layer during the second and thus uniform treatment is possible. During these operation relaxation and setting takes place.

1.1.5.2   Potting of woolen fabrics

This is also a relaxation and setting process followed immediately after crabbing. In this process the fabric is steamed or blown with steam and thus the fabrics are given surface stability without losing their shaping capacity in the making up process. The bonds that stabilise wool fibres in a particular configuration are disulphide cross-links, hydrogen bonds, electrostatic bonding between anionic and cationic side chains (see below in dyeing section) and hydrophobic side chains in the protein molecules. During setting process the material is relaxed and the previous cross-linking bonds may be broken and new bonds are formed in relaxed position. During steaming the existing bonds may be broken as shown below:

$$R - S - S - R + H_2O \rightarrow R - SH + R - S - OH$$

The formation of new bonds may be represented as:

$$R - S - OH + R - NH_2 \rightarrow R - S - NH - R + H_2O$$

Various bonds in wool and the reagents/treatments which can cause relaxation and rebonding are shown in table below:

Type of bond	Cross link/bond structure	Mobilizing agents
Disulphide links	$\overset{\mid}{C}{=}O \qquad O{=}\overset{\mid}{C}$ $H\overset{\mid}{C}{-}CH_2{-}S{-}S{-}CH_2{-}\overset{\mid}{C}H$ $NH \qquad\qquad HN$	Thioglycollic acid, reducing agents, high pressure steaming or steam at lower pressure for longer time
Hydrogen bonds	$C{=}O\dots\dots\dots\dots NH$	Water, steam, urea, dimethyl formamide
Electrostatic bonds	$\overset{\mid}{C}{=}O \qquad\qquad O{=}\overset{\mid}{C}$ $HC{-}R1{-}COO{-}\dots\dots H_3N{-}R2{-}\overset{\mid}{C}H$ $HN \qquad\qquad\qquad NH$	Water or steam
Hydrophobic bonds	$\overset{\mid}{C}{=}O \qquad\qquad O{=}\overset{\mid}{C}$ $H\overset{\mid}{C}{-}R_3\dots\dots R_4{-}\overset{\mid}{C}H$ $NH \qquad\qquad HN$	Organic solvents like alcohols, dimethyl formamide

R1 and R2, hydrocarbon units; R3 and R4, hydrocarbon side chains.

One can see, in almost all cases steaming is a common factor which can do relaxation and setting. Hence crabbing, potting, decatizing, etc. is done to set the woolen fabrics.

1.1.5.3   Scouring of woolen goods

Even though loose wool is given a scouring treatment, further additions in the spinning/weaving operations like lubricating oil, coning oils has to be removed. Hence soap/soda, detergent/soda scouring is done. During scouring attention is mainly paid to the removal of fatty substances applied in the process of oiling. When choosing the scouring method, it should be kept in mind that first of all the fibre must be released from fats which may be achieved in different ways. Accordingly, all existing scouring methods are classified into three groups which are based on (1) the extraction of impurities by special fat solvents; (2) the adsorption by the use of substances capable of adsorbing fats and mineral oils and(3) saponification and emulsification of fatty substances. The first of these processes envisages direct solution of fats in organic solvents like benzene, dichlorethane, carbontetrachloride, and others. This process possesses many advantages, particularly, quick action, absence of wool damage by alkali, the possibility of regenerating and making reiterated use of the main amount of solvents; easy recuperation of fats contained in the fabric and the discharge of sewerage waters.

Thus, special equipment must be used to ensure safety of operation. It is also necessary to take into account that only fat is extracted from the fabric while for cleaning it from all other impurities scouring with water or with soap/detergent and alkaline solutions is to be carried out.

When there is no excess fat on the fibre, the scouring process does not take long and presents no difficulty.

It should be used only in special cases, and particularly in scouring coarse woollen cloth oiled with mineral oil which is unsaponifiable and difficult to eliminate by treatment with soap and alkali solutions. For this reason, in older methods these solutions are sometimes replaced by water suspensions of fuller's clay; the particles of this clay coming in contact with the fabric and rubbing against its surface adsorb fats and mineral oil. Fuller's clays which by their chemical composition are aluminium silicates containing calcium, magnesium and iron oxides and others. Water suspensions of this clay are prepared in special clay breakers and contain from 60 to 100 g of dry substance per litre.

In modern methods, which is widespread the scouring method consisting in the saponification and emulsification of fats and oils with the use of sodium and triethanolamine soaps or synthetic detergents with additions of alkali.

Impurities can be eliminated only if the scouring solution penetrates inside the fabric, i.e. affects its wetting. It can be achieved by the detergents or if necessary additional wetting agents can be added. The scouring solution penetrates inside the fabric structure and thus conditions are created for impurities soluble in water to pass into solution. At the same time, the alkali contained in the wetting liquor neutralizes free fatty acids on the fabric and formed soaps are dissolved. Most of fatty substances contaminating the fabric are emulsified and eliminated in the form of emulsions.

These considerations on the essence and aim of scouring and its separate stages have led to the assumption that the initial solution of the detergent and soda ash is converted in the course of scouring into a composite system comprising: solutions of alkali and detergent, the concentration of which has been changed by the processes occurring in scouring; solutions of impurities eliminated from the fibre; emulsions formed by the action of the detergent liquor on fats and oil contaminating the fabric; suspensions of insoluble particles of impurities. If sized a separate desizing operation can be given. Generally, scouring is done in a rope washing machine shown below:

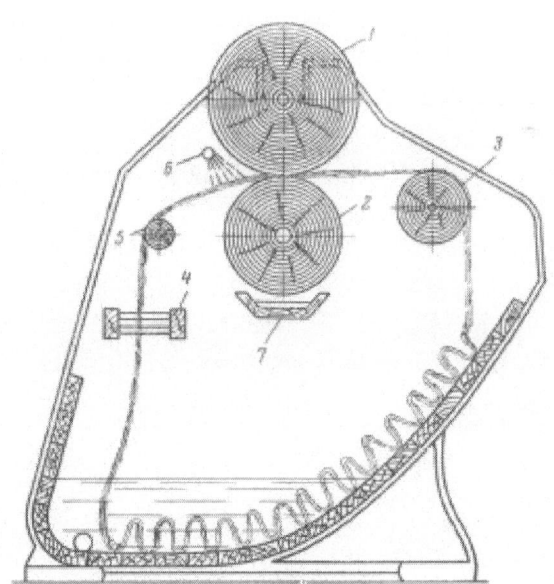

Schematic drawing of rope washer

Roll no 1 and 2 are used as squeeze rollers. After wringing is completed roll 3 shifts the ropes away from the rolls for their further displacement along the inclined bottom of the machine. In the front part, throughout the full width of the machine, there is a rope dividing frame 4 with a row of cross bars; between the frame and the main rolls there is a roller 5 directing the ropes into the roll nip. A sprayer, i.e. water supply pipe 6 with halos directed towards the fabric, is arranged above roller 5 (cold and warm water supply). Under the main rolls there is a suds box 7. The liquor squeezed out of the fabric is discharged directly into the suds box and depending on the degree of its contamination is either discharged through the side hole in the suds box into the drain or returned into the bowl through the hole in the bottom.

After filling the washer with the scouring solution "soaping" is started, which is the first stage of scouring. During soaping the fabric ropes are shifted inside the machine, squeezed many times between the main rolls, and soaked again in the bowl with the new portions of the scouring solution. During this operation, all the discharge holes of the bowl must be closed, and the liquid squeezed between the main rolls is discharged from the suds box into the bottom part of the bowl. At this stage of treatment all the dirt and impurities of the fabric are washed out by the scouring solution, and a great amount of foam is formed. In order to attain a better effect of the liquor on the impurities, soaping is carried out at a temperature of 35–40 °C. The increase of the temperature above this limit may cause deterioration of the wool.

The contaminated liquor is eliminated from the fabric surface by clean water supplied from the sprayer, while used-up liquor is gradually drained through the discharge openings.

Besides rope washers there are machines for washing the cloth in open width. However, these machines are less efficient and find a limited use, only for processing fabrics which form creases when treated in rope form.

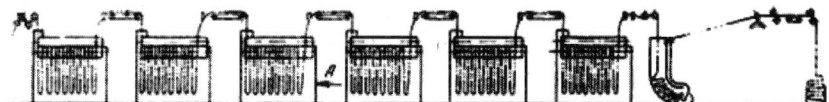

Woolen worsted rope washing and scouring machine

Modern scouring uses detergent, soda ash at a pH of 8.5–10, for 15–30 min at 40–45°C.

1.1.5.4   Desizing

Most wool yarns, particularly warp yarns, are sized; starch and gelatin being the principal agents used. Although gelatin may be readily removed in the ordinary washing operation, starch requires breaking down with enzymatic desizing agents. For this purpose the goods are impregnated on a mangle with a desizing enzyme solution at 50–60 ° C (120–140 °F) and allowed to lie overnight. Another method is to treat the goods for 6–8 h on a winch or steep overnight in an enzyme solution at 60–70 °C (140–160 °F). In both cases the best results are obtained at a pH of 5–5.4 (The manufacturers' instructions should be consulted when working with enzymatic desizing agents).

1.1.5.5   Carbonising of piece goods

The conventional carbonising process comprises five main stages : scouring, acidizing, drying and baking, burr crushing and dedustering and neutralising. This can be done either before or after dyeing. In recent years the carbonization of piece goods has been carried out as a continuous process, the impregnation with sulphuric acid 3–6 °Tw (2–4 °Be.), carbonizing in the hot chamber, washing and neutralizing being effected in one run. Carbonizing after dyeing lessens the danger, but the colours are often altered in shade unless dyestuffs fast to carbonizing have been used. Acid stable wetting agents (0.5–2 g/l) have also proved to be of considerable assistance in the carbonization of piece goods. The goods wet out more rapidly and evenly, thus lessening the danger of stains and faults. It is needless to say that the quicker the passage of the goods, the larger should be the addition of wetting agent. On no account should the goods contain residues of soap or lime-soaps. The material should, if possible, be impregnated, carbonised and neutralised in one continuous run. If this is not possible and impregnated material has to lie for some time before carbonizing, it should be plaited down and completely covered with a cloth impregnated with the carbonizing liquor, as otherwise stains may cause by direct sunlight falling on the goods, drops of water, draughts or uneven drying. After carbonizing, the goods must be thoroughly washed and neutralised with a soda solution of 3–4 °Tw (2–3 °Be). Unneutralised pieces sent for dyeing should be dyed at the commencement without acid. Dyestuffs which require acetic acid to commence dyeing tend to give unlevel results, it thus being essential that the goods be neutralised in this case. The goods to be carbonised have a cotton selvedge this has to be adequately protected. The selvedge is covered by brushing with sodium silicate 42–52 °Tw (25–30 °Be) or with a 20% soda solution thickened with chalk. This protective covering is removed in the subsequent washing operations.

In pad-dry bake method scoured wool fabric is padded, either in rope form or in open width, with a liquor containing dilute sulphuric acid (5–7% by mass (6–8 °Tw); approximately 65% wet pickup), and dried at 65–90 °C in order to concentrate the acid. Baking at 125 °C for 1min chars the cellulosic material. The goods are then neutralised immediately or run dry through a dolly or milling machine to remove the charred vegetable matter and then neutralised (Another method to remove the carbonised vegetable matter is to run through rollers to crush the brittle mater and wash off). Neutralisation of wool fabric can also be carried out in open-width scouring machine. Neutralisation with ammonia or ammonia/ammonium acetate mixtures removes acid from the fabric far more rapidly than either sodium carbonate or sodium acetate.

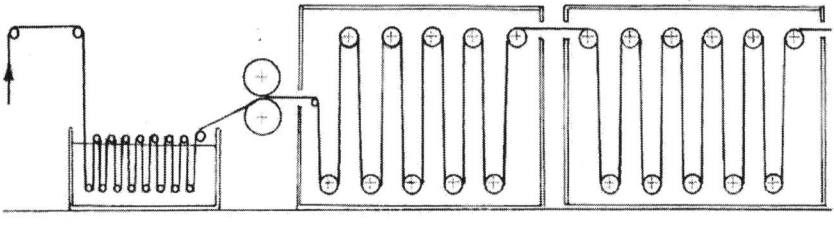

| Wetting vessel | Squeezing | Dryingunit | Carbonising unit |

Schematic drawing of a carbonising range

In a rapid carbonising process wool is treated in up to 8% (w/v) sulphuric acid solution for 30 s and time delay of about 10–15 min is introduced before drying. The locally damaged areas can greatly weaken the fibre if surface acid is not evenly distributed during drying. However, in the rapid carbonising method with a delay of 10–15 rain, the distribution of sulphuric acid between surface acid and total acid inside the wool changes. Free surface acid penetrate into the wool fibres and bonds to wool. Thus the amount of concentrated acid formed during drying and baking is minimised and as a consequence, less chemical attack occurs to wool in rapid carbonising.

Magnesium chloride, aluminium chloride (8–10 °Tw) or gaseous hydrochloric acid may also be used for carbonising of wool. Rags are carbonised to remove cotton stitchings prior to use in shaddy by the 'dry carbonising' process by exposing to hydrochloric acid gas in an enclosed chamber. Metal acid salts liberate hydrochloric acid at high temperature and attack the cellulose in wool. But owing to their high cost and higher temperatures required which cause injury to dry wool, are rarely used.

1.1.5.6   Solvent based carbonising

This method is a safer and environmental friendly process. Perchloroethylene, thanks to its low surface tension, can soak textile fibres deeper and faster than an aqueous solution. On the contrary, since vegetal impurities contained in the fabric are highly hydrophilic, their affinity for the solvent is lower than the fabrics; the solvent is contained only in the surface of vegetal particles. When the fabric soaked with solvent comes in contact with an aqueous solution of sulphuric acid, the aqueous solution cannot remove the solvent from the wool and replace it. On the contrary, the aqueous solution is absorbed by vegetal hydrophilic particles. In practice, with this system vegetal impurities absorb the acid solution selectively, and the acid carries out only a gentle action on wool. The benefits of this process are lower pollution, considerable reduction of

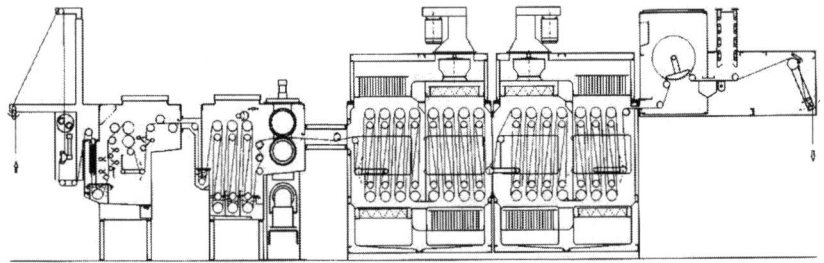

A solvent carbonising unit

damage to the wool and the possibility of by-passing the acid removal step (or if necessary, this step is considerably easier and faster).

1.1.5.7   Milling/fulling process

Milling is a mechanical treatment which causes the fibres to creep and tangle and form a more compact fabric. Felt is produced from loose wool by a somewhat similar process called "felting".

In order to ensure the best milling effects and to accelerate the process, it is necessary to use milling agents.. Felting is a specific property of wool, chiefly caused by its particular surface structure. As the flake cells overlap each other, as with roofing tiles, the friction resistance towards the fibre tip is lower than the reverse. The direction-dependent coefficient of friction (DFE) is obtained from this. The movement of a fibre swollen by moisture absorption under the effect of an external force, is steadily towards the root. In the opposite direction, the ratchet effect of the flakes hinders the migration of the fibre so that an irreversible shrinkage takes place. Internal parameters, flake structure, hydrophobia of the cuticle, longitudinal and transverse elasticity, swelling, fibre length, fibre fineness, grease content, crimping, yarn construction and external parameters like moisture content, mechanics, time, pH, temperature, lubricants contribute for felting.

Milling is the processing of piece goods under extensively similar conditions such as in felting to increase the density and strength of the fabric (normally with a moisture content of 90–120%of the raw weight and at 30–35 °C). Depending on the intensity of the milling process, a so-called pre-milled item is obtained, whose fabric weave can still be perfectly recognised, or a felt. In acore felt, the fabric is evenly felted in the cross-section; in a surface felt the top of the fabric only exhibits a random layer. In pre-milling the compression is achieved through relaxation and input of the piece goods in warp and weft direction. The number of warp and/or weft threads can be increased by a targeted compression to an extent, which cannot be achieved solely by fabric adjustment. Besides the compression of the fabric, more or less strong matting is carried out by the milling. This further increases the mechanical union through the existing yarn and fabric bonds and so the fabric is made more resistant to wear.

Traditionally milling operation is carried on woolen fabrics to allow it to felt so that you get fabric which are more or less shrink resistant. The woolen material is wetted with suitable wetting agent at around 40 °C and subjected to continuous pressure both in weft and warp direction. Under these conditions, wool fibres tend to felt, thus causing fabric shrinkage and a subsequent dynamic compacting. The conditions required for felting – moisture, mechanical stress and temperature has to be carefully controlled to get proper shrinkage without damaging wool or causing rope marks and irregular shrinkage and the required degree of felting is obtained in a short time. The process is done in open width or tubular form by stitching selvedges together (to favour the wrinkle movement, and avoid irregular tensions on both selvedges; in the process, an air pocket forms inside the cylinder of wet material thus favouring the wrinkle movement) on special machines called milling machine. Markers are put in the centre of the fabrics (often 1 m lengthwise) and checked periodically to ensure the required shrinkage has been achieved. Milling cycle usually followed by a washing cycle.

The milling processes are divided according to:

1. Dirt milling:
    * grease milling,
    * soap milling,
    * neutral milling,
    * acid milling,
    * special acid milling processes.
2. Milling of pre-washed fabrics:
    * soap milling,
    * neutral milling,
    * acid milling.

Natural wool chair fabric contains auxiliaries from the spinning and warp sizing in the weaving preparation, which must be removed in the washing and/or milling processes. The practical implementation of the milling process and the selection of milling method are therefore not only determined by the desired fabric result, but also by quantity and chemical structure of the spinning lubricants. Neutral oils, technical oleic acid (olein) and low-viscosity mineral oils are used as textile lubricants with and without the addition of emulsifiers.

Chiefly mineral oiled or yarn dyed carded yarn fabrics and neutral oiled worsted yarn fabrics are treated in soap milling. Neutral milling can be used should a non colour fast fabric be milled or only a low degree of felting is striven for. Carded yarn fabric, which contains an emulsifying spin oiling, can also be milled dirty at neutral pH. Synthetic surface-active substances are thereby used as milling agents. This, for example, includes fatty acid condensation products, fatty alcohol sulphates, alkyl naphthalene sulphonates, oil and sulphonated surfactants and/or sulphates.

Acid milling is carried out at pH 2–4.5, adjusted by formic acid, with the addition of acid-resistant, synthetic milling agents. This milling process is chiefly used in felt manufacture (hats, technical felt), as particularly high degrees of felting can be achieved in the acid range; also in fabrics with a lack of colour fastness or with a large proportion of short fibres. Acid milling is the best method for mohair additions due to the low fibre loss.

Milling cycle: The fabric is run in the machine by the squeezing rollers pulling the fabric from the bottom of the vessel and deliver to the backside of the machine where the bottom side of the vessel is slanting and the fabric slip to the lower portion of the vessel from where it is lifted to the squeezing roll. Fabric is loaded in rope form and rotates inside the machine till the process is completed. At the delivery end of the squeeze roll there is a box (square section tube) where the fabric is packed, slowed down by means of the adjustable plate (steel or wooden hinged plate on top of the box; it can be lowered by reducing progressively its section, slowing down the fabric) which can press the fabric and accumulate the fabric in box and when the pressure of the

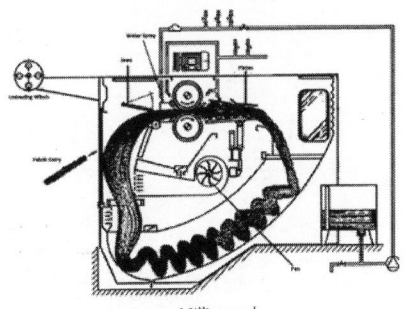

Milling cycle

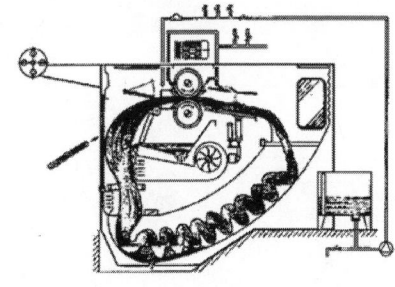

Wahing cycle

accumulated fabric increases some part of the fabric comes out lifting the plate which is free but is pressed with some weights. This action causes the lengthwise shrinkage of the fabric. Vertical parallel steel plates with adjustable-pressure cylinders and push the fabric inside the box, positioned in the front part of the machine that make the fabric shrink in the weft direction by squeezing the fabric. These machines can handle fabrics whose weight ranges between 80 and 800–1200 g/m.

Washing (gentle) step: The plate is lifted up while the liquor is uninterruptedly fed on the fabric and the air jets move the rope wrinkles. Since the plate and the top of the box is lifted the fabric freely runs without any pressing. As per the washing procedure the vessel can be opened to the drain, so that dirty water can be drained and fresh water can be sprayed on the cloth or partially drain and fresh water can be added. This type of machine can run at a maximum speed of 200–220 m/min (see figure above).

Fast washing step: The air jets can remain open. The fabric, drenched with liquor, moves at a speed ranging between 400 and 600 m/min with the plate open, and runs into the grid at the back part of the machine. Beating, combined with high speed, causes a slight felting on the surface and the yarn swells, as a result hiding the comb marks.

There are many variations to the machine with devices to improve felting action, washing efficiency, avoiding wrinkle marks by changing the rope position, increasing capacity, etc. On these machines too, the expertise of machine manufacturers and the application of the latest electronics have allowed the introduction of some devices that increase the output capacity and permit more accurate controls, ensuring excellent repeatability.

There are also open width milling machines which are mainly used for furnishing felts made from wool fleeces and are manufactured in a wide form, on plate or roller mills.

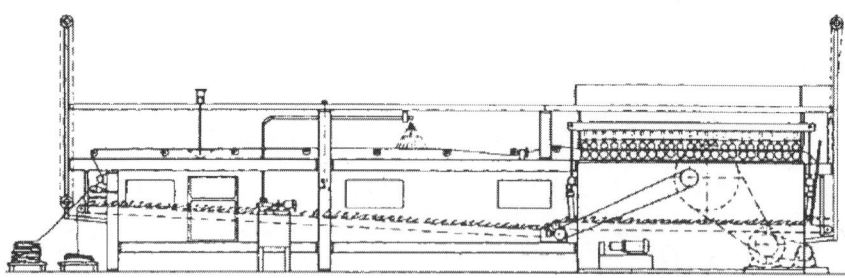

Open width milling machine

Notes:

1. In order to ensure the best milling effects and to accelerate the process, it is necessary to use milling agents.

2. According to the effect desired, the material is milled in a neutral, alkaline or acid medium.

3. Neutral milling: The goods are given a neutral milling if excessive felting is to be avoided, and all that is required is a more compact material. This method should also be applied to materials containing coloured effects which do not withstand treatment with alkaline milling agents.

4. Alkaline milling: This process is used for the production of heavy cloths and was originally carried out exclusively with soap plus soda or ammonia. In cases where saponifiable fats were present, milling was carried out solely with soda (milling in the grease).

5. In the last few years modern milling agents have replaced soap to a great extent. If mineral spinning oil is still present in the material, these milling agents are essential.

6. Acid milling with sulphuric acid, as employed in the manufacture of hats and felts, has a deleterious action on machine parts and cellulosic fibre effect threads. Organic acids such as formic acid for example, are also used in place of sulphuric acid.

1.1.5.8   Bleaching

Natural colour of wool can be cream, yellow, brown or brownish black depending on its origin, and breed. The natural colour of the wool also intensified its position in the fleece. The tips of the fibre is more heavily tinted than the rest of the fibre due to physical causes and exposure to light; certain other regions of the fleece also have a strong yellow colour. These differences in shade are taken into consideration when the fleeces are sorted. Wool with a natural brown to brownish black colour caused by melanin pigmentation is only bleached in exceptional circumstances. In such cases, since the bleaching process may damage the keratin, bleaching is commonly carried out with hydrogen peroxide using ferrous sulphate as a catalyst and formaldehyde as a protective, followed by a clearing operation with ammonium bifluoride.

The bleaching process should consider the fact that boiling and pH values above the isoelectric point shall enhance the yellowing of wool.

It is necessary to scour the wool thoroughly before submitting it to chemical bleaching processes. Any remnants of spinning oils due to the poor scouring can result in unlevel and unsatisfactory bleaching.

Wool can be bleached by a reduction process or oxidation process or a combination of both.

1.1.5.8.1 Reduction process

The possible reducing agents are sulphur dioxide and its derivatives like sulphurous acids (H_2SO_3), and its salts like sodium bisulphite ($NaHSO_3$), sodium hydrosulphite ($Na_2S_2O_4$). The effect of this treatment is to reduce the yellow impurities in the wool to colourless leuco compounds. However, the original colour will return on exposure to sunlight, oxidation or washing in alkaline conditions.

1.1.5.8.2 Sulphur stove bleaching

The goods are wetted out with water or 0.2–0.3% soap solution, hydroextracted and suspended loosely on poles in the sulphur stove where they stay overnight. Approximately 3–6 kg of sulphur is required per 100 kg. Care should be taken to burn sulphur as completely as possible avoiding sublimation. Any sublimed sulphur may be trapped by means of baffle plates or damp cloths placed over the burning pan. After bleaching the goods are thoroughly washed.

1.1.5.8.3 Bleaching with a solution of sulphur dioxide in water

When sulphur dioxide is passed through water sulphurous acid is formed. This also can be prepared by acidifying sodium bisulphate or sodium sulphite. The clean well squeezed or hydroextracted material is placed in a cold bath containing 4–6 l sodium bisulphite (38 °Be) or 2–3 kg sodium bisulphate crystals crystals and 300–400 ml sulphuric acid (66 °Be) per 1000 l. After working for a short time the material is left overnight in the bath. No iron must be present in the machines used or in the water. The goods are then rinsed and dried at room temperature if possible.

1.1.5.8.4 Bleaching with hydrosulphite

This is the most practiced method these days. The above two methods may be obsolete and more of academic interest. The bath is set at 45–65 °C with 3–5 g/l hydros at pH of 5.5–6.0 and the goods are entered and left in the bath for 4–6 h and sometimes even up to 24 h. The bath may be stirred or the goods are turned periodically. It is beneficial to keep the vessel covered to restrict the access to air and to delay the cooling down of the bath. Bleaching is followed by acidification with sulphuric acid.

Other reducing agents which can be used for bleaching wool are thiourea dioxide, sodium formaldehyde sulphoxylate and zinc formaldehyde sulphoxylate. Thiourea dioxide is more expensive than sodium dithionite, but it can be used for bleaching wool with 1–3 g/l at 80 °C and pH 7 for 1 h.

Another novel method of bleaching wool is by producing the active bleaching species sodium dithionite *in situ* by the reaction between sodium borohydride and sodium bisulphite to produce:

$$NaBH_4 + 8NaHSO_3 \longrightarrow 4Na_2S_2O_4 + NaBO_2 + 6H_2O$$

This combination of reagents is also used in the textile industry to strip dyestuffs from synthetic fibres. This process has been found much better in achieving robust whiteness than using the ready hydrosulphite.

1.1.5.8.5 Oxidation process – bleaching with hydrogen peroxide

The generally used oxidizing agents are hydrogen peroxide, sodium peroxides, per salts and potassium permanganate.

The oxidizing agent predominantly using wool bleaching is hydrogen peroxide. Wool is generally bleached at pH 9. Phosphates (e.g. tetrasodium pyrophospahte) have proved valuable in stabilizing the bleaching bath or a specialised stabilizing agents can be used. Ammonia is added to give a pH of 9.Trisodium phosphate and sodium silicate do not require such an addition since their reaction is sufficiently alkaline. They are, however, not so suitable for wool bleaching since their stronger alkalinity may have a tendering effect on the fibre. In addition to this, sodium silicate has a stabilizing effect only in hard water. Thus, the bleaching bath should contain acertain quantity of hard water. If this is not available,0.005–0.01 g/l epsom salt may be added. Sodium silicate is liable to render the wool harsh.

Other problems in bleaching wool with hydrogen peroxide are the presence of transition meta ions in the water and wool and the side effects due to this. The presence of transition metal ions can cause rapid decomposition of hydrogen peroxide to water and oxygen.

$$2H_2O_2 \xrightarrow{M^{n+}} 2H_2O + O_2$$

Transition metal ions can also lead to the generation of hydroxyl radicals, which can damage wool fibres.

$$H_2O_2 + M^{n+} \longrightarrow M^{(n+1)+} + OH^- + \cdot OH$$

There will be inherent transition metal ion traces in wool which are strongly complexed to the fibre, particularly iron and copper, which are not fully removed during scouring. Hence, it is necessary to add sequestering agents (e.g. tetrasodium pyrophosphate – TSPP) in the bleaching bath to prevent from above reaction taking place.

Bleaching is generally commenced at 50 °C and is continued for 5 h and often overnight in cooling bath.

1.1.5.8.6 Bleaching of woven piece goods, knitted goods and felts on winch

Recipe:

Quantity	Unit	Additions
15–20	ml/l	Hydrogen peroxide 35 Vol.
2	g/l	Tetrasodium pyrophosphate crystals
0.5	g/l	Nonionic detergent
0.75	ml/l	Ammonia (density 0.88) to pH 9

There are special peroxide stabilizers which can improve the performance of peroxide bleaching considerably.

The winch is run for 30–60 min h and the goods are then left in the bath (below the surface) for 1–2 h. The goods are again worked and then immersed for 5–6 h or, better, overnight. In order to ensure level bleaching, it is advisable to run the goods again for1 h at 45 °C before removal. They are then thoroughly rinsed. Great care should be taken to ensure that the goods are completely covered by the liquor when the winch is not running, otherwise salts will occur through oxidation.

1.1.5.8.7 Bleaching of loose wool, slubbing yarns

Recipe

MLR 1:10, circulating machine

Quantity	Units	Additions
2	g/l	Sodium pyrophosphate crystals
20–30	ml/l	Hydrogen peroxide 40%
0.5–1	ml/l	Ammonia 20%

Bleaching is carried out for 5–6 h at 40–45 °C during which circulation may be interrupted from time to time. Even better results are obtained if the liquor is allowed to circulate for about 2 h at 40–45 °C and the goods then left in the bath overnight. The wool should be rinsed thoroughly several times after bleaching.

Notes:

1. According to the bleaching effect desired and the quality of the wool, the baths should contain 10–30 c.c./l hydrogen peroxide 40%. The content is determined by titration with potassium permanganate.

2. During bleaching, the pH should not be lower than 8.2, since a neutral bath causes the wool to take on a reddish tinge which cannot be removed. The lowest permissible degree of alkalinity can easily be determined with phenolphthalein. If a test sample of the bleaching bath is not tinted red by the addition of a few drops of this indicator, it shows that the pH is below 8. This, of course, necessitates the addition of a little alkali.

3. In some cases of synthetic stabilisers is sodium pyrophosphate, the ammonia addition should be increased somewhat in the following recipes to conform to the above titration values.

4. Bleaching of wool also can be done at lower pH (slightly acidic, like pH 5) to reduce the wool damages. Bleaching is slow and it needs special peracid activator like citric acid or Prestogen W – BASF (a proprietary mixture of organic salts that generates percarboxylic acids) which are found to be effective activators at pH 5.5. Typical bleaching conditions are 0.75–1.00% w/v H_2O_2 with about 5 g/l Prestogen W for 1 h at 80 °C.

5. The level of whiteness achieved by peroxide bleaching under acid conditions is significantly less than is obtained by conventional alkaline peroxide.

1.1.5.8.8 Continuous bleaching

The peroxide bleaching also can be done continuously using special machines designed for the same.

Sequence: Pad steam – Wash – Neutralise – Dry, etc.

Padding

The material is padded at 30 °C with following recipe:

Quantity	Unit	Additions
30–60	ml/l	Hydrogen peroxide 35 Vol.
3–6	ml/l	Stabiliser
2–3	g/l	Detergent
4	g/l	Trisodium polyphosphate

Liquor pick up: 100–110%.

Steaming temperature: 85 °C.

Speed: 6 m/min.

The material is then washed, neutralised and dried.

1.1.5.8.9 Bleaching with sodium peroxide

Hydrogen peroxide can be replaced by more stable sodium peroxide. Aqueous solution of sodium peroxide react strongly alkaline and thus have to be neutralised with sulphuric acid.

100 l of water is set with

1.0 l of conc. sulphuric acid (168 °Tw).

The bath is then cooled, preferably with ice and

1:1.4 kg sodium peroxide is mixed while stirring.

The bath should react weakly acidic and if necessary little more sulphuric acid can be added to ensure all sodium peroxide has been converted. The bath is finally made weakly alkaline with ammonia/sodium silicate. The goods are entered at 500 °C and left for 4–6 h.

1.1.5.8.10   Whitening of wool

Wool is normally whitened (Optical Brightening Agent treatment) after bleaching, even though good whitening effect can be achieved by OBA treatment of unbleached wool also. Optical whitening agents are available in the market suitable for protein fibres like wool, silk, etc. These are applied as an acid dye in the presence of acetic acid.

Application of OBA increases the whiteness of the wool but reduces their photo stability and hence the whitening of wool is not very popular. A common whitener used for pure wool is bis-stilbene derivatives (see below), with light fastness ratings of only 2–3 which is the most stable products available for wool.

Example of bis-stilbene derivatives

Guide recipe

Quantity	Units	Additions
0.5–5	%	Suitable optical whitening agent
5	%	Acetic acid 40% or
2–4	%	Formic acid

The goods are treated for 20–30 min at 40–50 °C higher temperatures reduce the uptake of in case of some OBA, but in other cases higher temperature can be used (as per manufacturer's instructions). Rinsing before drying is not absolutely necessary.

1.1.5.8.11 Full bleaching

To make full white wool it is best to go with a two-stage process. First the wool is bleached with hydrogen peroxide under alkaline conditions, as described above, and then a reductive bleaching is conducted, usually using either sodium hydrosulphite or thiourea dioxide, as described above. Optical whitener can be applied in the hydrosulphite bleaching bath.

1.1.5.8.12 Application of OBA from hydrosulphite bath

The bleaching bath is set at 40–50 °C at a pH of 5.5–6 with

Quantity	Unit	Additions
2.5–5.0	g/l	Sodium hydrosulphite
2.0–0.5	%	OBA

and the goods are treated for 1–4 h. It should be noted that more the hydrosulphite in the liquor, the greater is the affinity of OBA for the fibre. The goods are finally rinsed.

Notes:

1. Some OBA can be applied to wool from a neutral, weak alkaline or acid bath, and can, therefore, be used in the peroxide or hydrosulphite bleach. It gives a neutral to reddish white.

1.1.6   Dyeing of fabrics

Since there are acid and basic functional groups on the side chains of wool fibre it is possible to dye wool with acid, metal complex, direct, chrome, and reactive dyes. Level dyeing of any wool dye will depend on its affinity, substantivity, migration properties and degree of dispersion of the dye. The exhaustion rate of dyes is mainly determined by their diffusion properties. Dyes with small molecular size usually have fast diffusion rates and hence faster exhaustion and vice versa. The affinity between dye and fibre also affects the exhaustion rate positively. As a general rule for the migration capacity of dyes, those with small molecules (levelling dyes), due to their lower bonding capacities arising from van der Waals forces and because of their high diffusion capacity, migrate relatively well and therefore compensate for the disadvantages of a high exhaustion rate. Dyes with large molecules (milling dyes and 2:1 metal complex dyes), on the other hand, hardly migrate at all or only migrate a little because of their high affinity and low diffusion capacity, and because of their relatively low rate of adsorption, can easily dye unevenly. The wet fastness of wool dyeings are directly related to the migration capacities of the dyes.

Keratin (wool protein) becomes positive in acid solution by the addition of a hydrogen ion

$$R\text{-}NH_2 + H^+ \rightarrow R\text{-}NH_3^+.$$

The possible acid and basic groups on wool fibre are given below:

Acidic Functional Groups	Basic Functional Groups
Aspartic acids	arginine
C terminal amino acids	lysine
Glutamic acid	N terminal amino acids
	histidine

The resulting electric charge must be counterbalanced by ionic coupling with suitable anion, for example the anion of the relieved acid, anions of salts or dye anions. But the fact that in this competition dye anions are preferred indicates that there are other bondings forces at work between the fibre and dye anion besides the electrostatic forces – these are non-polar van der Waals forces and hydroxyl groups, hydrophobic interactions, etc. The main bondings come into play during the dyeing of wool fibres are:

1. Ionic bondings: As indicated earlier the acid dyes dyeing bath in the presence of hydrogen ions charges the amino groups on the wool positively. The attraction between this positively charged $-NH_3^+$ and negatively charged dye ions forms the ionic bonds. The ionic bonds are relatively weak and can be removed by modifying the charge on the wool (by increasing the pH). Additional forces are therefore necessary to achieve an adequate fastness.

2. Coordinate bonding: In case of chrome dyes, other than ionic bonds, they can form coordinate bonds, formed between the chrome (metal) atom of the dye complex and the amino and imino groups on the wool.

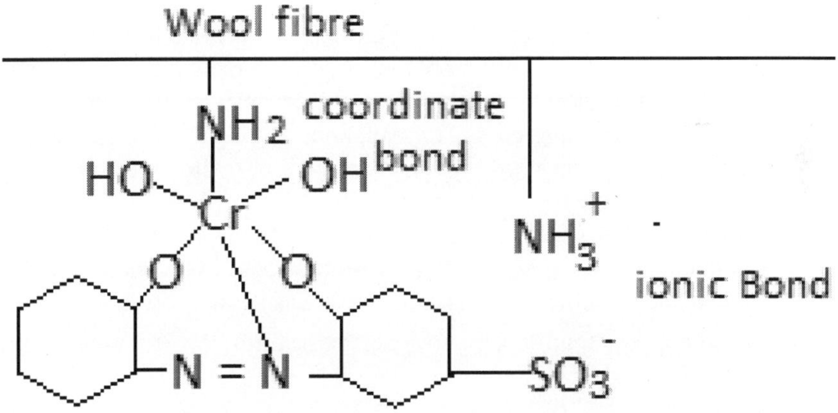

Bonds formed in Chrome dyeings

In case of 1:2 metal complex dyes case is a little different as the coordination sites.

The metal atoms are both taken up by the two associated dye molecules. Formation of

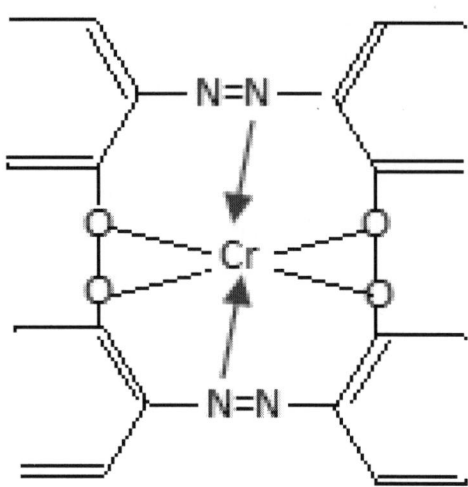

A typical structure of a 1:2 metal complex dyestuff showing the co-ordination sites taken up by dye molecules

metal complex does not take place until it is rinsed with weak acid, the amino groups which are capable of reacting, only being formed in the acid range where they exist as ammonium ions.

3. Solyophobic or hydrophobic interactions: These arise from the effect of non-polar parts of water-soluble solutes on the structure of water. When such molecules are brought into aqueous solution through a relevant solubilising group (e.g. sulphonate) then the water structure must change to accommodate the non-polar or hydrophobic residues. This change represents a gain in entropy for the whole system. Since most dyeing processes are restricted to aqueous systems, hydrophobic interactions are likely to play an important role in determining both dye uptake by fibres and subsequent wet-fastness.

4. Covalent bonding: Reactive dyes can also form covalent bonds with the reactive groups

Lysine: CH_2–CH_2–CH_2–CH_2–NH_2 (ε amino group),

Cysteine: CH_2–SH (β thiol group),

Threonine: CH_2–CHOH–CH_3 (secondary aliphatic hydroxyl),

Serine: CH_2–OH (primary aliphatic hydroxyl),

N terminal amine: CO–CHR–NH_2 (α amino group).

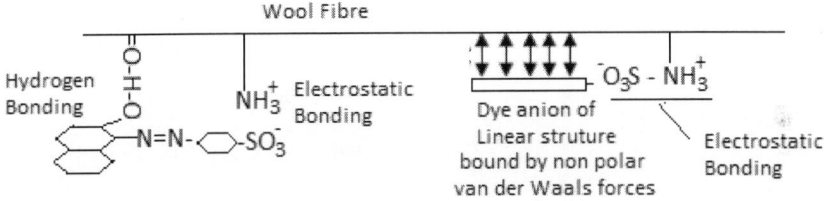

Hystidine: (imidazole group)

Tyrosine: —⟨ ⟩—OH (phenolic hydroxyl group)

on the wool. Under dyeing conditions, it is supposedly the thiol, amino and imino groups which become available as the reaction partner.

Different types of bondings

Main classes of dyes used in dyeing wool are:

Acid dyes,

Mordant dyes,

Acid metal complex dyes,

Reactive dyes.

Given below the main classes of dyes used for wool dyeing, their dyeing properties and fastness and the area of application. The dyes and dyeing method selection has to be done considering the fastness property requirements, levelling stages of dyeing, final usage, customer requirements, etc.

Class of dye	Levelling properties	Fastness properties	Applications
Equalising acid dyes	Very good migrating power	LF, moderate to good but poor fastness to wet treatments	Useful where high Light Fastness and less importance for wet fastness. Light and better quality dress fabrics, furnishing fabrics, plush, carpet yarns, knitting yarns
Moderately acid dyeing acid dyes	Good to very good	Wide variation in LF, wet fastness moderate to good	Bright shades of moderate LF and wet fastness. Used for dyeing blankets, dress goods, knitting yarns, carpet yarns, furnishing felts
Weakly acid or neutral dyeing dyes	Moderate to good	LF as individual dyes, wet fastness very good	Bright shades for more stringent LF and wet fastness requirements. Used for dyeing union articles, carpet yarns, knitting yarns for hose, sportswear
Mordant dyes	Moderate to good	LF good to very good, excellent wet fastness, potting fastness good in many dyes	Suitable where excellent LF and wet fastness required. Only for materials which are not affected by the fairly long dyeing process. Used for dyeing slubbing and loose stocks for suiting, knitwear, uniform fabrics, yarns for upholstery, heavily milled materials
Strongly acid dyeing 1:1 metal complex dyes	Very good to excellent, good migrating power	LF good to excellent, wet fastness inferior to mordant and milling dyes, potting fastness moderate	Suitable where good levelling and good all-round fastness are required. Use of strong acid may have an effect on the final appearance of the dyed goods. Used for piece dyed suitings, upholstery and furnishing fabrics, hosiery, heavier dress fabrics, carpet yarns, yarns and packages for lightweight wool articles
Weakly acid dyeing 1:2 metal complex dyes	Moderate to good. Poor migrating power	LE very good to excellent	As for mordant dyes, but also in cases where the long dyeing time affects the material. For woven and knitted goods, chlorinated yarns and tops
Weakly acid or neutral dyeing 1:2 metal complex dyes	Good, poor migration	LE very good to excellent. Potting fastness moderate	For worsted and woollen yarns, loose wool and tops, where maximum fastness along with good wash and milling fastness is required. Uniform fabrics, suitings, gabardines, carpet yarns, furnishing fabrics in pale shades
Wool reactive dyes	No migration after reaction	Excellent fastness	Extra ordinary bright shades with all-round fastness for shrink resist finished wool. Suitable for fabrics with "super wash" label. Can be used for dyeing wool at all stages

Following table shows the relative levelling and fastness properties:

Levelling properties ↑					
Acid Levelling					
	Acid Half Milling				
					Afterchrome
		1:1 Metal Complex			
			Acid Milling		
				1:2 Metal Complex	
					Reactive

Fastness propeties ————————————————————>

Relative Levelling and fastness properties of wool dyes

1.1.6.1 *Acid dyes*

Acid dyes are so called because of the acid groups in their dye molecules and also due to the fact that they are dyed in an acid bath. The presence of a sulphonic group ($-SO_2H$) in the molecule makes these dyes soluble in water.

Acid dyes possess affinity for protein fibres are a class of dyestuff primarily used to dye wool, silk, polyamide or protein fibres. Acid dyes are mostly sulphuric or carboxylic acid salts and are essentially applied from an acidic bath and hence the name 'acid dye'. They ensure that the dyes carry a negative charge (i.e. they are anionic). When strongly acidic conditions are used in the dyeing process, the protein molecules acquire a positive charge and this attracts the acid dye anions by ionic forces. As well as these ionic forces of attraction between dye and fibre molecules, van der Waals forces, dipolar forces and hydrogen bonding may also play a part in the affinity of acid dyes for protein fibres.

Usually, the dye anion is the active coloured component in these dyes; invariably synthesised as sodium salts – the most important among them are sulphonic derivatives of azoic dyes as well as of tri aryl methane and anthraquinone dyes. It is difficult to isolate them as free dye acids and hence is isolated as sodium salts. Under certain conditions they may be used for dyeing polyacrylonitrile fibres. Similarly to direct dyes, they are anionic dyes with the general formula RSO_3Na, but they do not dye or poorly dye cellulosic fibre without ensuring colour fastness, which is in the first instance connected with their structure, i.e., they do not have a sufficiently long chain of conjugated double bonds and their structure is not coplanar.

Fastness properties of acid dyes may vary with chemical constitution as well as application class of dyes. Acid dyes are water soluble and, being salts of a strong acid (the dissociation constant is within $10^{-1}–10^{-2}$), they give neutral solutions. They are strong electrolytes in aqueous solutions that are their dissociation is practically complete. Generally at usual dyeing temperatures (close to the boiling point) even in the presence of other electrolytes, acid dyes are found in the form of ions or small aggregates (aggregation degree not over 2–3).

As mentioned earlier, an acid dye can be simply represented as D $(-SO_3^-Na^+)_n$ where D is a chromophoric entity which may be substituted azo, anthraquinones, phthalocyanines or others, and n is an integer, usually 1–3. Ancient days wool has been dyed with these low molecular weight simple dyes but the wet fastness of these dyes were poor even though they used to dye bright shades. Modern manufacturers in an attempt to improve the fastness properties has been making dyes more sophisticated and these acid dyes were modified, usually by increasing their molecular size.

Acid dyes are a broad group of dyes. The dyes which are being explained in next sections are also acid dyes. But generally, by the name acid dyes covers only simple acid dyes with low molecular weights. In actual acid dyes classification is as follows:

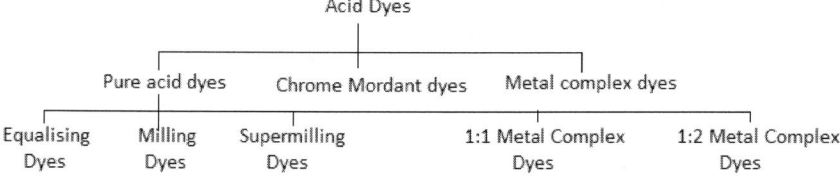

1.1.6.1.1   *Classification of acid dyes*

Acid dyes may be classified either by their chemical structure or on application method.

Based on chemical structure they may be classified into the following:

(i) Monoazo or Biazo

Monoazo dyes includes a number of much used wool dyes that are distinguished by brilliance of shade, very good levelling power, and particularly low cost, while their wash- and light fastness meet only low to medium requirements. Acid monoazo dyes are classified according to the type of coupling component like aromatic amines, naphthols and naphtho-l-sulphonic acids, aminonaphtholsulphonic acid, 1-phenyl-5-pyrazolones,etc.

Example:

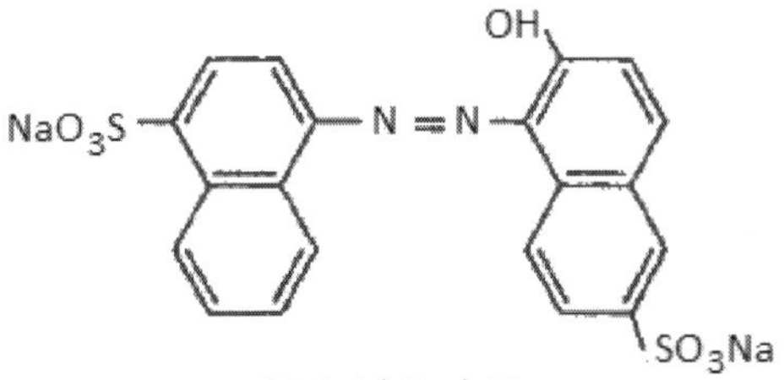

C.I.Acid Red 13

Monoazo acid dyes derived from aminonaphthol sulphonic acids like gamma acid and H-acid as coupling components exhibit very good light and wet fastness. Especially, acid coupling of gamma acid yields dyes with very good light fastness, which is presumably attributable to the formation of a hydrogen bond between the azo bridge and hydroxyl group adjacent to the azo bridge.

Dyes with two azo groups are divided on chemico structural principles into primary and secondary disazo dyes. Primary disazo dyes are manufactured either according to the scheme D1 → K ← D2 from a bifunctional coupling component K and two identical or different diazo components D1 and D2, or according to the scheme K1 ← D → K2 by coupling a bis-diazotized diamine D to the coupling components K1 and K2, which may be the same or different. Secondary disazo dyes are manufactured according to the scheme D → M → K (M = middle component) from a diazotized aminoazo dye (D → M) and a coupling component K (series coupling).

Example,

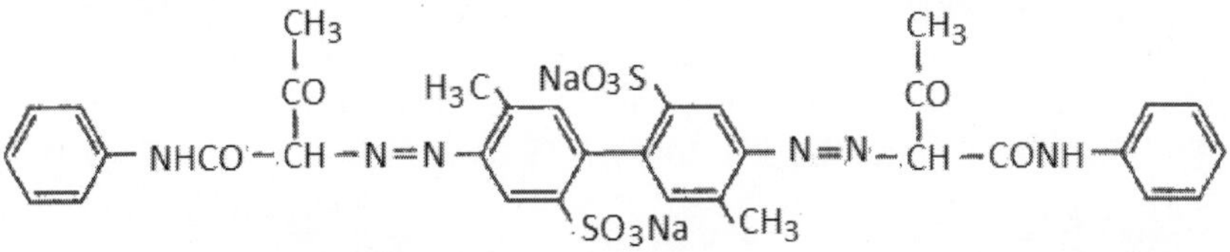

C.I. Acid Yellow 44

(ii) Nitro,

Acid dyes of nitro class are used for dyeing natural animal fibres such as wool and silk. They are nitro derivatives of phenols, e.g., picric acid or naphthols, e.g., C.I. Acid Yellow 1,

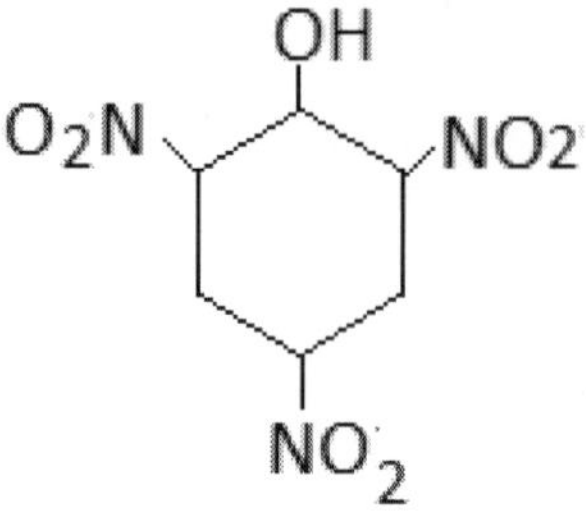

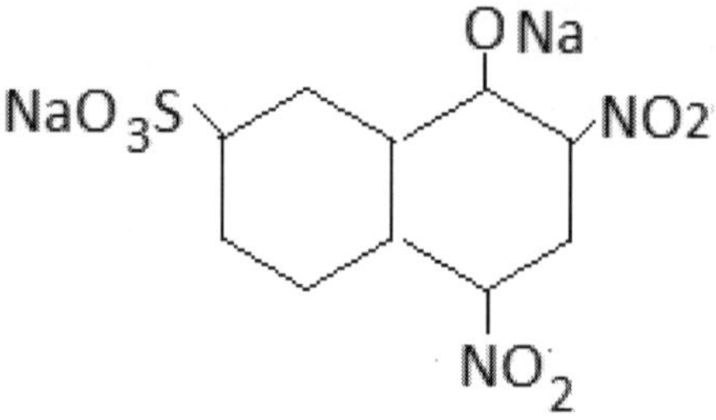

(iii)   Nitroso,

Nitroso dyes are metal-complex derivatives of o-nitrosophenols or -naphthols. The only nitroso dyes important commercially are the iron complexes of sulphonated 1-nitroso-2-naphthol, e.g., C.I. Acid Green 1,

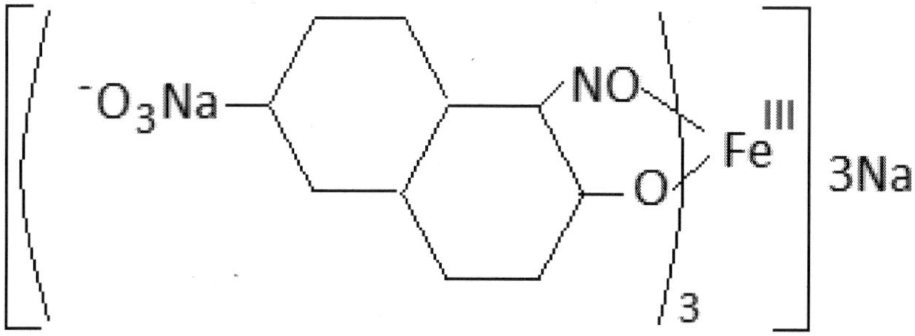

(iv)   Triphenylmethane,

This group is mainly bright blues and green dyes but of less importance as they have poor light fastness. These dyes can be manufactured by the sulphonation of the corresponding basic dyes. There are many other manufacturing possibilities like derivatives of N-ethyl-N-(3-sulphobenzoyl) aniline, naphthalene sulphonic acids, etc.

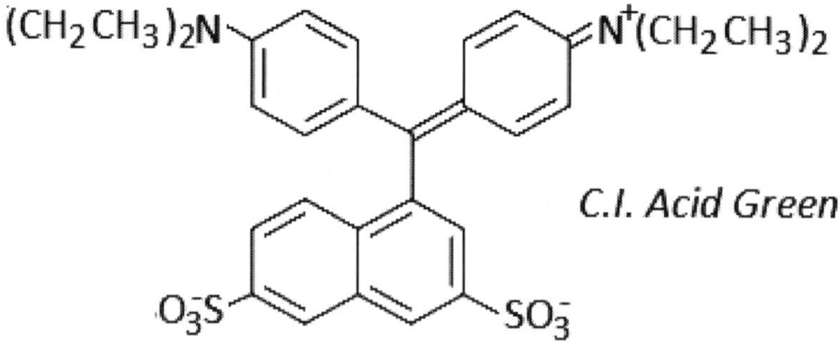

C.I. Acid Green 16

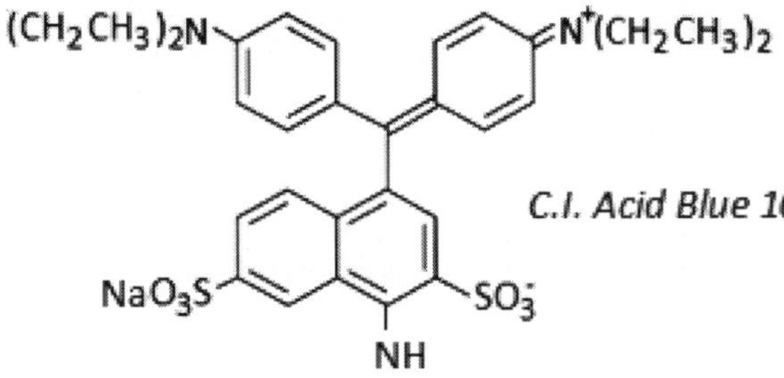

C.I. Acid Blue 108

Their dyeing properties of which are determined by the amino-, sulpho- and carboxylic groups attached to the molecule.

(v)   Xanthenes,

Xanthenes are derivatives of triphenyl methane carboxylic acid or triphenyl carbinol carboxylic acid, of the common structure as follows:

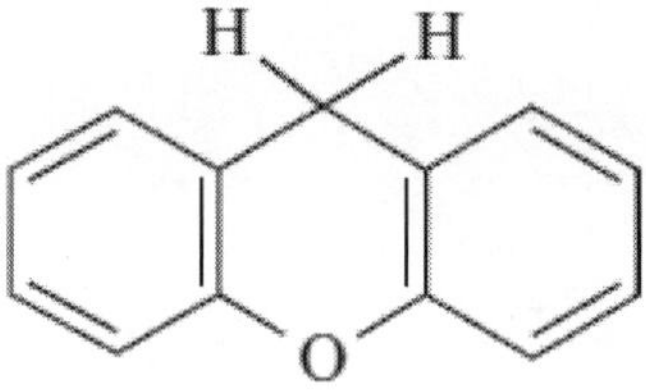

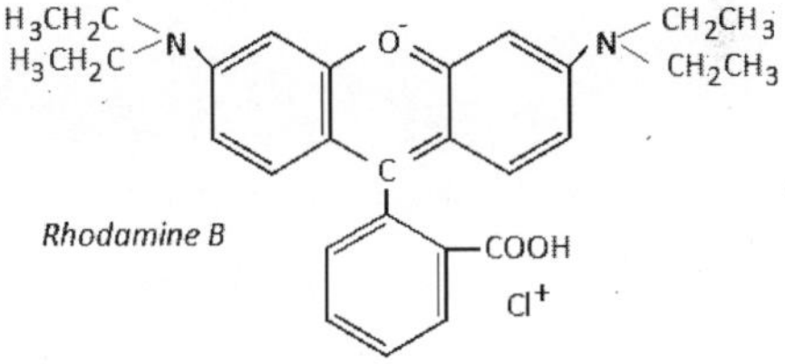

Rhodamine B

Examples are phenolphthalein, fluorescein, eosine and rhodamine.

(vi)   Azine,

It is a coincidence that the first synthetic dye manufactured mauveine belongs to this group and was an acid dye. Dome dyes of this class being still used as wool/silk dyes are C.I. Basic Red 2, aniline black, azocarmine and Wool Fast Blue, etc. Oxazine and thiazine dyes also belong to this group.

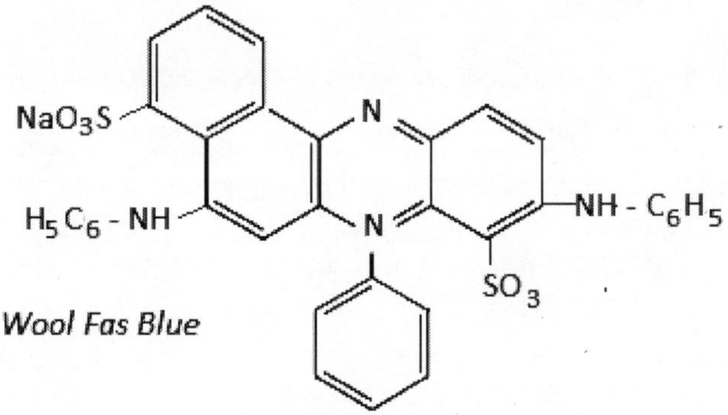

Wool Fas Blue

(vii)   Anthraquinone,

Anthraquinone dyes are derived from the parent compound anthraquinone, and belong to the quinonoid class of dyes. Simple anthraquinone derivatives are coloured when at least two further donor substituents (–OH, –NH$_2$) are present in addition to the basic anthraquinone structure. A group containing lot of vat dyes but there are a few acid dyes, especially certain blue and green types. They are, in some cases, are very difficult or impossible to discharge, so that they can be employed in the production of coloured discharges in printing. Examples:

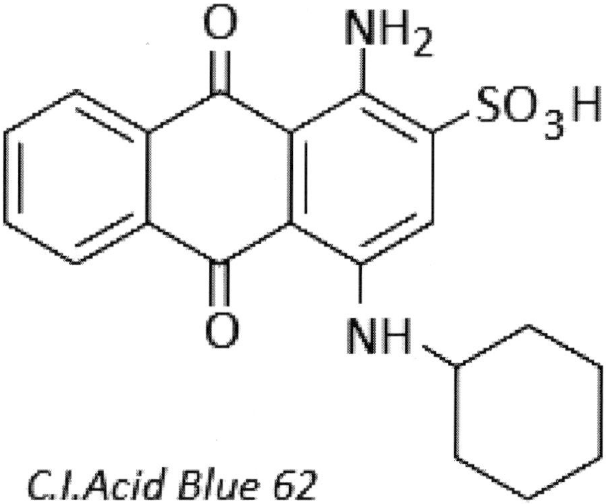

C.I.Acid Blue 62

(viii)   Phthalocyanine,

Phthalocyanines possess extraordinarily clear, luminous colour shades with extremely good light fastness and usually excellent wear fastness. It comes under general category of metal complex dyes.

1.1.6.1.1 1   Classification of acid dyes according to application

The acid dyes are more suitably classified according to application of three types, viz. (i) strong or levelling or equalizing acid dyes, (ii) milling or weak acid dyes and (iii) very weak or super milling or neutral or aggregated acid dyes.

Levelling acid dyes are generally monoazo dyes having one or two sulphonic acid groups and are held on to wool by electrostatic bonding. They are simple dyes, are markedly hydrophilic in character and this necessitates them being applied to the substrate from a strongly acidic dyebath.

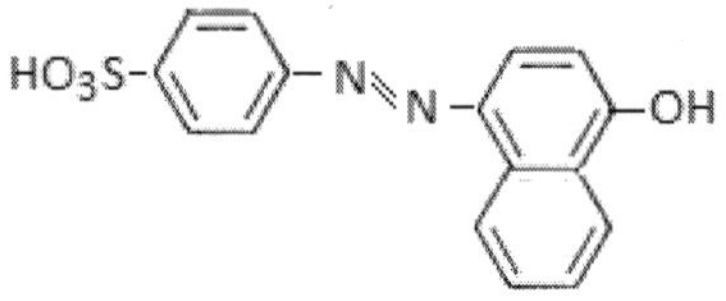

C.I. Acid Orange 20 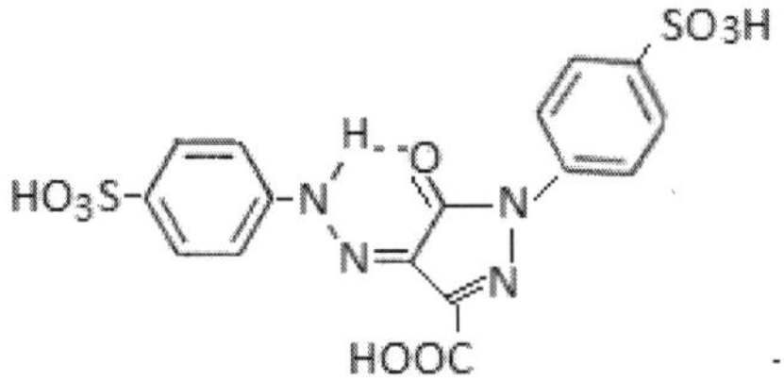   C.I. Acid Yellow 23

Milliing acid dyes are of better wash fastness due to their bigger molecular size and are usually disazo dyes. They usually have two sulphonic acid groups.

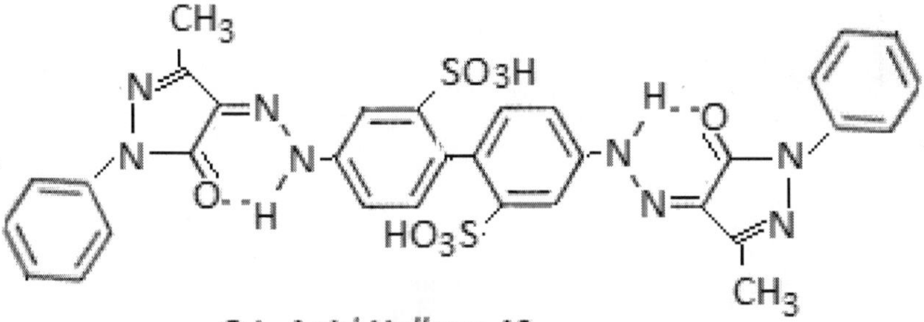

C.I. Acid Yellow 42

Super milling dyes have still superior wet fastness which is achieved by incorporating a highly hydrophobic group at the same time the molecular size is also bigger. These dyes are dyeable at neutral pH because of the non-polar bonding between this group and hydrophobic side chains in wool.

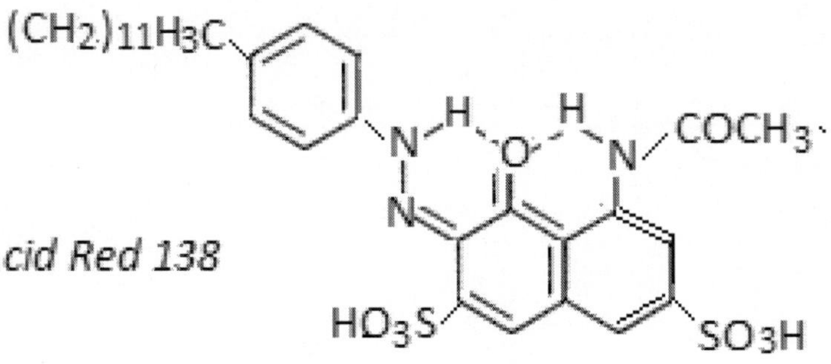

C.I. Acid Red 138

The dyeing conditions, levelling and wet fastness properties of these three classes of dyes can be summarised as follows:

Group	Acid system	pH in dyeing	Wet fastness	Levelling properties	Affinity of anions	Low molecular weight
Levelling or equalising acid dyes	H_2SO_4	2–4	Poor	Good	Low	Low molecular weight with very high solubility molecular solution
Milling acid dyes	CH_3COOH	4–6	Good	Moderate to poor	High	High molecular weight with low solubility colloidal solution
Supermilling acid dyes	CH_3COONH_4	6–7	Very good	Very poor	Very high	High molecular weight with very low solubility colloidal solution

This classification was originally devised for dyeing wool even though now it is applied for all protein fibres. The disadvantage of this classification is that there is no sharp distinction between the groups, and indeed some acid dyes are difficult to categorise.

As explained in the theory of dyeing right from the commencement of dyeing, acid dyes are seldom adsorbed evenly. The degree of unlevelness developed during dyeing is largely governed by the 'strike', the extent to which dye is initially adsorbed. With levelling dyes, which exhibit good migration, levelling can eventually be improved by movement of dye molecules from heavily dyed to more lightly dyed parts of the substrate. Milling dyes, however, exhibit poorer migration, and super milling dyes migrate only to a very limited extent.

Acid dyes are anionic in aqueous media, the anion being coloured. Most acid dyes are sodium salts of aromatic sulphonic acids.

1.1.6.1.2   *Theory of acid dyeing*

When a fibre is dipped in a dye solution, the external surface of the fibre absorbs the dye, and equilibrium is instantly established. The finer the fibres, the greater is their external surface absorbing the dye and the greater is the dyeing rate, which in the final analysis is determined by the speed of dye ion diffusion from the external layer inside the fibre. The diffusion speed vary depending on the size and shape of the dye ion, the ion charge, the dye concentration and its affinity for the fibre, the kind of fibrous material, the degree of its swelling, the dyeing temperature, and the presence of other electrolytes in the bath. Speed of wool dyeing at low temperature is very low, due the resistance of its scaly surface structure. It is needless to say that any violation of this layer (mechanical damage or chlorination) promotes an increase in dyeing speed. Wool swelling has a great influence on the dyeing process, and in swollen wool, micropores as large as 40 Å are formed through which the dye ions diffuse. When the temperature is increased to 50 °C the swelling of fibre in water is slightly reduced, but increases again with the increase in temperature and hence the acceleration of dyeing at a temperature over 50 °C is partially connected with this phenomenon. Dyeing speed is also promoted by greater swelling of wool in the presence of acids. In case of silk, since there is no scaly structure or resistance, the dye diffusion is much smoother and even as the temperature rises and in case of polyamide the hydration and the swelling of the fibre is much slower than natural fibres which explains the low speed of dyeing of this fibre.

1.1.6.1.2 1   *Mechanism of dyeing*

It has been found that when wool or silk is dyed with free acids pertaining to the class of acid dyes, maximum absorption values are in equivalent ratios and equal to about 0.08 g/equiv per 100 g of wool and 0.02 g/equiv per 100 g of silk. Polyamide fibres interact with acid dyes in the same way as protein fibres. They can bind acid dyes like simple acids by their terminal primary amino groups; however, the maximum absorption for these fibres is lower (0.005–0.006 g/equiv per 100 g of fibre) which corresponds to their lower acid capacity. This absorption ranges are same as sulphuric, hydrochloric and other acids bound by fibre. Thus, we can assume that acid dyes in the form of free acids, as well as simple inorganic and organic acids, are capable of forming compounds (salts) with wool keratin and silk fibroin.

Theory of the dyeing mechanism is studied with an experiment where the wool is dyed from a dye bath containing an acid dye (Crystal ponceau) and a strong dye (hydrochloric acid). It has been observed that there is an initial rapid absorption of both hydrogen and chloride ions. Over time, the slower diffusing dye anions displace chloride in the fibre, as demonstrated by first a rapid fall in chloride concentration in the bath followed by a more gradual rise again.

Wool retains each of $-NH_2$ and $-COOH$ groups at either ends which are capable of taking part in chemical reaction. Based on this chemical structure of wool may be better represented as $H_2N-W-COOH$, where 'W' denotes rest part of wool structure. When wool is immersed in water, H-atom attached to the carboxylic group at one end of wool is transferred to $-NH_2$ group at the other end of the macromolecules so that the two ends of wool chain acquire opposite electrical charges, called zwitter ions:

$$H_2N - W - COOH \; \overset{H_2O}{\rightleftharpoons} \; H_3N - W - COO^-$$

On addition of acid in bath containing wool with zwitter ions, some of the negatively charged carboxylate ions ($-COO^-$) take up hydrogen ions released by acid into solution and are transformed to electrically neutral carboxylic acid groups ($-COOH$). At the same time, the acid anions, released from acid in the bath are absorbed by the positively charged amino ends of keratin macromolecules.

$$H_3C - COO - H \; \rightleftharpoons \; H^- + {}^+OOC - CH_3$$

$$H_3N^+ - W - COO^- + H^+CH_3 - COO^- \; \rightleftharpoons \; NH^+_3 - CH_3 - COO^- - W - COOH$$

Due to this action, the cationic nature of wool increases with increase in time of acid treatment with generation of more and more positive sites ($-NH_3^+$) in wool through absorption of excess acid. When an acid dye

(R–SO$_3$Na) is added to this bath containing cationised wool, the dye anion gets attached with cation of wool through electrostatic force with liberation of salt.

$$H_3N^+ - W - COO^- + R - SO_3^- + Na^+ \rightleftharpoons R - SO_3^- \cdot {}^+H_3N - W - COOH + CH_3COONa$$

Some of the groups present in dye are bound to wool by hydrogen bonds too. Since the dye anion is held on the protein molecules more strongly than the acetate anion, the acetate ions taken up by the positive sites prior to addition of dye are continuously replaced by dye anions. As dyeing proceeds, transfer of different ions from solution to wool and from wool to solution takes place.

In solutions with a pH higher than the isoelectric point, i.e. in neutral and alkali liquors, keratin and fibroin acquire an excess negative ionization, and therefore conditions are not suitable for an electrostatical interaction of protein fibre with the anions in the solution. Dye uptake also depends on nature and concentration of acid (strong or weak) in bath. In acid solutions with a pH below the isoelectric point, keratin is positively ionised due to the addition of protons according to the following equation:

$$
\begin{array}{cc}
NH_3^+ & COO^- \\
COO^- & NH_3^+
\end{array}
\quad +2H^+ \rightleftharpoons \quad
\begin{array}{cc}
NH_3^+ & COOH \\
COOH & NH_3^+
\end{array}
$$

In a dyeing bath where wool (silk) + strong mineral acid + acid dye (in the form of a salt), an excessive positive ionization of keratin takes place caused by the action of mineral acids and conditions are created for the electrostatic interaction of keratin with the anions in the solution, including the dye anions (AC)$^-$, whose fixation compensates for the keratin charge:

$$
\begin{array}{cc}
NH_3^+ & COOH \\
COOH & NH_3^+
\end{array}
\quad + 2AC^- \rightleftharpoons \quad
\begin{array}{cc}
NH_3^+AC^- & COOH \\
COOH & NH_3^+AC^-
\end{array}
$$

In case of polyamide fibres also it behaves the same way as protein fibres, the amount of acid dye absorbed by polyamide fibres increases with a decrease in the medium pH,

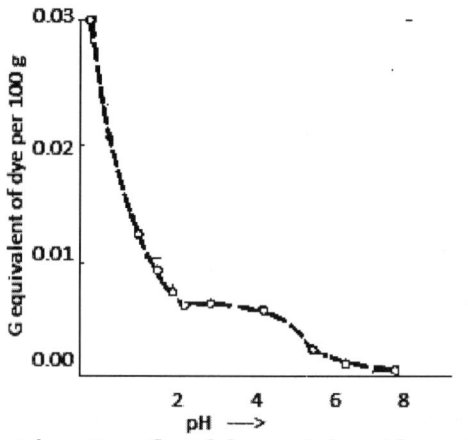

Exhaustion of acid dye on Polyamide vs pH

The high binding of acid dyes in strong acid medium is also due to the ionization of acid amide groups in the fibre and to the addition of protons retaining dye anions.

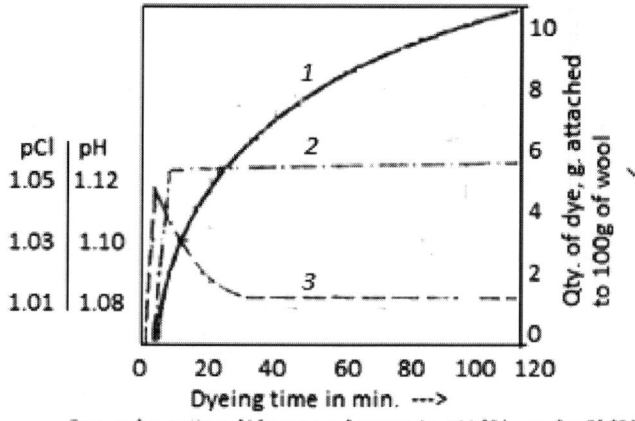

Dye exhaustion (1) as per change in pH (2), and pCl (3)

When the exhaustion of acid dyes is plotted against the changes of Cl⁻ and H⁺ ions in the bath it can be seen that acidity is reduced at the very beginning of dyeing due to the fixation of H⁺ ions by keratin, until a certain maximum is attained, after which the acidity remains the same owing to the binding of chlorine ions by the fibre, their concentration in the beginning is also reduced (pCl increases); however, this process is soon reversed with chlorine ions passing from the fibre into the solution. At the same time, the amount of fixed dye is gradually increased. Schematically this process may be illustrated as follows:

$$\begin{array}{cc} NH_3^+ & COO^- \\ COO^- & NH_3^+ \end{array} +2H^+ \rightleftharpoons \begin{array}{cc} NH_3^+ & COOH \\ COOH & NH_3^+ \end{array} +2Cl \rightleftharpoons$$

$$\begin{array}{cc} NH_3^+Cl^- & COOH \\ COOH & NH_3^+Cl^- \end{array} +2RSO_3^- \rightleftharpoons \begin{array}{cc} NH_3^+ \ ^-O_3SR & COOH \\ COOH & NH_3^+ \ ^-O_3SR \end{array}$$

Thus the acid dyeing process is only an exchange of ions.

1.1.6.1.2 2   *Effect of acids*

In wool (protein) dyeing, dye uptake also depends on nature and concentration of acid (strong or weak) in bath. With increase in concentration of acid in bath, exhaustion of dye bath also increases. H_3PO_4 has been recommended for dyeing of wool with acid dyes in place of H_2SO_4 [1% Conc. H_2SO_4=2% H_3PO_4(75%)] or CH_3COOH [1%CH_3COOH (80%) = 0.8% H_3PO_4 (75%)] as H_3PO_4 causes less damage to wool and any cellulosic fibres present in blend, if buffered with Na_2CO_3, H_3PO_4 also produces better levelled shades.

Acid dyes possess hydrophobic or non-polar head (benzene rings) and hydrophilic or polar tail ($-COOH$, $-SO_3H$). On the other side, wool contains a hydrocarbon or non-polar backbone (keratine) which feels higher affinity for the non-polar head of dye causing higher affinity of acid dye for wool; dyeing is thus analogous to extraction of non-polar solvent from an organic solvent. It also has been found that introduction of non-polar groups in dye structure, such as $-CH_3$, increase its affinity for wool many folds while introduction of polar groups ($-COOH$, $-SO_3H$) reduce affinity. In addition to an electrostatic bond (NH_3^+ D$^-$), the dye anion is attached to wool with co-ordinate linkage – the strength of which increases from equalizing dyes to aggregated dyes causing improved fastness when the latter is used. Kinetics of dyeing wool with acid dyes and nylon has been disclosed via existing explanations of wool–dye isotherm data, i.e. Gilbert–Rideal model, Donnan model and the law of mass action.

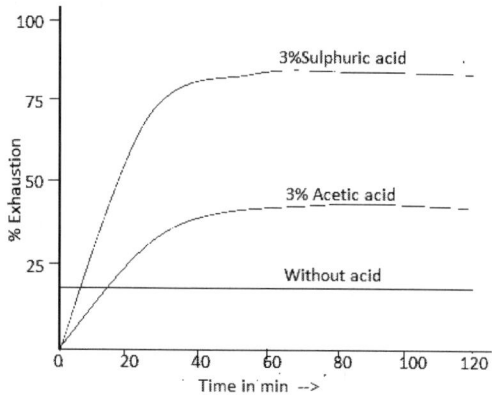

Rate of exhaustion of Solway Blue BS in the presence and absence of acid

Not only the total amount of the dye adsorbed is influenced by amount of acid, but the rate of exhaustion is also dependent upon acidity or pH of bath. When different acids are used in equivalent quantities, these produce similar extent of exhaustion on wool only if dyeing time is short; if dyeing is done below boil then weaker acid is preferred in larger quantities than the stronger ones at lower concentration.

1.1.6.1.2 3   *Fixation*

The aforesaid explanation does not fully explain the fixation of the dye on to these protein fibres. The comparatively high affinity of acid dye anions for protein, as compared with the anions of simple mineral acids, is the most important characteristic of acid dyes which determines the possibility of their practical application as dyes. Though the ionic electrostatic bond is essential for the transfer of the acid dye from the liquor onto the fibre, it is not the only way of interaction of the dye and the fibre. This phenomenon is not fully understood but there are various theories put forward by textile investigators. The main theories suggested regarding interaction of acid dyes with protein and polyamide fibres are van der Waals polar forces (dipole interaction) and the formation of hydrogen bonds and van der Waals non polar forces between hydrophobic parts of the fibre and the dye. Thus a so-called hydrophobic bond also may be playing a part in the interaction of acid dyes with protein and polyamide fibres. Hydrophobic bonding is simply the term used to describe the tendency of hydrophobic groups to associate together to avoid interaction with water, as in the formation of soap micelles and dye aggregates. The driving force in hydrophobic bonding is the collapse of the hydrocarbon–water interface that occurs when the hydrophobic groups in the dye molecule interact with hydrophobic groups in the wool. The involvement of non-ionic interaction for binding of acid dyes is further confirmed by the sorption of certain dyes by protein and polyamide fibres in amounts exceeding the acid capacity. For example when wool was with Carbolan blue B, 0.210 g/equiv of dye was absorbed per 100 g of wool. Acid dye fixation may be a total of dipole interactions, hydrogen bonding, and dispersion forces operating between groups in the dye molecule and appropriate groups in the protein. Wool, silk and nylon all have regions along the polymer chains that are quite hydrophobic, being void of ionic and polar groups. Dye–dye and dye–fibre hydrophobic interactions might be expected in such regions.

1.1.6.1.2 4   *Implications in practical dyeing*

When temperature of dye bath is increased from room temperature the dyeing rate is increases but the exhaustion is reduced. This may be due to an exothermic dyeing process. Also, stronger the dye bath acidity (pH), higher the rate of dyeing especially to the outer surface of wool fibre. As the temperature is increased the diffusion of the dye molecules from surface to inside of the wool fibre takes place. The implication of this in practical dyeing process is that one should see that the dyeing rate should be controlled in beginning stages of dyeing so that uneven colouration does not take place.

Thus there are three steps in the process of wool dyeing from an aqueous dye bath:

(i)   Adsorption of dye to fibre surfaces

(ii)   Transfer across the surfaces

(iii)   Diffusion of dye into wool fibre structures

Temperature control, quality dyeing equipment, controlling liquor circulation, should eliminate the delays and uneven access to dyes possible in the first step. For the second step, the wool scales offer a significant resistance to dye penetration. It is now accepted that dyes gain access to undamaged wool fibres mainly via junctions between wool cuticle scales. Lipids are present to varying degree at the surface and in intercellular interstices, and they present an obstacle to entry of dyes into wool fibres. In the second stage the dyes diffuse throughout all the non keratinous regions, and also the endocuticle and inter macrofibrillar material regions of the cell membrane complex. The final stage, of dyestuff diffusion inside wool fibres entails a progressive transfer of dye into the sulphur-rich matrix proteins surrounding microfibrils in cortical cells. Even after the exhaustion is completed in wool dyeing process, the diffusion into the inner layers of the fibre may not be completed. Hence, it is customary to give extended dyeing time for a proper and fast dyeing. It has been observed that in many cases the non-keratinous regions, which are so important in the early part of the dye cycle, are found to be almost totally devoid of dye after the diffusion into the inner portions of the fibre is completed. Wools with surface damage, or degraded by sunlight, chlorination or carbonising treatments for example, will respond differently when exposed to dye bath conditions.

1.1.6.1.2 5   *The effect of electrolytes*

The presence of electrolytes has a prominent effect in the dyeing of acid dyes. The neutral salts reduce dye exhaustion in acid dye baths. This peculiar effect is utilised in practice: such salts for example sodium sulphate and sodium chloride are used to regulate the dyeing process (Contrary to the general notion, that the electrolytes are added for exhaustion as in the case of many other classes of dyes). The addition of a neutral salt brings an increase in the concentration of simple anions in the dye bath and consequently a shift of equilibrium from right to left, as well as a decrease in the dye transfer onto the fibre.

$$H_3N^+ - W - COO^- + R - SO_3^- + Na^+ \rightleftharpoons R - SO_3^- \cdot {}^+H_3N - W - COOH + CH_3COONa$$

Considering the dyeing equation cited earlier, CH_3COONa is eliminated. In the starting of the dyeing process the forward reaction proceeds faster due to higher attraction between cation of fibre and dye anion, more and more salt. This fast strike rate can result in uneven dyeing. In order to reduce the strike rate excess salt is added in dye bath at the start of dyeing to ensure backward process, i.e. stripping and thereby achieving a more even dyeing; higher the concentration of salt added, higher is the rate of stripping. In other words, if an excess of salt is added to bath, stripping is favoured and it acts as retarding or levelling agent.

But in the case of super-milling (see below) dyes the electrolyte plays a different role. These dyes are dyed from a neutral bath they are actually used for exhaustion as in the case of direct and other dyes and not as a levelling agent. Salt promotes dye uptake on protein fibres by reducing zeta potential. Wool probably acquires a negative electrical potential in neutral solutions, which normally repels the negatively charged acid dye anions present in dye bath. Presence of an electrolyte reduces this surface negative charge difference between dye and fibre through absorption of positive sodium ions released form salt.

Basically, at higher ionic strengths, hydrogen ions are more readily absorbed because the repulsive electrical charge which develops at the surface boundary of the fibres is substantially masked. Exactly the same effects explain why salts, commonly sodium acetate or sulphate, are to be found in most acid dye baths. Less acid is required for a given uptake of hydrogen ions. This facilitates dye anion absorption at less extremely acid conditions and helps a dyer to reduce the acid quantity in the dye bath and avoid acid damage of wool.

In the dye bath there are many forces applicable like van der Waals (VDW) forces, electrostatic interactions, induction forces, charge transfer stabilisation effects and solvophobic interactions. But the major forces of significant size of importance may be:

i) Electrostatic interactions between static molecular charge distributions: This includes attractions such as those between protonated amino groups in polyamide fibres and sulphonated anionic dyes, but also, according to classical definitions, hydrogen bonding.

ii) van der Waals interactions: These non-covalent molecular interactions are the sum of dispersion and repulsive energies.

iii) Solvophobic or hydrophobic interactions: To have a high affinity for the wool, anionic dyes must have a hydrophobic character. The structure of the dye molecule must be such that the hydrophobic substituents are situated some distance from the polar groups.

In the presence of strong acid, the carboxyl groups on the wool are protonated first. The acid anions rapidly diffuse into the fibre to produce a state of electro-neutrality. Due to the affinity between the dye and fibre, the slowly diffusing dye anions replace the inorganic anions as the dyeing time increases, i.e. the dye is bonded to the fibre in an ion exchange reaction. These types of dyeing conditions are mainly to be found in the case of dyes with a lower affinity for wool, such as levelling acid dyes. The high concentration of cationic groups at pH 2 provides the necessary attraction for the dye anions. Under these conditions, the addition of salt, for example sodium sulphate, may have a levelling effect. Because of their superiority in numbers, the additional sulphate ions compete with the dye for the cationic ammonium groups on the wool. Thus the addition of an electrolyte is not for exhaustion in case of acid dyeing of wool but for a levelling action.

In the case of weakly acid dyeing dyes the levelling agent acts in a different way. During dyeing of a weak acid dyeing dye, the driving force for bath exhaustion is the potential difference between the dyes with negatively charged SO_3 groups and the positively charged wool. In the presence of a levelling agent (auxiliary) it forms a complex with it and an increased total negative potential which produces increased dye affinity. This again results in increased bath exhaustion, hence achieve level deeper dyeings and subsequent advantages. Once the dye is diffused in sides the fibre the levelling agent remains in the liquor.

- vely charged dye ion   positively charged levelling agent   Increased negetive charged dye ion

Dye auxiliary complex   wool   Dye inside the fibre   Auxliary left in the bath

Schematic representation of increased negetive chanrges of dye auxiliary complex and the affinity to fibre and subsequent entry into the fibre at higher temperature leaving the auxiliary in thebath itself

1.1.6.1.3   Classification of acid dyes according to application

Based on the above theories, the acid dyes used in wool dyes can be classified into three groups based on the acid requirement in the dyeing bath.

1. Strongly acid dyeing dyes.
2. Moderate acid dyeing dyes.
3. Weakly acid or neutral dyeing dyes.

The levelling properties and wet fastness properties of these dyes and application procedure are interrelated. We can generally say, strongly acid dyeing dyes have good levelling property or poor affinity and poor wet fastness and weakly acid dyeing dyes have poor levelling properties but high affinity and good wet fastness.

Sometimes another sub class is made in this class called *Fast acid dyes*. The difference of this group compared to equalising acid dyes is that they are of somewhat higher molecular weight and are usually monosulphonated ones. The dyeing method is also similar. They have reasonable migration at boil, and sometimes a levelling agent is recommended. Fast acid dyes are used when level dyeing is necessary with good washing and perspiration fastness.

Dyeing properties of acid dyes:

Dye bath	Process examples	Affinity for the dye	Levelling power	Wet fastness	Remarks
Strongly acid	10% Na$_2$SO$_4$	Moderate	Good to very good	Moderate	No levelling problems
	2–4 H$_2$SO$_4$				
	pH 2.5–4				
Moderately acid	10% Na$_2$SO$_4$	Fairly high	Fairly good	Fairly good	
	3–4% Acetic acid 40%		0.35		
	pH 3.5–5		0.25	0.06	
Weakly acid or neutral dyeing	4% NH$_4$COOH or (NH$_4$)2SO$_4$	6.3–6.5			
	pH 4.5–6.5	Very high	Moderate to poor	Good to very good	Level dyeing obtained by controlling adsorption

1.1.6.1.4   Dyebath constituents

A typical dye bath contains dyes, sulphuric acid (or sometimes formic acid), and Glauber's salt (hydrated sodium sulphate, $Na_2SO_4 \cdot 10H_2O$).

Function of dye bath assistants:

1.1.6.1.4.1   Sulphuric acid

When wool is immersed in strongly acid solutions, such as sulphuric acid solution, its carboxyl groups lose their negative charges and overall positive potential is established on the fibres (see Sect. 1.1.6.1.2). Anions, such as sulphate and dye ions are then attracted towards the positively charged amino groups of the wool (protein fibres).

1.1.6.1.4.2   Glauber's salt

If Glauber's salt is included in the dye bath, the number of sulphate ions present is raised. The competition between sulphate and dye anions for the positive sites on the wool increases. The sulphate ions, being smaller, reach the positive sites more easily, but dye anions are eventually preferentially adsorbed because of their much higher affinity. This competition leads to good migration and levelling, but also to incomplete adsorption from the dye bath.

1.1.6.1.5   Equalising acid dyes

Acid dyes are sodium salts of sulfonic acids, which are azo, anthraquinon, xanthene and tri-phenyl methane dyes. These are planar in structure. While dyes may be synthesised from many different chemical groups, they are usually classified into three groups on the basis of their technical properties such as dyeing method, levelling properties and wash fastness.

These dyes are also known as 'levelling acid dyes' (relative molecular mass (rmm) values of around 300–600) or 'equalizing acid dyes' due to their capability to produce mostly levelled shades, and the dyes are adsorbed rapidly at a pH 2–3 even at 40 °C. In terms of size and shape, often an important consideration in the design of dye molecules, these dyes may be described as a small to medium-sized planar molecule. This allows the dye to penetrate easily into the fibre and also permits a degree of movement or migration within the fibre as the ionic bonds between the dye and the fibre are capable of breaking and then re-forming. This type of dye is termed acid-levelling since, as a result of its ability to migrate, it produces a level or uniform colour. However, as the dye is not very strongly bonded to the fibre and because of their low molecular weight with higher water solubility they possess poor wash fastness, which can be improved through suitable post-dyeing treatments.

Acid dyes can be further subdivided into two groups (1) disulphonated and (2) monosulphonated. The monosulphonated types are of lower relative molecular mass (typically 300–500), slightly more level dyeing and have marginally better fastness than disulphonated products.

1.1.6.1.5.1 Dyeing process

Method 1: Dye bath is prepared at room temperature with sodium sulphate (5–10%) and formic acid or sulphuric acid to a pH of 2.5–3.5. After entering the material and equalizing the bath after 10 min

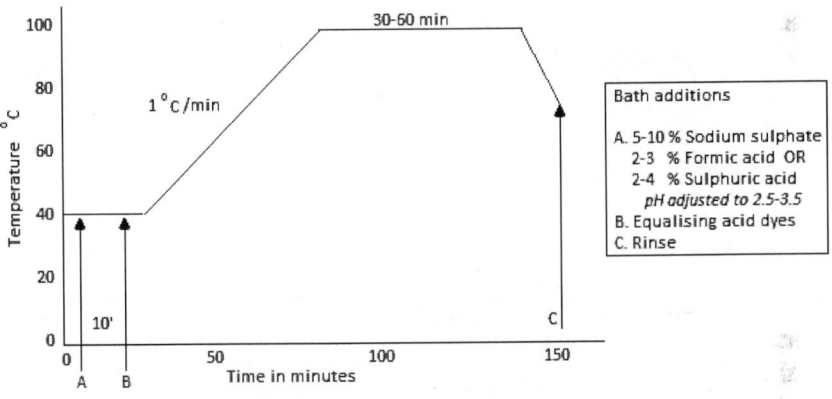

Dyeing process for Equalising acid dyes on wool

the dissolved dye is added to the bath. After working the material at this temperature for a while the temperature is raised to boil at 1 °C/min. Dyeing is continued at boil and completed in 30–60 min as per the depth of shade, after which the bath is cooled below 80 °C and drained. Rinsing and soaping is followed.

Method 2: At A set bath at 50 with 2% formic acid (85%) 5% Glauber's salt anhydrous,pH 3.5–4.5. At B add required amount of dye. At C add 2% sulfuric acid (96%) or 2% formic acid (85%). Thoroughly rinse after dyeing to remove loose colour.

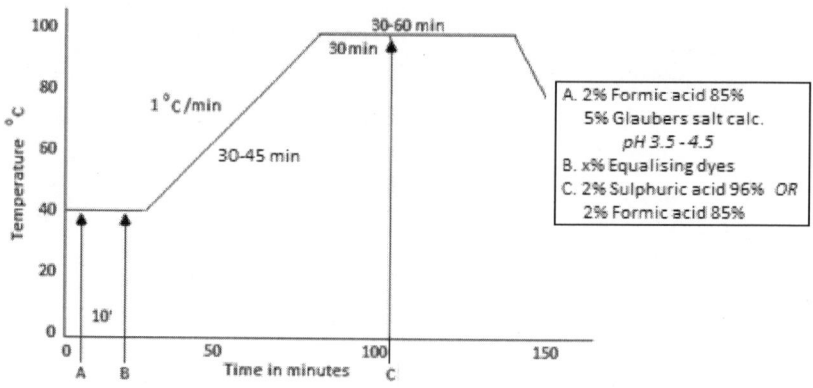

Dyeing process for Equalising acid dyes on wool (2)

Notes:

1. The level of equilibrium exhaustion is determined by the rmm, the level of anionic salt ions, dyebath pH and the number of solubilising groups on the dye molecule.

2. Under acid dyeing conditions, a majority of the substituent amine groups (e.g. arginine, histidine and lysine) in wool will be protonated. Control of uptake of the dyes is achieved with sodium sulphate (Glauber's salt).

3. For unlevel dyeing a further addition of sodium sulphate may be made to increase the concentration to 15–20% and continue boiling for another 30 min or till uniformity is achieved.

4. For shade corrections the boiling is stopped by turning of the steam for some time and the addition of dyes may be made. After running in the cooling bath for 5 min the temperature is raised to boil again and continued dyeing for another 30 min at boil.

1.1.6.1.6 *Fast acid or half milling acid dyes*

Fast acid dyes have a higher relative molecular mass (500–700) than acid levelling dyes. Dyes belonging to this group possess better wash fastness than that with strong acid dyes but do not migrate well and are called 'milling acid dyes' as these are fast to milling process imparted to woollen textiles and dyeing is carried out using acetic acid at around pH 5.2–6.2. Wash fastness of dyeings is good due to high molecular weight of dye rendering low solubility. Levelling is moderate as affinity of dye anion for fibre is quite high.

1.1.6.1.6.1   Dyebath constituents

A typical dye bath contains dyes, acetic acid and Glauber's salt.

Functions of dye bath assistants:

Acetic acid: Milling dyes, which generally have higher molar masses and lower solubility's in aqueous media than levelling dyes, are also more substantive to wool. Milling dyes, therefore, depend to a much smaller extent on ionic attraction for complete adsorption, and fewer positively charged sites on the wool fibres are required. Thus, a dye bath of higher pH is desirable, and this is conveniently obtained by using acetic acid rather than sulphuric acid. Indeed, if the pH is low, the attraction between dye and fibres is too high; the adsorption of dye is rapid and the dyeing is unlovely.

Glauber's salt: Glauber's salt still promotes levelling, migration, etc. as in the case of levelling dyes. However, its effectiveness is reduced, since the anions of milling dyes have higher affinity and there are fewer positively charged sites on the wool fibres.

1.1.6.1.6.2   Dyeing procedure:

Dye bath is set at 40 °C with CH_3COOH_3 (1–3% owf) and Glauber's salt (10% owf). Dyeing is started at 50 °C with pH around 4.5–5.0, raised to boil in 30–45 min and kept at boil for further 30–60 min at pH 5.2–6.2. If adsorption of dye is not sufficient by this time, some further acetic acid may be added.

Same dyeing profile given for acid levelling dyes can be used, the only difference being the pH is adjusted to 4.5–5.0 with acetic acid and if necessary a levelling agent is added.

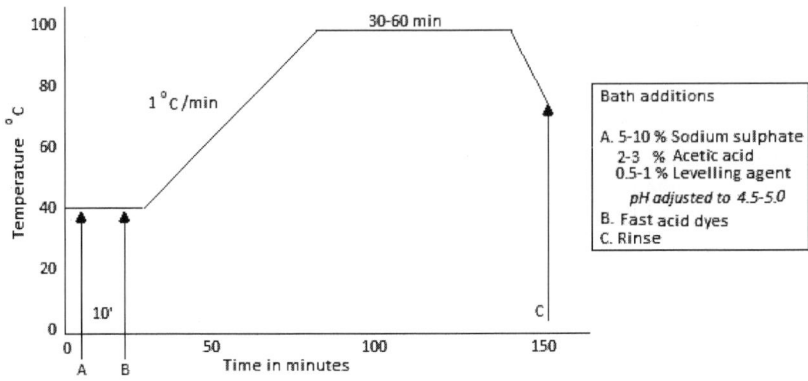

Dyeing process for half milling acid dyes on wool

Notes:

1. Wool has a problem of dye affinity difference between root and tip and can dye differently which is called tippy dyeing. This can be solved to some extent by using fast acid dye range of Clarient (Optilan MF range) and by adding 1.0–2.0% formic acid to the dyebath after 30 min at the boil and then boiling for a further 30 min.

2. Correction of dyeing can be done by cooling the dyebath to 60–70 °C, adding predissolved dyes, running for 5 min and again raising to the boil and boiling for 30 min.

1.1.6.1.7   *Super milling acid dyes*

Dyes are applied from a neutral bath with acid liberating agent, e.g. ammonium acetate, sulphate or phosphate which liberates respective acid beyond 80 °C to develop required pH. Dye anions possess higher affinity for fibre even at neutral pH requiring a minimum of acid. These are commonly known as 'super milling dyes' due to their high fastness to milling. Dyes possess poor levelling property due to very high affinity and consequent higher strike rate for fibre. Wash fastness is excellent due to larger molecular size of dye due to which remains insoluble in water. These are also known as 'aggregated acid dyes' due to their availability in the form of aggregates, necessitating dyeing at higher temperature by breaking down these.

1.1.6.1.7.1   *Function of acid liberating agent*

These are larger dye molecules with relative molecular mass (rmm) values of 600–1000 and owing to the higher rmm, two or more solubilising groups are often used to give enough solubility. Super milling dyes possess the highest affinity of the acid dyes. They require virtually no ionic attraction for good adsorption and so are applied from an almost neutral dye bath. Significant acidity in the dye bath leads to very rapid, unlevel adsorption of dye. The ammonium salt serves to maintain the pH of the dye bath, by neutralising any alkali in the wool residual from scouring. In addition, it can slowly liberate acid as dyeing proceeds, e.g.:

$$NH_2OOCCH_3 \rightleftharpoons NH_3\uparrow + CH_3COOH$$

The pH of the bath will fall as the ammonia escapes, and adsorption of the dye will be assisted. However, if the dyebath is enclosed, ammonia cannot readily escape and the liberation of acid is then inhibited.

General properties of acid dyes

Property	Self-levelling acid dye	Milling acid dye	Super milling acid dye
Molecular size	Very small	Relatively bigger	Biggest
Affinity	Less	High	Very high
Migration property	Excellent	Poor	Very poor
Levelling	Excellent	Moderate	Poor
Wash-fastness	Poor	Good	Very good
Dye bath pH required	2–4	4–6	6–7
pH adjusting agent	Sulphuric/formic acid	Acetic acid	Acid liberating salts
Water solubility	High	Low	Low
Light-fastness	Very good	Good	Good
Behaviour in solution	Molecularly split	Aggregates	Aggregates

1.1.6.1.7.2   Dyeing process

1.1.6.1.7.2.1 Dyeing at boil

Wool dyes for application by exhaust process are usually dyed at boil. The sequence being

Setting the bath → Heating up → Boiling → Cooling down

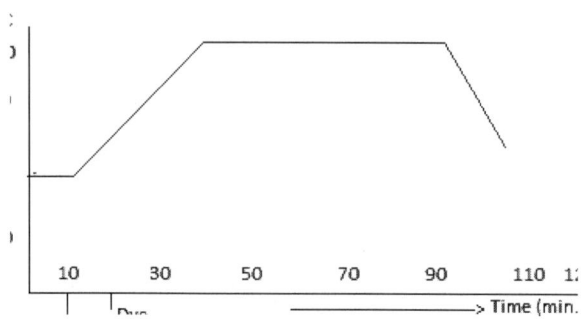

The duration of each case is determined by the substrate, dye, dyeing assistants and the machine.

Method I: Sulphuric acid and Glauber's salt –strongly acid dyeing equalizing dyes

The low molecular weight dyes of this group (e.g. Sandolan E) usually have only on sulphonic group. They are applied from a bath set with sulphuric acid (pH 2.5–3). The amino groups in the wool attract the hydrogen ions donated by the sulphuric acid, thus making the wool cationic. The dye anion is therefore able to enter into an ionic bond with the wool. Other intermolecular forces also operate at the same time. Since the dye molecule is very small and the bond is not very strong the dye anion can detach from a present position and again make another bond at another site. Thus,the dye is said to migrate from one area of higher concentration to another area of lower concentration. The absorption of the dye is rapid in the initial stages and hence to achieve a level dyeing the migration stage of the dyeing is very important.

The levelling properties of the dyes can be enhanced by the addition of Glauber's salt. One should understand that here the Glauber's salt does not function as an exhausting aid but as a levelling agent as the mobile sulphate anion tend to take up the cationic site on the fibre, thus preventing the dye anions being completely absorbed by fibre which result in unleveled dyeing. It is understood such dyeing are not very fast and perspiration, or lying wet for long time, washing can bleed the dye.

Dyeing is carried out with an addition of

Quantity	Unit	Additions
10–20	%	Glauber's salt cryst.
2–4	%	Sulphuric acid 168 °Tw (66 °Be)

The goods are entered and the temperature is slowly raised to the boil. Dyeing is continued at the boil for 45–60 min and the goods then rinsed.

Carbonised, unneutralised material is dyed with the assistance of Glauber's salt only.

Method II: Acetic acid and Glauber's salt, and a subsequent addition of sulphuric Acid – moderately acid dyeing equalizing dyes

Acid dyes of high molecular weight, are completely taken up by the fibre from a weakly acid or neutral bath (pH 6–8) and become permanently fixed to the wool. Strong non-polar van der Waals forces form the predominant link between these large dye molecules and the wool, while the ionic bond between the sulphonic group in the dye and the amino groups in the wool plays only a secondary role. For this reason, Glauber's salt has no levelling action with neutral-dyeing dyes. This explains why neutral-dyeing dyes migrate only to a small extent during dyeing at the boil. Producing level dyeing with dyes of this class is therefore not; as straightforward as with the acetic or sulphuric acid dyeing equalising dyes.

Another important point is that the equalizing dyes by virtue of their rapid diffusion penetrate the all wool fibres uniformly, the neutral dyeing dyes of high molecular weight, on the other hand, usually show some selectivity between the root and the tip of the wool fibre, and between wools of different origin, fineness or pretreatment, so that dyeings do not have such a "solid" appearance as those produced with equalising dyes, neutral-dyeing dyes are not freely combinable and they are therefore most suitable for self shades. Levelling is greatly improved by the use of levelling agents (Glauber's salt or specialised products)in conjunction with careful control of the pH, adsorption and temperature.

Dyeing is carried out in the presence of

Quantity	Unit	Additions
10–20	%	Glauber's salt cryst.
3–5	%	Acetic acid 40%

The goods are entered at approx. 50–60 °C, the bath is brought to the boil within t hour and kept at the boil for 1 hour. An addition is then made of 1–2% sulphuric acid 168 °Tw. (66 °Be), and boiling continued for a further 30 min.

Carbonised goods should either be neutralised before dyeing or treated with an excess of sodium acetate. The sodium acetate bath can subsequently serve as dye bath.

Method III: Acetic acid and Glauber's salt – Weakly acid dyeing acid dyes

The bath is set with

Quantity	Unit	Additions
10	%	Glauber's salt cryst.
3–5	%	Acetic acid 40%

The goods are entered at 50–60 °C, the bath brought to the boil within 20–30 min and kept at this temperature for 1 h. If the bath is not completely exhausted, a further addition of is made and boiling continued for a further 1 h.

Quantity	Unit	Additions
0.5–1.0	%	Acetic acid 40%
0.125–0.25	%	Formic acid 85%

Method IV: Ammonium salts – neutral dyeing acid dyes

This method is used when levelling difficulties are to be expected, even in an acetic acid bath.

The dye bath is set at 40–50 °C with:

2–5% ammonium acetate, or

2–5% ammonium sulphate.

The bath is then slowly brought to the boil and kept at the boil for about 1 h.

Note:

Follow manufacturer's directions. Certain dyes may not allow vigorous boiling.

Method V: Neutral Glauber's salt bath–neutral dyeing acid dyes

This method is chiefly used in the dyeing of anions. Since it is little used in all-wool dyeing, a general method only will be given here.

The dye bath is set at 50–60 °C with:

10–20 g/l Glauber's salt cryst.

The goods are entered and the bath is slowly brought to a gentle boil. It should be borne in mind that, in many cases, Method V does not give such a good yield as when dyeing is carried out in an acid bath.

1.1.6.1.7.2.2 High temperature dyeing

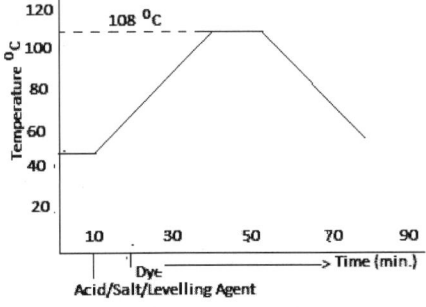

Dyeing with acid dyes at High Temperature

The acid dyeing can be done at temperature higher than 100 °C, usually 105–108 °C, where the dyeing time can be reduced to half compared to the dyeing at boil. In this method the fibre damage is minimal. The main advantage is that the total dyeing time is reduced by 35–40%. The main disadvantage being the shading additions or prolonged dyeing at high temperature is not advisable since it may be detrimental to the fibre.

Dyeing can be done as per the dyeing graph given above.

1.1.6.1.7.2.3 Dyeing below boil (80–85 °C)

Wool can also be dyed at temperature below 100 °C by using special wool dyeing assistants and selected dye. Dyeing at 80–85 °C gives the following advantages:

1. No yellowing of wool when dyeing brilliant pastel shades.
2. Minimal felting and less creasing on the winch when dyeing delicate pieces.
3. Improving spin ability of loose stock and slubbing, particularly in the case of deep chrome dyeing.

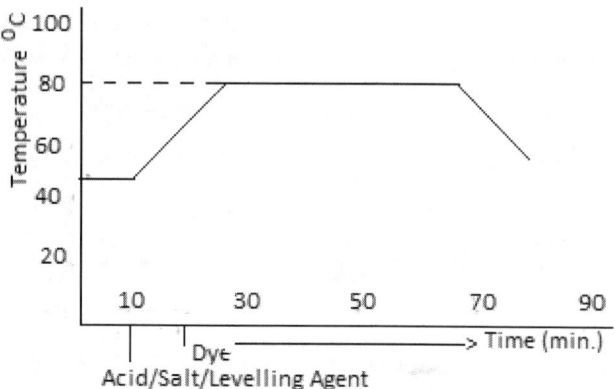

Dyeing with acid dyes below boil (80-85 ^{o}C)

The technique is not without limitations; a level, completely penetrated dyeing is not always obtained with woven wool fabrics since there is scarcely any migration of the dyes at temperature below 90 °C.

1.1.6.2   Chrome dyes (mordant dyes)

Before the introduction of synthetic dyes, natural dyes were applied to various fibres using a mordant. The material was first impregnated in a solution of a metal salt, this process was called mordanting, and then the dye was applied. These mordants used to help the dyes to improve fastness especially wash and light fastness. Many metal oxides and salts were used those days but now it is mainly chromium.

A mordant is a simple chemical which possesses affinity for both fibre and dye. If a dye has less or no affinity for fibre, the mordant previously applied improves affinity of that dye and makes a dye–mordant–fibre complex. Mordants are metal salts and are electrically cationic. In contrast, all acid dyes are anionic in nature, and their affinity is increased to large amount when mordant is applied on the fibre before application of the dye. Various natural dyes, such as logwood black or madder, are also used for dyeing wool with the help of mordants such as alum, chrome, iron and tin salts. Brighter colours and better fixation can only be obtained by pre-mordanting. Some substantive or non-mordant dyes, such as turmeric, berberis, dolu,annato and henna, contain tannin in the colouring matter itself as a natural mordant, and can produce fast colours when applied at the boil.

This process of mordanting and dyeing technique is seldom used now. But the principle is used in dyeing mordant dyes. Such dyes were basically acid dyes with the ability to form in addition a stable complex with chromium. All chrome mordant dyes are capable of forming coordination complexes with chromium which are called lakes, but the nature of the complexes is far from completely established. It is known that chromium exists in the trivalent state in these complexes and it has been thought probable that 1:2-complexes are mostly formed. (Such complexes would contain one equivalent of chromium to two equivalents of the parent dye.) However, it is likely that multiple bonds between the wool keratin, chromium and dye can also be formed. Most commonly, this takes the form of two hydroxy (OH) groups on either side of (ortho to) the azo group of a monoazo dye, as shown below.

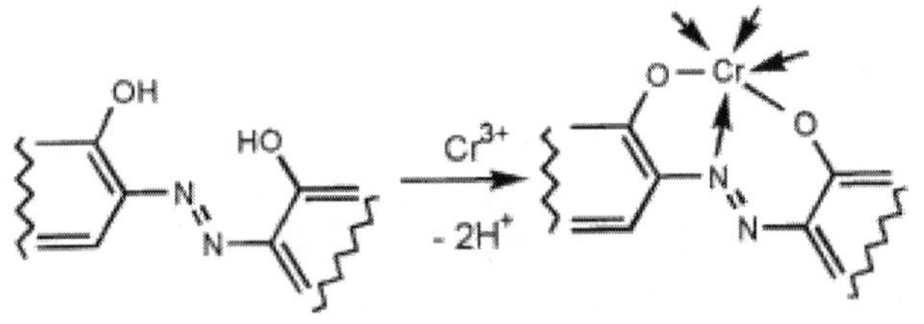

Mordanting process

The mordanting process makes a chromium complex of the dye within the fibre. A dye of this type acts as a tridentate ligand. This term implies that the dye molecule forms three bonds simultaneously with the chromium; in this case the chromium bonds with two oxygen atoms derived from the hydroxy groups and with one nitrogen atom of the azo group. The chromium atom in complexes of this type always bonds to six atoms and the complexes show octahedral geometry. It is not established with certainty how the remaining three valencies of chromium are satisfied in the mordant dyeing of protein fibres, although the possibilities include bonding with water molecules, or coordinating groups on the fibre or with a second dye molecule. Principal problem currently with chrome mordant dyes is environmental associated mainly with the undesirable presence of chromium residues in dyehouse effluent.

The coloured product formed within the fibres was largely insoluble in water. Thus, mordanting conferred a degree of wet-fastness upon the dye; it also tended to improve the light fastness. Although many metals were used as mordants at one time, nowadays only chromium is used to any extent. Dyes capable of combining with chromium are still commonly applied to wool, their mode of action relying on the high content of the aminoacid, cystine, in wool keratin chains. Chrome mordant dyes have generally good light- and wet-fastness properties, but they nearly all possess dull shades. However, in practice, chromium mordants have to be used with great care, to prevent undesirable discharges of chromium.

All mordant dyes are acid dyes, but the reverse is not true. In simpler techniques, chromium salts are invariably used for complex formation, the reactions are carried out during dyeing to obtain the dye–metal complex. These are often referred to as chrome dyes. Chromium salts act as efficient mordants, e.g. $Na_2Cr_2O_7$, $K_2Cr_2O_7$. Potassium chromate is sometimes used instead of dichromate, but in solution one species changes to the other on adjusting the pH in the range 6–8:

$$Cr_2O_7^{2-} + 2OH \rightleftharpoons 2CrO_4^{2-} + H_2O$$
$$2CrO_4^{2-} + 2OH \rightleftharpoons Cr_2O_4^{2-} + H_2O$$

As mentioned earlier, the mechanism of mordanting is highly complex and is still not yet fully understood. The first stage, the interaction of dichromate ions with amino groups in the wool, is similar to the adsorption of acid dye anions:

$$2\text{—}NH_3^+ \ ^-OOC\text{—} + Cr_2O_7^{2-} + 2H^+ \longrightarrow \left[\text{—}NH_3^+\right]_2 Cr_2O_7^{2-} + 2HOOC\text{—}$$

The dichromate ions, which contain hexavalent chromium, are then reduced by a complex route to trivalent chromium ions.

Mordant dyes do not only promote dye–metal complex formation to improve fastness, but they also enhance acid dye uptake due to the cationic nature of chromium salts when applied to wool before dyeing. These mordant dyeing methods are useful tools for cottage industries to dye wool with superior wash fastness. A single mordant with different mordant dyes, or vice versa, can produce a wide range of hues. Mordant dyes are classified chemically as azo, anthraquinone, oxazine, xanthene, triphenyl methane and nitroso and thiazine types, in which only azo dyes have a good spectral colour range and are subsequently more prominent.

These dyes are soluble acid dyeing dyes but have special atomic groups that enable these dyes to form a co-coordinating complex with heavy metals. Normally chromium salts are not environment friendly. They possess the highest fastness properties of any of the classes of dyestuffs used in wool dyeing. Hence these dyes can be used for dyeing wool at all stages like loose wool, slubbing, yarn and piece. They level well and are easy to apply and due to the method of dyeing it is not easy to shade the dyeings. Normally, metal complex dyes are the products most suitable for shading mordant dyeings. If the highest fastness requirements are demanded of worsted and woolen cloths, especially uniform and face cloth, dyeing is carried out with mordant dyes.

Some of the mordant dyes:

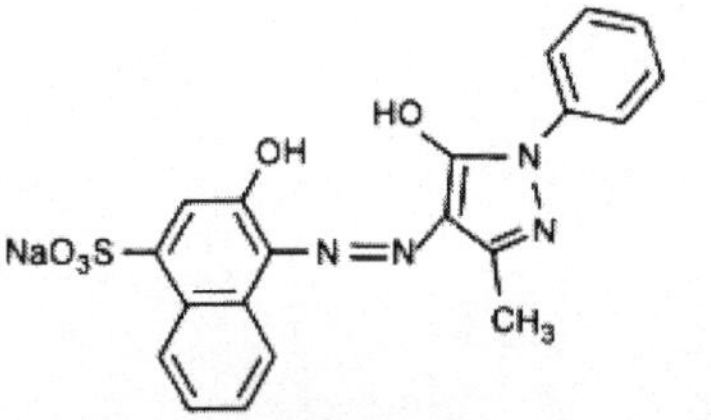

C.I.Mordant Red 7

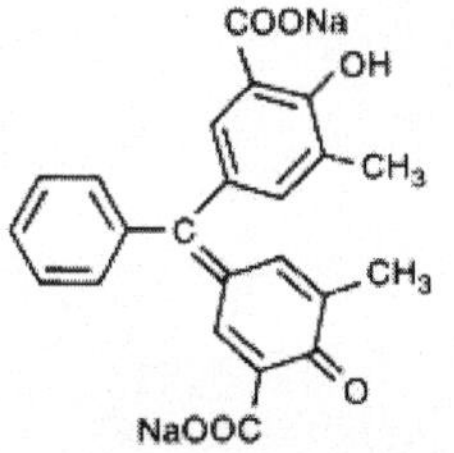

C.I. Mordant Blue 1

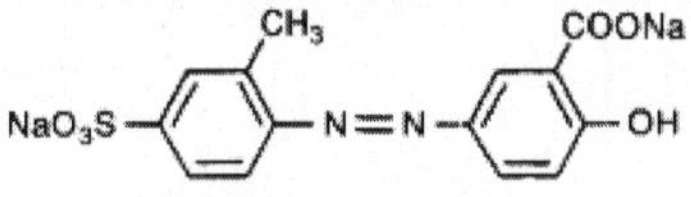

C.I.Mordant Yellow 60

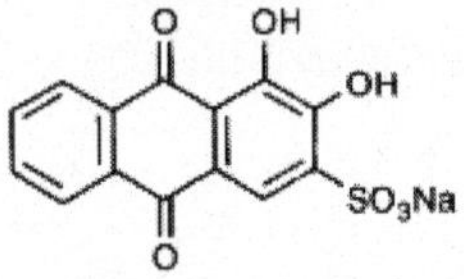

C.I.Mordant Red 3

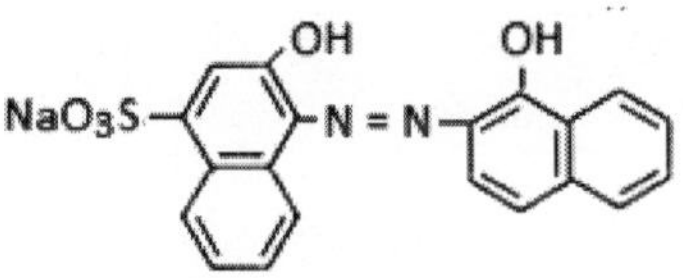

C.I.Mordant Black3

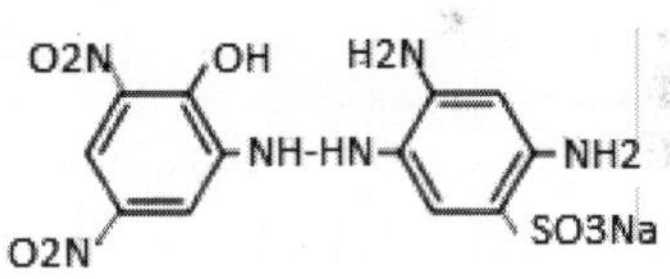

C.I. Mordant Brown 33

1.1.6.2.1 Theory of chrome dyes and dyeing

Metal-complex formation

The capacity of transition metal ions to form simple salts in aqueous solution by reacting with anions is used in the formation of complexes with dyes to increase its fastness properties. The simplex example is the formation of salts such as $FeCl_3$, which results from a combination of one Fe_3^+ and three Cl^- ions. They also combine with anions and neutral electron donors to form complexes involving coordinate covalent bonds. For example, potassium ferricyanide, $K_3Fe(CN)_6$, consists of three K^+ ions and one complex ion

$Fe(CN)_6^{3-}$. Coordinate bonds involve sharing of electron pairs that originate with the anion or electron donor.

$$\overset{+++}{Fe} + 6\overset{-}{CN} \longrightarrow Fe(CN)6^{---}$$

The species providing the electron pair is a ligand. The orbital containing the lone electron pair of the ligand donor overlaps with a vacant orbital of the transition metal ion acceptor. For a transition metal, the lowest energy vacant orbitals available are d orbitals or their hybrids. The maximum numbers of ligands that can attach to the metal ion are generally 4 or 6 and is called the coordination number.

In case of Cu, the coordination number is 4. When the free ions are in aqueous solution the available water molecules acts as electron donors and forms a complex. But if an ammonium salt is available in the aqueous solution, the ammonium ions which are stronger electron donors (stronger ligands) will replace the water molecules. These type of bonds are covalent bond it is difficult to severe this bond especially when the ligand is a strong electron donor. The action can be represented as follows:

$$\overset{++}{Cu} + 4H_2O \longrightarrow Cu(H_2O)^{++} \xrightarrow{4NH3} Cu(NH_4)^{++} + 4H_2O$$

If anions act as ligands, then the charge on the complex ion will be less than that on the hydrated metal ion and can be negative, as in the case of ferricyanide, or even zero. This can be understood if to a solution of the complex $CoCl_3 \cdot 4NH_3$ is added excess to silver nitrate solution. One expects all the chlorides has to be precipitated as silver chloride. But only one third of the chloride is only precipitated. The balance chloride ions are strongly bonded via covalent bonds and not available for forming silver chloride.

$$\overset{+++}{Co} + 4NH_3 + 3Cl^- \longrightarrow Co(NH_3)_4 \overset{+}{Cl_2} + \overset{-}{Cl}$$

Since the coordination number is 6 it combines with $4NH_3$ molecules and two chloride ions and thus Co^{---} ends up as +. The third chloride ion is the counter-ion to the complex cation and it is this that precipitates as silver chloride. Different metal ions have different coordination numbers, but in textile the most common metal ion used for mordanting purpose is chromium (to a lesser extent Cu and Co) whose coordination number is 6. Cr and Cu has the advantage that these resist demetallisation during process and subsequently. Most of the dyes suitable for mordant (ligands)acts as tridentate.

The above type dye (mordant dyes) will have generally, two hydroxyl groups or one hydroxyl and one amino group, or one hydroxyl and one carboxyl group ortho-oriented in respect to the chromophore group, as well as two hydroxyl or one hydroxyl and one carboxyl group ortho-oriented relative to each other in their structure. With this constitution, salt-forming groups (usually OH) and groups forming a coordinated bond with the chromium atom

(for instance $\underset{C=O}{\underset{\scriptstyle}{>}}$, $\overset{OH}{\underset{O}{<}}$, $-N=N-$) interact with chromium and may enter the composition of internal complex salts at the formation of five or six membered rings.

The trivalent chromium taking part in the complex formation may interact, for instance, with three dye molecules; by the forces of the primary valency it is attached to salt-forming groups and by coordinate bonds it is connected with corresponding co-ordinately unsaturated atoms of oxygen or nitrogen in keto-, carboxyl or azo groups.

This process where hydroxyl group is involved may be represented by the following way

or for salicylic acid derivatives where carboxyl group is involved.

The great majority of known chrome dyes compose the following anthraquinone and azoic groups:

1. o,o'-dihydroxy azo dyes, for instance, Acid fast chrome blue,

2. o-amino-o'-dihydroxy azo dyes for instance Acid chrome brown R,

3. derivative of salicylic acid for instance Acid chrome yellow,

4. Azoic dyes which originally have no lake – forming groups, but acquire the latter when they pass into a quinoid form as a result of oxidation at the interaction with the salts of sexivalent chromium; for instance, Acid red 2C belongs to this group,

and its oxidised quinoid form can be represented as:

5. derivatives of hydroxyanthraquinones , for instance, Acid bluish black anthraquinone C,

examples of commercially available mordant dyes and complexes

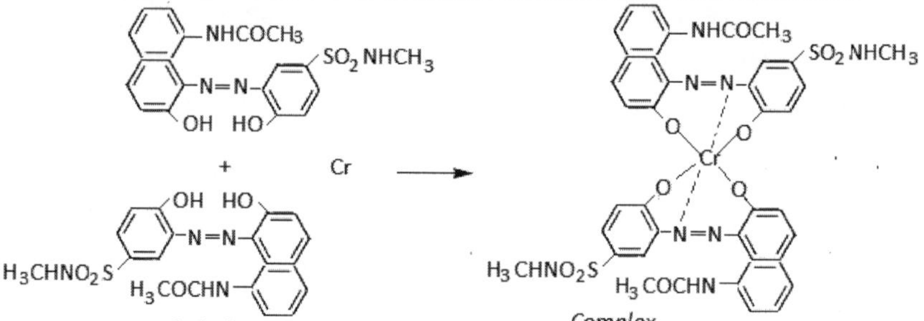

C.I. Acid Black 60

Complex

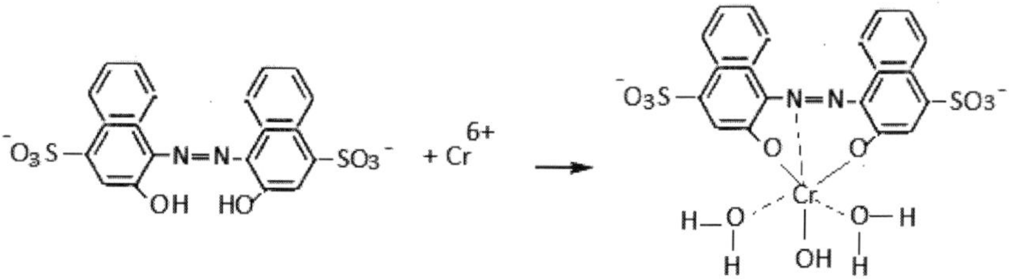

C.I. Acid Blue 158

Chromium complex

1.1.6.2.2   Application of chrome dyes

Both dye and mordant have to be applied under conditions which promote their combination within wool fibres. The form of chrome used is nearly always potassium (or sodium) dichromate.

Dyeing methods:

There are three ways of applying chrome dyes:

(a) Chrome mordant or Onchrome method where dichromate is added before dyeing.

(b) After chrome method where dichromate is added after dyeing.

(c) Metachrome method where dichromate and dye are applied simultaneously.

In all cases it is necessary to avoid lake formation in the dye bath and on the fibre surface.

1.1.6.2.2.   Onchrome method

This is the oldest method. However, it is nowadays least used, for it requires two baths, one for mordanting and the other for dyeing. It is, therefore, often uneconomical in comparison with other methods, but some chrome dyes can be applied only by this method. After preliminary chroming, wool is thoroughly washed to eliminate completely the bichromate solution to avoid any precipitation of the complex in the bath or on the fibre and then dyed in a separately prepared bath.

The mordant or dichromate can be applied first on the material in the neutral or acidic medium. When applied in neutral medium the process is called a 'sweet chrome' process and if acidic it is called a 'sour chrome' process. There is also a reduced chrome method is used for dyes that are susceptible to oxidising agents.

In the 'sweet chrome' method, wool is treated with $K_2Cr_2O_7$ solution (2%) in neutral pH at 60 °C for 30 min. Chromium in a higher valency state is absorbed by the wool, forming a wool–chromium complex consuming only half the amount of chromium.

The 'reduced chrome' method is especially useful for dyes susceptible to oxidation. The pre-treatment is carried out using $K_2Cr_2O_7$ (2%), formic acid (2%)/lactic acid (3–4%) at 50–60 °C for 15–20 min.

In 'sour chrome' method the goods are first added at about 60 °C to a mordanting bath, containing dichromate solution (1–2%) and often either sulphuric (1%) or an organic acid like formic acid (2%). The temperature is raised to 100 °C over 45 min and held there for a further 60–90 min when all the mordant is absorbed by the material. In all cases, chroming is succeeded by dyeing with selective mordant dyes.

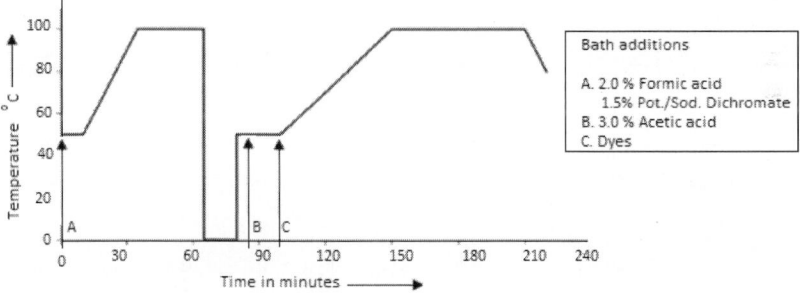

Dyeing profile for prechrome (sour chrome method) Mordant dyeing process

The mordanting bath is set with

Quantity	Unit	Additions
1.5	%	Pot. bichromate
2.0	%	Formic acid 85%

The goods are entered at 50 °C and the bath is slowly brought to the boil and kept at this temperature for 1½ h. After mordanting, the goods are rinsed and dyed in a fresh bath containing

0.25–1% formic acid 85%, or

1–3% acetic acid 40%,

according to depth of shade, and the requisite amount of dyestuff. After running the material in the bath for 10 min at 50–60 °C for homogenization, the well dissolved dyes are added. The temperature of the bath is slowly raised to the boil and dyeing continued at the boil for 1–2 h. A further addition of formic acid may be made, particularly when dyeing full shades, to assist exhaustion of the dyebath.

Notes:

1. The following mordanting liquors also may be used in place of that given above:

Quantity	Unit	Additions
3	%	Pot. bichromate
2.5	%	Cream of tartar

or

Quantity	Unit	Additions
3	%	Pot. bichromate
1	%	Sulphuric acid 168 °Tw (66 °Be)

or

Quantity	Unit	Additions
2	%	Pot. bichromate
3	%	Lactic acid 50–60%
1	%	Sulphuric acid 168 °Tw (66 °Be)

The chrome consumption is greater with these three mordanting liquors than with the formic acid/potassium dichromate process, and the chroming bath is not so well exhausted. Thus, the liquors cannot be used as dyebath.

2. When mordanting according to the above method, the chroming liquor is fairly well exhausted, thus making it possible to dye in the same bath. For this purpose the bath is cooled down after the mordanting operation to 60–70 °C.

3. 1% Formic acid 85%

4. and the dyestuff is then added, the bath is raised to the boil, and dyeing continued in the usual manner. However, for better results and fastness before dyeing, the mordanted wool is thoroughly rinsed. It is then added to a dyebath containing acetic acid at 50 °C. The bath is brought to the boil over 45 min and maintained at the boil for 60–90 min longer.

5. All of the 'on chrome' methods are two-bath processes; this is expensive and usually only one batch of material can be chromed per process and there is always the danger of impairing wool during chroming and subsequent dyeing. The methods are also only suitable for light and medium shades. The advantages of this processes are matching of colour is relatively easy; regular building of shade occurs on tone with little change in shade during dyeing. Pre-mordanting increases dye uptake substantially because of electrostatic interaction between the mordant and the dye.

1.1.6.2.2.2   Metachrome method

Some chrome dyes (milling and super-milling acid dyes with chelating sites) can be applied simultaneously with mordant. In spite of being electrically opposite in nature, the dye and the mordant do not react initially, rather both are exhausted on the substrate after which the dye-mordant reaction occurs *in situ* under favourable conditions to develop the hue. If conditions are not maintained properly during exhaustion, the dye and the mordant react in the bath and the metal complexed dye is precipitated from the dyebath.

Chromium is deposited on, or combines with, the fibre, followed by reduction of CrO_3 to Cr_2O_3 at the boil. At the same time, dye combines with the wool in the same way as with a milling acid dye. The dye and the reduced chromium then combine with fibre to form a dye–chromium–wool complex.

It is likely that all three reactions occur simultaneously in bath as follows:

$$(NH_4)_2SO_4 \rightarrow H_2SO_4 + 2NH_3$$
$$2Na_2CrO_4 + H_2SO_4 \rightarrow Na_2Cr_2O_7 + Na_2SO_4 + H_2O$$
$$Na_2Cr_2O_7 + H_2SO_4 \rightarrow Na_2SO_4 + 2Cr_2O_3 + H_2O$$
$$Cr_2O_3 + dye \rightarrow Cr\text{–dye lake}$$

The chances of the formation of another metal–dye–fibre complex, rather than with chromium, arising from the presence of other metals that may be present in the supply water used for dyeing is prevented by chromates which form a protective oxide film on the surface of other metals and do not allow them to take part in reaction.

The bath is made with chromate (Na_2CrO_4 (sodium chromate) or $Na_2Cr_2O_7$ – 2–5%) and ammonium sulphate (5–10%) and a surfactant (0.5%) and is made very slightly alkaline with ammonia. The pre-treated wool is added at 50 °C and worked for 20 min; the temperature is then raised to 100 °C over 45 min or so and held there for a further 60–90 min. Exhaustion may be enhanced by applying 0.5–1% of CH_3COOH (40%) or half the amount of HCOOH (85%), preferably 30 min before dyeing is completed. They exhaust well at a pH of 6.0–8.5.

Quantity	Additions
2–5%	Chrome mordant (potassium chromate)
2–5%	Ammonium sulphate/acetate
10%	Anhydrous sodium sulphate

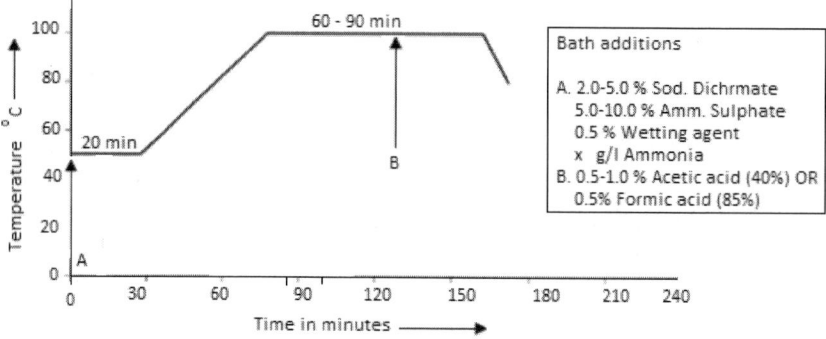

Dyeing profile by Metachrome method of dyeing Mordant dyes

$(NH_4)_2SO_4$ promotes deposition of CrO_3 on the wool without damaging it and generates a pH around 6.0–8.5 and CrO_3 is reduced to Cr_2O_3 by the wool during this process. Due to generation of NH_4OH, the bath is likely to become alkaline with time, in a closed dyeing machine.

$$Na_2CrO_4 + H_2O \rightarrow 2NaOH + CrO_3$$
$$2NaOH + (NH_4)_2SO_4 \rightarrow Na_2SO_4 + 2H_2O + 2NH_3$$
$$Na_2CrO_4 + (NH_4)_2SO_4 \rightarrow Na_2SO_4 + H_2O + 2NH_3 + CrO_3$$

$(NH_4)_2SO_4$ may be substituted with CH_3COONH_4 when using dyes known to cause uneven dyeing.

The use of the metachrome method is restricted by several factors. The 'metachrome' process is a single-bath process; it is simple to apply with a higher rate of production/time. Shade matching is easier because the final shade is produced during dyeing. Chromium does not form a complex with the dye in chromate form, permitting the application of sodium chromate and dye from the same bath; pH of the dye bath is maintained at around 6.0–8.5 for maximum conversion of dichromate to chromate. The dulling of shade due to presence of iron or copper does not occur, as it often does in other chrome dyeing methods. One prerequisite is that only dyes which do not react with hexavalent chromium (Cr_6) can be used in this method. In spite of all this, the range of shade is limited, as only a few dyes are suitable for this method and unlevel dyeing cannot be rectified.

Considering that any prolonged treatment intensifies the deterioration of keratin structure, it should be assumed that dyeing with simultaneous chroming, being the shortest technique, is the most efficient of these three methods. If for some reason the interaction of the dye with chromium cannot be sufficiently retarded, the lake is partially formed on the fibre surface which reduces the colour rubbing fastness, especially in case of deep shades.

Due to relatively higher pH of the bath, exhaustion remains incomplete, especially with deeper shades; blacks, navy blue and other deep shades are seldom produced in this method. The dye must, for example, resist oxidation by the chromate in the bath, and it must have good substantivity up to pH 8. Thus, the range of shades is limited, especially in comparison with the range which can be produced by the after chrome method.

These dyes are also used for the same application as after chrome dyes. These dyes have following advantages:

Shorter dyeing time,

Lower steam consumption, hence economical,

Unlimited combinability,

Can be shaded easily.

Special process for high twisted yarns:

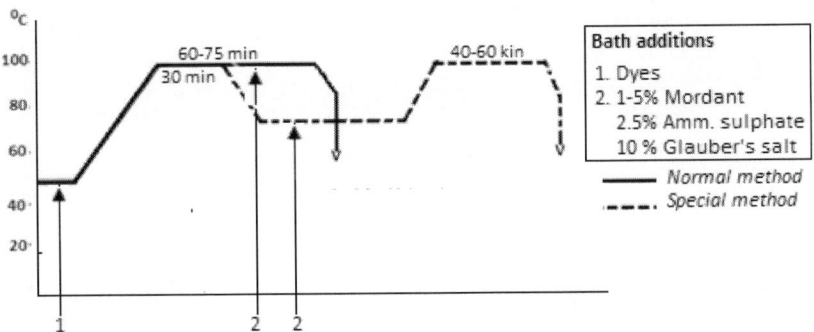

Special method for tightly twisted yarns and fabric using such yarns

In case of tightly twisted yarns which are difficult to penetrate, dyeing is started without mordant and is continued for 30 min at the boil. The bath is then cooled to 75 °C and the mordant is added. The temperature is then raised to boil again and the dyeing is continued for another 45–60 min.

Notes:

1. At the start of the dyeing the pH of the bath is 6.2–6.5. The pH should never be allowed to go below 6.2. Otherwise there will be premature salt formation in the bath, resulting in the precipitation of the dye which can cause poor rubbing fastness of the dyeing. These dyes are not suitable for low temperature dyeing.

2. If machines are available for high temperature dyeing then it can be dyed at 102–105 °C, whereby the dyeing time can be reduced to 40 min.

3. The amount of the mordant used is generally equal to the weight of the dye but it should never be less than 1% or more than 5%.

4. The amounts of ammonium sulphate or ammonium acetate required depend largely on the liquor ratio, the nature of the water, the quality of the wool, and the type of dyestuff used. Hard or alkaline water, long liquor ratios and deep dyeing s require large quantities of ammonium sulphate.

5. When dyeing medium depth and full shades, it is advisable to add
0.5–1% acetic acid 40%, or
0.25–0.5% formic acid 85%,
to the dyebath 1 h before dyeing is complete.

6. The addition of dichromate at 70 °C before any formic acid provides an opportunity for the chroming agent to exhaust uniformly. At low pH, dichromate ions have a strong affinity for the wool fibre, and if added at pH 3.5–3.8, they might exhaust too rapidly, leading to unlevelness. By delaying the addition of formic acid, this issue is overcome and shade levelness is more readily achieved, which is particularly relevant in yarn and piece dyeing.

7. The issue of levelness in chroming may also be overcome by chroming in a fresh bath, starting at 40–50 °C. This route has the additional benefit of producing greatly reduced chromium residues.

8. The main disadvantage of after chrome dyeing is that the final shade is not developed until the chroming stage is completed. It is not possible to add further chrome dye to adjust the shade, unless

the addition is followed by yet another chroming process. As this would be impracticable, except for very major adjustments, shading additions are often made with milling or 1:2 metal-complex dyes.

1.1.6.2.2.3   After chrome method

This is the most popular and efficient way of applying chrome dyes. The dye is applied like an acid milling dye in the presence of an acid. Once the dye has been adsorbed, the dye bath is cooled to 70 °C, whereupon dichromate ($K_2Cr_2O_7$ 1–2%) is added or as a separate bath at 60–80 °C and the material is transferred to it and worked. The lower temperature helps to distribute the dichromate ions evenly within the wool fibres, as the chromium is not significantly reduced to the trivalent state at this temperature. The temperature is then raised to 100 °C and held there for 30 min, while the dye is mordanted.

The mechanism of the reactions in the after chrome method is as follows:

1. At a pH below 3.5, dichromate anions are absorbed and interact with protonated amino groups in wool:

$$2H_3N^+ \text{-(Wool)-COOH} + Cr_2O_7^{--} \longrightarrow \text{HOOC-(Wool)-} N^+H_3\ Cr_2O_7^{--}\ H_3N^+\text{-(Wool)-COOH}$$

2. Cystine crosslinks are oxidatively decomposed and helps to reduce the absorbed chromium(VI) to chromium(III):

| Cystine | Dehydroalanine | Perthiocysteine | Cysteine - S - sulphonate |

3. The chromium(III) cations then combine with the carboxylate groups in the wool fibre:

4. At the final stage, dye ligand interacts with free or complexed chromium (III) to form 1:1 and 1:2 metal complex dyes, predominantly the more stable 1:2 complex. These coloured complexes are bound to the wool primarily through van der Waals and electrostatic forces. Any excess chromium (III) will remain linked to carboxylate sites in the wool.

The method has a number of advantages. It requires only one bath. The dye can be initially applied in a level manner, yet after mordanting, it does not migrate. The dye, however, must be completely adsorbed before the addition of dichromate. If dye and dichromate are in the bath together, a chromium complex of the dye may be precipitated and can affect the fastness of the resultant dyeing. Dyeing with after-chroming has acquired a most widespread use, as in this technique there is less danger to impair wool (the presence of dye on the fibre increases keratin resistance to the bichromate oxidizing action). However, using this method, it is difficult to obtain the required shade, as it is necessary to take into account the changes caused by chroming; moreover, to avoid the formation of a lake in the liquor, chrome dyes cannot be entered for colouring into a bath containing bichromate.

The dyeings by this process show superior fastness to milling and potting. Shade matching is not easy and any change in shade can be difficult to rectify. However, the concentration of dichromate required is remarkably less than other mordant dyeing methods. This has led to this method becoming more popular, although the dye uptake is lower compared to that obtained using the onchrome method.

Dyeing with simultaneous chroming (Metachrome method – see Sect. 1.1.6.2.2.2) may be performed only with a certain group of dyes, i.e. metachrome and monochrome dyes which interact with the chromium salts only after passing onto the fibre. This is promoted by the adjustment of the dye bath pH within the limits of 6 as well as by the replacement of usually used bichromates by ammonium chromate which when heated is slowly hydrolysed into acid ammonium chromate. In this way conditions are created for a slow interaction of the dye and chromium on the fibre. The best mordants are sodium or potassium bichromates and chromic fluoride. Upon direct interaction of wool with bichromates, chromium is partly reduced on the fibre due to keratin oxidation. Wool treated by the chromate method oxidizes to a high degree during subsequent storage, particularly under the action of light, which may cause damage to the fibre and impair its fastness. Therefore, a reducer is usually entered into the bath for preliminary chroming in addition to the bichromate.

All chrome dyeing methods suffer from the increased stiffness of the dyed fibre, due to precipitation of a part of the complex at the interstices as well as the formation of the complex on the surface of the substrate. The formation of dull shades with change in hue, in comparison to that produced with parent acid dye, poses problems in dyeing wool. The reproduction of hue and shade becomes difficult, and partial tendering of the wool keratin occurs due to prolonged dye bath treatment.

Mordant colours are also can be used for silk, if required. Acid-mordant (chrome) dyes find little use in silk dyeing, for they produce dull colours while chroming imparts rigidity to the fibre and deprives it of its characteristic scroop.

Quantity	Unit	Bath additions
20–30	% owf	Lactic acid 40%
6–8	% owf	Chromium lactate containing 11–14% Cr_2O_3

Run the material in blank bath for 15 min (wet out). Add dye liquor and continue dyeing without heating for 10 min. Add chromium lactate without heating and run for another 14 min. Add half the amount of lactic acid and heat to a temperature of 35–40 °C and run for another 15 min and add the balance of lactic acid and heat to 90 °C in 60 min. Continue running for 30 min again for another 30 min closing the steam. Fabric is washed with warm water and then for 15 min at a temperature of 90 °C with a solution containing 5 g/l of 60% detergent, then again with warm water and finally with cold water. After this the fabric is subjected to brightening in a solution of acetic acid.

When dyeing wool according to the after chrome process, only those dyes can be considered which contain functional groups capable of forming ultra-complexes. In most cases, these are hydroxyl groups which are situated in the ortho position in relation to the azo group on the aromatic ring.

In dying process, the dye first adsorbs on to the wool fibre like an acid dye. The fastness of the wool dyeing is significantly improved by fixing the acid dye with Cr^{3+} ions. When after chroming using $K_2Cr_2O_7$, a redox reaction takes place with the wool. As the potassium dichromate is a strong oxidizing agent, it is capable of reacting with the amino acid, cystine. Because of their anionic nature, sulphonic acid groups can either act as a source of surface charge or, in their role as charged functional groups, soften the scale structure by absorbing water into the outer regions of the wool fibre by osmosis.

Oxidation products of cystine on afterchroming (with potassium dichromate)

The Cr^{3+} ions produced are complexed by the dye ligands and the functional groups of the wool. Direct use of Cr^{3+} salts prevents complex formation with the dye molecules as the hydrolysed Cr^{3+} ion is relatively inert to substitution and adsorbs very poorly on to the wool.

Generally this method yields the fastest results, but has the disadvantage that the final shade is only developed on chroming, and levelling difficulties are experienced when shading to pattern with chrome dyes.

These are acid dyes which are applied in the presence of acetic acid. They have good levelling property. Initial shade is not the final one as the colour changes when developed with potassium dichromate when the complex is formed. After adding dichromate to the exhausted bath the liquor is maintained at boil for 60–90 min. The special constitution of chrome dyes enables them to form a very stable insoluble coloured complex or salt. The chrome is also able to form a co-ordination bond with the amino group in the wool as well. Hence, these form the fastest dyeing on wool.

The after chrome process was the most widely used because it gives the highest wet fastness properties and the best levelling properties. The after chrome dyestuffs give good coverage of different wools but are difficult to strip or reprocess due to the stability of the fibre bonds.

Dyeing process

Application: Menswear, uniform cloths, heavy suitings, coatings, etc., carpet yarns, knitted yarns, loose wool and tops.

Machines: Winches, Beam Dyeing, Jet Dyeing, Package dyeing Machines.

Method I: Dyeing at the boil and 105 °C

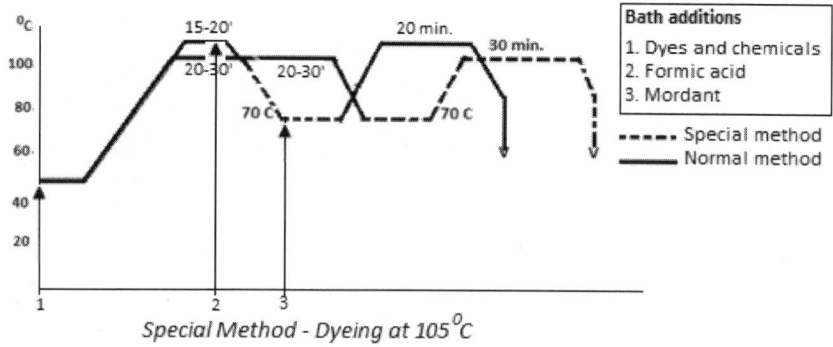

Special Method - Dyeing at 105 °C

Set the dye bath at 50 °C with

Quantity	Additions
10%	Sodium sulphate
1–3%	Acetic acid 80%
x%	Dye

After entering the goods at 50–60 °C, raise to boil in 20–50 min and the dyeing is continued at boil for 20–30 min 1–2% formic acid (sulphuric acid) is added (with steam turned off) and dyeing is continued for a further 20–30 min at boil. Then the dye bath is cooled to 70 °C and sodium or potassium dichromate is added. The amount varies according to the depth of shade, but it should not exceed 2% and should not be less than 0.5%; usually an amount equal to the half of the weight of the dye is sufficient. The temperature is again raised to boil and the boiling continued for 30 min to develop the dyeing. If the bath is not completely exhausted, chroming is carried out in a fresh bath.

If dyeing is carried out at 105 °C the time required for exhaustion and chroming may be reduced to almost half (see the dyeing graph)

Notes:

1. This method is time consuming and the process is having an adverse effect on the wool.

2. Since the chromium salts are discharged into the effluent, the use of chrome dyes is now limited.

3. Loose wool and slubbing are often dyed without an addition of Glauber's salt.

4. It should be ensured that the exhaustion of the dye liquor is complete otherwise the dye remaining in the bath will after chrome as well and then deposit onto the fibre surface. This will result in diminished rubbing fastness for the dyed goods.

5. In the after chrome dyeing process, cross linking is also possible between the protein chains due to the effect of potassium dichromate (for example, by complexing two dye ligands bonded to two different protein chains). As potassium dichromate is a strong oxidizing agent it can yield a large number of oxidation products by oxidation of amino acids like cysteine.

6. Using after chrome dyes is significantly cheaper than using reactive dyes and at the same time better fastness results.

1.16.2.2.4   *Special low temperature dyeing method*

This method is developed for improving the damage on wool by the previous method. A dyeing assistant (e.g., Lyogen SMK, Lyogen MS, Ekaline S of Clarient) is used in dyeing.

Dye bath is set at 50 °C with

Quantity	Additions
10%	Sodium sulphate
3–4%	Ammonium sulphate
0.5–1%	Acetic acid 30%
0.5–1%	Dyeing asst.
x%	Dye

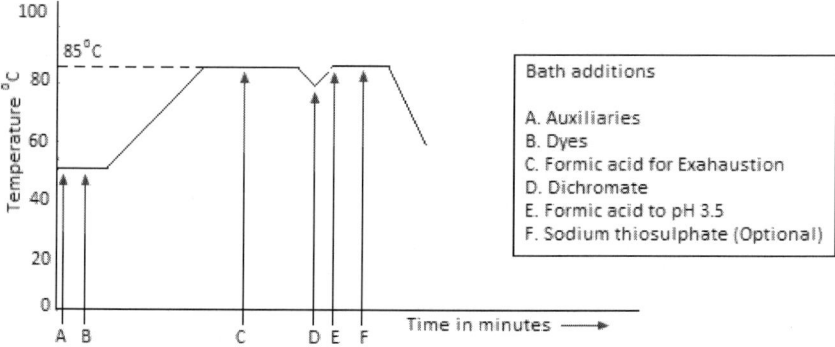

Special chrome mordant dyeing method at lower

The temperature is raised to 85 °C in 25 min and is held at this temperature for 30–60 min. Cool the bath to 80 °C and add

0.2–2% Sodium dichromate,and

2% Formic acid 85% (diluted),

and again raise the temperature to 85–90 °C and the dyeing is chromed at this temperature for 30 min drain and soap, etc.

Notes:

1. The addition of formic acid ensures good development of the shade.
2. For pale and medium dyeings an amount of chromium salt equal to half of the weight of dye is used, though it should not be less than 0.2%. Heavy shades require min. 1.5–2%.
3. Developing with aluminium salts

The dyebath is set with

Quantity	Unit	Additions
10–20	%	Glauber's salt cryst.
2–3	%	Sulphuric acid 168 °Tw (66 °Be) or
3–4	%	Oxalic acid

The goods are entered at approx. 30 °C, the bath is slowly raised to the boil and kept at this temperature for 1 h. Developing is then carried out at the boil with 5–10% alum (free from iron).

1.1.6.3   Metal complex dye

Metal complex dyes have been developed for increasing the wet fastness of the then known acid dyes, instead of doing after treatments for improving fastness properties. Hence, most of the metal complex dyes are having the starting unit as the acid dyes. They are synthesised through coordination of bi- or polyvalent transition metal ions with selective acid dyes as mentioned above, these dyes may be produced on the substrate, during dyeing, through a one- or two-step process, which is known as mordant dye. Overall increase in the size of the acid dye molecules after coordination with metal is at the root of improved wash fastness. The extent of exhaustion, diffusion and levelling through migration, is utilised to promote build-up of shade. The disadvantage of the inclusion of metal in the dye molecule is the dullening of the shades of the original acid dye and the environmental implications. For example, chromium salt, the most commonly used mordant, increases waste-water load.

As explained in Sect. 1.1.6.1, during the dyeing process, the strong acid in the dye bath (low pH) cause more and more amino sites to protonate and along with it higher dyeing time, lead to higher number bonds being formed between dye and fibre. But these fibre–dye bonds are very weak and the dye molecules being small, the bond can break and the dye molecule can come to the surrounding bath, and hence the acid dyes are having poor wet fastness properties.

$$H_3N^+ - W - COO^- + R - SO_3^- + Na^+ \rightleftharpoons R - SO_3^- \cdot {}^+H_3N - W - COOH + CH_3COONa$$

Introduction of milling and super-milling dyes, since these are larger dye molecules has largely improved the wet fastness problem. This migration of acid dye can be reduced or halted, if the dye structure is made significantly large. The alternate method to improve the fastness property is to mordant the dye with metal salts to make the molecule bigger. This is basically a process done on the shop floor and was possible for those dyes which are having the possibility of accepting a metal ion. There are many disadvantages for this method like all dye baths used in dyeing with mordant dyes are mostly two-bath or two-stage processes, the shades produced lack brightness, shade change is evident causing matching problems, maintenance of pH, poor rubbing fastness due to deposition of a part of complex on dyed fibre, etc. Given below are examples of mordant dyes:

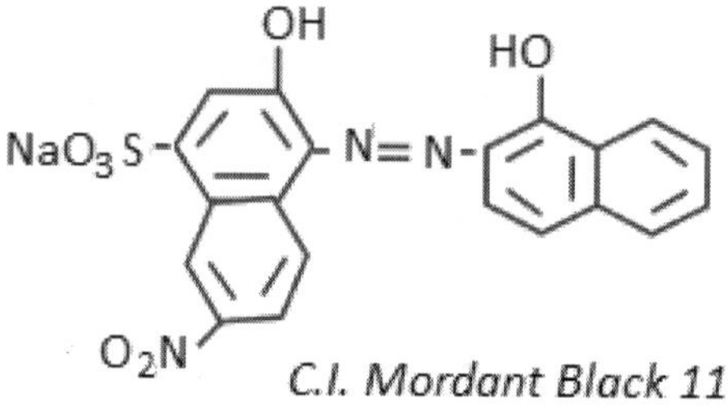

C.I. Mordant Black 11

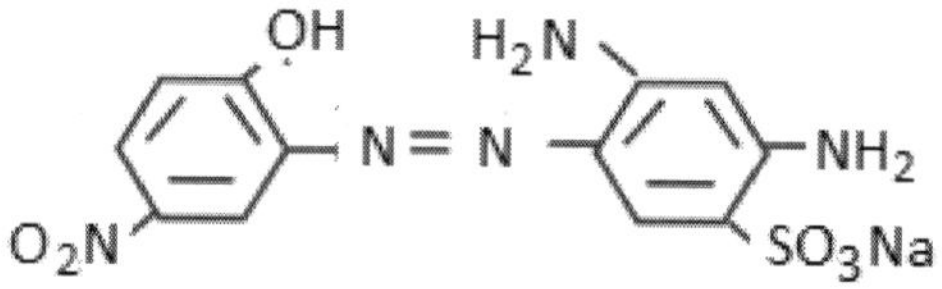

C.I. Mordant Brown 33

To avoid obvious complications associated with reaction of metals with dye possessing chelating sites in shop-floor level, in many cases, the reaction is carried out in the dye manufacturing plant itself and the dye–metal complex is supplied to the dyeing sector, known as pre-metallised dye. This has necessitated development of eco-friendly formazan and other dyes through complex formation with iron and aluminium. In spite of this, metal-complexed dyes, especially 1:2 chromium premetallised dyes, are commercially important to produce bright, wash-fast shades on natural protein and polyamide fibres without the risk of over dyeing and degradation of fibre.

Metal-complex dyes are selective acid dyes possessing chelating sites to coordinate with metal atoms to develop bigger dye–metal complexes through coordinate and covalent bond formation as shown below:

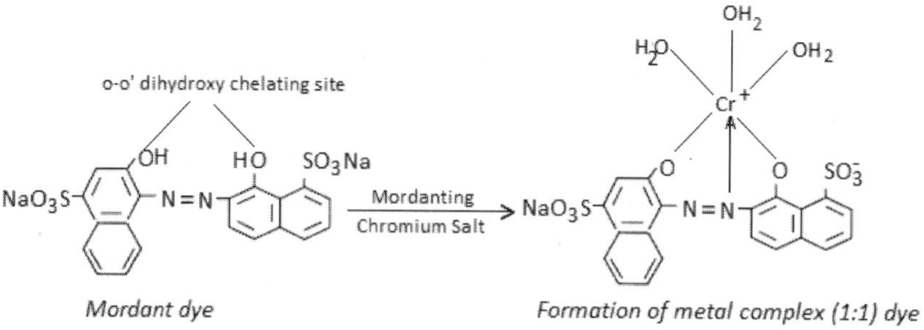

Mordant dye   Formation of metal complex (1:1) dye

In the application of mordant dyes, chromium and dye are combined in the wool fibres to form larger molecule. Alternatively, chromium–dye complexes can be prepared before dyeing by the dye manufacturer; these are the premetallised (metal-complex) dyes. Thus there is no much difference between a mordant and premetallised dyes. When produced on a substrate, the dye–metal complex is called mordant dye and, if manufactured before dyeing, is called premetallised dye.

The evolution of premetallised acid dyes are basically to avoid obvious complications associated with reaction of metals with dye possessing chelating sites in shop-floor level. Thus the mordanting reaction is carried out in the dye manufacturing plant itself and the dye–metal complex is supplied to the dyeing sector, known as pre-metallised dye. Selective acid dyes are complexed with copper, chromium, cobalt or nickel at suitable proportions to develop these dyes.

The incorporation of metal in the acid dyes changes the characteristics of the parent dye. Metal-complex dyes have low water solubility. Colour of the parent dye becomes duller due to the introduction of the metal ion in the dye molecule (But the incorporation of some metal ions does not affect the brightness as much as in the case of mordant dyes). Other disadvantages are the environmental pollution due to the presence of heavy metal in the dyes and in the effluent solubility in aqueous media of these dyes are improved by the presence of one or two solubilizing groups.

There are two main types, the 1:1- and 1:2-chromium complexes, but the latter are commercially far more important.

1.1.6.3.1   1:1 Metal complex dyes

The 1:1-complexes are derived from monoazo dyes, whose molecules also contain chelating groups. When the dye have two sulphonate groups, the dye has an overall negative charge; if there is only one such group, the dye behaves as a zwitterion, in which different parts of each molecule possess equal and opposite charges.

Chemistry of 1:1 metal complex dyes

Many monoazo type acid dyes having o-o' chelating sites such as –OH, –COOH, $–NH_2$, etc. are complexed with selected metals to make 1:1 metal complex dyes. The electron donating ligand or ion combines with a transition metal ion to form the complex. In order to form a complex, there must be at least two ligands in the dye structure. However, –tridentate or –tetradentate ligands in the parent acid dye structure occupy three or four coordination sites of the transition metal ion. The combination of ligands with metal ions depends on the coordination number of the metal. In 1:1 complexes, three coordination sites are occupied by trivalent metal, whereas the remaining three coordination sites are occupied by solvents, like water (see below). The synthesis of this dye will give an idea of the molecular structure of the dye better. It is prepared by diazotization of 5-amino-2-chlorohydroquinone dimethyl ether and coupling onto 1-hydroxynaphthalene-5-sulfonic acid. The reaction product and Cr_2O_3 in formic acid are heated in an autoclave at 130 °C.

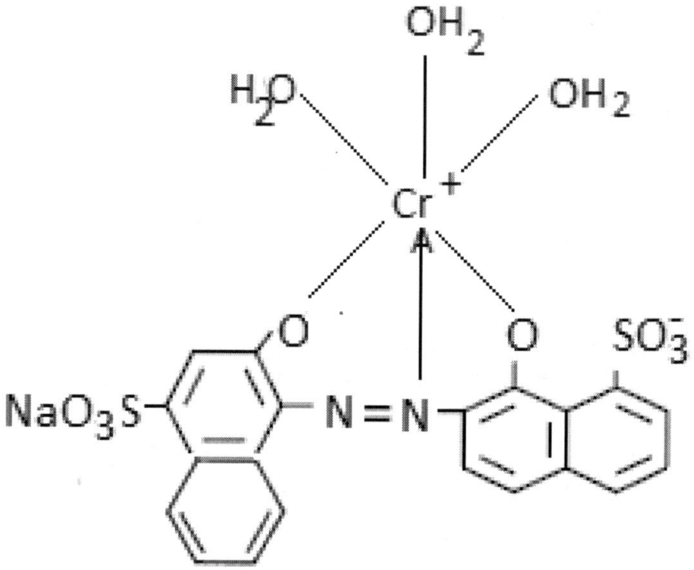

Formation of 1:1 metal complex dyes

By introducing one or more sulphonic acid groups or amino sulphate (– SO_2NH_2) groups the solubility of these dyes were improved. As discussed earlier, depending on the nature and number of the solubilising groups and the nature of the monodentate ligands present, the dyes are either effectively uncharged (see above structure) or carry an overall negative charge (see below A) (because of the presence of the Cr(III) cation); dyes that contain no ionic solubilising group (see blow B) have an overall positive charge.

The electronic configuration of Cr^{III} gives stability to the complex formed by dye molecules. The stability of the complex is such that it resists demetallisation during dyeing or subsequent processing and use. In chromium, 3d orbitals and 4s orbital are filled with single electrons. It becomes a Cr^{3+} ion by releasing three electrons and can then form a complex with a dye molecule and water.

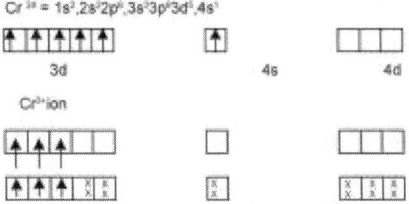

$Cr^{3+} = 1s^2, 2s^2 2p^6, 3s^2 3p^6 3d^5, 4s^1$

3d   4s   4d

Cr^{3+} ion

Cr^{3+} ion gained six pairs of electrons in $[Cr(H_2O)_6]^{3+}$, X = donated electrons.

In a 1:1 metal complex dye, the metal ion Cr3+ matches with a single monoazo dye ligand and three other ligands, usually water or colourless hexafluorosilicate ligands (see below). Depending on the nature and number of the solubilising groups and the nature of monodentate ligands present, the dyes are either effectively uncharged or carry an overall negative charge (because of the presence of CrIII). These dyes exhibit good even dyeing, good penetration characteristics, are suitable for acid milled wool, cover irregularities in substrate and produce uniform shade, and have very good light fastness and moderate to good wet fastness.

(A)

C.I. Acid Red 183

(B)

C.I. Acid Orange 76

1.1.6.3.1.1 Application

1:1 Metal dyes needs very strong acid (8% H_2SO_4) for dyeing, which imposes certain limits on their range of applications. The 1:1 metal complexes are not suitable for polyamide, which is partially decomposed under the dyeing conditions for these products. Their main area of application is in the dyeing of wool, but they are also suitable for leather dyeing. Shades of any depth can be dyed by these dyes. The quantity of acid, which can cause hydrolytic damage of wool and serious impairment of its soft handle, can be reduced by the addition of cationic levelling agents (e.g. Palatine fast salt O and Neolan Salt II). These levelling agents are usually fatty alcohol–ethylene oxide condensates having general formula $CH_3-(CH_2)_x-O-(CH_2-CH_2O)-H$.

Levelling agents form loose complex with dye as these are of opposite electrical nature. The complex is adsorbed by the fibre and at boil; the complex breaks to release levelling agents. The acid is removed by proper neutralisation washing off after dyeing. Some dyeings can be used for discharge printing also. Examples of this type of dyes are Neolan (Ciba) brand and Palatine (BASF) brands. 1:1 Metal complex dyes are sensitive to metal-complexing agents. Use of higher pH values in the dyebath produces uneven dyeings due to unprotonated nucleophilic groups in the wool keratin, which displace ligand water from the 1:1 complex and thus form coordinative bonds to the chromium atom. Nevertheless, because of high migration and levelling capacity, the brighter hues compared with mordant dyes, and the single-bath dyeing process, the 1:1 chromium complexes are of commercial interest to the present day. Moreover, they are essential intermediates for manufacture of unsymmetrical 1:2 chromium complex dyes.

Wool dyeing with acid metallised dyes of the type of 1:1 complex is carried out in the same manner as the common acid dyeing, but with an increase of sulphur acid amount (up to 10% instead of the usual 1–4%). Therefore, in order to achieve a more complete elimination of the acid, dyeing should be followed with washing in a solution containing up to 5% of ammonium acetate or soda up to 2%. Besides, to attain even dyeing with a reduced consumption of sulphuric acid (up to 4–5%), it is better to enter into the bath 1–2% of the levelling agent (quaternary salt of diethyl aminomethyl polyglycolic ester of alkylphenol).

Wool dyeing with acid metallised dyes of the type of 1:1 complex is carried out in the same manner as the common acid dyeing, but with an increase of sulphur acid amount (up to 10% instead of the usual 1–4%). Therefore, in order to achieve a more complete elimination of the acid, dyeing should be followed with washing in a solution containing up to 5% of ammonium acetate or soda up to 2%. Besides, to attain even dyeing with a reduced consumption of sulphuric acid (up to 4–5%), it is better to enter into the bath 1–2% of the levelling agent (quaternary salt of diethyl aminomethyl polyglycolic ester of alkylphenol).

$$\left[\text{Alkyl} - \underset{}{\overset{O(CH_2CH_2O)_nCH_2CH_2OCH_2N(C_2H_5)}{\bigcirc}} \right]^{+} \quad C_2H_5SO^{-}$$

Leveller A

The interaction of this preparation with the dye at the creation of hydrogen bonds leads to the formation of slowly diffusing complexes with low mobility, which decompose very slowly and thus ensure colour levelling.

1.1.6.3.1.2 Strongly acid dyeing metal complex dyes (1:1 metal complex dyes)

These are premetallised chromium complex dyes which have been used for dyeing wool for long time. The metal atom and the dye molecule form a complex in the ratio 1:1. The solubility in water is taken care of by the two sulphonic acid groups. They are having high fastness, good levelling, excellent penetration and easy application. The original type used a dye molecule capable of forming a complex with a chromium atom in the ratio 1:1. o,o'-dihydroxy azo dye was combined with a single chromium atom that carried a cationic charge, and was partially quenched by water. The dye usually contains an ionic sulphonic acid group, which imparts solubility and neutralises the cationic charge on the chromium atom. This class of dye has very good levelling properties when applied to wool at pH levels of around 1.8–2.0 and is ideal for the level dyeing of acid milled and pre-carbonised fabrics. Carbonising is an acid bake process, which tends to make wool fibre surfaces more anionic in character. The means by which such dyes are absorbed by the fibre is unclear but the most likely mechanism is that the dye attraction is initially ionic, via the sulphonic acid group. This would explain the high migration properties. It is also thought likely that the dye may form larger 1:2 metal complexes within the fibre. There is no evidence that this class of dye can form coordinate links with the fibre and, as such, their fastness properties are not as good as after chrome dyes. Indeed, the use of 1:1 metal complex dyes on washable wools is restricted because their shade is unstable when the wool dyeings are washed with modern detergents containing perborate.

Since the dyes are good levelling dyes, these can be used for piece dyeing and also there is possibility of dyeing directly after carbonizing without neutralizing. They give dyeing of high wet fastness and perspiration fastness. The main disadvantage of this group of dye is that, there are chances of wool damage on prolonged treatment in strongly acid bath.

There are two methods of dyeing:

Using optimum amount of sulphuric acid at a pH of 1.9–2.1, with at least 1½ hat boil. The acid addition is calculated on the weight of the material and the volume of the bath. A dyeing in medium depth for example would require 4% (plus 1 ml/l sulphuric acid 90%). See table below.

1. Using a lower quantity of acid in conjunction with an auxiliary product having affinity for the dye at pH 2.2–2.4 (e.g., Ekaline F liq. of Clarient, etc.).

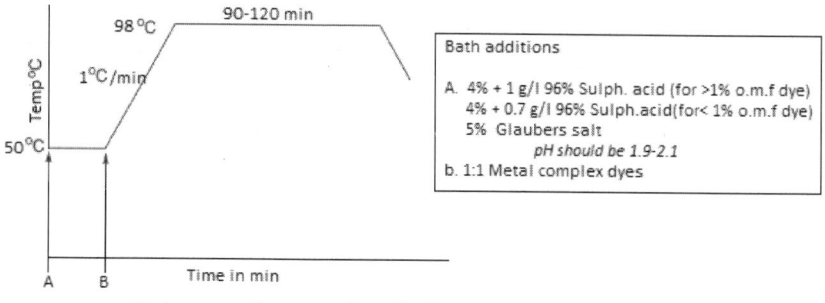

Dyeing process for 1:1 Metal complex dyes on wool

Usually 1:1 metal complex dyes are commonly applied to wool using a strongly acidic (pH 2) dye bath (hence dyes are sometimes referred as 'acid dyeing metal complex dyes'). Under these conditions, the dyes possess excellent transference and thus are good levellers. Since wool absorbs approx. 4% W.O.W (Weight on weight). of H_2SO_4 (96%), excess acid is required to maintain the pH. The amount of acid depends upon the liquour ratio and dyeing method used as shown below:

Material Liquor Ratio	<1% Dye o.w.m (on weight of material)	>1% Dye o.w.m (On weight of material)
10:1	0.7	1
20:1	1.4	2
30:1	2.1	3
40:1	2.8	4
50:1	3.5	5

Notes:

1. When acid milled or carbonised wool is taken for dyeing without neutralising the sulphuric acid on the material should ascertained and the addition of H_2SO_4 in the dyeing bath has to be adjusted accordingly.

2. In view of the high concentration of sulphuric acid used, it is necessary to either neutralise or buffer the residual acid in the fibre at the end of dyeing.

3. Excessively hard water may affect the quantity of H_2SO_4 required; chelating agents are unsuitable due to demoralisation of some dyes.

4. As prolonged boiling at such low pH can impart fibre damage, either reduced amounts of H_2SO_4 or a proprietary levelling agent can be employed.

5. In case of certain blacks the following method of dyeing used:

For a 1:40 MLR dyeing the dye bath is set at 70 °C with the dyes and 2% sulphuric acid (168 °Tw)and carefully brought to the boil. After boiling for a short time and when the bath is exhausted a further addition of 3% sulphuric acid (168 °Tw) is made. The total boiling period should be at least 1½ h.

In all the three methods the amount of acid required does not depend only on the weight of the wool fabric but also on the MLR. The following table shows the quantity of sulphuric acid for all the three methods for ready reference A general formula also is given under each method.

Liquor ratio	Method I		Method II		Method III
	Sulpuric acid 168 °Tw		H_2SO_4	Spl. auxiliary*	H_2SO_4
	Dyes <1%	Dyes >1%	168 °Tw		168 °Tw
1:10	4.7	5	4.3	2	4.3
1:20	5.4	6	4.6	2.5	4.6
1:30	6.1	7	4.9	3	4.9
1:40	6.8	8	5.2	3.5	5.2
1:50	7.5	9	5.5	4	5.5
1:60	18.2	10	5.8	4.5	5.8
pH	2–2.1	1.9–2.0	2.2–2.4	2.2–2.4	2.2–2.4
Gen. formula	4%+0.7 g/l	4%+1 g/l	4%+0.3 g/l	1.5%+0.5 g/l	4%+0.3 g/l

Notes:

1. The wool to be dyed often contains certain amount of alkali and this has not been taken into consideration in the above table.

2. For tightly woven wool goods or felts, it is better to use some anionic wetting agent to facilitate penetration.

3. After dyeing the goods should be thoroughly rinsed and in certain cases the acidity should be adjusted with sodium acetate.

Advantages:

1. Good levelling and penetration, even in the case of closely woven goods,

2. Dyeing of carbonised, non-acid-free goods,

3. Better wet fastness than acid dyes for middle to dark shades.

Disadvantages:

1. Dyeing with 1:1 metal-complex dyes at pH 2 and at temperature 80–120 °C leads to significant damage to the wool fibre due to hydrolysis known as peptides hydrolyse, N-terminal acetyl groups split off and protein fragments leach out of the wool in the form of wool gelatin and extraction of parts of the cell membrane complex. These results in a negative effect on fibre properties such as wear resistance and stretching.

2. In the case of dark shades, residual sulphuric acid in the goods has to be buffered.

1.1.6.3.1.3   Weakly acid dyeing metal complex dyes (1:1 metal complex dyes)

Modified 1:1 metal complex dyes have been introduced by many manufacturers which can be dyed at pH 3.5–4 for all colour depths and make-ups, adjusted using formic acid. Examples are some dyes like 1:1 copper-azo dyes derived from I-(3-N-benzenesulphonamido)phenyl-3-methyl-S-pyrazolone, which, owing to the presence of sulphonamide groups, were suitable for dyeing wool from a weakly acidic dyebath. Some of the modified 1:1 metal complex dyes contain fluorosilicate. The modification is produced when the dye is dissolved in almost boiling water (minimum final temperature 80 °C) in relatively concentrated form. 1:1 Metal complex dyes such as these can used for the gentle dyeing of fibres at pH 3.5–4 in the presence of special levelers (with formic acid). Commercial examples are Neolan P (Ciba) which can be dyed at a pH of 3.5 which need only formic acid. These dyes are suitable for dyeing wool/PAN and wool/PA blends with excellent levelling character, very good fastness properties and high exhaustion. Since it is dyed at higher pH a neutralisation step in the dyeing process can be avoided. These particular dyes are 1:1 chromium complexes of sulphonated azo dyes that contain colourless hexafluorosilicate ligands and are applied in conjunction with Albegal Plus (Huntsman), an amphoteric levelling agent that contains an ethoxylated fatty amine and ammonium hexafluorosilicate ($(NH_4)_2SiF_6$). The dyeing process is as follows:

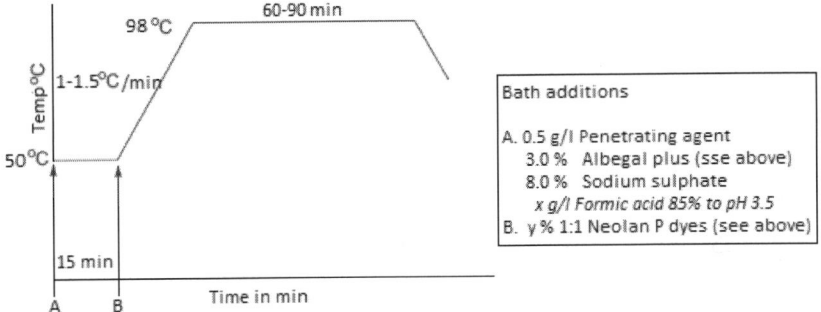

Dyeing process for Neolan P type 1:1 Metal complex dyes on wool

1.1.6.3.2 1:2 Metal complex dyes

The introduction of 1:1 metal complex dyes have solved the problems inherent of acid dyes like poor wet fastness, but it has posed some other anxieties. As explained earlier the 1:1 metal complex dyes have to be dyed under strong acidic bath which has complications of acid damaging of wool, silk and polyamides. 1:1 Metal complex dyes exhaust too less from a neutral bath with poor rubbing, milling and light fastness. Strong acidic pH which damages wool, poor exhaustion due to higher solubility of $-SO_3H$ groups (oxygen in $-SO_3H$ is electron donor to H of water) of 1:1 metal complex dyes and prolonged boiling of dye bath have paved the way for the evolution of 1:2 metal complex dyes with the pre-condition that (i) the complex should carry a negative charge, (ii) any residual valency of the metal atoms must be fully or nearly satisfied, (iii) solubilising groups should be less hydrophilic, may be by introducing sulphonamide groups in place of $-SO_3H$, (iv) complex should retain no orminimum hydrophilic groups and (v) all dyes in range should have identical exhaustion property.

The 1:2-complexes are also mostly derived from monoazo dyes containing chelating groups. A few dyes possess azomethine ($-CH=N-$) groups instead. Because of their structure 1:2 metal-complex dyes exhibit anionic character. Sulphonate groups are generally absent. Instead, low, but adequate solubility in aqueous media is conferred by incorporation of highly polar nonionic groups, such as methylsulphonyl (SO_2CH_3) and sulphamoyl (SO_2NH_2) groups. The absence of sulphonate groups avoids tippy dyeing, to which the corresponding sulphonated dyes often give rise. The problem is less serious with deep shades and indeed some sulphonated 1:2-complex dyes are now marketed specifically for producing deep colours. The introduction of 1:2 metal-complex dyes which are applied from a neutral to weakly acid bath, represented a significant technical advance over the strong-acid-dyeing 1:1 chrome complex dyes. It has led to better protection of the fibre material, simplification of the dyeing process, and improvement of the fastness properties.

1.1.6.3.2.1   Chemistry of 1:2 metal complex dyes

Bivalent Cu^{2+} ion, possessing coordination number four, can be complexed with two bidentate ligands in an acid dye or a trivalent or a tetravalent one. When coordination of Cu^{2+} ion occurs with a tridentate ligand, the fourth coordination site is generally occupied by a solvent molecule. Trivalent chromium (Cr^{3+}) or cobalt (Co^{3+}) have coordination number six and hence can easily form 1:1 and 1:2 metal-complex dyes. 1:2 Metal complex dyes can be made as chromium complexes with one or two sulphonic acid groups. By definition, 1:2-metal complexes with one sulfonic acid group must be unsymmetrical, because only one of the two azo dye ligands bears a sulfonic acid group. The dyes with one sulphonic acid possesses the same favourable properties as the unsulfonated 1:2 complex dyes and is moreover distinguished by higher brilliancy, better solubility and simpler preparation. Dyes with two or more sulphonic acid (or any other hydrophilic group), because of its strongly hydrophilic character this dye type needs specific dyeing auxiliary agents to perform even and deep dyeings.

Unsymmetric Complex

C.I. Acid Red 296

1:1 Metal complex dyes can be manufactured at lower temperature whereas the 1:2 metal complex dyes has to be manufactured at elevated temperatures which limits its manufacture.

When two dye molecules are complexed with the metal atom a larger dye molecule is formed and it can form strong links with the fibre and so they tend to have low migration properties. Molecular weight is usually 500–1000. Chromium is 2–5% by weight of the dye molecule for the commercially-traded product. There are typically three distinct types of 1:2 metal complex dye, unsulphonated, monosulphonated and disulphonated. The disulphonated types are the most soluble and have greater pH dependence. This means they have a low neutral affinity and require more acid to achieve exhaustion, but in practice they have the advantage of the highest fastness to domestic washing since any desorbed dye has only limited affinity for adjacent fibres at the pH conditions experienced with domestic detergents. Disulphonated dyes are, however, more fibre selective than the mono or unsulphonated dyes and care must be taken to avoid skittery dyeing.

Modern brand wool dyes like Lanaset (Ciba) and Lanasan CF (Clariant), consist of optimised mixtures of the three different types of 1:2 metal complex dye, and are formulated to give good combinability, good coverage of fibre irregularities and good overall fastness properties. Acid milling and 1:2 metal complex dyes have similar application requirements and are often applied in combination to brighten the shades obtainable with metal complex recipes. Since 1:2 metal complex dyeings usually exhibit dull shades, the 1:2 metal-complex dyes are combined with small amounts of acid dyes or suitable reactive dyes to brighten the tint.

1:2 Metal complex dyes are dyed onto wool, polyamides and silk. They have excellent light fastness, very good wet fastness (better than 1:1). Potting fastness better than 1:1, but not as good as after chrome dyes. It has a tendency to streaky dyeing on polyamides. Their too little affinity for fibre and being salts of strong acids, they ionize in water with lesser solubility and applied from mild acidic or neutral baths. Their solubility in bath can be improved by introducing $-SO_2-CH_3$, $-NHCOCH_3$, $-SO_3-CH_2$, etc., groups in the molecule. See below:

C.I. Acid violet 78

Metal complex dyes with more solubilising groups

The solubilised metal complex was a commercial success and many dyes were introduced with non ionic hydrophilic substituents, such as mono- and dialkylated sulfonamide, ethylsulphonyl (see above), ureido, alkylsulfoxide and cyclic sulfone groups besides the above groups. 1:2 Metal complexes with hydrophilic groups can be applied under neutral to weakly acidic conditions to give even dyeings with outstanding wet- and light fastness.

Of late 1:2 metal-complex dyes are made with two sulfonic acid groups. Contrary to the opinion frequently expressed in the literature that such dyes would exhibit inadequate levelness when applied in a weakly acid bath and inadequate stability to acid in a strongly acidic dye bath,they are very well suited for dyeing wool and polyamide if certain pH conditions are observed. These products are very soluble in water, have high tinctorial strength and are very high yield dyes with a relatively simple structure.

1:2 Metal-complex dyes can be broadly divided into two types as per the nature of the solubilising groups present in the dye ligands: weakly polar dyes, which contain no water-solubilising groups, and strongly polar dyes, which contain water solubilising groups.

1.1.6.3.2.2   *Weakly polar 1:2 metal complex dyes*

They are dyes without solubilising groups like $-SO_3H$, $-COOH$, in them commercially introduced by Geigy in 1951 under the brand name Irgalan dyes. However, water solubility is conferred by the inherent anionicity of the 1:2 structure (arising from the loss of four protons from the two dye ligands) and the presence of nonionic, hydrophilic substituents (e.g. $-SO_2CH_3$) or mono- or di-alkyl-substituted sulphonamide, ethylsulphone and cyclic sulphone groups. Examples of this types of dyes are C.I. Acid Violet 78,C.I. Acid Black 58, C.I. Acid Black 60, etc.

C.I. Acid Violet 78   C.I. Acid Black 58   C.I. Acid Black 60

1.1.6.3.2.3   *Strongly polar 1:2 metal complex dyes*

These dyes contain water solubilising groups like $-SO_3H$ groups. SO_3H group can be mono or di, symmetrical or asymmetrical. Examples of these types of dyes are C.I. Acid Orange 148 (asymmetrical), C.I. Acid Orange 142 (symmetrical), etc.

C.I. Acid Orange 148   C.I. Acid Orange 142

1.1.6.3.2.4   *Application*

These dyes need very short dyeing time and dyeing cycle is also simple. Dyeing can be started at room temperature, with proper agitation or in machines like jet/soft flow machines as the migration power and rate of dyeing is low below boil to achieve level dyeing. Proper pH control is necessary during the dyeing process. The migration is maximum at 82 °C beyond this temperature migration decreases but the rate of dyeing increases. Low rate of heating is essential at higher temperature as the rate of dyeing is increased near boil. Rate of exhaustion depends on pH, temperature and agitation of bath. An optimum pH around 6.5 at boil may be sufficient for level dyeing with maximum dye uptake; to increase rate of dyeing, pH may be lowered down to 4 or below it. Any correction of shade can be done by adding the dye at the boil itself after raising the temperature little by adding an alkali like ammonia which reduces the rate of exhaustion and as the dyeing goes on the pH automatically decreases as the ammonia evaporates and the rate of exhaustion increases. For proper levelling needs a special levelling agent (Lyogen SMK Clarient) even though addition of a salt partially helps.

As the dyeings are of excellent wet fastness, high light fastness, and good fibre levelness, the 1:2 metal-complex dyes are of major importance in dyeing wool and polyamide.

Sometimes an extra group of 1:1:1 is named where mixed complexes made from one coloured, one colourless component and one metal central atom (variation from 1:2 complexes). When 1:2 metal complex dyes are made without any soluble groups in it, can be used as disperse dyes for polyamides with some advantages and disadvantages –good light and wet-fast, building up to the deepest tones. Tendency to streaky dyeing in brighter shades, usually dull tones.

As already mentioned, in strongly acid conditions, sodium sulphate takes on the role of a levelling agent. In the case of dyes with high affinity, to achieve even dyeings, it is appropriate to add levelling agents which

(i)   Temporarily reduce affinity,

(i)   Teduce the exhaustion rate,

(ii) Improve the migration capacity,

(iii) Equalize differences in affinity on the wool.

By means of their ionic character and affinity, levelling agents are classified into either fibre-affinity or dye-affinity auxiliaries. Non-ionic and amphoteric auxiliaries which have a high affinity for dyes are important for 1:2 metal-complex dyes. The dye affinity of the auxiliary is less important. A surfactant micelle is produced which incorporates the dye. Due to its hydrophilic properties, this micelle is oriented towards the liquor (the micelle is very soluble) and has limited mobility in the bath due to its large volume.

With the advent of weakly acid dyeing metal complex dyes the scenario of wool dyeing was changed. With their ease of dyeing and fastness properties, they successfully challenged almost all classes of dyes used in wool dyeing. There are two types of 1:2 metal complex dyes:

1.1.6.3.2.5   *Weakly polar dyes without sulphonic groups*

They are better levelling dyes and exhibits good penetration properties. Even in pale and medium shades they have very good/excellent light fastness and very good fastness to wet treatments on wool. They can be used for loose stock, slubbing, yarn and piece goods dyeing. Examples of commercial dyes are Isolan range of Dystar.

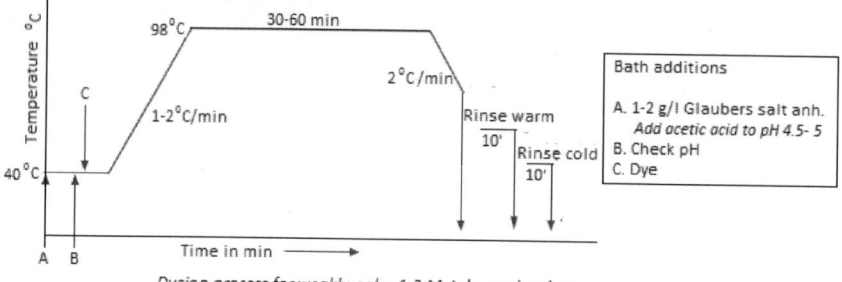

Dyeing process for weakly polar 1:2 Metal complex dyes

Strongly polar 1:2 metal complex dyes with sulphonic groups

These dyes due to the neutral dyeing character preserve the fibre quality, for improved spinning yield and elasticity compared to the fibre dyed with other classes of dyes.

1:2 Metal complex dyes have the following advantages:

1. Wide range of applications,
2. Very good all round fastness,
3. High light fastness even in pale shades,
4. Simple reliable methods of dyeing to get level shades,
5. Good combinability,
6. Dyeings can be easily shaded.

Dyeing method

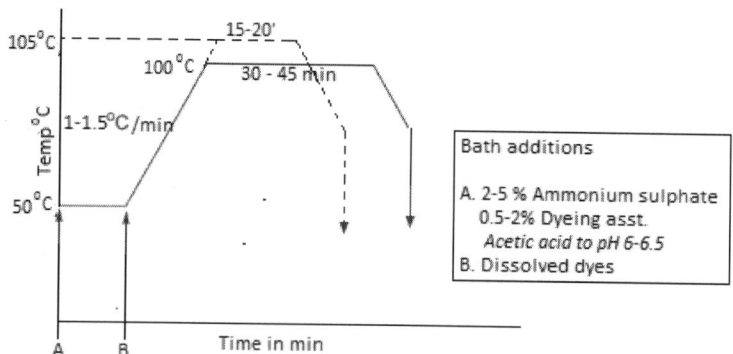

Dyeing process for strongly polar 1:2 Metal complex dyes

Dye bath is set at 50 °C with

Quantity	Additions
2–5%	Ammonium sulphate
0.5–2%	Dyeing asst. (e.g., Lyogen SMK, Clarient)
	Acetic acid of pH 6–6.5
x%	Dissolved dyes

Material is entered at this temperature and treated for 10 min and pH is checked and corrected if necessary before the dye is added. The temperature is then raised to boil in 25–30 min and the dyeing is continued for 30–45 min at boil.

Notes:

1. Better levelling can be achieved by adding 10% Glauber's salt calc. and pH adjusted to 5.5 with about 2% acetic acid.

2. A dyeing time of 15–20 min at 105 °C produces the same effect as 30–35 min at boil if the provision is available in the machine.

1.1.6.4   Direct dyes

The main usage of direct dyes are in the area of union dyeing i.e., in the dyeing of wool/cotton (cellulosic) where the direct dye dyes both components.

The direct dyes are applied from a neutral bath containing

10–20% Glauber's salt crystals.

Improved exhaustion can be effected by setting bath with

2–5% ammonium acetate or sulphate,

10–20% Glauber's salt crystals.

The goods are entered at 60 °C, the bath is brought to boil and kept at the temperature for 45 min.

In order to ensure complete exhaustion

2–4% acetic acid 40%

should be added carefully towards the end of dyeing.

1.1.6.5   Basic dyes

Basic dyes on wool give exceptionally bright shades but with poor fastness properties. Basic dyes are not very popular on cotton. However, this class of dye is only used when very bright shades are required where the fastness is not very important, e.g. for the dyeing of botany and fancy yarns, and sometimes as toning colours to give brightness to the dyed shades with other colours.

Methods of dyeing:

The dyebath is set with

1–3% acetic acid 40%,

and the goods are entered at 50–60 °C. The bath is slowly brought to the boil and dyeing continued at this temperature ½–¾ h(Some maynot be dissolved or dyed at temperatures exceeding 70–75 °C). If necessary, the dyestuff can be added to the bath in portions.

1.1.6.6   Wool reactive dyes

Wool reactive dyes have been introduced long back in 1930 even before the actual reactive dyes (for cotton) has become popular. ω-Chloroacetylamino, sulphatoethylsulphone (eliminates to vinylsulphone (VS) group while dyeing) reactive groups were patented by many manufacturers (Clingenstein and IG Farben – Supramino Orange R see below), but the market was not caught

Supramino Orange R (IG Farben)

up well till the reactive dyes were developed for cellulosics. Another problem with these dyes were it gave considerable problems of unlevel dyeing, especially in piece dyeing and yarn hank dyeing; thus their usage was mainly in the fields of loose-wool stock and wool top dyeing. Further development in this line was the introduction of dyes based on the α-bromoacrylamido group and difluoro-monochloro-pyrimidine dyes. These dyes with the special levelling agent was able to achieve even dyeing of machine washable wool.

Wool reactive dyes are the reactive dyes which are able to react with the amino group in the wool or protein fibres and establish a covalent bond. Covalent bond, being firm and cannot be easily broken gives a good fastness to these dyeing but fixation depends on the reactive group employed. The disadvantage of metal complex dyes achieving bright of shade is solved by the introduction of these dyes. However fastness will depend on the effective removal of unfixed dyes from the fibre.

The importance and acceptance of wool reactive dyes increased due to concerns of using premetallised and chrome dyes regarding environmental pollution. In this context dye manufacturers have increased their efforts to offer wool dyers ranges of attractively priced reactive dyes. Most of the dyes have sulphatoethylsulphone residues, which activate to vinylsulphone on boiling at pH 5.5.

There are various functional groups in the wool keratin which can react with these dyes like amino ($-NH_2$), hydroxyl ($-OH$) and thiol ($-SH$) groups. However, alkali is not used to induce reaction, lest the wool be degraded. Instead, the pH is maintained at 4–6, and reaction is induced by an increase in temperature. Silk, however, is more resistant to alkali, and alkaline conditions can be used to fix reactive dyes to silk. Much of the fixation is through the hydroxyl groups on some of the constituent amino acids. A wide range of reactive dyes is commercially available for application to protein fibres, including the chlorotriazinyl dyes, α-bromoacryloamido, the fluorochloropyrimidine (FCP) and vinylsulphone dyes, whose general formulae are given below:

$$D - NH - CO - CBr = CH_2 \quad \cdot \quad \textit{a-bromoacrylamidoamodo dyes}$$

$$D - SO_2 - CH_2 - CH_2 - \underset{\underset{CH_2}{|}}{N} - CH_2 - CH_2 - SO_3 Na \quad \textit{Vinyl sulphone dye}$$

Dye–fibre attachment resembles the same with that of acid dye on wool, i.e. chlorine of dye forms ionic bond with amino group of fibre with relatively higher bond energy.

Reactive dyeing gives good bright shades with all round fastness (LT 5–7, W.F. 4–5). Since the affinity of reactive dye for wool is very high, it cause problem on production of level shades. The initial strike off has to be avoided using suitable levelling or retarding agents. One such levelling agent is supposed to have the following structure:

$$(CH_2CH_2O)_n\ SO_3$$
$$C_{18}H_{37} - N - (CH_2CH_2O)_m\ H$$
$$CH_2$$
$$C = O \qquad m+n = 7$$
$$NH_2$$

Example of an amphoteric levelling agent for reactive dyeing

Dye-fixing agents are only applied during last wash to avoid staining and to improve wash fastness.

Some of the commercial wool reactive dyes:

Brand	Company	Reactive groups
Lanasol	Ciba	- NHCO - $\overset{Br}{\underset{}{C}}$ = CH$_2$ (α - Bromoacrylamido)
Drimalan	Sandoz	- HN⟨N⟩F ... Cl, F (2,4-difluoro-5-chloro-pyrimidyl (FCP)
Hosatalan	Hoechst	- SO$_2$CH$_2$CH$_2$- $\overset{CH_3}{\underset{}{N}}$ - CH$_2$CH$_2$SO$_3$H (N-methyltaurine-ethyl sulphone) and some β -sulphato ethyl sulphone dyes (see below)
Realan	Dystar	- SO$_2$CH$_2$CH$_2$OSO$_3$H (β -sulphato ethyl sulphone) and some FCP dyes

1.1.6.6.1   Reactions of the wool reactive dyes

The wool reactive dyes are capable of reacting with amino, thiol and hydroxyl groups in wool. There can be two types of reactions – neucleophilic substitution and Michael addition.

In neucleophilic substitution reactions, attraction of an electron-deficient carbon atom for the free lone pair of electrons on the nucleophile takes place. Generally, this reactive centre on the carbon atom is activated by electron-withdrawing groups adjacent to it (usually SO$_2$ or C=O).

The reactive carbon atom is also attached to a leaving group, usually halogen, sulpho or quaternary nitrogen. For example reactions of wool reactive chloroacetyl reactive dye with an organic amino group in wool fibre is shown in figure below.

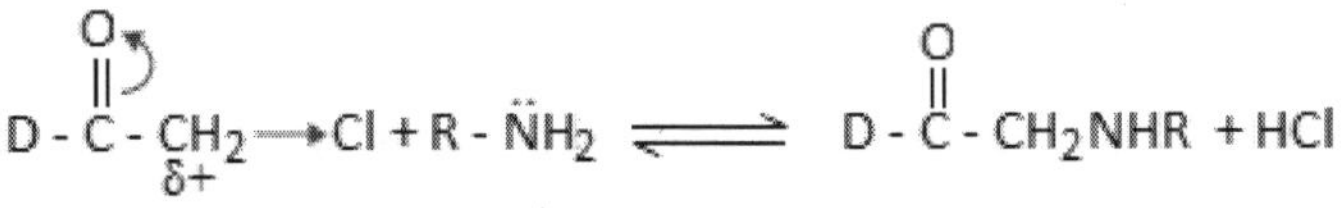

$$D-\overset{\overset{\displaystyle O}{\|}}{C}-\underset{\delta+}{CH_2} \longrightarrow Cl + R-\ddot{N}H_2 \rightleftharpoons D-\overset{\overset{\displaystyle O}{\|}}{C}-CH_2NHR + HCl$$

Example of Neucleophilic substitution reaction (Reaction of chloro acetyl dye with amino group in wool)

The reaction of a VS dye with an amino group in the wool is a Michael addition reaction. The polarised, unsaturated carbon–carbon double bonds are activated by the presence of an electron-withdrawing substituent, such as a carbonyl or sulphonyl group and reacts with the amino group in the wool fibre. It is a 1,2 trans addition reaction as shown in the figure below.

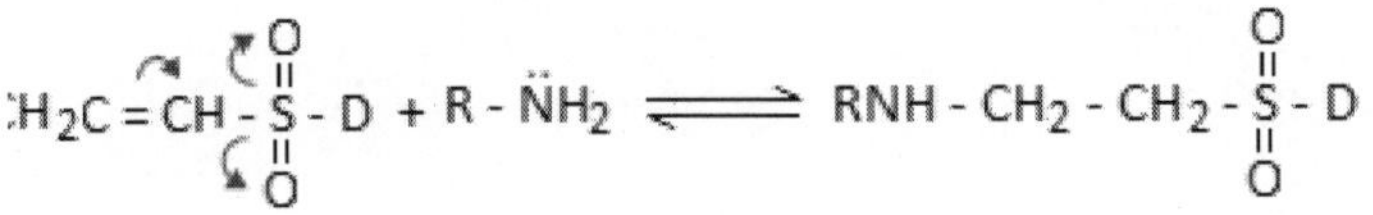

$$H_2C = CH-\overset{\overset{\displaystyle O}{\|}}{\underset{\underset{\displaystyle O}{\|}}{S}}-D + R-\ddot{N}H_2 \rightleftharpoons RNH-CH_2-CH_2-\overset{\overset{\displaystyle O}{\|}}{\underset{\underset{\displaystyle O}{\|}}{S}}-D$$

Example of Michael addition reaction (Reaction of vinyl sulphone dye and amino group in wool)

Application

Even though wool reactive dyes are introduced earlier, there was problem to in practically useing them due to the skittery dyeing results, especially in trichromatic shades, until a new Ciba–Geigy introduced the a new amphoteric auxiliary product, Albegal B (and other equivalent products by other manufacturers like–Lyogen FN (Sandoz), Eganol GES (Hoechst), Avolan REN (Bayer), etc.). The auxiliary may be amphoteric or weakly cationic of the typical following structure:

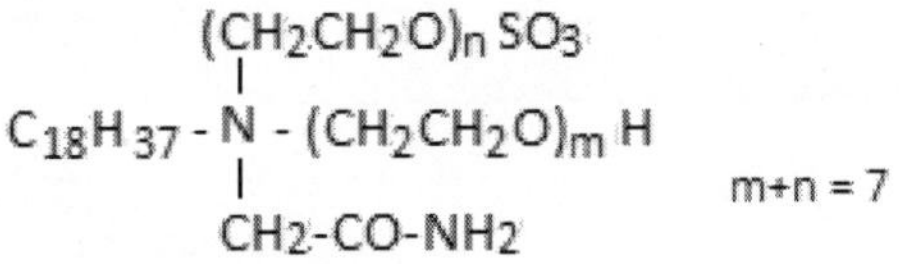

$$C_{18}H_{37}-\overset{\overset{\displaystyle (CH_2CH_2O)_n SO_3}{|}}{\underset{\underset{\displaystyle CH_2-CO-NH_2}{|}}{N}}-(CH_2CH_2O)_m H \qquad m+n=7$$

Bath is prepared with dye, 30% CH_3COOH (2–5%) and a non-ionic levelling agent (2%). (Instead of acetic acid ammonium acetate or ammonium sulphate (1–3%) may be used for light shades.) Dyeing is started at room temperature and raised to boil in about 45 min. Dye fixing agent may be used after washing off the unfixed dyes as in the case of reactive dyes on cellulosics.

The application involves the same three stages required for dyeing cellulosic textiles, although fixation of the dye is achieved by an increase in temperature, instead of addition of alkali. The conditions of dyeing are initially arranged so that the dye is evenly adsorbed under weakly acidic conditions (pH 4–6), usually at a temperature well below boiling point. Glauber's salt is also present in the dyebath. At this stage, the dye behaves like an acid milling dye. A levelling agent is often included amongst the dyebath ingredients.

The temperature of the dyebath is then gradually raised to the boil and subsequently kept at this temperature. The adsorbed dye now reacts with the wool to form covalent linkages. Thus, provided fixation is sufficiently slow, the dye can migrate in the early stages of dyeing below the boil. As the temperature increases, however, so the rate of reaction increases, and the dye consequently becomes fixed. Finally, unattached dye is removed by soaping.

Wool reactive dyes are used to produce dyeings of especially high wet fastness in any depth of shade. Their inability to migrate is compensated for by even adsorption which results in level dyeings. They are primarily used for self-shades on untreated i.e. unchlorinated, wool to produce full, bright shades with higher wet fastness.

An important field of application for Drimalan F dyes is in dyeing wool rendered resistant to shrinking by chlorination. Wool dyes, even the fastest of them, are known to produce dyeings of' unsatisfactory wet fastness in medium and deep shades on wool of this kind. With wool reactive dyes, however, fast, bright or muted shades can be dyed. Furthermore, since chlorination largely eliminates skitteriness, the dyes are freely combinable..

These dyes have chemical structure like acid dyes but can make a chemical bond with keratin under normal dyeing conditions (pH 4.5–6, 98 °C). Because of the chemical bond they give dyeing of high wet fastness and bright shades. They are very much suitable for dyeing shrink resist wool(Chlorination –see finishing of wool).

1.1.6.6.2   Reactive dyeing by Ciba special Lanasol method

Since the environmental regulations restrict the use of chrome dyes, which has good all round fastness otherwise, the importance of metal free reactive dyes have been increased. Ciba has developed a faster and cost effective dyeing method with selected Lanasol dyes. This method gives high wet cross dyeing fastness and deep shades which are necessary especially for suitings.

Dyeing diagram

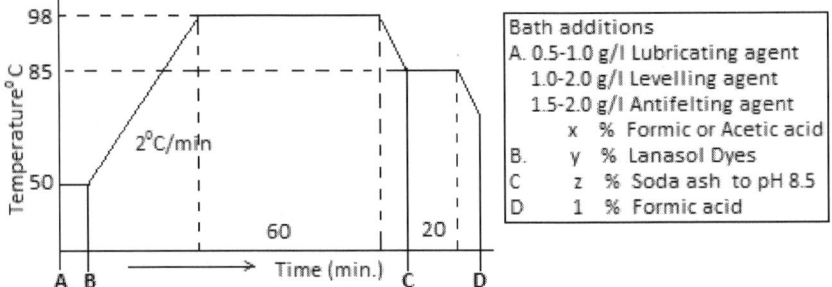

Reactive dyeing by Ciba Special Lanasol Method

Dyeing procedure

Dye bath is prepared at 50 °C with dyeing auxiliaries and material is introduced and run for 10 min and add the well dissolved Lanasol dyes. After running 5–10 min at this temperature is raised to boiling at 2 °C/min. Run at this temperature for 60 min and cool to 85 °C and add the alkali (soda ash) to adjust pH to 8.5. Hold at this temperature for 20 min and then cool to 70 °C and drain. Rinsing and soaping as usual.

By this method the dyeing can be completed by around 140 min whereas a chrome dyeing method can take up to 160–175 min.

1.1.6.6.3   Dyeing of chlorinated wool (treated wool)

Recipe

	Upto 1%	Upto 2%	Above 2%
Initial pH (Optimum)	6–5.5	5.5–5	5–4.5
Glauber's salt calc.	10%	10–5%	
Ammonium sulphate	4%		
Acetic acid 80%	0.5–1.0%	1.0–1.5%	1.5–3.0%
Dyeing asst.	1%	1%	1–1.5%
Dye	x%	y%	z%

1.1.6.6.3.1   Medium to deep shade

Set the bath with above recipe enter the goods at 20–40 °C and treat for 20 min as at this stage 40% of the dyes are absorbed rapidly and loosely attached to the fabric, to ensure even

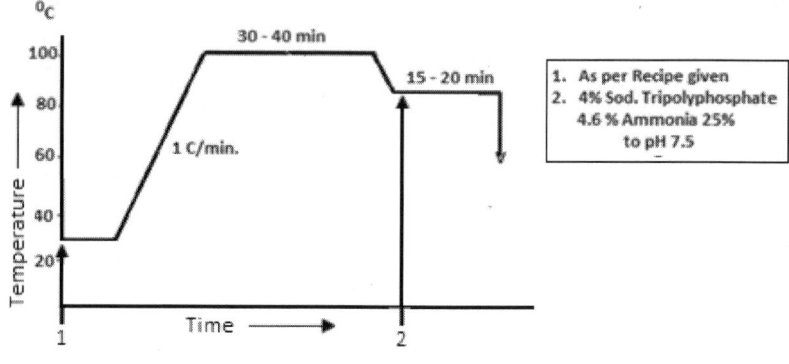

Dyeing on chlorinated wool

distribution of the dye. Raise the temperature to boil at the rate of 1 °C/min and allow at least 30–60 min for fixation at the boil depending on the depth of shade. Cool the dyebath to 90 °C and add

4%	Sodium tripolyphosphate
2–4%	Soda ash or
4–6%	Ammonia 25%

and treat for 15–20 min at pH 7.5.

This removes all the unfixed dye. After this treatment rinsing etc. can be done as usual.

1.1.6.6.3.2   Pale shades

Set the bath with above recipe enter the goods at 20–30 °C and treat for 20 min as at this stage 40% of the dyes are absorbed rapidly and loosely attached to the fabric, to ensure even distribution of the dye. Raise the temperature to 65 °C at the rate of 1 °C/min and continue dyeing at this temperature for 15–20 min. This pace in temperature rise is especially important for dyeing hanks and yarns. Cool down to 85–90 °C and allow at this temperature for 30 min and add

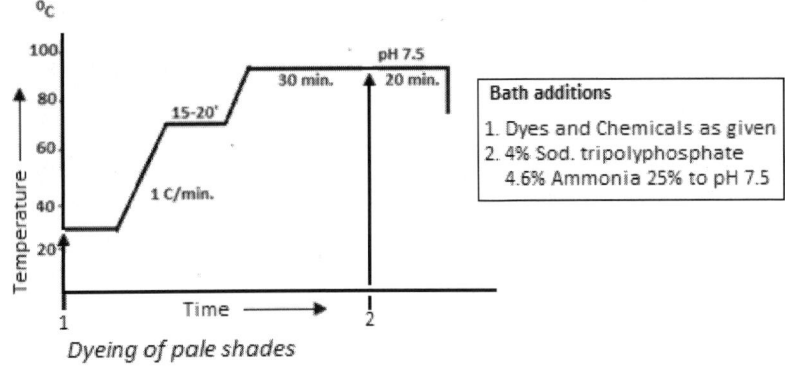

Dyeing of pale shades

4%	Sodium tripolyphosphate
2–4%	Soda ash or
4–6%	Ammonia 25%

and treat for 15–20 min at pH 7.5. Rinse, wash, etc. can be followed as usual.

Notes:

Levelling asst. (e.g., Lyogen F Liq.–Clarient) is a must to ensure even distribution of the dye in the beginning stage of dyeing.

1.1.6.6.4   Dyeing of untreated wool

In untreated wool the strike rate being lower especially at lower temperature when the optimum pH is kept. Adsorption commences at 50–60 °C and the shade is built up slowly and evenly.

Additions	Upto 1%	Upto 2%	Above 2%
Initial pH (optimum)	6–5	6–4.5	4.5–4
Glauber's salt calc.	10–5%	5%	
Acetic acid 80%	Upto 1.5%	1.5–2%	2–4.0%
Dyeing asst.	1%	1%	1–2%
Dye	$x\%$	$y\%$	$z\%$

Procedure

The goods are treated for 10 min at 40–50 °C in a bath containing the chemicals. The pH of the bath is determined and adjusted if necessary and then the dye is added and the bath is run for 5 min until the dye is evenly distributed and the temperature is then raised to boil at the rate of 1–2 °C/min depending on the goods and machine. For machines with poor circulations a pause for 20 min can be introduced at 65 °C. Then the temperature is raised to boil and held at boil for

% Of dye	Time for fixation at boil
Upto 1%	20 min
1–2%	20–30 min
Over 2%	30–60 min

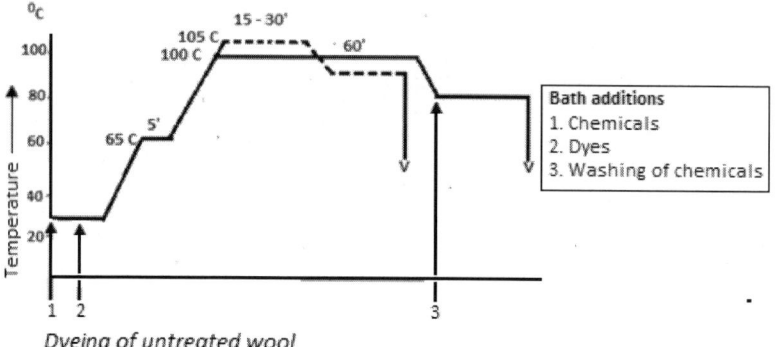

Dyeing of untreated wool

Notes:

1. These fixation times can be reduced to half by dyeing at 105–106 °C.

2. For pale shades, to avoid yellowing of the shades dyeing can be done at 85–90 °C, where the fixation time is doubled.

3. Shading addition can be made at 60 °Cand run for 30 min at 85–90 °C.

4. After treatment can be done as for treated wool.

5. Glauber's salt in these methods acts as a powerful levelling agent and not helping in exhaustion.

6. There are special auxiliaries available which can be added in the dyeing bath to improve the brightness of the shades.

1.1.7   Continuous dyeing of slubbing

1.1.7.1   Pad steam process

Recipe for pad liquor

Quantity	Unit	Additions
4–8	g/l	Thickener
x	g/l	Dye

10	g/l	Dyeing asst.(e.g., Lyogen V)
10–20	g/l	Levelling agent (e.g., Lyogen CW paste)
10–20	ml/l	Formic acid 85% (pH 2.8–3)
		For wool reactive dyes 6–6.5 with acetic acid

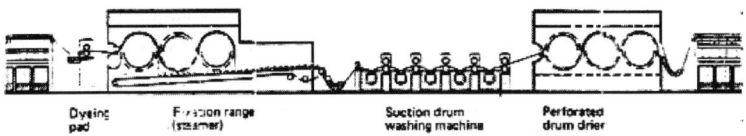

Dyeing pad   Fixation range (steamer)   Suction drum washing machine   Perforated drum drier

Pad at temperature 25–30 °C.

Liquor pick up 100–120%.

Fixation by steaming at 1000 °C for 15–45 min.

Washing.
1st Bath – 60 °C with 5 ml/l ammonia.
2ndBath – 60 °C with water.
3rdBath – 60 °C with water.
4th Bath – 60 °C with acetic acid at pH 4.5.
5thBath – 30 °C with 10–5 g/l softener.

Notes:

1. Nowadays acid dyes are rarely used due to its poor fastness properties.

2. Normally 1:2 metal complex dyes are used for slubbing especially for duller shades like greys, browns and navy blues. In addition some wool reactive dyes are also used for this method.

3. Whenever possible concentrated dyes usage is preferred for this method since they give a better stable pad liquor.

4. Dyeing assts. are necessary to avoid uneven dyeing.

1.1.7.2   Cold pad batch method

Normally wool reactive dyes are used for this method.

Recipe for pad liquor

Quantity	Unit	Additions
350	g/l	Urea
8	g/l	Thickener
x	g/l	Dye
10	g/l	Dyeing asst. (e.g., Lyogen V liquid)

2-Jan	g/l	Levelling agent (e.g., Lyogen CW Liq)
		10 g/l sodiumbisulphite for some class of dyes
		pH adjusted to 6 with acetic acid

Pad at 25–30 °C.

Liquor pick up 100–130%.

Batching time with rotation for pale to medium shades 24–48 h at 20–30 °C.

Dark shades 48–72 h at 20–30 °C.

Washing and after treatment as per pad steam process

Notes:

1. This method is usually used for larger batches.
2. Shrink resisted wool present considerable problems with this method with regards to levelness.
3. This method can be followed without any specialization, provided required machinery is available.
4. The handle and appearance of the fabric is not affected by this method.
5. If procedure is followed correctly, the method gives good level and fast shades.
6. After padding the goods are batched and fixed for 1–3 days according to the room temperature and depth of shades. The rolls should be turned periodically and to prevent the outer layers drying out the batch should be wrapped with polythene film.

1.1.7.3   Vigourex printing or melange printing

Vigoureux printing is a special technique in which slivers of raw, bleached or dyed wool are printed with stripes, steamed, run through a backwasher, and then dried.

Advantages of vigoureux printing are:

*   Minimum fibre damage.
*   Exceptionally regular melange effects.
*   Virtually no discoloration of the wool.
*   Low steam and water consumption.
*   Low effluent pollution.

Melange printing is an alternative to slubbing dyeing and allows mixture effect (mélange) yarns to be produced with a much more uniform appearance than by mixing dyed and undyed slubbing.

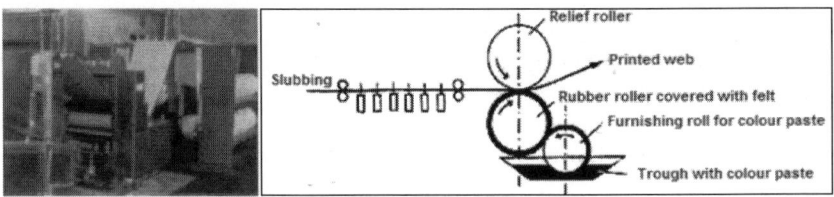

Vigourex printing system   Schematic diagram of vigourex printing system

Greys can be produced by printing chrome black. The use of chrome dyes is prohibited in many countries because of the danger of fluorochrome pollution. Hence nowadays 1:2 metal complex dyes are used.

Recipe for printing paste:

1. Chrome colour

Quantity	Unit	Additions
6–8	g/l	Thickener
5–10	g/l	Formic acid 85%
5	g/l	Penetrating agent (e.g., Lyogen CW Liq)
30–50	g/l	Chrome dye
20–30	g/l	Chromium fluoride
15	g/l	Wetting agent

2. 1:2 Metal complex

Quantity	Unit	Additions
6–8	g/l	Thickener
5–10	g/l	Formic acid 85%
5	g/l	Printing asst.
1	g/l	1:2 Metal complex dye
15	g/l	Levelling agent

The procedure for making the printing paste:

40 l Cold water is taken in a preparation tank. A high speed stirrer is used to agitate the liquid and while stirring add800 g thickener are sprinkled in, followed by1000 g formic acid 85% diluted with 9 lwater and500 g Lyogen CW paste dissolved in10 l water. The chrome dye dissolved in20 l water is

then added. After stirring for 30 min add15.5 l cold water. Ensure that the temperature of the paste is not exceeding 50 °C,3 kg chromium fluoride are sprinkled in and finally1.5 kg Lyogen V liq is added.

Temperature of printing paste 25–35 °C.
Steaming with saturated steam at 98–103 °C for 60–90 min.

The steaming time is governed by the working temperature in the steamer, the amount of material and coverage of print. Steaming should be carried out while the material is still moist. Localised drying should be prevented.

Washing

1st Bath water at 50–60 °C + 1 g/l ammonia,

2nd Bath water at 50–60 °C + 1 g/l ammonia,

3rd Bath 2 g/l acetic acid 40%,

4th Bath x g/l softener.

3. Wool reactive dyes

Quantity	Unit	Chemicals
x	g	Wool reactive dyes
y	g	Cold water
z	g	Boiling water
100–300	g	Thickener (2% solution)
15–25	g	Acetic acid 80%
2	g	Wetting agent
10–30	g	Padding assistant
1000	g	Make upto

Printing and steaming is done as above. The steaming times tabulated below are for guidance only. In each specific case the precise length of the treatment will depend on depth of shade, steaming temperature, density of material and type of steamer used.

Depth of shade	Semi-continuous	Continuous
Pale shades	30–60 min	15–30 min
Deep shades	About 60 min	20–45 min

Saturated (not superheated) steam should be used. If dry steam is used, it is good practice to add the necessary moisture by injecting water, to interrupt steaming after half the time has elapsed, and to allow the slubbing to cool down. The steaming temperature is 100–102 °C (212–216 °F), depending on the type of steamer.

Thickener

The amount of dye applied can largely be controlled by the viscosity of the print paste or padding liquor. The right thickening is the key to good printing. In principle, only thickeners should be used which are easy to wash off and resistant to acid. When present in the print paste or padding liquor, the thickener should be well dissolved, otherwise problems can arise, e.g. agglomeration of dye and thickener, sticking on the back washer, wrapping round rollers, dusting during gilling or spinning. To be on the safe side, therefore, many mills prefer to use stock thickenings.

1.1.7.4   Importance of machinery and equipment

Machinery and equipment plays a very important role in determining the dyeing programme and therefore the levelness of the dyeing, than the substrate and dyes, etc. A large open hank dyeing machine, for example, requires great care in heating up, while dyeing in an enclosed package dyeing machine of more modern design is reliable and rapid. Correction is costly and non productive exercise and hence the optimisation if dyeing process with respect to substrate, products and machinery is very important.

One of the most important ways to obtain level dyeings is to control the adsorption. This is possible when the relationship between the number of contacts and the exhaustion rate is linear,further the amount of dye exhausting on to the fibre per contact or cycle should not exceed a given maximum. This maximum varies according to the form in the wool (any material) is dyed, as follows (assuming that dyeing is carried out in a machine with circulating liquor, without special level ling agents and without reversing the direction of the flow):

Loose stock: Maximum value of D= 2–3% per cycle.

Slubbing: Maximum value of D = 1.5–2.5% per cycle.

Yarn, pieces: Maximum value of D= 0.755–1.0% per cycle.

The differences are due to the fact that the levelness of loose stock is not depending on such stringent demands as that of slubbing or pieces. The permissible maximum values D can be increased in two ways: by changing the direction of flow of the liquor or by using levelling agents.

Like the exhaustion of the bath, the reversal of flow has to be based on the bath revolution or cycles:

No. of bath cycles per flow reversal	Factor f
1.0–1.9	4
2.0–3.9	3
4.0–7.9	2
–8	1

Maximum values D, can be increased by the factor f as a function of bath cycles per reversal of liquor flow. This is of particular interest in the case of machine with low pumping efficiency, for a package dyeing on an older type of machine with a circulation rate of only 1/2 cycle per minute, for example, D is 1% per cycle without reversal of liquor flow. However if the direction of flow is altered every 4 min, i.e., after every two cycles then $f = 3$, hence $D=$ 3% per cycle, i.e., the rate of dyeing can be three times higher than when the liquor flow is not reversed. If in addition a level ling agent is added, the D can be doubled.

If the rate of adsorption is to be controllable in this sense, the pumping capacity of the machine must be known. If the liquor throughput in presence of the material is known in litres per minute, the number of cycles achieved per minute is:

$$\frac{\text{Litres per minute}}{\text{Total liquor volume in litres}}$$

The pumping capacity in litres per minutes per kg material of a fully loaded machine is often known. In this case the number of cycles is calculated as follows:

Specimen calculation for a 100 kg machine with a capacity of 1500 l and a pumping rate of 30 l/min/kg.

Number of cycles per min = $(100 \times 30)/1500 = 2$ cycles/min.

If the pumping efficiency of the machine is known, the minimum exhaustion time for a dyeing can be easily determined.

Assuming that the maximum permissible exhaustion on to the fibre $D = 1\%$ of the total amount of dye applied, then by definition 100 contacts of cycle are required to achieve linear dyebath exhaustion.

Thus the minimum exhaustion time

$F=100/D.C=$ minutes.

Chapter 2
Processing of silk

2.1   Silk fibre

The term silk is used for filaments exuded by the glands of larvae of certain species of moth. The larva is called silk worm. Silk is the only continuous yarn existing in nature. The main two types of silk worms are:

1. *Bombyx mori* silk worm: Cultivated silk. They give almost white silk after degumming even though slightly yellowish brown before degumming.
2. Wild silk worm, also called tussah: Wild silk filaments are coarser, more irregular and beige/brownish in colour even after degumming.

Since both these silks are produced by the glands on the head of the silk worm they consists of a trilobular filament, the fibroin and sericin which is a glue like protein which sticks these two fibroin filaments. Sericin will be about 20–30% in cultivated silk and 8–15% in case of wild silk.

The cultivated silk is raised and fed on mulberry leaves. The home of the mulberry tree is Asia. Even today most cultivated silk comes from this part of the world. Today China dominates with about 90% of worldwide exports of silk and silk products.

In the 19th century China was the greatest producer of silk as well, but during parts of the 20th century Japan was in the lead. Although the production of silk in Japan has now declined very much, this country is its greatest consumer; in the early 1980s,about 80% of all Japanese silk requirements went into the making of kimonos.

Silk production is also important in India, Thailand and South Korea.

In Europe however cultivation of the silkworm has declined very considerably and is today practically unimportant compared with 100 years ago. Only in the Soviet Union and Bulgaria is the production of silk worthy of mention. In recent years Brazil has also become a silk producer. With the aid of the Japanese, great efforts are being made to increase production. Wild silk also comes from Asia, with India and China as the major producers. They are grown in the wild usually feeding upon oak tree leaves. The tannin and the iron salts in the oak leaves are the reason for the brownish colour of the wild silks. The cocoons are coarser, harsher, and more and more uneven than the silk from the cultivated silk worm.

As silk is a natural product, differences in self-colour, fineness and regularity of the filaments can be seen, depending on the origin and region of cultivation. Cocoons, yarns or woven fabrics are therefore often traded under the name of the corresponding region of origin.

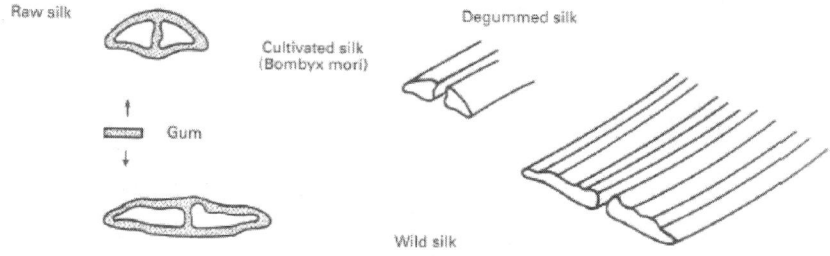

Cross-section, longitudinal view and perspective of silk filaments

Cross-sectional (a) and longitudinal (b) views of silk fibres

2.1.1   Bombyx mori silk

The female moth lays about 400 eggs the size of the head of a pin. After 10–14 days at a temperature *above* 13 °C the shells are broken through and the caterpillars creep forth. On hatching, the caterpillars are about 3 mm long and weigh 0.47 mg. Eating (mainly on mulberry leaves) voraciously, they grow *very* quickly. After 30–33 days they weigh 3.65–4 g and are 9 cm long, having increased their length about 30 times, their size 400–500 times and their weight 7–10 thousand times. During this development the *larvae* change their skin 4 times, or *every* 5–6 days. These figures make it clear that silkworm *larvae* require an enormous amount of fresh mulberry leaves for their food. For example, 50,000 silkworms require 1.5 metric tons of fresh mulberry leaves. According to an old Chinese tradition, at the early stage, the silkworm must be fed *every* half hour.

When they reach maturity the *larvae* are selected and gathered. This point is reached when the head of the caterpillar, for example, becomes lighter, more transparent and begins to take on colour. Here the timing and correct selection require considerable experience. To facilitate spinning, the silkworms are placed in bamboo spinning nests which are heated from below and *covered* with straw. The silkworm seeks the most *favourable* position for spinning and begins making its pupa bed. Then, moving its head in a figure-8 motion, it begins to spin the silk filament, forming the cocoon itself. The two spinning organs are located on either side of the lower jaw. The two silk glands contain the fibroin in dissolved form as a concentrated substance (fibrinogen) which on emerging and being drawn in the air becomes insoluble in water. The silk gum (sericin) which envelops and binds together the two strands is added at the moment of spinning. It is produced in the stomach of the worm. These substances are the two components of the protective cocoon which the silkworm spins round itself.

From the spinning of the cocoon (3 days) to the *larva's* transformation to the chrysalis stage (3–4 days) and the pause for development (14 days) to the emergence of the moth takes about 20 days. The cocoons from which the silk is taken must be gathered at exactly the right moment. Equally important is the selection of suitable silkworms for further breeding. This requires an especially great amount of experience, since suitable males as well as female moths must be selected. With some types of silkworms the chrysalis are often hardly distinguishable; the female may either be more rounded at one end or its cocoon somewhat indented. Only skilled breeders can recognise these fine differences, thus also contributing to the production of high-quality silk by careful cross-breeding of good moths.

2.1.2   Tussah silk

The most important representative of this type of silk spinning moth belongs to the family of *Antheraea* or wild silkworm. There are also an enormous number of different species of this type of moth, only few of which are used for producing silk. In India there are also other wild silks such as the traditional 'Tasar'(*Antheraea mylitta*), 'Eri'(*Philosamia ricini*) and 'Muga' (*Antheraea assamensis*) silk.

2.2   Silk preparation

Silk cocoons are mostly white and yellow, but may be orange, green or *even* dark brown. For high-quality silk, the cocoons should be sorted into batches of the same colour. The cocoon colour results from the pigments which form during the development of the silkworm, e.g. green is derived from chlorophyll (the pigment issoluble in water and not in organic solvents). Yellow, derived from karotin, is basedon a pigment which is insoluble in water but soluble in organic solvents such as alcohol or ether. The colour pigment is added to the sericin just before the filamentis spun.

A silk cocoon

Cocoons *vary* considerably in their shape. Some are spherical while others are egg shaped; still others *have* a peanut-like indentation or are elongated and with pointed ends. The exterior is *uneven* and gritty while the interior – after about ¼ of the entire filament has been reeled – is smooth. The rough exterior form of the surface is important for the processing and quality of the silk. Thus, for example, the number of spherules per square centimetre is defined. The quality of the silk is assessed on the basis of the surface graininess. The occurrence of spherules is explained by the fact that the silkworm lays the filament on the wide-meshed cocoon structure in figure-eight patterns which extend into the wide meshes, thus forming grains or spherules.

2.2.1   Reeling

Reeling is mostly done in the country where the silkworms are cultivated. The chrysalises in the cocoons are killed either by hot air or soaking in hot water and the cocoons are then placed in water in a steam heated basin. By being turned regularly with brushes the cocoons are kept submerged so that the silk gum softens. The flock silk, i.e. the outer layer, falls away and permits the outside ends of the filament to be located for reeling. This operation demands considerable skill on the part of the reeler. The treated cocoons are now placed in the reeling basin and drawn out together, depending on the fineness number desired (e.g. 5–10 filaments), dried and reeled into one thread. All these operations require *very* skilled hands. In a large Chinese silk processing mill, as many as several hundred women may be employed in this important operation. The *raw or reeled silk* produced in this manner is called 'greige'. The total length of the filament spun by the silkworm for attaching and building its cocoon is ca. 2500–3000 m for the cultivated silkworm. Only about 1200 m of this can be reeled, the rest being used for valuable by-products such as shape and burette silk.

2.2.2   Schappe and burette silk

As already mentioned, about 2/3 of the cocoon filament is not highest quality greige silk, the processing of unreelable filament is also important. This material, as well as scraps from reeling, twisting and weaving, can be further processed. For a long time these valuable materials were simply pulped or used for other purposes in the country of origin, due to the intensive work involved and the high investment costs for machinery. The cocoon residues often serve as animal feed, e.g. for chickens or fish.

Specialist terms for processing by-products of greige filaments:

- Silk wadding (Spelaia, Blaze): loose silk mesh around the cocoon.
- Flock silk (Frison, Stursa, Kibizzo): the outer, irregularly spun cocoon layers (ca. 25% of weight of material), regarded as best raw material forschappe silk.
- Cocoons perces (cocoons piques): pierced, damaged cocoons which cannot be normallyreeled.
- Bassins (Pelettes, Ricatti, Galettame, Bisou): the inner, parchment-like layer of the cocoon skin.
- Bourre (Strazza):scraps produced during reeling the skeins, twisting andcleaning the threads.

Unlike cotton spinning, where the small amount of impurities can be removed in the dry state, silk residues require an entire series of dissolving and cleaning operations such as degumming, washing, drying, beating and cutting (opening).

2.2.2.1   Schappe silk

Yarns spun from silk waste are usually termed "schappe" or "floret" silks. As is well known, a raw silk or greige thread results from reeling cocoons not pierced by the silkworm. Pierced or damaged cocoons cannot be utilised in this manner and yield the raw material of the silk spinning industry, which also processes the wastes fromthe reeling, winding, cleaning and throwing operations in the treatment of greige silk.

Thus, spun or schappe silk is pure natural silk. Spun silk yarns are high-grade products which differ from greige yarns solely in that they consist of shorter fibres more difficult to process. The outstanding properties and the value of the silk fibre early led to the wastes from silk manufacture being put to full use too. Man first pulled apart the filaments of perfect cocoons for spinning-as waste is spun to-day-before hitting on the idea of reeling the continuous fibre. As a textile raw material, silk waste may, therefore,be considered to be older than reeled silk.

In sorting, which precedes reeling, the following types of cocoon are eliminated as waste:

(a) Pierced cocoons of standard white or yellow colour from which the silk moth has escaped for mating, provided the races concerned are of European, Levantine, Iranian(Persian), Caucasian, or Chinese origin. The yield is 48–5% of dressed or combed fibre suitable for spinning;

(b) Doubles ("doupions"), i.e. cocoons formed by two worms spinning close together so that their tangled filaments cannot be unreeled (yield: 24–25%);

(c) Stained and pierced cocoons from diseased silkworms, or damaged by pests (yield:22–24%);

(d) Canton, Tonkin, or Mysore pierced cocoons, i.e. the cocoons of brown, golden yellow, or green races from rearing areas in South China or India. As there may be from two to three annual crops of cocoons in these districts, quality and fibre yield vary widely, depending on seasonal weather conditions. Yields are between 30% and 36%.

The types of wastes formed in reeling the cocoons are extremely numerous. According to the standard sorting procedure, the outside fluff is first removed and put aside; it is formed by the filaments first spun by the silkworm as a support for the cocoon proper (="blazes", or floss silk). The yield is 26–30%. The cocoons, which float in a reeling basin, are worked with a revolving brush and the filaments caught by it reeled. The waste is straightened out and yields strands of flock silk known as "knubs" or "frisons" about3–5 ft. long and held together by the drying gum (yield: 58–60%). Flock silk, which comprises from 25% to 30% of the entire cocoon, *is* the premier raw material and in the silk spinning industry serves as the standard for all other types of waste. Cocoons reeled only in part and discarded from the reeler's basin as being too difficult to process enter the market as "bas sines" or "galettame" silk (yield: 15–20%).

2.2.2.2   Bourrette

Short fibres of 10–50 mm which are produced when combing are used for bourrette spinning. These yarns are characterised by the dark, irregular chrysalis residues which even after bleaching and dyeing are still visible as small black specks irregularly distributed over the fabric surface. Nowadays the fineness of schappe and bourrette silk is expressed in metric figures (Nrn), while that of free thread is still mostly given in denier (den).

Bourrette yarns which are to be woven must be sized before hand. This is different from greige silk, where the natural gum of the fibres, i.e. the sericin, mostly provides sufficient abrasion resistance on the weaving loom. Either starch products or synthetic sizes are used. The latter require particular care that substances are not introduced into the processing stage which cause problems in desizing, dyeing and printing.

2.2.3   Structure of silk

The silk fibre is a continuous double filament consisting of fibroin (72–80%) surrounded by sericin (17–28%). The fibroin is highly oriented. Crystalline domain and is insoluble whereas sericin is soluble in hot soap solution. The crystallites are contained in microfibrils packed together in bundles form a single strand of fibroin. Details of silk structure can be found in the book. The Substrates – Fibres, Yarn and Fabric Chapter 4 by the same author.

2.3   Silk manufacturing

As explained earlier, silk processing begins reeling of silk the production of the fibre is performed by the larvae. The fibre is used for making the cocoons. When the larvae are transformed into moth inside the cocoons it will break the cocoon to come out. Hence the cocoons are harvested before the moth is formed and put in the boiling water to kill and dissolve the larvae and the cocoon is separated.

This cocoon is the reeled to separate the fibre (reeling is the process where the end of the fibre is separated from the cocoon surface and wound on a reel and it is made in the form of the hank. The silk is sold in this form. The processing of silk starts from this hank form.

2.3.1   Hank to spool winding

As the yarn to be wound is extremely delicate, special hank to spool winding machines are used. The slipperiness of the silk yarn requires the use of tapered bobbins. Take up packages may be flanged bobbins, bottle bobbins or cops, depending on the subsequent use. The reel on which the hank is positioned can be rotary (deroule winding) or fixed (defile winding).

On these machines the yarns are controlled by means of mechanical flanges with pressure variable according to spool diameter and desired density. Doubling machines are specially designed which can double four ends and optionally six ends. They are usually provided with magnetic sensors on each end to be doubled which senses a yarn break or unavailability. There are some doubling machines are provided with purpose designed tensioner. Here the sensors have been replaced by a spiral throwing device and thus the last end is twisted around the other already doubled ends. The sensors for the three ends already doubled are electronic, whereas for the fourth end there is a diapason device. Tension is controlled by a ceramic split yarn clearer.

There are other machines which couples two, three or four ends prior to throwing, ensuring an excellent yield. The machine can couple ends by special process that imparts a cohesion effect. On the two-for-one twister, this fact improves yarn unwinding from flanged bobbins. The machines are equipped with either electronic or non-contact optical system stopping the end in case of break or lack of yarn and raise the single take up positions when the preset diameter has reached. The gear box operates through sophisticated system to produce flanged bobbins without hard edges.

The two-for-one twister has spindles in oil bath, specifically designed for silk. Energy consumption has been minimised. The take-up speed can be adapted to both high and low twists. Two types of flyers are available for unwinding low and high twist yarns.

2.3.2   Warping, weaving and circular knitting

Silk yarn warping can be performed on electrically controlled section warpers which allow to handle the delicate silk yarn. For the weaving process the looms must be suitably adjusted bearing in mind that silk is not a "speed-friendly" material. The same goes also for circular knitting machines.

Classic silk woven fabric, however, is made by twisting the greige threads, in accordance with the number of filaments, fineness, the number and direction of the twists; factors which have a decisive influence on the appearance, handle and fall of the fabric.

General names of greige silk threads come from the old silk school of Lyon at a time when the creation of specific articles of silk made the city and region famous:

1. Voile individual greige threads with 600–3000 twists per metre. Similar to crepe, but finer and lighter. Used in the raw state as warp and weft material.

2. Organzine: Silk yarn from two or more twisted greige threads. Single strands are given a preliminary twist and then two or more are doubled and twisted again in the reverse direction. The preliminary twist consists of about ca. 600–700 and the subsequent one 500–600 twists per metre. Organzine yarn has a stiff handle and produces a tightly closed effect. In the boiled off state a soft lustre is produced. It is a preferred warp thread for plain and colour-woven fabrics but is also used for knits, which have recently made increasing headway into fashion collections.

3. Grenadine: Silk yarn from two or more threads, similar to organzine, but much more strongly twisted: 1200–1800 twists per metre for the pre-yarn, 1000–1500 for the subsequent yarn. The threads have a much closed effect, show high abrasion resistance and are used as both warp and weft material for special weaves as well as ladies hosiery.

4. Tram: Yarn of 2 or more greige threads which have been doubled without preliminary twisting and then twisted with 150–200 twists per metre. A lofty, voluminous yarn, soft and supple. When boiled off it has an elegant lustrous effect. Dyed as yarn, it is a preferred weft material for colour-woven fabrics. More strongly twisted, it can also be used for knits.

5. Crepe: Yarn from one or more greige threads without a preliminary twist and then twisted with 2500–3500 twists to the metre, producing a hard, strong, tightly closed effect. When boiled off it becomes more supple and softer, with a dull lustre. The appearance of the fabric resembles that of a woven crimp effect.

6. Cordonnets: Coarser double twists consisting of several greige threads with a preliminary twist of 400–800 twists per metre and subsequent doubling of 2 or 3 threads followed by 300–600 twists per metre. Cordonnets yarns are mostly used for sewing threads but sometimes for knitting yarn as well.

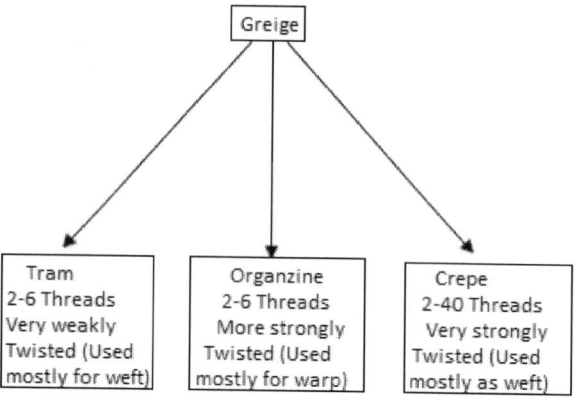

2.3.3   Raw silk processing from hanks

a.   From hank to weaving

Well known fabrics with plain weave

1. Habutae and pongee: Greige warp, greige or tram weft. Fabric has discreet lustre. Habotai silks are mostly light fabrics of 30–ca. 60 g/m², often known as pongee.

2. Taffeta: Organzine warp (possibly slightly weighted, degummed). Fabric has a 'papery' handle, with typical rustling. As a rule it is woven with dyed yarns, as further treatments, especially at higher temperatures, may cause the fabric to lose its characteristics. In some cases these fabrics are very crease-prone.

3. Organdy or organza: Warp and weft of non-degummed organzine yarn, woven in a lattice structure. A light, very transparent fabric, characterised by a rather stiff handle ('Organza' is a registered trademark).

4. Honan or Henan: Fabric of tussah yarns woven on a hand loom. Warp of raw tussah greige and weft of either raw tussah greige or tussah tram. The predominant general impression is of linen, as relatively strong warp and weft bars appear. This fabric is also often called raw silk. In accordance with the new designation of the province, this quality is now no longer called 'Honan' but 'Henan'.

5. Antung: Warp and weft of tussah greige. Unlike Honan silk, this fabric is produced on a mechanical loom. The fabric has a firm, lustrous appearance.

6. Shantung: Warp and weft of tussah greige from province of the same name; today often (wrongly) applied to fabrics of natural silk or other types of fibre with e.g. tram warp and doupion tram weft. A light fabric with a nep structure in the weft; the effect is called 'Shantung', the name applied to this typical woven fabric, which today is found on the market in practically all fibres.

7. Toile de Soie Warp and weft of schappe twisted thread; a soft, rather matt fabric.

2.3.4   Names of crepe fabrics

Fabric made in some cases of high to very high-twist yarns, can be in plain weave.

1. Crepe de Chine: Probably the best known silk woven fabric. Warp of greige threads, weft of crepe threads with *2S/2Z* twist. A soft, flowing fabric with good draping effect.

2. Crepe georgette: Warp of crepe threads with *2S/2Z* twist, weft also of crepe threads with *2S/2Z* twist. Fabric is soft and slightly kinky.

3. Chiffon warp of down silk or greige, weft of crepe with an S twist. Slightly transparent article with a bark effect along the length.

4. Muslin: Warp of crepe threads with an S or Z twist and weft of crepe threads with *1S/1Z* twist. Slightly transparent, flowing fabric.

5. Crepe marocain: Warp of greige threads, weft of crepe multi twisted of threads with *2S/2Z* twist.

6. Crepe satin: Warp of greige threads, weft of crepe twisted thread with *2S/2Z* twist and in a satin weave. Soft, flowing fabric with an elegant lustre.

7. Crepe satin/Faconne: Threads as above, but with a Jacquard weave. The patterning produced by the weaving method lends itself to rich effects which are very much in demand, especially when overprinted.

Names of fabrics with a satin weave:

1. Satin duchesse: A thicker, richly lustrous satin.

2. Damask: Warp and weft satin with different motifs, alternately matt and lustrous.

3. Satin envers Doupion (Shantung): Warp satin fabric with flame or nep effects in the weft.

Fabrics with a twill or serge weave:

1. Twill warp and weft of greige threads with a 2:2 twill weave (diagonal weave structure). Light, flowing fabric, also called 'foulard silk'.

2. Surah: Thread-dyed warp of organzine and weft of tram with a 45° diagonal weave structure. Even-sided twill with strongly marked slant-ribbed effect.

Classic woven fabrics with special effects:

1. Velour or velvet: Three dimensional fabric with upright thread nap, i.e. open end (cut velour), or nap with thread loops (crimped velour).

2. Plisse tisse (pleat-woven): Made with very strongly twisted threads which are shrunk in pretreatment to produce a pleat- or relief-like fabric appearance (bark structure).

3. Brocade: A fabric mostly produced by the Jacquard technique, with interwoven gold or silver threads.

2.3.5   Silk manufacturing and processing routes

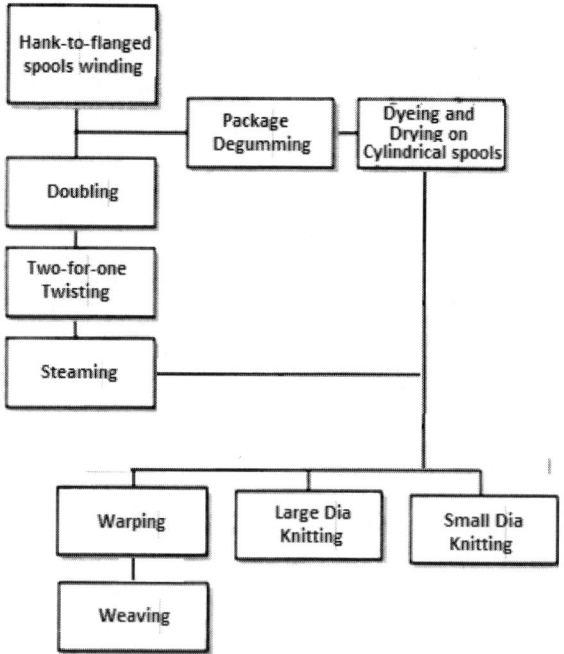

Silk preparation and Manufacturing process

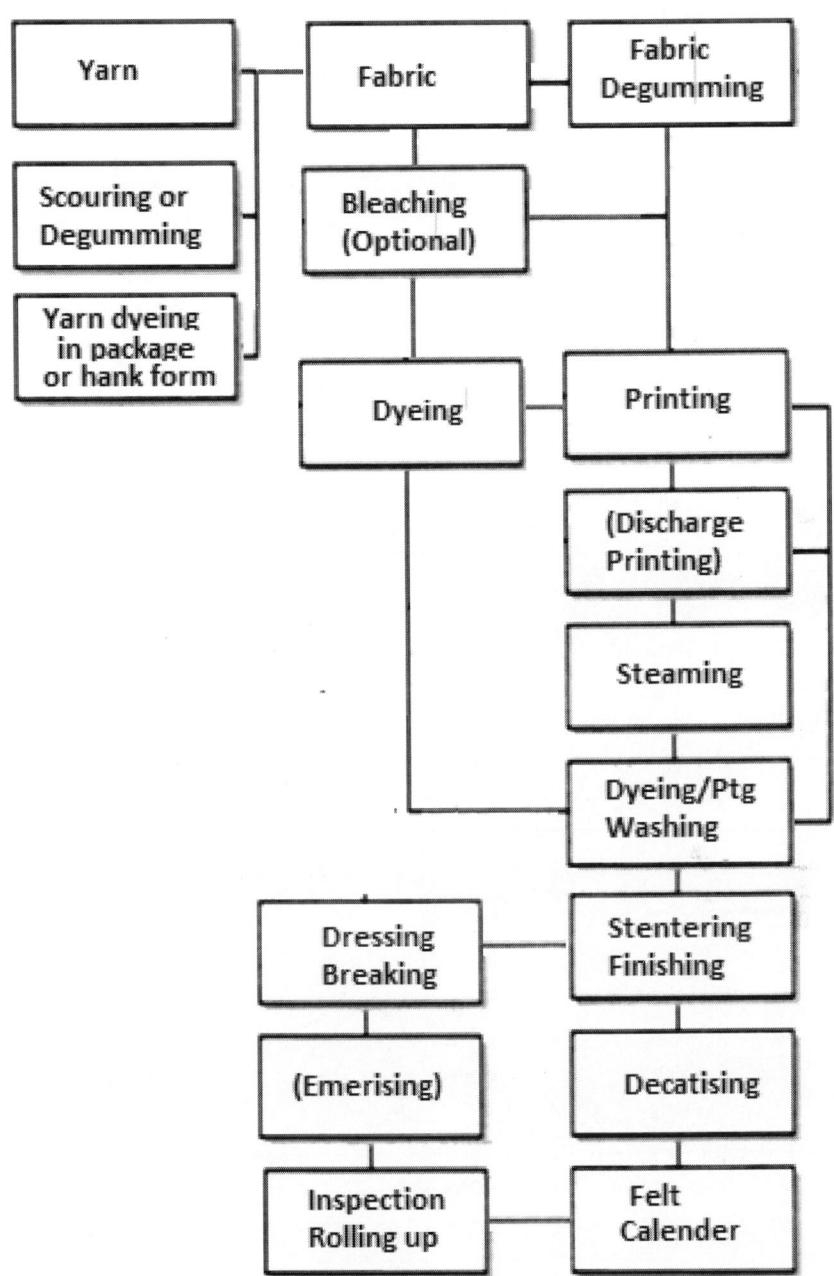

Silk Fabric processing routes

2.4   Pretreatment of silk

As explained earlier, silk is a protein fibre formed by firm and continuous filament coated with a gummy layer called sericin. To make the protein fibre bare the sericin is to be removed. Hence pretreatment mainly involves in removing sericin.

Pretreatment of silk consists of either degumming only or degumming and bleaching. In rare cases size may be applied on silk, the pretreatment involves desizing also. Even there are part degumming processes for special purposes.

Degumming is a process where the sericin is totally or partially removed by a suitable treatment. There are different types of degumming processes:

1. Traditional soap boiling.
2. Degumming with synthetic detergents.
3. Enzymatic process.
4. Degumming at high temperatures.

2.4.1   Degumming by traditional method

Traditionally soap solution is used. It is still being followed even though it has some disadvantages.

1. Conventional working method

   a. Soaking overnight or 6 h at 40–50 °C in a bath containing

3–5 g/l	Soap
1–2 g/l	Dispersing agent

   b. Boiling off for 2–6 h at 90–95 °C in a bat with8–10 g/l soap at pH 9.5 with soda ash.

   c. Rinsing (which is very important) First with hot water at 50 °C containing 1 ml/l ammonia and then two times once with warm water (40 °C) and then with cold water

2. Rapid method

Recipe

8–10 g/l	Soap
1–2 g/l	Suitable scouring agent (e.g. Imerol XN liq.)
1 g/l	Soda ash (optional)
0.5–1 g/l	Sodium tri polyphosphate
	Adjust the pH to 9.5–10

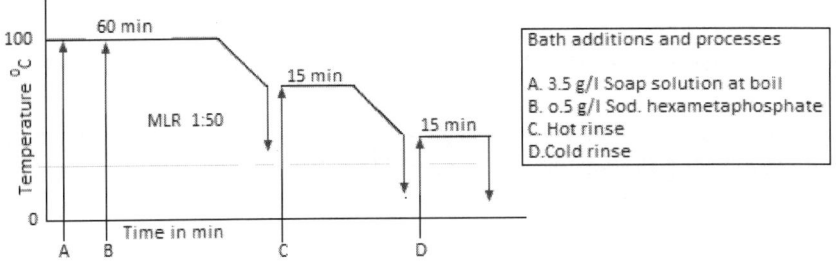

Degumming profile (Traditional method - Rapid)

Treat for 1–2 h at 90–95 °C.

Repeat treatment if necessary.

Add 1–3 g/l hydros (stabilised hydros preferred).

2.4.2   *Degumming in Yarn form*

Usually degummed in hank form – The hanks are hung on polished wood or metal rods which are submerged in the bath and turned by hand. The hanks are turned on a four-sided or double rod of porcelain or stainless steel (the rotation of the rods can be regulated) in hank dyeing machine. Hanks are also degummed in spray dyeing machine –The hanks are hung on perforated arms and the bath is circulated by spraying while the arms turn. foam boiling off process: The hanks are not submerged in the degumming bath but remain above a concentrated soap bath, the foam of which comes in contact with the silk and causes degumming.

2.4.3   Degumming in fabric form

Machines used

2.4.3.1   *Star machine – batch process*

This type of degumming is indicated for delicate articles and cases where the primary consideration is perfect production quality, especially for goods which are to be plain dyed. Star machines consists of:

Vertical single star (goods hang from above only) or vertical double star (goods are attached from below on hooks as well as hanging from above) and a cylindrical vessel height and diameter enough for the star and the hung fabric to freely hang inside the vessel. There will be steam heating coil inside the vessel (both perforated and non-perforated) to heat up the bath. Water is taken inside the vessel and first directly heated via perforated steam coils (water taken considering the volume increase due to condensation of steam). Bath additions are done in between and when the bath is near boil the volume is made up and heating is further done by closed coils. The material is hooked on to the star frame concentric and entered from the above by means of a hoist which is having a facility of inching up and down. The star along with the cloth is kept hung inside the hot soap bath for the time required for degumming, and taken out and rinsed and washed in separate cylinders kept in line using the same hoist which can move horizontally in the same line in which vessels are fixed. In case of single star the shrinkage of the fabric is not controllable.

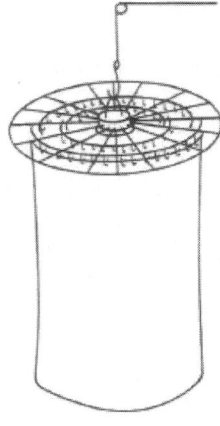

Vertical star

Horizontal star (double star), used for very sensitive articles such as heavy taffetas and tightly woven fabrics; used mostly for dyeing; also known as velvet star. Here both the hooks are held firmly on the horizontal vessel in which the star frame is dipped fully

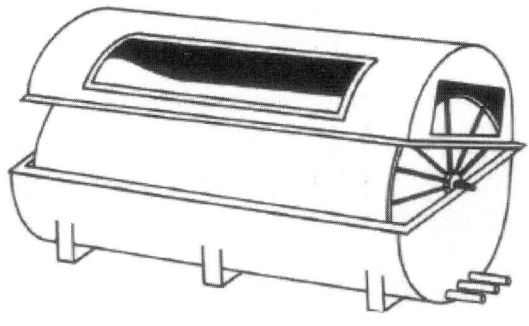

Horizontal star

helps in controlling the shrinkage of the fabric.

2.4.3.2   Continuous degumming machines

The goods are hung in loops on rods which remain fully submerged in the bath and move continuously forward. It is very important that the goods are correctly fed in, so as to avoid creases, and that any protruding threads are automatically cut off so that they do not catch on the rods and wind round them.

Continuously operating machines may contain one or two large baths, depending on their capacity. As already mentioned, it is now possible to reduce treatment times considerably by using appropriate synthetic detergents. As little as 30–40 min may suffice for readily

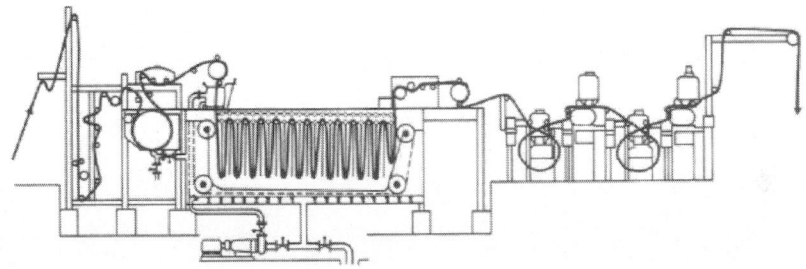

removable types of gum, but 1 h is required in most cases. It is of course very important to check the baths and make relatively rapid checks of the degree of degumming. Silk qualities suitable for this type of processing generally exhibit a uniform final appearance and do not display chafe marks.

2.4.3.3   Jigs

Tensionless jiggers can be used for degumming, but is not a perfect machine for this purpose. Pre-degummed materials like spun materials can be treated on this machine.

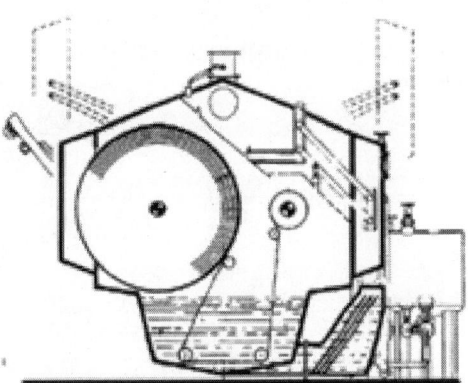

Degumming on Jigs

2.4.3.4   Winch and overflow machines

Main problem of these machines are that they are treated in the rope form which may leave the fabric unevenly degummed and with creases. Fabric which are crease resistant due to construction, for e.g., knitted fabrics can be degummed on these machines especially overflow machines.

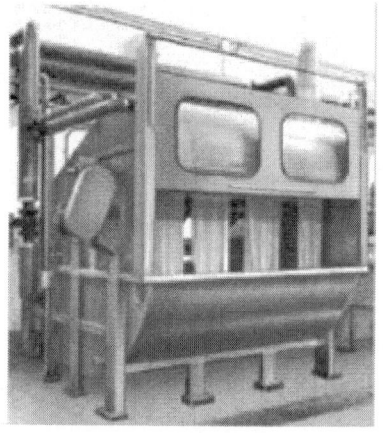

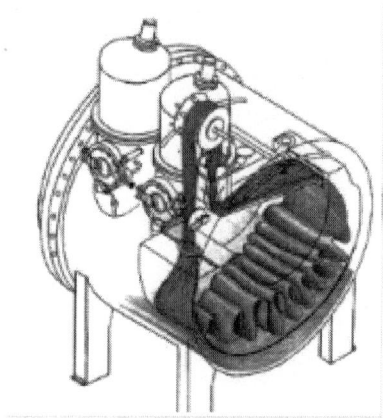

Winch   *Soft flow*

2.4.4   Special effect degumming

1. *Ecru silk:*

Here the main objective is to remove fats and waxes, however during the treatment 2–5% of sericin also may be removed. The thread remains in the raw state and will be stiff.

The raw silk is immersed overnight in a weak (4–5%) solution of formaldehyde at about 20 °C.

Alternatively, the material also may be soaked for 1 h in a 15% formaldehyde solution at 75 °C.

After this the silk is treated in a weak soap solution. Care being taken that the sericin is not attacked.

2. *Souple silk/half boiling*

In this method the partial degumming does not take place in an alkaline medium, but in an acid medium. The weight loss is between 6% and 16% on the original weight (before degumming).

Process:

    a. Degrease at 30–40 °C in a weakly alkaline bath.

    b. Treat at boiling temperature either in sulphuric acid/magnesium sulphate/sulphuric acid soap bath for 1–3 h.

    c. Rinse thoroughly.

    d. If still sticky, treat again briefly in a soap bath at 30–40 °C.

Any further process where acid is used the balance gum will not be lost whereas alkaline treatment tends the gum to be lost at least partially.

2.4.5   Pretreatment of tussah silk

2.4.5.1   Conventional method

(a) Soaking

Treat for 30–60 min at 40 °C with1 g/l wetting and dispersing agent.

(b) Desizing

Treat for 1 h at 70 °C with

5 ml/l	Sericin splitting enzyme (e.g. Bactosol MTN Liq.)
4 g/l	Common salt
0.5 g/l	Calcium chloride
1 ml/l	Wetting and dispersing agent

(c) Soda bath

Treat at 80 °C for approximately 1 hour with3–4 g/l soda ash.

(d) Boiling off

Treat at 95 °C to boiling for 2–8 h with 8–10 g/l soap adjusted to pH 9.5–10 with soda ash.

The amount of soda ash to be increased depending on the gum. Care must be taken to avoid any damage due to the increased amount of alkali.

(e) Rinsing

Rinse at 50–60 °C with1 ml/l ammonia followed by one or two cold rinses.

Since the tussah silk is produced by a wide variety of moths, the procedure has to be adopted depending on the gum in the silk and experience.

2.4.5.2   Discontinuous degumming scouring and washing (summary)

Form	Yarn and piece goods			
Machines/MLR	**Yarn dyeing machines/winch MLR 10:1–30:1**			
Silk quality	Degummed	Souple	Schappe and bourette	Tussah
Procedure				
Wet out				2–3% Wetting agent at 40 °C 30–45 min
Soak		5–10% Soap, 5% wetting and dispersing agent, 50 °C, 1 h (e.g. Imerol XI)		
Desize				3–5% Enzyme, 1–3% NaCl, 2–3% wetting agent, 70 °C, 1 h
Intermediate rinse		Soft water, 40 °C, 20 min		
Acid treatment (Souple)		12.5%H_2SO_4, 5% Na_2SO_4, 80 °C, 1 h		
Boil with soda ash				3–4 g/l soda ash, 80 °C, 1 h
Intermediate rinse		Soft water, 40 °C, 20 min, soft water + 2 ml/l ammonia, 40 °C, 20 min		
Boil off with soap bath 1	20–30% Soap+0.5 g/l soda ash, 90–95 °C, 2 h	20–30% Soap, 90–95 °C, 20 min	20–30% Soap, 90–95 °C, 45 min–2 h	20–30% Soap, 90–95 °C, 2 h, if necessary 10% sod. perborate
Boil off with soap bath 2	15–20% soap, 90–95 °C, 1 h			
Rinse	Soft water, 1 ml/l ammonia, 50 °C, 20 min, soft water 40 °C, 20 min, soft water cold, 20 min	Soft water, 1 ml/l ammonia, 50 °C, 20 min, soft water 40 °C, 20 min, soft water cold, 20 min	Soft water 40 °C, 20 min, soft water cold, 20 min	Soft water 40 °C, 20 min, soft water cold, 20 min

2.4.6   Degumming with synthetic detergents

Synthetic products are increasingly more popular due to the non-availability of soap and its cost. Their principal advantage is that they permit continuous processing of piece goods. It can also be applied in traditional machines. Another advantage is that they permit considerably shorter treatment times; degumming in under an hour (for example in 40 min) depending on the origin of the silk.

In synthetic product degumming the final handle of the degummed silk is not always same as one degummed with soap. But by selecting the degumming agent now we can get almost the same effect. Degumming takes place in most cases between pH 11.2 and 11.5. If these values are not adhered to, the degumming effect may decrease rapidly.

Guide recipe and procedure:

1st bath: 2.5–5 g/l synthetic degumming agent (usually a modified polyglycol ether)treat at pH 11.2–11.5 (with caustic soda) at the boil (depending on the gum) for 30–120 min.

2nd bath: Rinse thoroughly with water at 80 °C.

Processing may be discontinuous or continuous.

Notes:
1. Maintaining the pH is very important. The most effective degumming takes place at pH 11.2–11.5.

2. Soap boiling gives a softer feel.

3. The time of treatment and the feel required has to be set by trials in the lab, especially when continuous machines are used.

2.4.7   Degumming with enzymes

1st bath: Soaking

Treat at boil for about 1 hour with

Quantity	Unit	Additions
0.5	g/l	Soda ash
0.5	g/l	Glauber's salt calc.
3	g/l	Wetting and dispersing agent
2	g/l	Sodium silicate

2nd bath: Degumming

Treat at 40–45 °C for about 1 h with

Quantity	Unit	Additions
x	g/l	Suitable protein enzyme*
1	g/l	Soda ash
2	g/l	Glauber's salt calc.

*1–5 g/l Depending on the product.

3rd bath: Boiling

Treat at boil for about 1 h with

Quantity	Unit	Additions
0.5	g/l	Soda ash
2	g/l	Sodium silicate
2	g/l	Wetting and dispersing agent
0.5	g/l	Glauber's salt calc.

Notes:

1. The fastness of degumming by enzyme method varies as per the product. Generally it is not as fast as the detergent degumming but faster than soaping method.

2. Soaping is important, since the soaking helps the enzyme to act on protein/Sericin.

3. When setting a recipe for a sensitive articles it is also advisable to test it for the damage to the material, since it is observed that long degumming may attack the fibrous filaments as well (since it is also protein).

2.4.8   Bleaching

As explained earlier the colour of *Bombyx mori* silk is a light yellow (called in different names like ecru, soft white, off white, etc.). Even though this is widely accepted in the market but there are requirements for a better white also. For such requirements the material has to undergo bleaching treatments and in many cases optical whitening also. In case of tussah silk the natural colour is a pale brown or so and it is difficult to make a white fabric by even by bleaching or optical whitening, even though by bleaching it can be made still lighter natural shade.

Silk can be bleached oxidative or reductive or both.

2.4.8.1   Oxidative bleach

Generally done by hydrogen peroxide in the presence of a stabilizer. Guideline recipes are given below. Care should be taken to see that there areno heavy metal ions in the bath to avoid any catalytic damages on the fabric (pin holes). Sometimes peroxide is added in the degumming bath to save time especially in case of schappe and bourette silk.

2.4.8.2   Discontinuous bleaching of silk with hydrogen peroxide

Material: Silk yarn, woven and knit goods.

Machines: Hank Yarn machines, winch, etc.

Liquor ratio: 10:1–30:1.

Types of treatment	One bath degumming and bleaching	Bleaching after degumming	Bleaching of tussar silk
Soap (owf) (%)	25–30		
Hydrogen peroxide 35% ml/l	10–30	10–20	30
Stabiliser for H_2O_2(ml/l)	0.25–0.5	0.5–1	2
Ammonia or trisod.phosphate	to pH 10	to pH 10	
Soda ash (g/l)			5
Working method			
Treatment temp. (°C)	90	70	90
Treatment time (h)	2–5	2–4	4–6
After treatment	Rinse warm/ cold	Rinse warm/ cold	Rinse warm/ cold, then treat with 2–3 g/l pot. permanganate for 30 min at 20 °C

2.4.8.2.1   Alternate recipe for Bombyx mori silk

Quantity	Unit	Chemicals
0.5	g/l	Stabiliser
2	g/l	Suitable wetting washing and dispersing agent
15–20	ml/l	Hydrogen peroxide 35%
		pH 9 adjusted with trisodium phosphate

Enter the goods at 40 °C, treat at 90 °C for 1–2 h, then rinse hot and cold.

2.4.8.2.2   Alternate recipe for Tussah mori silk

Quantity	Unit	Chemicals
0.5	g/l	Stabiliser
2	g/l	Suitable wetting washing and dispersing agent
20–30	ml/l	Hydrogen peroxide 35%
		pH 9 adjusted with trisodium phosphate

Enter the goods at 40 °C, treat at 90 °C for 4 h or more, then rinse hot and cold.

Alternative method for tussah silk and other silk with darker natural shade.

Treat with 2–3 g/l potassium permanganate for 30 min cold.

Remove the goods from the bath, allow to drain, then treat with

Quantity	Unit	Chemicals
15–25	g/l	Stabilised hydros e.g. Arostit BLN, Rongolite C
0.4	ml/l	Sulphuric acid conc.

In cold

Then rinse thoroughly in warm and cold water, acidify cold with hydrochloric acid and rinse.

2.4.8.3   Reductive bleaching

Reductive bleaching done with about 0.5–1 g/l zinc sulphoxylate formaldehyde (e.g. Arostit BLN granulated) which is added in the soaping bath. It is also normal to combine a reductive bleach with a fluorescent brightener in a separate bath.

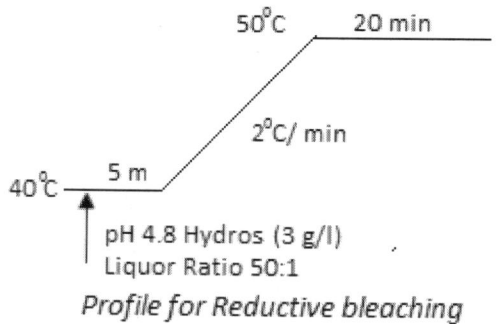

Profile for Reductive bleaching

A typical bleaching procedure with normal hydros in separate bath is given above.

2.4.8.4   Optical whitening

In exhaust method an optical whitening agent suitable for protein fibre is exhausted from a bath containing hydros or stabilised hydros (2–5 g/l).

In padding method following recipe can be used

Quantity	Unit	Chemicals
5–20	g/l	Suitable OWA
1	ml/l	Nonionic wetting agent

Pad at room temperature at a pickup of about 95–100% and dry at 110–120 °C.

2.4.8.5   Weighting of silk

There was a time when silk was traded in weight. The processors used to buy the silk fibre by weight without degumming and by the time it is degummed and made into fabric and dyed it used to loss about 25% of the original yarn weight. This loss in weight led the silk dyer to compensate this loss by special weight increasing process and in certain cases actually making the silk heavier than its original weight. This process is called weighting. Thus the weighting silk has become a process requirement.

The word 'par' has been introduced to express the degree of weighting; thus e.g. 10% lighter than the original silk is termed 10% 'below par', par being the weight of the original silk. Thus 'above par' means that the silk has been weighted to more than its original weight. Depending on the process used, the material may be weighted up to 300% of its original weight. Recently with polymer weighting it has also become common to give the percent content, calculated on the present weight of the degummed silk, as e.g. 10%, 30%, 50%.

Weighting can also improve certain properties of the thread such as fullness and better handle more elegant lustre (e.g. for necktie silk) better flame retardant properties better wash and wear behaviour (crease resistance, less splitting). But nowadays the silk is weighted mainly to improve lustre, handle, drape, and fullness for specific finished products like neckties, trimmings, embroidery, etc.

2.4.8.5.1   Chemical weighting

As a rule the silk is weighted in the hank form before dyeing. Suitable weighting agents are:

1. Metal salts (tin phosphate, silicate weighting)
2. Vegetable substances (tanning agents)
3. Resins (e.g. methacrylamide)

Procedure

Metal salts like zinc posphate gives a weight increase of about 10–15%. Silk is treated in cold for 1–1½ h in tin (IV) chloride solution of 48–58 °Tw, hydro extracted and rinsed cold. It is a must to use soft water for weighting treatment as well as cold rinse as calcium salts may cause precipitation impairing the lusre of the fibre. After rinsing the silk is further treated with disodium hydrogen phosphate (Na_2HPO_4) solution of 7–13 °Tw. These two operations are repeated 2–4 times according to the required weighting. After the final weighting treatment the silk is immersed in a 3–10 °Tw sodium silicate bath at 50–60 °C to convert the basic tin phosphate so far formed into wash fast tin sodium silicate. This tin sodium silicate makes silk bulkier, gives more lustre and improves dye uptake.

A typical procedure can be summarised as follows:

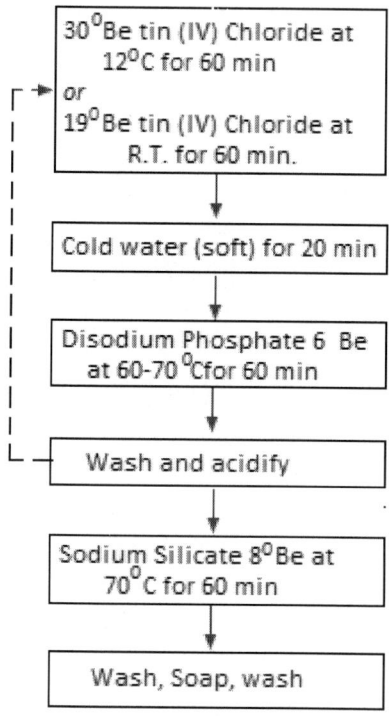

Route 1–If required weighting is achieved.

Route 2 – If required weighting is not achieved repeat the process.

2.4.8.5.2   *Vegetable weighting*

Vegetable weighting with tannins or special plant extracts (Persian berries) is easy to perform and does not damage the silk, but large amounts are needed to restore the original weight. It was often combined with colours such as logwood for black.

The maximum weighting feasible with these agents is 10% increase. This process is hardly used nowadays.

2.4.8.5.3   Polymer weighting

It has been found that resin treatment with methacrylamide can increase the weight of silk. The material is treated with a bath containing

- prepare a 20% aqueous methacrylic amide (MMA) solution,
- add 5% (on the MMA amount) ammonium persulphate 10% solution,
- set the bath, e.g.1–2 parts methacrylic amide solution 20% to9–8 parts bath (water),
- treat the silk at a liquor ratio of 4:1–10:1 at ca. 50–70 °C for 2 h.

This method gives reproducibility under strictly controlled conditions. The reaction between silk and methacryamide does not proceed to completion. An equilibrium system is probably set up, but the individual steps involved are not clearly understood. No exhaustion takes place and hence the bath cannot be recycled. Thus there is a lot of wastage of methacrylamide thus environmental pollution. Other disadvantages are possible fibre damage, reduced dye affinity and impaired wash fastness of dyed goods.

Example of a bulk trial for polymer weighting of woven fabric:

- Machine:Jig,
- Fabric: 10 kg silk muslin, degummed,
- Bath: 200 l,
- Liquor ratio:20:1.

The amount of methacrylic amide required was calculated as follows:

(a) 30% of weight = 3 kg,

(b) additional 15 g/l corresponding to liquor ratio for 200 l=3 kg,

(c) total 6 kg methacrylic amide cryst.

Catalysts used:

(a) 200 g (1 g/l) ammonium persulphate,

(b) 600 ml (3 ml/l) formic acid 85%.

Raise temperature to 90 °C, treat at this temperature for 1½ h and then rinse.

Results:

- Weight increase of 20%.
- Slight yellowing was improved by oxidative or reductive bleaching without impairment to the grafting.

2.4.9   Silk dyeing

Due to the inherent chemical structure of silk dyeing is similar to wool but since silk has a better stability toward alkali it can be dyed in a slightly alkaline conditions which gives wider dyeing chances. But alkaline baths at higher temperature and mechanical forces can tend the fibre to fibrillate and damage and give fluffy or hard feel. Silk also has less acid binding power and hence less affinity than wool. The affinity towards dye also depends on yarn quality (tram/organzine), percentage of degumming, weighted or not weighted, etc. generally it is easier to dye level silk than wool. However, in any dyeings it is important to keep dyeing temperature below 90 °C as higher temperature than 90 °C can reduce the lustre and hamper handle of the fibre.

The silk can be dyed in any form but generally dyed in yarn (Hank) for or piece form. It can be dyed by any machinery used for dyeing textile but the machines has to be modified in such a way as to give minimum tension and least mechanical forces taking the sensitivity of the silk toward these forces.

The table below give the suitability of various machines for dyeing silk:

Type of machine	Effect on material	Output	Levelness	Suitable for	MLR	Remarks
Yarn dyeing machines						
Hank dyeing machines	No hank deformation	Low	Level	Hanks	High	No strand splitting
Spay arm dyeing machine	Hank deformation	Low	Level dyeing but with foam formation	High twist yarn and threads	High	Strand splitting
Cheese dyeing machine	No strand slitting	Low	Level dyeing but winding has to be even	All type of yarn	Low	Specially used for fine yarns for neck ties, flags, banners. Gentle treatment
Piece dyeing machines						
Star frame	No chafe marks	Low	Level dyeing	All fabrics	As high as 100:1 to 300:1	For heavy and delicate fabrics such as shantung, velvet, fabrics for bed coverings. Gentle
Winch	High risk of chafe running marks	High output	Level dyeing possible	All fabrics	30:1–60:1	Specially for crepe, shantung. A driven rear guide roll reduces the risk of chafe marks
Jig	Risk of moire due to tension	High output	Level dyeing possible	Silk twill, Pongee (habutae) and satin	Short	Listing, tailing chances, suitable for qualities which are insensitive to tension

Jet/soft flow	High risk of chafe running marks	High output	Level dyeing possible	Suited for tricot, crepe, etc.	Short to medium MLR	Tensionless dyeing. Higher depths are possible but with higher risk of creases and chaffe marks
Beam dyeing	High risk of moire with longer yardages	High output	Level dyeing possible. winding is very important	Suitable for heavy and smooth fabrics	10:1 to 20:1 or even lower	Structured fabrics like crepe, Jacquard will have flat appearance. Now almost obsolete

2.4.9.1   Dyes and dyeing methods

The classes of dyes mostly used for dyeing silk are:

1. Direct,
2. Acid,
3. Metal complex,
4. Cationic dyes,
5. Reactive,
6. Mordant dyes.

2.4.9.2   Dyeing with acid dyes

All but a few acid dyes can be applied simply from slightly acid bath and good levelling properties on silk. All the shades including those specific to silk can be produced by judicious dye selection, but concessions has sometimes to be made as far as fastness properties. Generally speaking, bright shades will have lower light fastness and darker shades will have lower wet fastness like perspiration, water, even wash at 40 °C. In this context 1:2 metal complex dyes are more important. Shades are duller than acid dyes but overall fastness properties are good.

Acid dyes, due to their high affinity for wool, it is difficult to dye level shades. But, in the case of silk the affinity being less, can be dyed rather level shades following the dyeing process judiciously but the fastness properties are not very impressive even though shades are bright, especially light fastness. Super milling dyes show very good fastness due to a long alkyl side chain in the molecule.

2.4.9.2.1   Theory of acid dyeing

The number of amino sites in silk, are less than that of wool. Molecular weights of acid dyes range from 200 to 900. Most have one or two $-SO_3Na$ groups which are water-soluble and capable of bonding with fibres having cationic sites. The acid dye anions in solution experience a strong attraction towards the silk fibre because of their opposite charge and forms ionic bonds or salt links between the dye molecules and the fibre. In addition to ionic links, when silk fibres are dyed with acid dyes, hydrogen bonds as well as van der Waals forces will be formed between the dye molecule and the fibre polymer system. However, the ionic links between the acid dye molecules are stronger compared to the hydrogen bonds and van der Waals forces. Electrolytes are used as retarding agents when dyeing self-levelling acid dyes, but electrolytes act as an exhausting agent in the case of milling and super milling types of acid dyes on wool.

2.4.9.2.2   Exhaust dyeing with acetic acid

Material is entered the bath at 40 °C and run for 10 min. Next the pH of the bath is adjusted by adding acetic acid and Glauber's salt. After running for 10 min for the equalization of the pH, check the pH once again and add the well dissolved dye. After running in the dye for 10 min the temperature of the bath is raised to 85 °C at 1 °C/min. Run at this temperature for 30–60 min as per the depth of the shade to complete the exhaustion and to promote diffusion of dye into fibre improving wash and rubbing fastness. Dyeing at higher temperature beyond this reduces lustre. Drain.

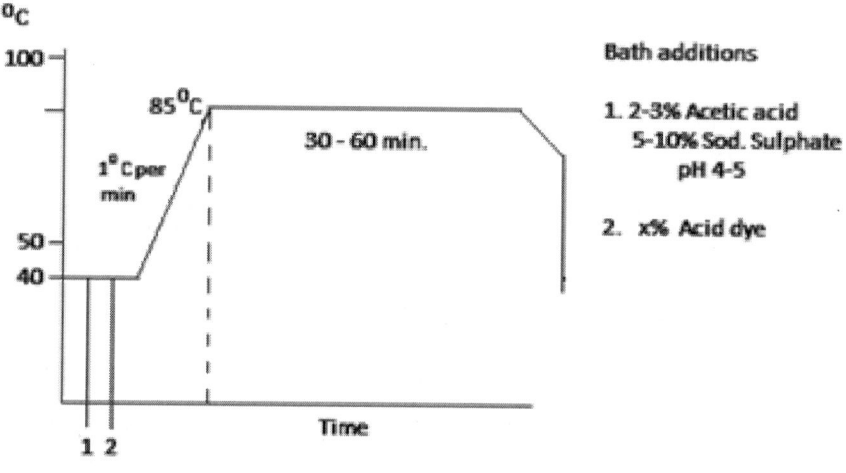

Exhaust dyeing with acid dyes

A thorough wash removes superficial dyes. Strong acid dyes can also be applied at pH~3–4 using 1–3% HCOOH (85%) and 5% Glauber's salt. wash fastness which can be improved by treatment with cationic dye-fixing agent (2–4 g/l) at 40–50 °C for 20 min.

2.4.9.2.3   Exhaust dyeing with sulphuric acid

Set the bath with acid dyes, 2–5% sulphuric acid (depending on the depth), 10% Glauber's salt, and MLR 1:50 at 40 °C. The material is introduced and after wirking for some time (15 min) and homogenisation the temperature is raised to boil at 3–4 °C according to the available provisions of heating. The dyeing is further continued for about 45–60 min at boil and the exhaustion is completed. The bath is dropped and rinsed with water and dried.

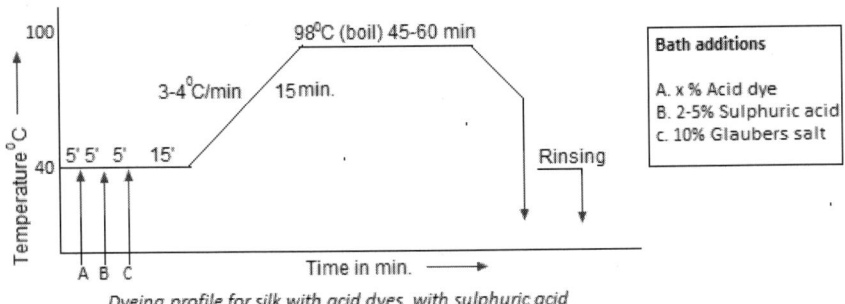

Dyeing profile for silk with acid dyes with sulphuric acid

2.4.9.3 Dyeing with metal complex dyes

2.4.9.3.1 Acid and 1:1 metal complex dyes

In acid solutions the silk fibro intakes on a positive charge through the adsorption of hydrogen ions (protons); the resulting electric charge can be counter balanced with negatively charged counter-ions. The coloured anions of the dyes present in the dye bath

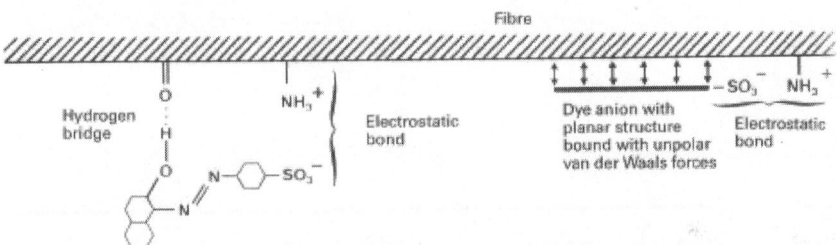

Three types of bond formed by acid dyes (Schematic)

serve this function. The fact that other bonding forces than electrostatic ones come into play is shown by the frequent necessity of dyeing silk above its isoelectric point (pH 5.0), depending on the type of dye, in order to obtain level dyeings. These forces are mostly hydrogen bridges and unpolar bonds formed by van der Waals forces.

Quantity	Unit	Dye/chemical
1	g/l	Sodium acetate
3–5	%	Acetic acid 80%
1	%	Levgen (Retarder)
5–10	%	Galuber's salt anh.
x	%	1:1 Metal complex/acid dye

Adjust the pH 4–5. Enter the goods at room temperature and increase the temperature

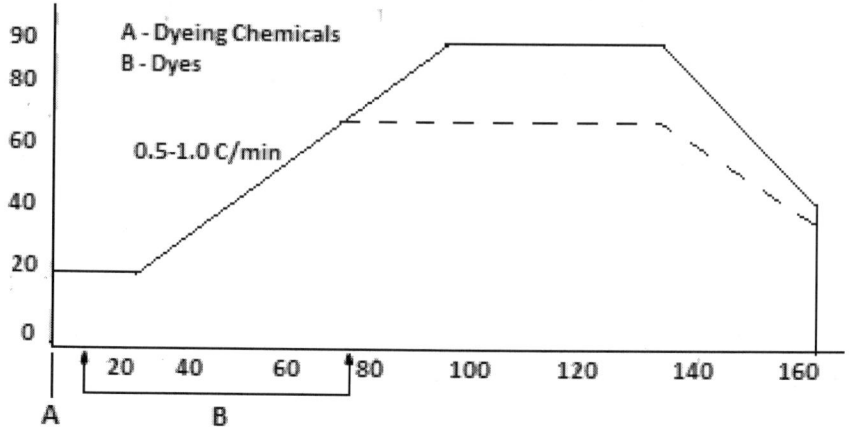

Dyeing Program

slowly (0.5–1 °C/min) to boil with proper movement of the fabric or yarn or liquor circulation or both as per the machines involved. Run at boil till exhaustion is complete.

Notes:

a. Since the strike rate of these dyes are very high it is better to start the dyeing at temperatures as low as possible and if necessary add a retarding agent to avoid any unlevelness. Degummed bath is a good levelling agent.

b. Light to medium shades can be dyed at slightly acidic pH of about 5–7 and dark shades at 4–5.

c. For sensitive piece goods the dyeing temperature can be kept low as 60–70 °C but for dark shades one can dye at boil.

2.4.9.3.2 1:2   *Metal complex dyes*

Quantity	Unit	Dye/chemical
1–3	%	Formic acid
5	%	Galuber's salt anh.
x	%	Dye

Adjust the pH 3–4. Enter the goods at room temperature and increase the temperature slowly (0.5–1 °C/min) to boil with proper movement of the fabric or yarn or liquor circulation or both as per the machines involved. Run at boil till exhaustion is complete.

2.4.9.4 Dyeing with cationic dyes

Recipe

Quantity	Unit	Bath additions
x	%	Dye
2	%	Acetic acid (owf)
1	%	Tannic acid
0.5	%	Tartarematic

Dyebath is set at about room temperature with basic dye, 2% acetic acid (MLR 1:30). Material is entered and dyeing can be started at 30 °C after homogenization of the bath (15 min). Next the temperature is raised to 90 °C in 30 min; dyeing is continued at this temperature for a further 45 min. Mordanting can be done by treatment with tannic acid (1% at 60 °C for 30 min) and fixed with tartermatic 0.5% at 90 °C to improve the fastness by anionic tannic acid complexing

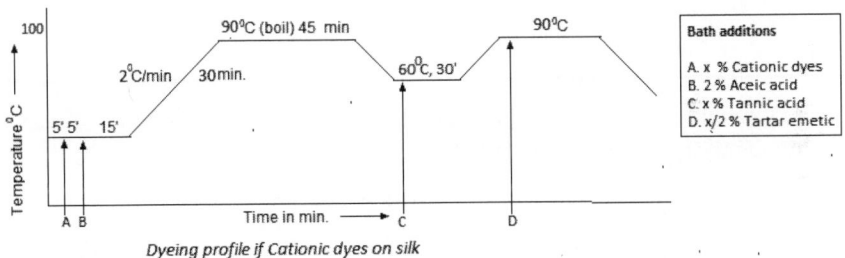

Dyeing profile if Cationic dyes on silk

with cationic dye thereby increasing the molecular size of the dye–tannic acid complex. This results in the improvement in fastness toward washing and consequently the fixation of the dye on the fabric.

2.4.9.5 Reactive dyes in dyeing silk

Lately, reactive dyes have gained more importance in silk dyeing because of the fastness properties of acid dyes and dull shades of metal complex dyes even though fastness properties are acceptable. There areno reactive dyes specially made for silk. Theoretically all reactive dyes

Reaction between silk fibroin and reactive dye

can dye silk. Reactive dyes give bright shades with good fastness properties as the dye reacts with the fibre and fixed to material by covalent bonds. All this is possible because silk can withstand slightly alkaline bath which is required for the reaction of reactive dyes with the fibre. Usually, highly reactive dyes such as dichlorotriazine, difluoro monochloro pyrimidine,2-methoxy4chlorotriazine and vinylsulphone dyes which can be dyed at low temperature with weak alkali are used.

As mentioned earlier, all reactive dyes are not preferred for silk. The reactive dyes fulfilling following conditions are usually selected.

1. They should give brilliant shades and be comparable to acid dyes especially on mulberry silk.

2. Should have high reactivity so that it can be dyed in a comparatively neutral, acidic or slightly alkaline bath. There are chances of silk getting damaged in an alkaline medium at high temperature.

3. Should have good storage stability as the consumption of dyes for the batch wise dyeing of silk is small, so the dyes should be highly stable to storage.

Generally selected dyes for silk are cold dyeing dichlorotriazine, dichloroquinoxaline or difluoropyrimidine classes. Exhaust dyeing at 60–70 °C and pH 5–6 gives satisfactory results, especially if a mildly alkaline after treatment is given to enhance fixation. Dichlorotriazine dyes can also be applied by pad-batch dyeing with bicarbonate and a batching time of 4–6 h. The relatively low reactivity of aminochlorotriazine dyes, however, results in moderate to poor build-up on silk. Tertiary amine catalysts such as DABCO (see below) can be used to accelerate the dye–fibre reaction and increase the fixation substantially, but it is difficult to achieve satisfactory compatibility in mixture dyeings by this method.

$$N-CH_2CH_2-N \underset{CH_2CH_2}{\overset{CH_2CH_2}{\Big\langle}}$$

DABCO

2.4.9.5.1   *Exhaust dyeing method*

2.4.9.5.1.1   *Dyeing with cold brand dyes*

In cold brand reactive dyeing the whole process is done at room temperature (25 °C) set the bath with the dyes and the Glauber's salt (20 g/l) and run exhaustion for 30 min and the alkali is added

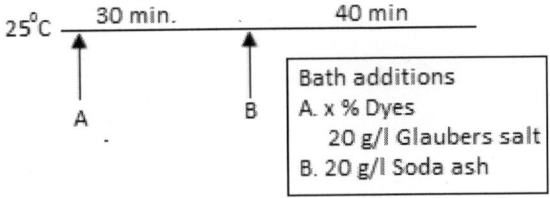

(20 g/l soda ash) and fixation is done for 40–60 min and rinsed and after treated.

2.4.9.5.1.2   *Hot brand dyes*

Dyeing is started at 50 °C with dyes and Glauber's salt and after 5 min the temperature is raised at 1–60 °C, after the temperature has reached the material is worked for another 30 min to complete the exhaustion and the alkali is added (2 g/l soda ash) and further the dyeing is continued for 40 min and fixed. Rinsing and after treatment is continued.

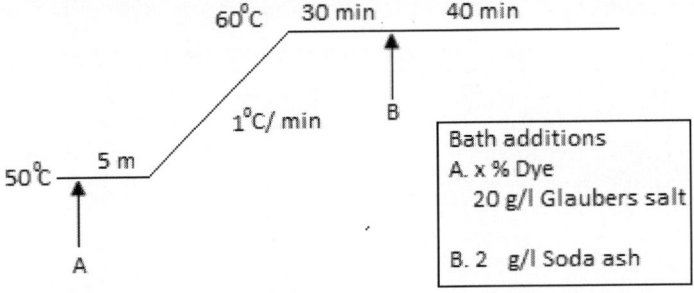

Alternatively, 1–2 g/l soda ash and 2–3 g/l trisodium phosphate can be used as alkali.

2.4.9.5.1.3   *Warm dyeing dyes (vinyl sulphone dyes)*

Vinyl sulphone dyes have proved highly suitable for silk, yielding brilliant hues of high wet fastness by application at 80 °C and pH 7–8 in the presence of Glauber's salt, usually followed by an alkaline fixation treatment to ensure optimum fixation. They are predominantly used for discharge grounds.

The dye is first exhausted and the bath is made alkaline to facilitate the dye fibre reaction. Since the substantivity is not very high the dyeing has to be done at the lowest liquor ratio possible. Start the dyeing at 20–30 °C and raised to 60–70 °C (90 °C for Black)in a neutral bath containing Glauber's salt. In the last 15–20 min the pH adjusted to 8–8.5 with sodium bicarbonate to fix the reactive dye. After rinsing, the dyeing has to be soaped to remove the dye that is not fixed.

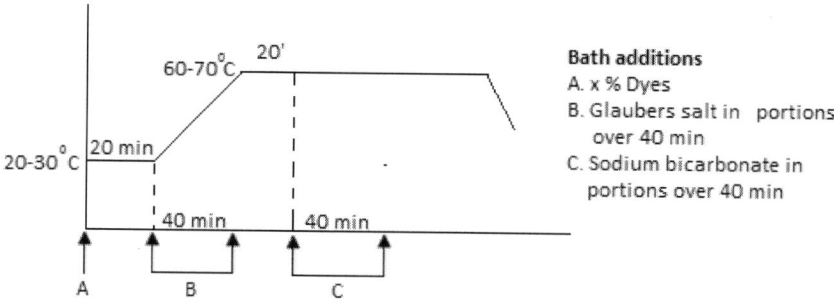

2.4.9.5.1.4   *Cold pad-batch method*

These days cold-pad batch has become more popular due to its versatility, economics and mild treatment.

Dyeing method

The machinery consists of a padding mangle with lowest pad box capacity, a rolling and unrolling device, and a storage (rotating) unit. Storage at rotating station for about 24 h for completion of the fixation of the dye. Once the fixation is over the material is rinsed, soaped and neutralised on an open width washing machine which can run without tension or dyeing machines (like soft flow).

Quantity	Unit	Dye/chemical
x	%	Dye
10–40	g/l	Galuber's salt/common salt
1–2	g/l	Soda ash

Cool, rinse thoroughly, soap with a suitable soaping agent, cool and rinse thoroughly, finally neutralise with 1 ml/l formic acid 85%.

Notes:

1. The padding mangle should have even squeezing over the full width if the dyeing is to be level. The expression should be around 80–100%.

2. If necessary thickener can be added in the dyeing bath to avoid any migration during storage.

3. Since the time of treatment in the dye bath is only seconds addition of wetting agent helps in rapid penetration of the dye into the fabric.

4. Liquor pick up is affected only very slightly by the auxiliaries and chemicals in the dye bath, but it can be affected a great deal by the nip pressure and the number of passes.

5. Fixation on tussah silk is lower than mulberry silk under same conditions. Higher storage time and higher temperature of fixation.

Alternative recipe for Pad batch

Quantity	Unit	Chemicals
x	%	Reactive dye
50–100	g/l	Urea
10–20	g/l	Sodium bicarbonate
0–2	g/l	Wetting agent

Even though by Pad batch method dark shades can be dyed, the colour yield is low.

2.4.9.5.1.3   Pad-steam method

Reactive dyes can be applied on silk can also applied continuously by pad-steam method. This method can achieve fixation levels up to 80%. This method does not need strong alkali and consequently the pH of the dye bath can be adjusted by sodium bicarbonate only.

Guideline recipe

Quantity	Unit	Dye/chemical
x	%	Dye
10	g/l	Alginate thickening
2	g/l	Wetting agent
20	g/l	Sodium bicarbonate

pH approximately at 8.5. Pad at liquor pick up of about 80–100%. The padded material is treated with saturated steam at 100–102 °C. Sopaing and neutralisation can be done on an open width washing machine as in the case of Pad batch method.

2.4.9.6   Dyeing with Indigosols

Indigosols also can be applied on pure silk and silk polyester blends, in light to medium shades. While applying the dye the pH should be only slightly alkaline. The following recipe can be used as guideline for padding:

Quantity	Unit	Bath additions
x	g	Indigosol dyes
50	g	Sodium nitrite
5	g	Soda ash
2.5	g	Wetting agent (non-ionic)
2.5	g	Dissolving asst.
100	l	Boiling water to make upto

After padding the Indigosol dyes are developed with 2 g/l conc. sulphuric acid and neutralised with soda ash and thoroughly soaped (preferably using non-ionic detergent) and finally neutalised with acetic acid. In case of polyester/silk blend after dyeing the material may be heatset at 170 °C for 2–3 min.

Indigosols can be dyed by steaming and bicarbonate methods explained under cotton dyeing (Dyeing of Substrates I – Cotton dyeing by the same author).

2.4.9.7   Dyeing with mordant dyes

As in the case of wool, silk also can be dyed with mordant dyes to achieve high fast shades.

Chrome pretreated silk is dyed with mordant acid dyes. The methods given under wool dyeing can be followed. The dyeing can obtain deep and level dyed shades with excellent fastness due to the formation of dye-chrome complex and its good penetration in the fibre structure. Mordant dyes gave better shades than acid dyes with attractive hues such as pinkish blue to red with very good depth and fastness properties.

2.4.9.8   Making of silver or gold zari (lace)

Gold zari is commonly used in Indian saree and other dress materials. The core of the gold zari is normally silk yarn of good quality 20/22 denier of 'S' twist with 6TPI, or sometimes equivalent synthetic filaments. The yarn is degummed with a bath 5 g/l soap and 2–2.5 g/l soda ash and washed well and dyed in golden yellow, orange or red. The golden yellow shade can be dyed with naphthol, reactive, vat or metal complex dyes. The dyeing should be fast to degumming. Two ply yarn is used for 2000 m quality and 3 ply yarn is used for 1600 m quality lace.

Silver with a copper content of 4% is beaten and drawn into a fine thin foil by passing silver thread through metal rolls to make into fine ribbons and wound on reels.

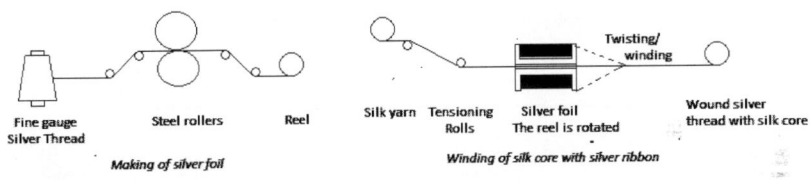

Next the silver foil (ribbon is wound over the previously dyed silk yarn.

Gold plating

Plaiting of gold over the silver covered yarn is done in an electroplating bath containing double cyanides of gold and potassium made by dissolving 40 g of gold and 60 g of potassium cyanide in a litre the sequence of application is as follows:

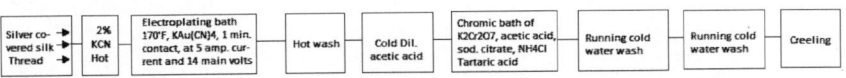

The bath volume used to be kept lowest as the gold used is very costly and cyanides are poisonous.

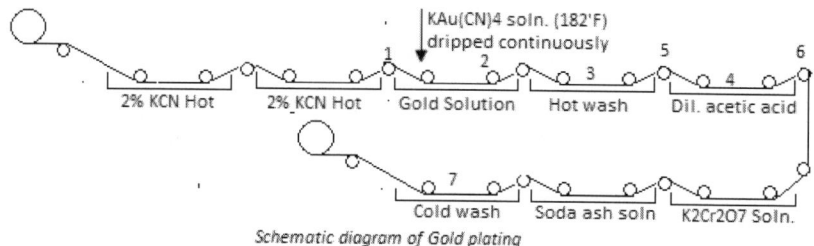

Schematic diagram of Gold plating

40 Bobbins (5 kg) of silver coated silk yarn is taken for electroplating at a time with a speed of 28–30 m/min. For every 175 kg of lace or 140 h of electroplating running the bath is taken fresh. While running, the electroplating solution (KAu (CN)$_4$) from reservoir, should be drip fed into the bath to maintain a uniform bath concentration. The bath of KCN used for pretreatment of the yarn should be changed every day. The bath concentration and the ampearage are adjusted to get a gold content of 0.4–0.7% in the final lace.

Generally a silver zari will have approximately 62–65% silver, 3–4% copper and 33–34% silk. In gold zari other than silver, copper and silk there will be around 0.4%. The percentage of gold used to be 0.7–0.8% when the gold was cheaper. Now a days the zari will have either very low percentage or made of other alloys or synthetic since the gold has become much costlier.

Part II

Dyeing of Regenerated Cellulosic fibres

Chapter 3
Dyeing of viscose

Since viscose is a regenerated cellulosic fibre, it is also a cellulosic material and the methods of dyeing cotton are applicable to viscose also. Certain precautions have to be taken since the dye-ability of viscose is higher. Given below the comparative rating of the properties of viscose compared to the most common natural and synthetic fibres, viz. cotton, polyester.

Properties	Cotton	Viscose	Polyester
Moisture regain	Good	Very good	Poor
Thermal protection	Good	Very good	Poor
Air permeability	Very good	Good	Poor
Softness	Good	Very good	Poor
Smoothness	Poor	Good	Very good
Dissipation of static electricity	Good	Very good	Poor
Drape	Good	Very good	Poor
Lustre	Poor	Very good	Very good
Crease recovery	Poor	Poor	Very good
Uniformity	Poor	Very good	Good
Resistance to pilling	Good	Very good	Poor
Wash and wear	Good	Poor	Very good

The higher dyeability of the viscose is due to the lower crystallinity of viscose fibres. In other words the degree of polymerization (DP) is lower. The average DP of cotton is 3000–5000 while that of viscose is 500–700. This proves the crystallinity of viscose is much lower than cotton. This is one reason for higher dyeability of viscose in comparison to cotton.

The average degree of polymerisation is measured by osmotic methods or viscosimetry. It provides information on polymer hydrolytic breakdown (damage). The specific viscosity (η_{spec}) of the solution of a polymer, the average degree of polymerization of which is to be determined, depends on the concentration, chain length (average degree polymerization) and temperature of the polymer (η = polymer solution viscosity) and the solvent (η_0 = solvent viscosity). The solution of the fibre is made in a solvent and the viscosity if determined using Ostwald's capillary viscometer. A blank is also done with the solvent. Since passage time is directly depending on the viscosity of the liquid the specific viscosity can be found using the time of passage of the solution (t) and time of passage of the solvent (t_0)

$$\eta_{spec} = \frac{\eta - \eta_0}{\eta_0} = \frac{t - t_0}{t_0}$$

The specific viscosity obtained in this way is connected with the average degree of polymerization as follows:

$$\eta_{red} = \frac{\eta_{spec}}{c} K_m \cdot DP \text{ (Staudinger equation)}$$

Thus,

$$DP = \frac{\eta_{spec}}{K_m \cdot c}$$

where,

c = polymer concentration in g/l,

K_m = constant, depending on polymer structure and solvent,

η_{red} = limit viscosity.

The limit viscosity η_{red} can be red by plotting η_{spec} against c and estimating viscosity for $c = 0$. Schulz and Blaschke equation gives a more accurate average degree of polymerisation which is given below:

Thus,

$$DP = \frac{\eta_{spec}}{K_m \cdot c \left(1 + K_\eta \cdot \eta_{spec}\right)} \cdot$$

where,

K_m = constant for polymer effect,

K_η = constant for solvent effect.

The average DP of cotton is 3000–5000 while that of viscose is 500–700. This proves the crystallinity of viscose is much lower than cotton. This is one reason for higher dyeability of viscose in comparison to cotton.

Dyeability of any fibre will mostly depend on the following factors, other parameters being kept constant:

- Chemical composition of the fibre.
- Dye class.
- The geometrical form of the fibre.

The relationship between fibre accessibility $A\%$ and the degree of crystallinity $K\%$ can be derived as follows:

$A\% = (100 - K) + \alpha K$

where α is a constant which is related to the crystalline part of the fibre. It can have values from 0–1. An ideal substance with hundred percent accessibility would be represented by $\alpha = 1$. In the case of high tenacity viscose fibres K can vary, for example, between the limits of 33–49%, $\alpha = 0.09$–0.67. The accessibility A represents an appropriate numerical criterion of dyeability.

The energy required to transfer dye molecule from solution to fibre throws light on the compactness of the fibre. This energy is one of the lowest in case of the viscose, showing it can be easily dyed. The energy for transferring dye from solution to fibre for various fibres are as follows (kcal/mol): for polyester-disperse dye, it is ~30; nylon-disperse dye ~22; wool-acid dye ~22; cellulose acetate-disperse dye ~20–24 and viscose-direct dye ~14 and so on.

The major proportion of the fibre is present as an amorphous intermicellar region. Due to these structural deviations in comparison with cotton, viscose reveals certain differences in its range of properties. The conventional viscose fibre thus exhibits lower strength properties, a higher water absorption, a greater creasing propensity and a higher extensibility than cotton. Its easy-care properties are likewise inadequate by today's standards.

Viscose fibres are particularly valued for their high degree of lustre in textile fabrics.

By newer manufacturing methods the morphology, degree of order and orientation of the fibre are improved in the regenerated cellulosic fibres like polynosic, modal, Lyocell, etc. Modal fibres have a higher degree of polymerisation (DP 350–600), higher resistance, wet modulus, dimensional stability and alkaline resistance compared with normal viscose fibres. These fibres have properties very similar to those of cotton. The polynosics have good form stability, increased linear strength properties and can be readily mercerised. The high wet modulus (HWM) types have an approx. 50–70% higher transverse strength; as a result they can also be mercerised under certain conditions, their bending rigidity is lower, and they are not subject to fibrillation. Their water retention capacity of 60–70% lies within the same range as that of cotton. The above properties predestine these fibres for blending with natural and synthetic fibres. Resistance to alkali for the high wet modulus fibres are less than that of polynosic fibres. The new fibres with higher molecular mass or chain length of the individual molecules and the more compact internal structure of the fibres play a crucial role in the processing of the viscose materials. The local and temporary deviation in the structure due to delay and orientation processes, chemical decomposition of the xanthate as well as subsequent thermal effects, etc. will cause uneven dye uptake, which can often end up as colour streakiness.

Mercerisation process: Mercerisation can be done on viscose material with proper understanding of its solubility behaviour in caustic soda and taking other precautions. Basically, viscose can with stand the caustic soda strength for mercerising but it disintegrates during subsequent washing. Following things has to be kept in mind when we are planning to mercerise viscose.

1. Solubility of viscose in caustic soda at higher temperatures is less than at lower temperature.

2. Mercerization with caustic potash solution is less critical, as viscose fibres swell significantly less in that than in caustic soda liquor. However, as the effect on cotton is less, one works with mixtures of both liquors.

3. Addition of 40–50 g/l sodium chloride in the mercerising liquor can protect the viscose fibres to some extent.

4. Blends of cotton with viscose can be mercerised without much problem.

Viscose and most other regenerated cellulosic fibres dissolve in caustic soda liquor at about 6.5% by mass (70 g/l), so causticisation rather than mercerisation is carried out using 3.3–5.5% NaOH by mass (35–60 g/l) to enhance wetting or dyeability. Polynosic fibres or HWM fibres are more resistant to alkali than regular viscose.

3.1   Mass colouration

The regenerated cellulose fibre is also dyed by mass colouration method. This is done by addition of pigment dispersion to the cellulose xanthate solution prior to extrusion and then coagulation by means of sulphuric acid treatment is no longer of much practical significance (see viscose manufacturing flowchart in 'Textile substrates – Fibre, Yarn and Fabric' in the same series Vol. 1). Selection of the pigment has to be done which can withstand the alkaline medium and following processes after the addition: strongly alkaline solution and strongly acidic coagulation processes.

3.2   Dyeing of viscose yarn

Without salt, swelling of viscose increases with temperature. This must be remembered in package dyeing where high swelling and high pump pressure can lead to flattening of the yarn cross section at cross over points. Flow should be predominantly IN to OUT so that the yarn layers float apart and the liquor can circulate more easily. With OUT to IN flow the yarn layers are compressed and flow is reduced. OUT to IN flow cycles should therefore only be 30 s or so.

3.2.1   Conical cones

The cones must be homogeneously wound to ensure an even dye result and the bobbin edges should be carefully rounded off (bumped). The winding hardness should be approx. 25 Shore. Taking the relatively high swelling of the material into account this equals a volume of approx. 360–380 g/l. Due to the higher swelling of the material we recommend that the winding diameter is limited to 160 mm. Dyeing problems have been experienced where the yarn tube diameter has reached 170 mm (approximately 850 g/cone).

3.2.2   Cylindrical cones

Lighter dyeing results on the edges can be avoided by using winding hardness's of approx. 25 Shore and a pressure rate of 20%.

3.3   Dyeing of fabric

When dyeing viscose in alkaline condition, mainly two factors affecting the dyeing has to be considered. One, under alkaline conditions the substantivity and reactivity of the fibre is very high. Secondly in alkaline bath the fibre is weaker. The most suitable dye class is reactive among the alkaline dyeing dyes.

The relationship between pH and fixing temperature and the possibility of dosing with alkali can be used to reduce the swelling properties in case of reactive dyeing.

In exhaust dyeing, it is better to choose lower MLR machines. When using jet/soft flow machines following precautions may be taken.

Short-liquor capabilities, special lifter reel, jet nozzle design and employing hydraulic (water) and/or pneumatic (air) fabric drives, minimum lift length, etc.

3.3.1   Dyeing with reactive dyes

There are five methods of dyeings which is practical for viscose dyeing:

3.3.1.1   Method I – Migration method

In migration method the viscose dyeing it is slightly different than the traditional method on cotton. As explained earlier precautions has to be taken to control the exhaustion and fixation. Hence the salt and alkali is dosed. In spite of the slow increase in substantivity, it is possible that the flow through the wound packages will be disrupted due to the swelling of the viscose filaments, particularly in water at lower temperatures. Consequently, dosing salt delays the absorption of the dye but the requirement that the dye be distributed uniformly before fixing begins is probably not fulfilled because the flow of liquor through the package is impaired. The bath is made at 50 °C with dyes and auxiliaries and the dyeing is started at this temperature. After 5–10 min the salt is dosed in 30 min. After dosing the temperature is raised to 800 °C in 10 min

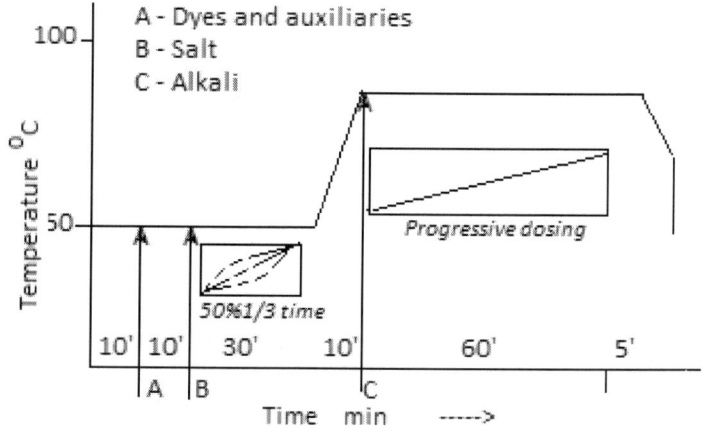

Migration method

or at 2.5–3 °C/min. Alkali is dosed at very slow rate for 60 min so that fixation takes place at the same time the pH in the bath is controlled. After dosing the dyeing may be continued for 5–10 min and drop the bath after cooling and continue after treatment.

3.3.1.2   Method II– High temperature salt dosing method

In the case of method II, the salt is dosed at a higher temperature. Swelling is limited at higher temperatures and the diffusion and migration of the dye improved. The probability of uniform dye distribution during the substantive phase is greater than it is for method I. Cooling and fixing in the cold bath could have a negative effect on the final levelling. The substantivity

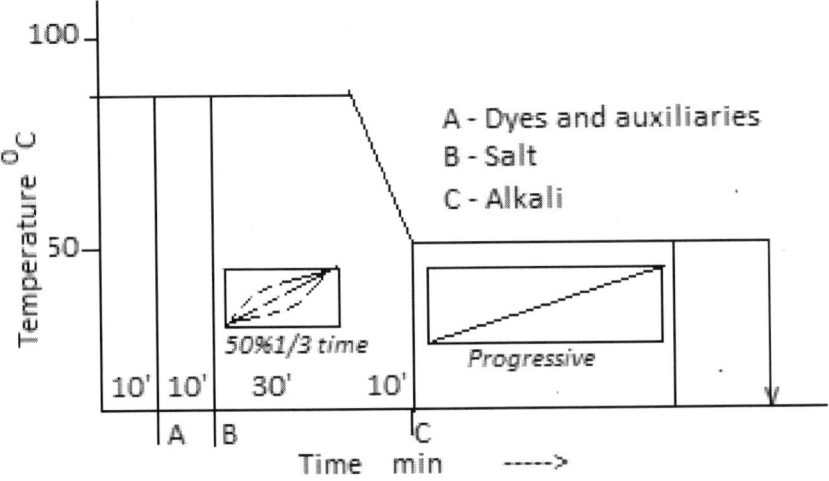

Salt Dosing at higher temperature (80°C)

of the dye increases as the temperature of the total quantity of electrolyte in the dye bath decreases. The dye fills up as the equilibrium shifts in the direction of the fibre. By adding alkali to the cold bath, three factors affect each other and cause increased swelling of the material and, therefore, impair the flow of the liquor through the packages:

- an increased tendency to swell due to the low temperature,
- an increased tendency to swell in alkaline solutions,
- lower fixing temperatures require stronger alkaline conditions to achieve the final fixing value. Swelling increases with increasing amounts of alkali.

3.3.1.3   Method III – Isotherm method

Judging by the dyeing profiles, the isotherm method III is the most suitable for achieving level dyeing, taking into account the lowest degree of swelling for the material and associated even flow through the packages. Based on uniform temperatures, non-uniform filling of the dye (as assumed in the case of method II due to cooling of the bath) is not to be expected. The fixing conditions are milder in line with the higher bath temperature, and this has a positive effect on the swelling state.

3.3.1.4   Method IV – Milder alkali method

In this method the following properties of viscose is taken in to consideration – at the selected temperatures, the reactive dyes have a high diffusion rate, where as the substantivity only increases slowly after the bath has cooled in a defined manner and salt has been added. The dye is also fixed at high temperatures. Thus in this process milder alkalis or with smaller amounts of alkali is used for fixation. The main factors (temperature and pH) which affect the swelling of the material are chosen to favour reduced swelling.

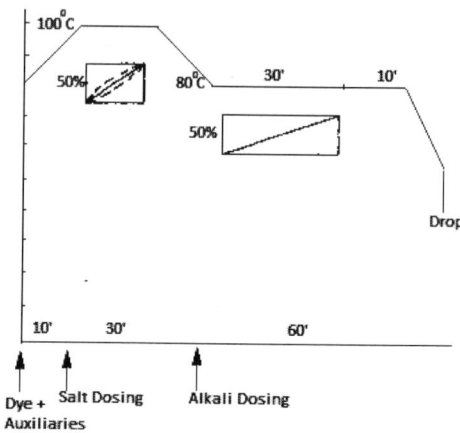

In the case of dyeing method IV, high temperatures are selected mainly to ensure better flow of liquor through the dye packages due to the reduced tendency to swell at high temperatures. When using reactive dyes, particularly in the case of the high-reactive type, these high temperatures cannot be achieved as dye fixing and increased hydrolysis takes place within this temperature range even in the absence of alkali (pH around neutral).

3.3.1.5   Method V – Lower temperature fixation method

The dyeing profile of method V is suitable for reactive dyeing. At the selected temperatures, the reactive dyes have a high diffusion rate, whereas the substantivity only increases slowly after the bath has cooled in a defined manner and salt has been added. The dye is also fixed at high temperatures. Thus, it is possible to work with milder alkalis or with smaller amounts of alkali. The main factors (temperature and pH) which affect the swelling of the material are chosen to favour reduced swelling and, therefore, uniform flow of liquor through the packages.

As viscose is a high affinity fibre, special consideration must be given to the choice of dyestuff and exhaust application technique. Hot dyeing reactive dyes are preferred for exhaust application in order to secure the highest migration and diffusion through the high affinity fibre when dyeing in rope form. Dye suppliers will thus generally recommend a migration technique to give levelness and reproducibility. This technique offers high dye mobility at temperatures as high as 110 °C before cooling to 80 °C for optimum fixation of the reactive dye to the cellulose chain. For higher substantivity dyes, the salt should be added gradually (over 30 min) at 95 °C, or at the highest temperature at which additions can be made.

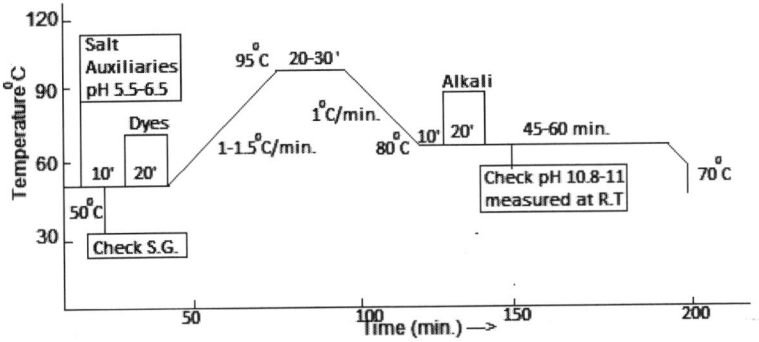

Reactive dyeing Method V

The bath is set at 50 °C with the required amount of electrolyte, auxiliaries and the pH adjusted with acetic acid to around pH 6.0. Pre dissolved dyes are then added in a linear manner over 15–20 min. The temperature is raised to 95 °C (or even up to 110 °C) at 1.5–2 °C\min and held for 20 min at 95 °C before cooling back to 80 °C at 1 °C\min. Hold at 80 °C for 10 min before adding the alkali in a linear manner over 15–20 min. Continue for 45–60 min.

After treatment

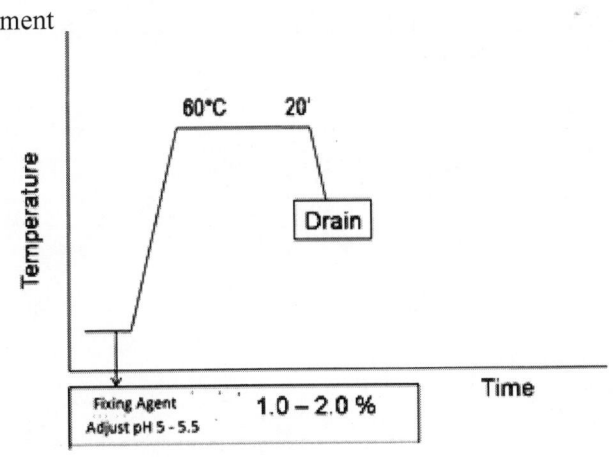

Take fresh bath and adjust the pH to 5–5.5 with acetic acid. Run the fabric for 5 min to equalise the pH and check pH again, if necessary adjust again.

Add the fixing agent as per the manufacturer's recommendations and run for 5 min at room temperature and raise the temperature to 60 °C and hold at this temperature for 2 min and drain.

3.3.1.6   Pad-jig develop process for viscose materials

When dyeing viscose rayon by this method, the quality of the dyeing and the colour yield are generally much improved by the introduction of a batching period between padding and developing. The batching time should be at least 1 h, and preferably should be 2 h.

The batching treatment is most beneficial when using the slower-diffusing dyes which are preferably dyed by the warm bicarbonate process and which are indicated in table below. It is important in this process to adjust the temperature of the pad liquor as closely as possible to the room temperature, in order to avoid migration effects during batching. The batch should be wrapped during storage to prevent migration, drying at the selvedges, and water spotting.

Development:

Set the jig at 30 °C with all the salt and soda ash, determined from table below for the cold soda ash process. Fixing may be done at 40–45 °C for 45 min or 60–90 min at 30 °C (the longer time for heavy shades).

Rinse and wash off at the boil for 15–30 min.

In case of hot brands the bath is set up at 40 °C with salt, soda ash and caustic soda (see table below)and the material is run increasing the temperature to 70–80 °C and run at this temperature for 45 min with minimum tension on the machine.

Dyeing recipe and stepwise condition for Pad-dry jig process:

Recipes and conditions	M Brands (% Shade)				H Brands (% Shade)			
	≤0.5	0.5–2.0	2.0–4.0	≥4	<0.5	0.5–2.0	2.0–4.0	>4
Starting temperature (°C)	30	30	30	30	40	40	40	40
Salt (g/l)	20	30	40	50	50	60	75	100
Soda ash (g/l)	5	6	8	10	6	6	8	10
Caustic soda (g/l)					2	2	2	2
Temperature of fixation (°C)	40–45	40–45	40–45	40–45	70–80	70–80	70–80	70–80
Time of fixation (min)	45	45	45	45	45	45	45	45
Washing	Thorough and repeated				Moderate			

3.3.1.7   Dyeing with HE colours

Dyeing can be done by almost the same method as above, the technique being somewhere in between M and H dyes. The padded material is worked in the salt bath while the temperature is raised to 80–85 °C followed by addition of alkali.It is recommended to heat up before each turn and keep the steam valve open slightly so that an overall constant temperature of dyebath for each turn is ensured. Dyebath pH should be in the range of 10.8–11.3 throughout the period of fixation at 85 °C; salt and alkali required for these dyes and their fixation times are listed in table below:

Viscose rayon	HE brands (% Shade)						
	Up to 0.1	0.11– 0.3	0.31– 0.5	0.51– 1.0	1.01– 2.0	2.01– 4.0	Above 4
NaCl/Glauber's salt (g/l)	5	10	20	30	40	55	65
Alkali –only soda ash	10	10	10	15	15	20	20
Mixed alkali – soda ash	5	5	5	5	5	5	5
Caustic soda	0.2	0.2	0.2	0.2	0.2	0.2	0.2
Time for fixation	30	30	45	45	45	60	60

3.3.1.8   Continuous methods (M and H brands)

In many continuous dyeing methods described under cotton dyeing, pad-dry, pad-dry-steam methods are suitable for viscose materials also. The material is padded with the following recipe:

Quantity	Unit	Additions
x	g/l	Dye
15–20	g/l	Sodium bicarbonate
50–100	g/l	Urea
2–5	g/l	Wetting agent

Padded goods are dried at 100–110 °C when water evaporates and transports dye at the interior of fibre for fixation. $NaHCO_3$ turns to Na_2CO_3 at higher temperature in presence of moisture raising dye bath pH to facilitate fixation. Urea acts as potential solvent, activates moisture to retain soluble form of dye and is essential for deeper shades. Urea also helps in pick up and where the quantity has to be of the level of 100–120 g/l in case of deep shades but since viscose is very absorbant this extra addition of urea may not be necessary when compared to cotton. (Dye uptake is also influenced by accessible free volume in fibre. Viscose has more free volume than mercerized cotton and so dye uptake will be according to these free volume data.)

In pad-dry-pad-steam method the same recipe can be used but with addition of 5–7 g/l resist salt to avoid oxidation, if any, during steaming.

Since the pad-dry method gives satisfactory results, pad-dry method is more suitable for viscose fabrics. The migration during drying has to be avoided. Float driers with predrying facility is most suited.

3.3.2   Dyeing with direct dyes

Direct dyes are very important in the case of rayon since it can be dyed in neutral medium. The fastness properties of direct dyes on viscose is better than on cotton. Only disadvantage in this case is that direct dyes can give only duller shades. Special direct dyes are available with ternary elements for high light fastness, low salt, low energy and water consumption, simple washing off and good reproducibility which makes them more suitable for cellulosic dyeing.

3.3.2.1   Exhaust dyeing

Most of the processes used in the case of cotton can be used for viscose. All kind of exhaust dyeing machines like jets, winches, package and beam dyeing machines, jiggers can be used for exhaust dyeing. In case of exhaust dyeing on jig it is better to follow pad-jig process.

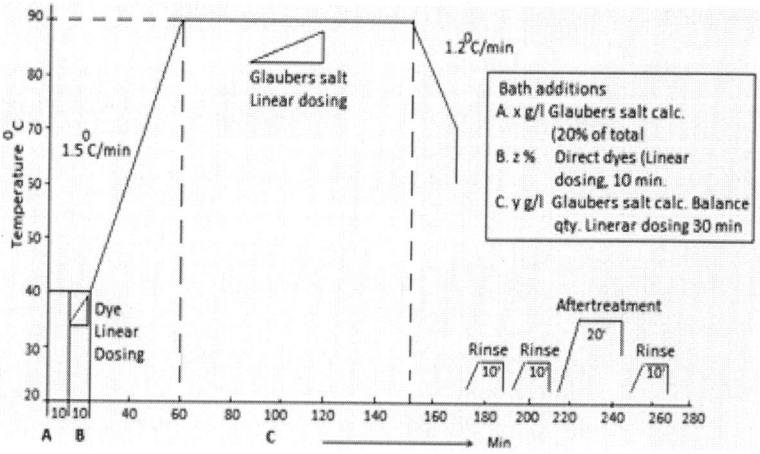

Exhaust dyeing with direct dyes

Bath can be set at 40 °C with necessary auxiliaries, 20% of the total required Glauber's salt calc. and the material is introduced. After 10 min running the required amount of dyes may be added in 10 min, by linear dosing. Next the temperature is raised to 90 °C (100 °C in case of cotton) at 1.5 °C/min temperature gradient. After running for 30 min at this temperature the balance Glauber's salt is added in 30 min by linear dosing. After running for 30 min at this temperature, the bath is cooled at 1.2 °C/min to 70 °C and drained.

Rinse 2 times at 30 °C, after treat and again rinse.

3.3.2.2   Pad roll method

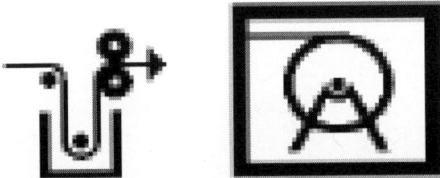

Recipe for padding

Quantity	Unit	Additions
x	g/l	Direct dyes
100	g/l	Urea
20	ml/l	Lyocol BC liq– Clarient
1	ml/l	Wetting agent
5	g/l	Sodium acetate and pH 6.5 (no less) with acetic acid

Padding: 20–25 °C

Batching time: 3 h at 80 °C

Washing off: 2 bath with cold soft water, 1 bath with cold hard water (4–6 °e)

Fixation as usual.

3.3.3   Dyeing with sulphur dyes

Sulphur dyes are used when excellent washing fastness is essential. Sulphur dyes also give a high degree of fastness to milling and to cross-dyeing. In dyeing with sulphur dyes on viscose also pad-jig process is most suitable. Dispersed C.I. sulphur dyes and the C.I. solubilised sulphur brands are the most suitable for application by this process owing to their lack of substantivity during impregnation and padding. Pad batched rolls can be directly loaded on to the jigger made ready with reducing agent and salt at 80–90 °C, then run for four to six ends before rinsing and reoxidising. In case immediately loading on the jig is not possible the batch may be covered and kept rotating till it is loaded.

3.3.4   Pigment dyeing

Pigment dyeing on viscose can be practiced in the same way as in the case of cotton (explained earlier in this book) by pigment pad process or cationisation process for light and medium shades. In pigment pad process pigment colorants, together with a binder and a catalyst, are fixed on textile fabrics by a heat treatment. The process may be combined with resin finishing as a single-bath application, or with other finishes and softeners. Only disadvantage being, the softer feel of viscose is negatively affected by the binder system which to some extent can be de better by the addition of softening agents or by using the softer binder system introduced by many manufacturers..

3.3.5   Precautions for dyeing viscose

Under alkaline conditions, in enclosed machines, viscose is especially prone to cause problems of dye reduction. It is essential that viscose is thoroughly desulphurised, otherwise appreciable colour value will be lost. Mild peroxide bleaching and addition of reduction inhibitor from the start of dyeing is recommended. When dyeing in rope form, there must be sufficient displacement and this is best achieved for woven fabrics in jets with aerodynamic systems. The liquor temperature should not fall below 50 °C and a suitable running crease inhibitor should be used in all hot baths. In order to guard against the danger of formation of running creases, the size of the load should not exceed approx. 80% of the maximum load. Fabric circulation speed should be set to between 80 and 120 s.

To prevent abrasion marks, the slippage (difference between winch speed and fabric speed) should be as low as possible.

If dyed viscose fabrics are allowed to begin to dry in patches – top of trolley – edges of folds – these areas may be visible as a physical/optical patch after finishing. Keep fabric thoroughly wetted until drying machine/ stenter is available.

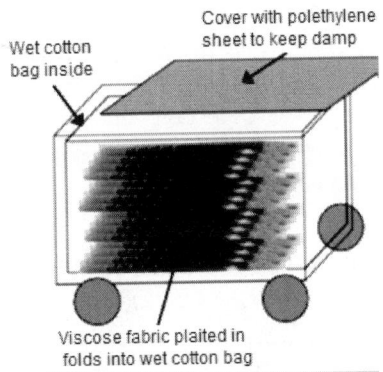

Viscose fabric plaited in
folds into wet cotton bag

Preventing pre drying of the fabric is important, but it is far better to ensure that production flows quickly and fabric does not stand wet for a long time. Unfixed dye, residual hydrolysed dye, will tend to migrate more in dyed viscose than in cotton. If dyed yarn or fabric is kept standing wet (bottleneck at drying machine), a cationic fixing agent can prevent this migration. With viscose/elastane blends, pre-setting is necessary. The dyer could run into creasing problems if attempting to pre scour and dye without setting. If pre-setting greige fabric – beware of temperature variation across the stenter, – temperature variation can have a marked effect on dye pick up on the viscose. Whether stentering before dye or after dye, where a viscose fabric hangs down heavily at the entry to the stenter, this can result in a gross variation in weight per square metre from the bottom of the sag to the edges near to the pins, especially on a non-support stenter, and this can lead to side-centre-side shade variation. Take cuttings across the piece and determine weight per sq.m. The results can vary greatly.

Viscose fabrics will absorb relatively high amounts of water and will swell to a greater extent than cotton. Fabrics made of viscose show a low level of dimensional stability in their swollen state and a higher propensity to crease than in their dry state. To give woven and knitted viscose dimensional stability and increase the fabric performance values it is beneficial to reduce the swelling capacity of the fibres. This can be achieved through the use of cellulose cross-linkers (low formaldehyde or zero formaldehyde) and additives. To ensure the efficiency of the cross-linker the fabric should not contain alkali and there should be sufficient time for the cross-linker to diffuse into the fibres and exchange with the swelling water already present. Similarly, it is important not to dry the padded fabric at too high a temperature in order to prevent migration of the cross-linkers to the fibre or fabric surface. Resin migration leads to a higher surface concentration of the cross-linker which is contrary to the desired effects and can reduce abrasion resistance. Viscose and Modal require a higher amount of cross-linker for easy-care finishing than cotton fabrics to achieve a comparable effect. The higher amount of cross-linker required can lead to a corresponding reduction in abrasion resistance. Compared to cotton, articles made of viscose and Modal fibres display much lower losses in tenacity due to cross-linkage. No losses in tenacity result from cross-linking providing the goods are given low-tension treatment in the preceding stages.

3.3.6   Resin finishing on viscose giving dimensional stability and 'bounce'

Quantity	Unit	Bath additions
50–60	g/l	Cross linking agent (low foaming)
12–14	g/l	Magnesium chloride
0.2	g/l	Sodium fluor borate
20	g/l	Polyethylene emulsion
20	g/l	Silicone softener
0.5	g/l	Wetting agent

Liquor pick-up: approx. 80%

Drying: 110–130 °C

Curing temperature: 155 °C

Curing time: 4 min

Final compressive shrinkage of the fabric for example on a Tope W or TurboTax leads to better fabric shrinkage values, to an attractive handle variation and visual fabric appeal.

Chapter 4
Dyeing of modal and lyocell

Modal is a cellulose regenerated fibre of the high wet modulus type – soft, silky and supple even after many washes. The fibre is manufactured according to a special viscose spinning process from high quality wood pulp Modal contains more crystalline structure(crystallite) as a highly ordered structure in contrast to non-crystalline of viscose. It is particularly absorbent absorbing up to 50% more moisture than cotton, and comfortable to wear and takes up colour fast, deep and permanently bringing out the natural brilliance of the colours. Its supple texture, soft look and brilliant silky sheen make it particularly attractive for top-quality fashion or home textiles. It is the ideal fibre for soft and comfortable woven and knitted fabrics. Because of its outstanding breath ability and easy-care properties, it is an apt material for ladies' outerwear, lingerie, sportswear and home textiles. The proportions of crystalline material in regenerated cellulosic fibres are about 40% in regular viscose, 50% in modal fibres and 65% in polynosics, compared with 70% in cotton. As crystallinity increases the water imbibition and dyeability decrease accordingly. Thus direct dye uptake under a given set of conditions generally increases in the order: cotton<polynosics<modal fibres<regular viscose.

Comparison of tenacity – viscose, modal, cotton

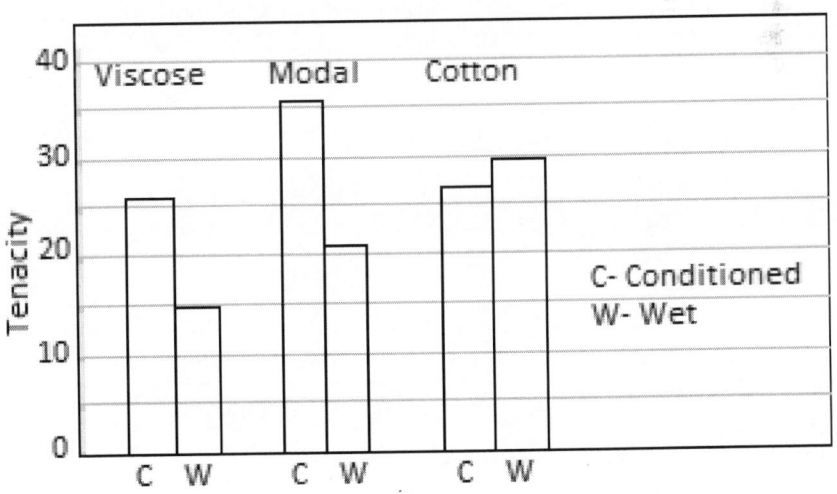

Modal is also ideal for blending with other fibres improving their natural properties and increases strength and improves the fluffiness absorbency and easy-care properties of blends. The common blended fabrics are:

Modal/cotton –Makes cotton even softer and suppleer. This blend higher-quality yarns an improved appearance, silky sheen, brilliant colours –the classic among the blends.

Modal/wool – Modal/wool blend improves climate control, increases comfort, gives a silky sheen and feels good against the skin.

Modal/silk –Gives silk an even more luxurious appearance and easier to care for.

Modal/polyester (and other synthetic fibres –Increases wearable comfort and breath ability. Blends of synthetics and modal improve the performance of sportswear making it more functional and kinder to the skin.

Modal/elastane –For stretch articles with a low proportion of elastane. Their sheen appearance outstrips even that of mercerised cotton.

4.1   Processing of pure modal fabrics

The machines used in processing should be selected in such a way as they exert a low mechanical influence on the fabrics.

Since modal is regenerated cellulosic fibre it does not need much preparation. The fibre as such is absorptive except spinning oils and any additive given during the fabric manufacture. The fibre may have approx. 0.3% of fibre finish and the 0.15–0.20% paraffin wax, which can be removed in a preliminary wash.

Preliminary washing:

Wash with

Quantity	Unit	Additions
0.5–1.0	ml/l	Non-ionic detergent
0.5–1.0	g/l	Soda ash

At 70–80 °C for 30 min and rinse with hot water.
A pure modal fabric may not need a bleaching as such.

4.1.1   Dyeing

Reactive dyestuffs are used mainly which show a good affinity with modal and lead to good yields with the exception of phthalocyanine types. For 100% modal fibre goods and modal/elastane blends reactive hot dyes are recommended.

Or selected reactive dyestuffs can be used for modal/cotton blends. As a result cold dyeable reactive dyes are also used depending upon the shade. Turquoise and green shades are generally dyed at temperatures above 80 °C to the lower affinity of modal fibres. A tried and tested basic dyestuff for shades of green is for example Drimaren Brilliant Green X-3G (Clarient).

4.1.2   Drying

When it comes to knitted fabric in cutting open the fabric high tensions and tensile loads should be avoided so as not to over-stretch the wet fabric.

Drying the fabrics can take place directly on the pin stenter and relax driers whereby the over feed and frame widths are set on the basis of preliminary trials and washing tests. But the adjustments at each stage have to be controlled to least any tension on the fabric. Band driers are ideal for fabric relaxation, in this respect one should, however, take care that e.g. the nozzle band drier —as long as it is still wet —is respectively not moved too much in order to avoid weight-related elongation (higher wet shrinkage).

4.1.3   Finishing

Finishing with cellulose cross-linking agents is to be recommended for washable fabrics in particular since cellulose fibres change their dimension in wet\dry treatments as a result of swelling. In order to reduce the extent of this change a reduction in the swelling potential is necessary which is aimed at with the help of finishing with cross-linking agents. Apart from the stabilising effect of finishing with cross-linking agents there are other advantages with respect to the durability of the fabric visual appeal and the reduction in the pilling tendency of fabrics of this kind. In practical terms finishing is performed by the use of medium amounts of cellulose cross-linking agents with no or a low level of formaldehyde with corresponding softening agents.

Guideline recipe

Quantity	Unit	Additions
50–60	g/l	Low formaldehyde resin
12–14	g/l	Magnesium chloride
0.5	g/l	Sodium fluoborate
10	g/l	Cationic softener suitable for cross linking finish
5	g/l	Non-ionic polyethylene wax emulsion
0.5	g/l	Wetting agent

Liquor pick-up: approx. 80%

Drying: 110–130 °C

Curing: 4min at 155 °C

4.2   Processing of modal/cotton

Blends of modal/cotton are mostly pre-bleached whereby the following recipe can be used depending upon the cotton quality.

Quantity	Unit	Additions
2–3	ml/l	Hydrogen peroxide 50%
1	g/l	Organic stabiliser for peroxide
0.5–1	g/l	Caustic soda 50%

Temperature 90 °C, time 45–60 min and rinse hot.

Wash and wear finish

Quantity	Unit	Additions
0.5–1.0	ml/l	Acetic acid 60%
50–60	g/l	DMDHEU resin (e.g., Knittex FEL–Ciba)
15–18	g/l	Catalyst (e.g., Knittex catalyst MO –Ciba)
20	g/l	Turpex CAN
20	g/l	Softener (e.g., Megasoft FMG –Ciba)
0.5	g/l	Wetting agent

Liquor pick-up: approx. 80%

Drying: 110–130 °C

Curing: 4′ at 155 °C

4.3   Processing of modal/elastane

The *dye* bath ratio is selected at a high rate for modal and modal/elastane fabrics by comparison and should equal between 1:12 and1:15 depending up on the fabric structure.

With elastane blends we recommend the preliminary setting of the fabrics to reduce the material density and thus the danger of the formation of creases.

With blends with elastane the chemicals used are selected in accordance with information from the elastane manufacturer.

4.3.1   Further precautions

Under alkaline conditions, in enclosed machines, viscose is especially prone to cause problems of dye reduction. It is essential that viscose is thoroughly desulphurised, otherwise appreciable colour value will be lost. Mild peroxide bleaching and addition of reduction inhibitor from the start of dyeing is recommended.

When dyeing in rope form, there must be sufficient displacement and this is best achieved for woven fabrics in jets with aerodynamic systems. The liquor temperature should not fall below 50 °C and a suitable running crease inhibitor should be used in all hot baths. In order to guard against the danger of formation of running creases, the size of the load should not exceed approx. 80% of the maximum load.

Fabric circulation speed should be set to between 80 and 120 s. To prevent abrasion marks, the slippage (difference between winch speed and fabric speed) should be as low as possible. Addition of lubricating agents also helps.

4.4   Processing of Lyocell (Tencel): fabric

Due to the many disadvantages of normal viscose, in the late 1970s and early 1980s, researchers, principally at the leading rayon producer Courtaulds Fibres and at an American firm American Enka, began investigating a new method of producing a cellulosic fibre through a solvent spinning technique. In this method, the cellulose is softened and then spun into fibre. Conventional rayon manufacturing, involves longer process - the cellulose is first chemically converted into xanthate, then dissolved in caustic soda, then regenerated into cellulose as it is spun. The solvent spinning technique is both simpler and more environmentally sound, since it uses a non-toxic solvent chemical that is recycled in the manufacturing process. The solvent-spun cellulosic fibre lyocell was first produced commercially in the United States in 1992 by Courtaulds. The company used the brand name Tencel. The fabric was given the generic name lyocell in the United States in 1996. Many other manufacturers like LenzingAG, American Enka has further introduced lycocell fibres.

Manufacturing process

1. Production of wooden pulp: The selected wooden logs are fed into a chipper where it is chopped into squares little bigger than postage stamps. It is loaded into vats of chemical digesters and softened into a wet pulp, which is further washed with water and may be bleached. It is then dried in a huge sheet, and rolled onto spools of about 500 lb (227kgs).

2. Dissolving the cellulose: The spools are unrolled and broken into 1"squares and loaded into a heated, pressurized vessel filled with amine oxide.

3. Filtering: After a short time soaking in the solvent, the cellulose dissolves into a clear solution. It is pumped out through a filter, to insure that all the chips are dissolved.

4. Spinning: The solution is pumped through spinnerets, when the cellulose is forced through it, the long strands of fibres are produced which are immersed in another solution of amine oxide, diluted this time. These sets the fibre strands are washed with de-mineralized water.

5. Drying and finishing: The lyocell fiber next passes to a drying area, where the water is evaporated from it. The strands at this point pass to a finishing area, where a lubricant may be applied. This may be a soap or silicone or other agent, depending on the future use of the fibre.

6. Drying and finishing: The dried, finished fibers are at this stage in a form called tow. Tow is a large untwisted bundle of continuous length filaments. The bundles of tow are taken to a crimper, giving it texture and bulk. The crimped fiber is carded by mechanical carders, which perform an action like combing, to separate and order the strands. The carded strands are cut and baled for shipment Recovery of the solvent

7. The amine oxide used to dissolve the cellulose and set the fibre after spinning is recovered and re-used - the dilute solution is evaporated, removing the water, and the amine oxide is routed for re-use in the pressurized vessel in step 2. Ninety-nine percent of the amine oxide is recoverable in the typical lyocell manufacturing process.

Details of the characterisics, processing, fibrillation etc are given in Vol. I in the Handbook of Textile Processors series- Textile substrates – Fibres, Yarn and Fabric - Chapter 5.4

4.4.1   Knitted fabrics

Lyocell yarn can be knitted in the normal way taking certain precautions to avoid fibrillation of the yarn. Special care has to be taken in waxing, yarn tensions, machine gauge/yarn count, etc. Generally, the machine settings used for other cellulosic fibres, including cotton, will apply. Positive feed mechanisms, yarn storage units and low yarn tensions (e.g. 2 g), waxing by hard wax, using fine gauge machines for low twist yarns, avoid steaming of the yarns are advantageous. Since the yarns are leaner the stitches can be clearly seen in the fabric. It is better to use higher twist to minimise the fibrillation.

4.4.1.1   Pretreatment

When the fabric is subjected to mechanical action in the wet state in preparation and dyeing the surface hairs receive the majority of the abrasive action, therefore fibrillation will occur predominantly on these surface fibres. The fibrils formed are relatively long and are able to become entangled leading to an extremely matted appearance. It is important that the fabric is 'worked' until the surface fibres are fibrillated to their maximum extent. Failure to achieve this will result in incomplete hair removal and a nuns table surface appearance in the finished fabric.

A cellulase enzyme is used to clean the fibrillated hairs from the surface of the fabric. Fibrillation of the surface hairs reduces the fibre diameter and this in turn enhances their vulnerability to enzymatic attack. Thus complete removal of the surface hairs can be effected, with minimal effect on the bulk of the fabric. The hydrolytic fibre degradation involved in this process results in a loss of weight, typically 4%, and is accompanied by some reduction in fabric strength. Preliminary local trials are therefore strongly advised. The use of an enzyme on greige fabric is not effective in complete hair removal unless excessive chemical treatment is used which leads to unacceptable loss of fabric strength. Dyeing can be carried out either before or after the fibrillation process, but certain dyes may affect the primary fibrillation characteristics and enzyme efficiency.

The choice of processing route and the type of equipment used have been of vital importance in the processing of Lyocell. Garment processing gives an ideal combination of the physical conditions for processing lyocell so this route processed most of the early commercial fabrics. Simple processing of lyocell fabrics by open width/ resination routes gave fabrics with a good technical performance but they did not have the 'peach touch' or the bulk of fabrics made by the garment route. Preparation, hair removal, dyeing and tumbling can all be carried out continuously,but the machines used were not sufficiently aggressive to generate fibrillation. The air- jet type machines for piece processing woven fabrics are found suitable for Lyocell. These rope processing machines must be able to re-orient the fabric frequently to ensure uniform fibrillation across the surface and to prevent crease damage marks appearing. The air jet rope-processors most easily meet the serequirements for woven fabrics but knitted fabrics are less prone to damage marks and a wider range of soft flow jets can be used successfully.

4.4.1.2   Scouring

As with other viscose fibres, since the lyocell is a clean cellulosic fibre, it needs only mild scouring treatment.

Guide recipe

Quantity	Unit	Bath additions
2	g/l	Soda ash
2	g/l	Wetting/scouring agent

Treat at 70 °C for 30 min.

The use of stronger alkali at higher temperature has to be restricted. In case a stronger alkali is needed for the scouring for any reasons one can use upto 20g/l soda ash at 70 °C or 5 g/l caustic soda at 80 °C or 1 g/l caustic soda at 95 °C.

4.4.1.3   Bleaching knitted fabric on soft-flow machines

Quantity	Unit	Bath additions
0.5	g/l	Contavan FLA
5	g/l	Peroxide 35%
1	g/l	Stabiliser
2.5	g/l	Wetting/scouring agent
1	g/l	Caustic soda 100%

Treat at 80 °C for 60 min.

Full white bleaching of knitted fabrics:

Quantity	Unit	Bath additions
0.5	g/l	Chelating agent
3	g/l	Stabiliser
2.5	g/l	Wetting/scouring agent
0.5	g/l	Caustic soda 100%
7.0	ml/l	Peroxide 35%
0.5–0.7	%	Optical whitening agent

Treat for 60 min at 85 °C.

4.4.2   Processing woven fabrics

The greige fabric made of lyocell has to be handled properly to avoid any undue fibrillation and creasing. Many precautions have to be taken from weaving to finishing, taking into consideration the wet strength of the lyocell. The slack or tight selvedges in weaving as it will giveun even tension on the fabric or selvedge and breakage during wet processing. Selvedges should also be robust enough to with stand aggressive wet processing.

4.4.2.1   Singeing

Singeing is strongly recommended and is best carried out on the loom state fabric. Best results are achieved with high intensity singers with pre-brushing facilities. Singe is important to minimise the amount of loose fibres and fibrils released from the fabric into the dye bath and to reduce the load on the dyeing machine filters.

4.4.2.2   Desizing/scouring

Desizing and scouring should always be carried out in open width. Rope ranges, jets, winches, etc. can cause permanent creasing in the fabric due to the stiffness of fabric when wet.

All the standard methods for open width desizing/scouring are applicable e.g. on the jigger, by pad-batch, pad-steam, etc. If starch is present, desizing using either an amylase or an oxidative desize will be necessary. During the process, it is essential to avoid creasing on rollers, nips, etc. as creases may not be possible to remove in later processes. 'Uncurlers', 'straighteners' and bow rollers should be adjusted by trials.

4.4.2.3   Causticisation

Causticisation is not a must for lyocell fabric. However, in special cases causticisation can be done with special care. Lyocell can acquire some properties by causticisation which are due to the swelling of the fibres as they absorb caustic soda, due to the reshaping and subsequent setting of the fibres while they are in a plastic state and changes in fibre structure. The main changes in properties are reduced fibrillation, increase in dyeability, permanent creasing, wet flexibility, reduced weft slippage, increased bulk, reduction of fabric growth. Normally, causticisation is done at caustic soda strength of 8–10% and concentrations higher than this offer no advantages. The process can be done by different methods:

1. Causticisation under tension: Caustic soda is applied in a system where tension is controlled in both widthways and lengthways directions, e.g. on a traditional mercerising machine. The effects of caustication can be controlled by the tensions applied within the system. For example, by tensioning width wise the degree of change in the weft yarn crimp can be controlled and the fabric can be 'set' by this process. Resultant physical properties can also be controlled. The effect on dyeability is relatively minor, hence concerns over evenness of application are considerably reduced. The effective setting of the warp yarns makes this a particularly effective system for reduction of weft line creasing that can be seen in garment processing.

2. Relaxed causticisation: It is obvious that the properties achieved by tensioning cannot be got in this method. The process will allow shrinkage and contraction freely. Since there is no tension in warp or weft way the fabric construction can be 'set' but only to the dimensions governed by the swelling of the yarns and the shrinkage induced. High shrinkage can lead to weft line creases being generated if warp shrinkage is uncontrolled. Extremely high increases in dyeability are possible as are significant improvements in abrasion performance, mainly due to the on siderableshrinkage taking place. However, tensile properties may be reduced due to the uncontrolled increase in yarn crimp. If the process is carried out in the relaxed batch wise form concerns exist over the vulnerable state of the swollen fabric and its susceptibility to abrasion damage.

3. Pad-batch causticisation: The performance of this system lies somewhere in between the above two process. Batch to batch variation is inevitable due to lack of tension control. Due to the difference in layer to layer differential shrinkage chances of moire effect is high. Variations in tension through the batch may also give rise to variations in dyeability and fibrillation performance. Caustic soda is applied in an

intermediate system, padding and batching onto a beam or 'A' frame. Some tensions are introduced in the batching operation.

4.4.2.4   Bleaching woven fabric in open width

Lyocell as such is a clean fibre and hence may not need bleaching for optic white and even for pale shades. If the basic white of the yarn is not sufficient for full white a mild bleaching may be given. When bleaching is required, this can be carried out by any of the normal techniques applicable to cellulosic fabrics e.g. jigger, beam, pad-batch and pad-steam.

Guideline recipes

Discontinuous (batch) process

Guideline recipe (1)

Quantity	Unit	Bath additions
0.5	g/l	Stabiliser
0.5	g/l	Wetting agent
1	ml/l	Lubricating agent
3.0	ml/l	Peroxide 35%
0.4	g/l	Caustic soda 100%

Treat at 85 °C for 45 min.

Guideline recipe (2)

Quantity	Unit	Bath additions
0.5	g/l	Chelating agent
0.5	g/l	Wetting agent
1.0	ml/l	Lubricating agent
3–5	ml/l	Peroxide 35%
0.4	g/l	Caustic soda 100%
4.0	g/l	Sodiumsilicate 38 °Be
1	ml/l	Stabiliser

Treat at 80 °C for 45 min.

Pad-batch recipe

Quantity	Unit	Bath additions
25	ml/l	Peroxide 35%
4	g/l	Caustic soda
5	ml/l	Stabiliser

Pad-steam

Quantity	Unit	Bath additions
15	ml/l	Peroxide 35%
2	g/l	Caustic soda
3	ml/l	Stabiliser

In all the above recipes wetting agents/scouring agents (0.5–2 g/l) and lubricating agents in case of jet process may be added. For full white, any optical whitening agent suitable for cotton can be incorporated suitably.

In case of lyocell, dyeing process and machine has to be selected as per the finish requirements since fibrillationis one of the most important physical properties of lyocell fibre. Its control and manipulation during dyeing and finishing are keys to a wide range of fabric aesthetics. It is caused by wet abrasion and occurs during wet rope processing and domestic laundering. Fibrillation does not occur during open width processing. If a clean surface is needed in the final finished fabric it is advisable to do the processing in the open width where mechanical abrasion is very less or if any fibrillation occurred has to be removed by processes like bio polishing. Thus any of the following two process route shown below can be followed.

Open width exhaust dyeing can be done on jigs and beam dyeing, as in case of viscose, but care has to be taken in selection and addition of chemicals, auxiliaries, temperature gradient, etc.

Reactive, vat and sulphur dyes may be also applied continuously using the pad-steam process. Reactive dyes may also be applied semi-continuously using the cold pad-batch process.

Rope dyeing can be done wherever necessary and peach skin finish by fibrillation is required (see procedure under fibre, yarn and fabric section under Lyocell). But where fibrillation is not required at the same time rope dyeing has to be done, care has to be taken to reduce fibrillation

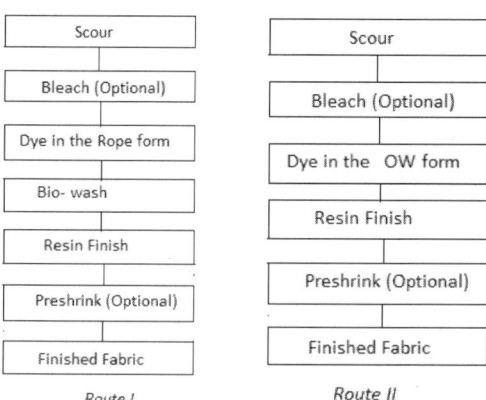

Route I   Route II

like, using the highest possible liquor ratio, using a suitable dyebath lubricant, using reactive dyes which can be dyed at lower temperatures and with low levels of alkali. However, some fibrillation will occur which may be removed by biowash using an acid cellulase enzyme. To reduce the tendency due to machine, it is advisable to use airflow type machines like the AF or AFS, Thies Luft-roto, Thies Airstream, Longclose Airsoft,etc.

Guide recipe

2–4% Cellulase enzyme at a pH of 4.5–5, for 45–75 min at 55 °C. After the treatment, the fabric should be rinsed and the residual enzyme should be deactivated by increasing the pH of the bath to 10 with sodium carbonate and run for a further 15 min.

4.4.3   Using fibrillation characteristic of lyocell for achieving peach skin finish

In the previous paragraph we have explained to produce dyed fabric with clean surface. But the fibrillation characteristic can be used for making peach skin effect on the finished fabric in the normal dyeing procedure. The microfibers generated in the wet processing of lyocell can be used to create a variety of interesting touch and feel aesthetics. A peach skin effect, which can also withstand repeated domestic washing at 40 °C is possible, providing the fibrillation is developed such that the fibrils cannot become long and entangled. This is the key. The production of a peach skin effect has three discrete stages: primary fibrillation, enzyme cleaning and secondary fibrillation.

4.4.3.1   Primary fibrillation

Lyocell grey fabric will be hairy due to the abrasions in fibre, yarn and fabric production stages. Singeing reduces the hairiness but a significant amount of hairs will remain. Singeing is important to minimise the amount of loose fibres and fibrils released from the fabric into the dyebath and to reduce the load on the dyeing machine filters.

When the fabric is subjected to mechanical action in the wet state the surface hairs receive the majority of the abrasive action, therefore fibrillation will occur predominantly on these surface fibres. The fibrils formed are relatively long and are able to become entangled leading to an extremely matted appearance (see figure below). It is important that the fabric is 'worked' until the surface fibres are fibrillated to their maximum extent. Failure to achieve this will result in incomplete hair removal and an unstable surface appearance in the finished fabric. Effective singeing of the fabric prior to the primary fibrillation step has been shown to be beneficial. At this stage the fabric is given treatment to increase the fibrillation.

The rate of fibrillation increases with increasing pH and temperature. Lubrication will decrease the rate of fibrillation but a lubricant should be included to minimise creasing damage.

Hence the following recipe can produce fibrillation.

Quantity	Unit	Bath additions
1.0–5.0	g/l	Soda ash
0.5–4.0	g/l	Lubricant

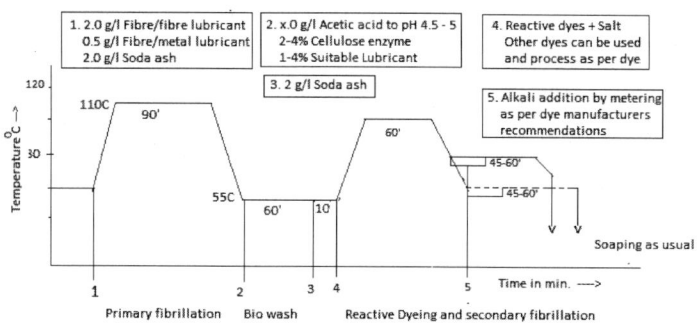

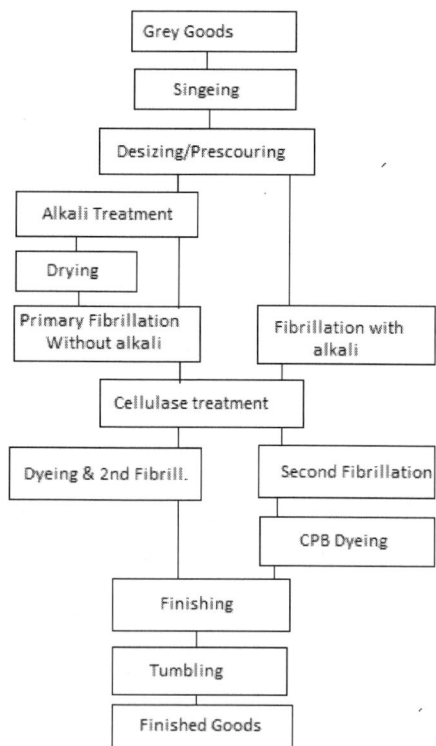

Run in a jet or soft flow and run for 60–90 min at 90–120 °C, at predetermined running parameters like speed of running, jet pressure, jet opening (where provisions are available).

Rinse and neutralise. Since it is difficult to ascertain the level of fibrillation in undyed state, we can do a laboratory dyeing (A sample from the fibrillated fabric should be immersed for 1 min, cold, in a 1 g/l solution of Indosol Black SFRL (Clariant), rinsed in cold water, hydroextracted and dried. The fibrils will stain much darker than the main body of the fabric.) to find out the fibrillation is up to the mark by visual assessment. (This time of running also can be standardized). Wet tumbling machines like Biancalani Airo 1000can be utilised for the fibrillation and enzyme processes.

4.4.3.2   Theory of action of caustic soda

In lyocell the cellulose chains are bound together by hydrogen bonds. The area where the chains lie parallel they come closer and the hydrogen bond at these places are stronger than the places where they are not parallel. The former portions of the fibre are called crystalline regions and latter amorphous portions. Water enters through the amorphous portions and swells the fibre whereby the weak hydrogen bonds are broken at these regions but not at crystalline regions and thus in water the fibre split into discreet crystalline and amorphous regions in water. But the caustic soda solution can further break more hydrogen bonds in the crystalline region also and swell the fibre fully. With a 12% caustic soda almost all hydrogen bonds are broken and the fibre is in a plastic state. On washing off the caustic soda the hydrogen bonds are again formed.

This behaviour of the lyocell is used advantageously in many processes. During causticisation the fabric is held under tension which allows the cellulose chains to form more parallel regions and thus more crystalline areas which makes the fabric/fibre more 'set'. At the same time if the fibre is deformed, say by bending or stretching, while the bond is broken, the new shape will be set into the fibre when the bond is reformed, and the new position will be maintained until the fibre is exposed to the same, or stronger, solution again. In weaker caustic soda, in swollen state mechanical agitation is given the partially split chains on the surface of the fibre forms microfibers which is called fibrillation.

Greige Fabric   Primary fibrillation

After biopolish   Secondary Fibrillation

Schematic diagram of fibrillation at different stages
of processing of Lyocel woven fabric

Schematic diagram of fibrillation at different stages of processing of
Lyocel woven favric

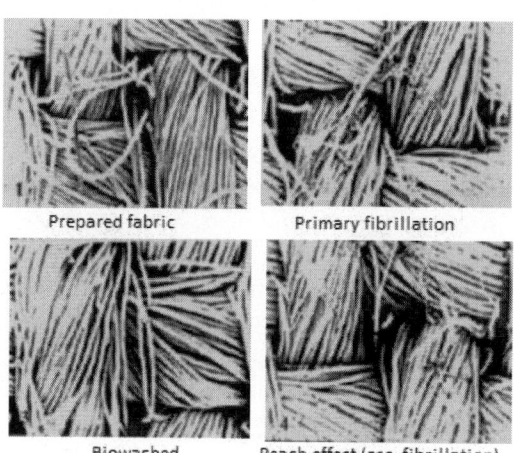

Prepared fabric   Primary fibrillation

Biowashed   Peach effect (sec. fibrillation)

Photo mocrograph pf the surface of a tencel woven
fabric at vrious stages of processing

To achieve maximum benefits from caustic soda treatment, the caustic soda should be neutralised by thorough washing of the fabric and the fabric should be dried prior to further processing. (Some of the new hydrogen bonds are not reformed, and the fabric is not properly 'set', until the fabric has been dried.)

4.4.3.3   Biowash

A cellulase enzyme is used to clean the fibrillated hairs from the surface of the fabric.

Fibrillation of the surface hairs reduces the fibre diameter and this in turn enhances their vulnerability to enzymatic attack. Thus complete removal of the surface hairs can be effected, with minimal effect on the bulk of the fabric. The hydrolytic fibre degradation involved in this process results in a loss of weight, typically 4%, and is accompanied by a reduction in fabric strength. Preliminary local trials are therefore strongly advised. The use of an enzyme on greige fabric is not effective in complete hair removal unless excessive chemical treatment is used which leads to unacceptable loss of fabric strength.

4.4.3.4   Secondary fibrillation

In order to generate a stable finished fabric appearance, it is necessary to fibrillate the fabric a second time. The characteristics of this 'secondary' fibrillation are, however, very different to the 'primary' fibrillation. With the absence of surface hairs, fibrillation is confined to the yarn cross-over points and the high points of the fabric construction (see figure below). Furthermore, the positioning of these fibrils means that they are physically unable to entangle and therefore pilling does not occur. A small pile is created on the surface of the fabric which gives the special touch and feel characteristics known as 'MillWash' or 'Peach Skin'. The fibrils are only a fraction of the size (diameter) of the fibres, therefore they appear much lighter in colour, even though they contain the same amount of dye. It is this optical effect which gives the dusted 'millwash' appearance.

Secondary fibrillation can be produced by either a simple washing treatment or by jet dyeing the fabric.Even though dyeing can be carried out either before or after the fibrillation process, stillcertain dyes can affect the primary fibrillation characteristics and enzyme efficiency. For this reason, therefore, application of dyes after the fibrillation and enzyme treatment stages in rope dyeing on jets is preferred to improve reproducibility of the process. Sufficient secondary fibrillation should be generated to a level that will not increase on subsequent washing. Dye bath lubricants should always be included. If dyeing by open width process, a separate process for fibrillation has to be followed, because in the open width process secondary fibrillation will not be developed (see flow chart above).

4.4.3.5   Cost effective special methods

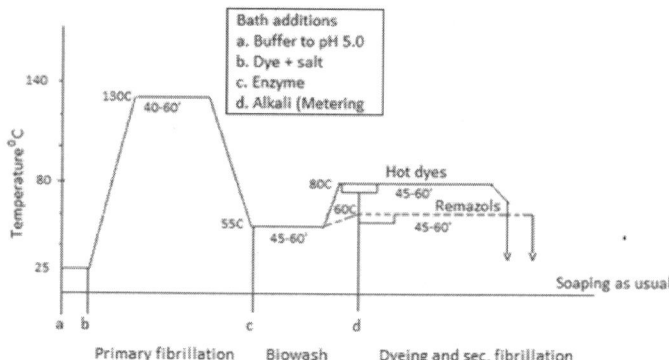

Notes:

Run with the maximum linear speed.

Run with the blower on maximum(In airflow machines).

Ensure the reel speed and the fabric speed are matched.

Use overflow rinsing to minimise lint deposits on the fabric.

Open Variojet to maximum(if provision is available in the machine).

Use a spray at the end of the plaiter (if provision is available in the machine).

Use a self-cleaning filter (if provision is available in the machine).

Use a more powerful blower (with silencer) (in air flow machines).

4.4.3.6   Dyeing of lyocell by reactive dyes by pad-batch method

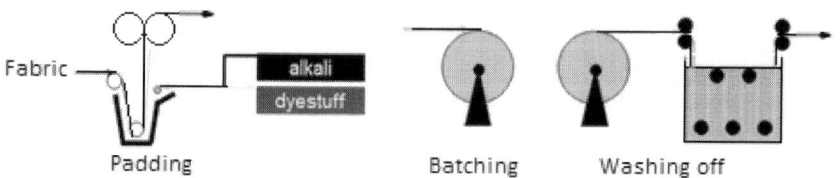

Padding is done as usual (recipe given below) batched for 12–24 h. And washed off. Padding should have following facilities:

Small trough,

Temperature control,

High padding speed (to avoid tailing)/short immersion time (~1 s),

85–100% pick-up.

After padding enough time has to be given for swelling of the fibres. PU-foam wrapping on the batch (provide space for possible swelling on the beam). Long batching time (to cover differences in wettability-migration). Take care to use dyes with similar substantivity behaviour.

Alkali requirement for silicate method (VS dyes)

Type of sodium silicate		Sodium silicate required		Quantity of VS dyes used upto					
°Be/°Tw	Mixture ratio Na2O:SiO2	g/l	ml/l	20 g/l	30 g/l	40 g/l	50 g/l	60 g/l	70-100 g/l
				Quantity of NaOH 50 °Be (50%) in ml/l required in addition to sodiumsilicate for each of above dye conc.					
37–40/69–77	1:3.3	65	50	11	14	14	14	17	20
40–42/77–82	1:3.3	55	40	11	14	14	14	17	20
48–50/100–106	1:2.6	50	35	6	9	9	9	12	18
58–60/134–142	1:2.1	45	30	16.5	11	11	16	16	15

Advantages of dyeing lyocell by cold pad-batch method:

High productivity,

Low energy cost,

Low water cost (<60 l/kg fabric; against exhaust > 100 l/kg in LR 10:1),

Salt free cpb methods,

Low effluent amount/effluent load,

Fibrillation is limited,

No problems with abrasion marks and crease marks in the dyeing process.

Disadvantages:

Secondary fibrillation cannot be developed during dyeing,

Long fixation time (hours) è disposition,

High repair cost/reproducibility,

System losses,

Fabric appearance, tailing.

4.4.3.6.1   Silicate free method or caustic soda/soda ash method

Alternative sodium silicate free method and hence more environmental friendly. Medium pad liquor stability. Dosing pump required.

Padding, batching and washing off are done as usual.

Alkali requirement for 20–25 °C.

Dyestuff (g/l)	20	40	60	80	100	>100
Caustic soda 50% (ml/l)	4	6	8	10	12	14
Soda ash (g/l)	30	30	30	30	30	30

4.4.3.6.2   Soda ash method

Higher reactivity dyes like Leva fix dyes can be dyed using only soda ash.

Sodium silicate free,

High dye liquor stability (2–3 h),

No dosing pump required (a dyestuff selection),

Long migration time/outstanding dyestuff penetration,

Process reliability in tropical conditions,

Long fixation time.

Padding and fixation temperature: 20–35 °C.

Dyestuff (g/l)	<5	6-9	10	15	20	30	>40
Soda ash (g/l)	10*	10	10	15	20	30	40

*For pale shades and padding temperature more than30 °C 1–2 g/l sodium bicarbonate is recommended.

Normal silicate free cystic soda/soda ash method for higher reactivity dyes (e.g., Levafix dyes)

Highlights

Alternative method sodium silicate free,

Medium dye liquor stability,

Medium fixation time,

Padding and fixation temperature: 20–30 °C,

Dosing pump required.

Dyestuff (g/l)	<5	6–9	10	15	20	30	>40
Soda ash (g/l)	10	10	20	20	20	20	20
Caustic soda 50% (ml/l)	–	0.5	0.6	1	1	2	2.5

4.4.3.6.3   Continuous dyeing on pad–steam range

The lyocell also can be dyed on pad-steam almost same way as cotton.

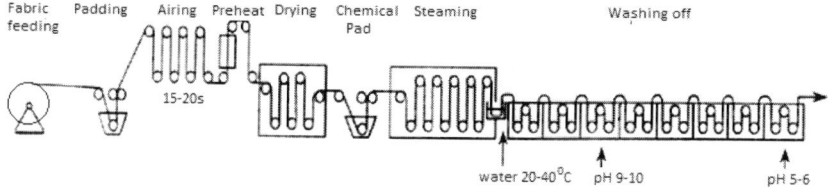

Dyeing processby Pad - Dry - Chemical Pad - Steam Method

4.4.3.7   Process for Biancalani machine

1. Development of primary fibrillation

Liquor 800 l water.

Recipe

Quantity	Unit	Bath additions
2	g/l	Soda ash
2.5	g/l	Sequestrant

2	g/l	Nonionic detergent
4	g/l	Lubricant

Bath temperature – 85–95 °C, air temperature –140 °C, air level – 60%, reel speed –90 m/min, cylinders closed, 2 bar pressure, lightening, time–30–60 min.

2. *Wash-off*

Drain. Fill 800 l. Add acetic acid to neutralise. Run 5 min.

Overflow rinse 5 min.

3. *Enzyme treatment*

   800 l water.

Recipe

Quantity	Unit	Bath additions
2.0–4.0	g/l	Cellulase enzyme
x	g/l	Acetic acid to pH 4.5–5

Bath temperature– 55 °C (Heat exchanger set point 61 °C), air temperature –60°C.

Air level –60%, reel peed –90 m/min, cylinders – open, time –60 min.

4. *Enzyme deactivate and wash-off*

   Drain, fill, add 2 g/1 soda ash, run 5 min, drain, Fill 600 l, run 5 min. Neutralise with acetic acid, overflow rinse 5 min.

5. *Dry In Airo*

   Squeeze with rollers for 1 rotation, nips closed –5 bar pressure.

   Air temperature –140 °C, air level –60%, reel speed –70 m/min, cylinders – open, time – until dry.

6. *Dry beat*

   When the fabric is completely dry.

Run at air temperature –55 °C. Air level –100%, reel speed –250 m/min, cylinders –open time –5 min. It is important that the fabric is dried in the Airo so that the beating action can assist in the removal of fibrils loosened by the enzyme treatment.

4.4.3.8   Pilling of lyocell due to fibrillation

In the swollen state lyocell has an extensive fibrillation tendency owing to linear high crystalline fibrillar morphology. On the other hand, the fibrillations induce e.g. rope marking defect in hank finishing, graying of dyed fabrics and a change of handle of clothes that spoils garments features. Efforts to control the fibrillation tendency in lyocell fibers include dyeing with reactive dyestuffs and treating fabrics with crosslinking agents.

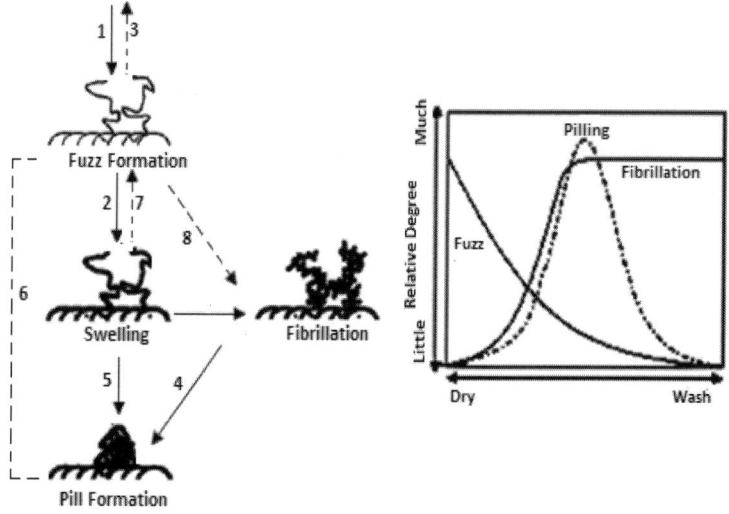

Schematic drawing of pill formation due to fibrillation.

In case of this fibre a treatment with 0.5–0.8 gpl NaOH or KOH will help and further anti fibrillation treatment will further reduce the fibrillation chances.

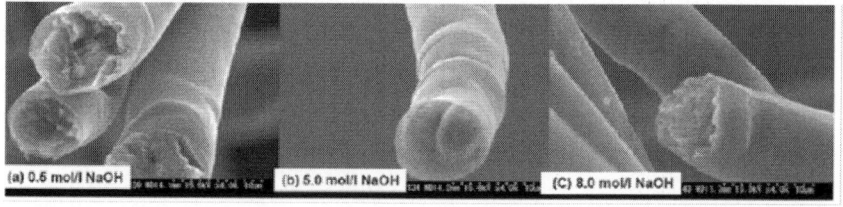

Next, the lyocell woven fabric can be given a resin treatment which can contain the pilling tendency (see below).

Part III

Dyeing of Synthetic fibres

Chapter 5

Dyeing of polyester

The name "polyester" refers to the linkage of several monomers (esters) within the fibre. Esters are formed when alcohol reacts with a carboxylic acid. Polyester fibre is a "manufactured fibre in which the fibre forming substance is any long chain synthetic polymer composed at least 85% by weight of an ester of a dihydric alcohol (HOROH) and terephthalic acid (p-HOOC–C_6H_4COOH)". The first polyester fibre was Terylene. In 1946 second polyester fibre prepared was Dacron. In 1958 another polyester fibre called Kodel was developed by Eastman. Today, polyester is still widely regarded as a "cheap, uncomfortable" fibre, but even now this image is slowly beginning to change with the emergence of polyester luxury fibres such as polyester micro fibre. Most polyesters are made from petroleum from which the constituent acids and alcohols are derived. The types of processes that manufacturers use vary, and little is known about specific manufacturing processes, because the companies want to keep them a secret in order to remain competitive.

The reason that these fibres have since enjoyed such widespread use is attributable to the large number of excellent properties they possess. The main ones are:

- very good crease recovery and dimensional stability,
- very good bulking characteristics,
- good tensile strength and abrasion resistance,
- good weathering and sun-light fastness properties,
- good resistance towards acids, reducing and oxidizing agents,
- and towards the majority of organic solvents, and
- very good resistance towards insects and micro-organisms.

Polyester textile materials are light, dry rapidly, are easy to look after, and their physiological characteristics are good.

5.1   Manufacture of polyester

The production process consists of:

5.1.1   Polymerization

Polyethylene teraphthalate (PET) is a condensation polymer and is industrially produced by either terephthalic acid or dimethyl terephthalate with ethylene glycol. While direct esterification of PTA is the preferred method of PET synthesis, ester interchanges between DMT and ethyleneg lycol is still utilised in some PET manufacture, partially because of local choice and partially because DMT is a product of polyester recycling by methanolysis or glycolysis. The second monomer, ethylene glycol, is a major material of commerce, produced by the oxidation of ethylene followed by ring opening with water. The large-scale production of all PET monomers assures low-cost polymers and makes competition from new compositions of fibre-forming polymers very difficult.

A. Terephthalic acid (PTA), produced directly from p-xylene with bromide-controlled oxidation.

B. Dimethyl terephthalate (DMT), made in the early stages by esterification of terephthalic acid. However, a different process involving two oxidation and esterification stages now accounts for most DMT.

C. Ethylene glycol (EG) initially generated as an intermediate product by oxidation of ethylene. Further ethylene glycol is obtained by reaction of ethylene oxide with water.

The first stage of PET polymerization is, in essence, the production of bishydroxy ethylterephthalate (BHET). In the direct esterification of PTA, this reaction

$$2CH_3COH + HOOC-C_6H_4-COOH \rightarrow H_3COOC-C_6H_4-COOCH_3 + 2H_2O$$

150–200 °C

$$H_3COOC-C_6H_4-COOCH_3 + 2HOCH_2CH_2OH + Catalyst \rightarrow$$
$$HOCH_2CH_2OCO-C_6H_4-COOCH_2CH_2OH$$

Catalyst 280–290 °C

$$HOH_2CH_2COOC-C_6H_4-HOH_2CH_2COOC \rightarrow$$

Vacuum

$$H[(OCH_2)_2O-OOC-C_6H_4-CO]n-O(CH_2)_2OH + (n-1) HO(CH_2)_2OH$$

The reaction actually results in a mixture of low amounts of free BHET with a variety of PET oligomers. The removal of the by product water, is critical to the ultimate achievement of high molecular weights.

$$CH_3OCO-C_6H_4-COOCH_3 + 2HOCH_2CH_2OH \rightarrow HOCH_2CH_2OCO-C_6H_4-$$
$$COOCH_2CH_2OH + 2CH_3OH$$

The introduction of ester interchange catalysts requires the killing of these catalystslater in the polymerisation sequence as they are equally effective as depolymerisation catalysts.

But larger quantities (e.g. tyre cord) are produced in a continuous process combining synthesis and melt-spinning according to the process of "polyesterification".

Catalyst 280–290 °C
$$nHOOC-C_6H_4-COOH + HO(CH_2)_2OH \rightarrow H[(OCH_2)_2O-OOC-C_6H_4-$$

Vacuum
$$CO]_n-O(CH_2)_2OH + (n-1)H_2O$$

Further process is to build polymers of high molecular weight. In this reaction, an ester interchange reaction occurs between two molecules of BHET to split off a molecule of glycol and build polymer molecular weight. The reaction is usually catalysed, and antimonytrioxide (Sb_2O_3). High vacuum is applied to push the reaction to high molecular weights. Typical melt polymerisation temperatures are 285 °C or higher, and viscosities are on the order of 3000 poise, making uniform stirring and the imparting of a constant shear history across the polymerization mixture difficult, although the power requirement to the stirrer thus becomes a useful QC tool. There are many variations of this process is used by different manufacturers.

5.1.2   Spinning

Polyester is a "melt spun" fibre, which means that it is heated, extruded through the spinnerets, and cools upon hitting the air. From there it is wound around cylinders. Once the above reaction is carried out and achieving molecular weight targets, the polymer may be extruded into strands and cut into chips for subsequent melt spinning (batch process) or fed directly into a spinning machine and converted to fibre (continuous process—CP spin-draw).

In the case of chipped polymer, the molecular weight can be further increased through solid-state polymerization. In this process, thoroughly dried PET chip is first crystallised at about 160 °C to prevent the amorphous as-polymerized chip from sticking together (sintering), and then heated just below the melting point under high vacuum and extreme dryness to advance the molecular weight upward to values of inherent viscosity of 0.95 (textile grade chip has an IV of about 0.65).

During the spinning operation, due to the temperature some side retain can take place which can cause change in the characteristics of the polyester. There can be the formation of di ethylene glycol through the coupling of two hydroxyl ends from the glycol ends (or BHET ends) by dehydration, forming a di ethylene glycol (DEG) unit in the chain, is especially troublesome.

DEG is a foreign unit in the backbone, although it does not directly affect the polymer chain length. This unit reduces crystallinity and lowers the glass transition, thermal stability, and hydrolytic stability of the polymer. It is impossible to completely eliminate DEG formation and about 1.0–1.5 mol% of DEG is always present. Depression of the polymer melting point is easily measured by differential scanning calorimetry (DSC), and this parameter provides an accurate measure of DEG content. Finally, any melt-processed PET always has some cyclic trimer content, which, while not a direct problem for polymer performance, does tend to exude during processing and may cause process upsets. In reality, commercially produced PET is always made by a continuous process involving a number of linked vessels between which the polymer is continuously pumped until the final product specifications are achieved. There can be variations in process which are not published due to trade secrecy.

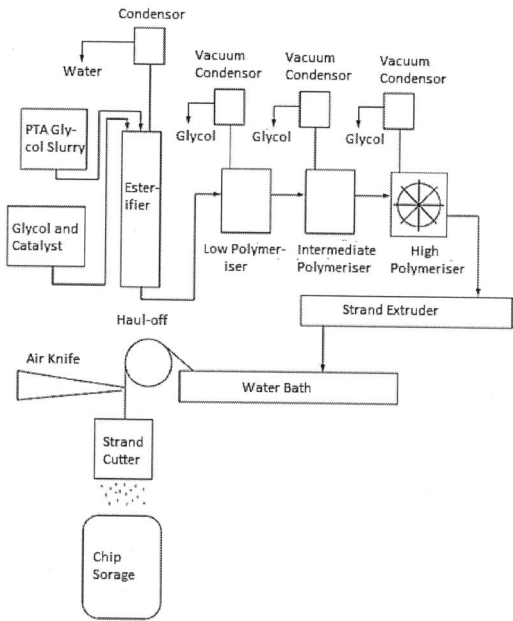

Polyester production process

Either continuous spinning or chip melt spinning the melted liquid is passed through as many as 200 separate spinnerets having five (hosiery yarn) to several thousand holes ranging from 180 to400 μm in size and the polyester melt streams are directed in such a way as to pass through all the holes uniformly in shear and thermal history. The temperature 280–300 °C may go by 10–15 °C more due to the friction while passing through the spinnerets.

The melt spinning process chart is as follows:

$$\text{Polymer chips} \xrightarrow{\text{Conversion by heat}} \text{Dope} \xrightarrow{\text{Spinning}}$$

$$\text{Viscoelastic filament} \xrightarrow{\text{Solidification by cooling}}$$

Solid filament fibre

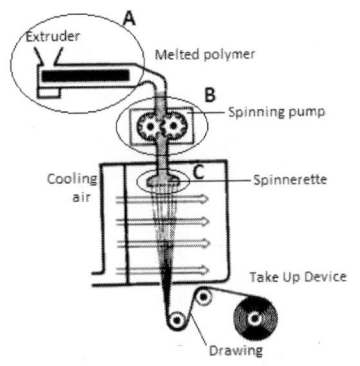 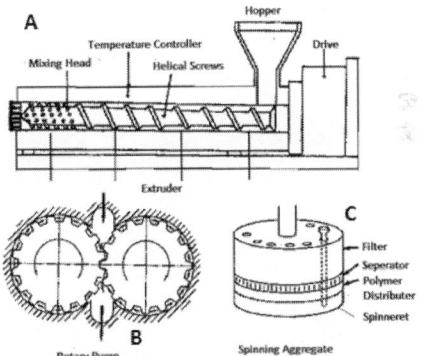

Schemotic diogram of melt spinning   Detailed diagram of marked parts A,B,C

5.1.3   Drawing

The polymer extruded by the spinnerets in form of filaments has not yet the properties which are typical of a textile fibre: in fact the polymer mass (solidified through cooling or solvent removal) is characterised by a mass of disorderly placed molecular chains (in amorphous state) which provides the material with poor thermal and chemical stability, low resistance to ageing, high plasticity and deformability and consequently insufficient physical/ textile properties. If we take natural fibres as models, we need to orientate the molecular chains (orientation phase) in the direction of the fibre axis and at the same time or successively activate or increase the ordered arrangement of the intermolecular structure (crystallisation phase). This process can be partly activated during spinning by increasing the ratio between the take-up speed and the extrusion speed (spinning ratio) but, except the case of high speed spinning of continuous filament yarns, the process needs to be completed by an additional operation of mechanical drawing.

During melt spinning, a lot of orientation of the molecular chains occurs. The combination of molecular entanglements and the presence of polymer chain crystallites lock this orientation into place. This, in turn, affects such parameters as tenacity, modulus, elongation at break, and heat-shrinkage. But further orientation and crystallinity is achieved by a process called drawing. The drawing process is carried out at temperature above the glass transition temperature (80–90 °C) with draw ratios R (1.5:1–6:1). There are two types of drawing.

1. *Single stage drawing*

Earlier this was done by a heated metal snubber pin around which the yarn was passed. The pin temperature was set to about 10 °C above glass transition temperature, i.e., about 85–90 °C for

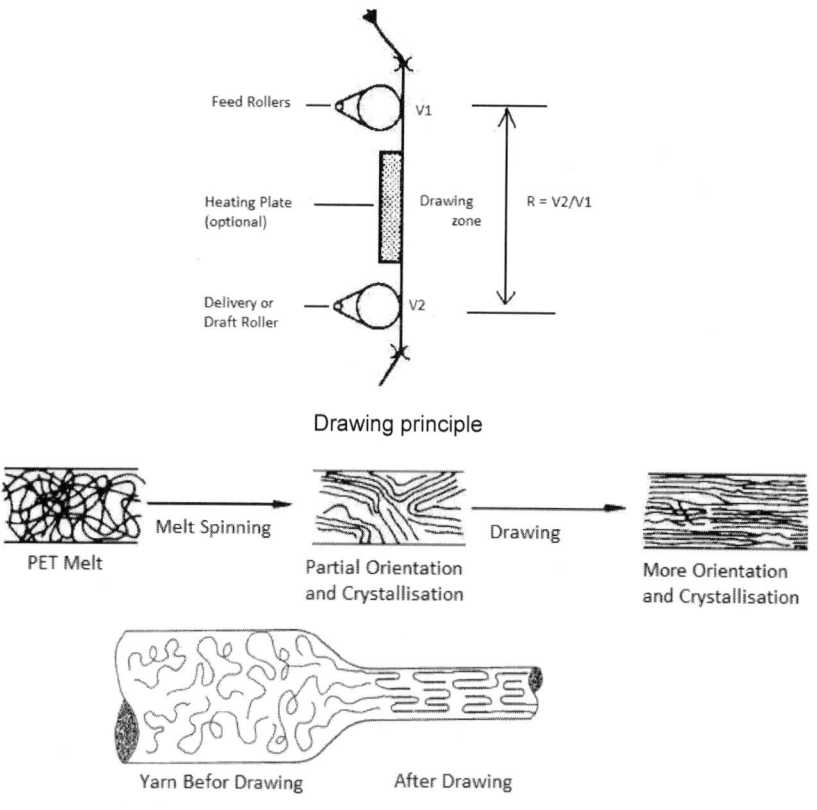

Drawing principle

Molecular orientation during drawing

PET process. However, this alone was not sufficient and the drawn yarn had an unacceptable degree of heat shrinkage. The latter defect was prevented by heat-setting the fibre by passing it over a long hot plate at about 130–140 °C, well above the effective Tg (~125 °C) of the drawn, crystallised yarn. This simple system was adequate when draw speeds were low (500 m/min), but, as draw speeds rose considerably, it was necessary to use separately heated feed rolls and draw rolls to achieve the same effect at much higher speeds. The heated rolls allowed for longer yarn contact times for thermal transfer, with the yarn wrapped several times around the roll and over an attendant idler roll. The draw ratio has a major effect on yarn elongation and tenacity. High draw ratios give high-tenacity yarns with higher yarn moduli and lower extensions to break as expected; low draw ratios give lower tenacities with higher extensions. A consequence of high-speed spinning is to shift the load supporting of the network chains of the fibre structure from noncrystalline to crystalline regions of the fibre morphology. This limits the draw ratio available to fully orient these fibres, resulting in fibres with nearly equivalent tensile properties, but significantly lower shrinkage at an elevated temperature.

In order to provide the drawn fibres with thermal stability, usually these fibres undergo also a treatment at temperature higher than drawing temperature, under controlled tensions or in a freestate, with the objective of eliminating internal tensions through readjustment of inter molecular chemical links and of the crystallization degree.

2. *Two-stage-process*
This technique can be applied low oriented yarn (LOY) or medium oriented yarn (MOY) or partially oriented yarn (POY). With the old traditional plants, the bobbins of LOY yarns were drawn in a suitable department by machines named "draw-twisters".

On these machines the yarn runs along a vertical path composed by the feeding system (with bobbins in upper position), by the draft zone and by a winding device similar to the one of a ring twister (Fig. A). During the winding on a stiff tube, the yarn is provided with a light twist originated by the rotation of the ring around the spindle. The speed of current machines can range between 600 and 1500 m/min depending on the yarn type, and the weight of the yarn packages (cops) can reach up to 4 kg; in order to increase productivity and to reduce costs, the machines can be equipped with automatic doffing device. The 1980s recorded the development of a new type of drawing frame (draw-winder), in which the winding on spindles was replaced by a take-up system on bobbins with cross-winding (Fig. B). This system permits a higher

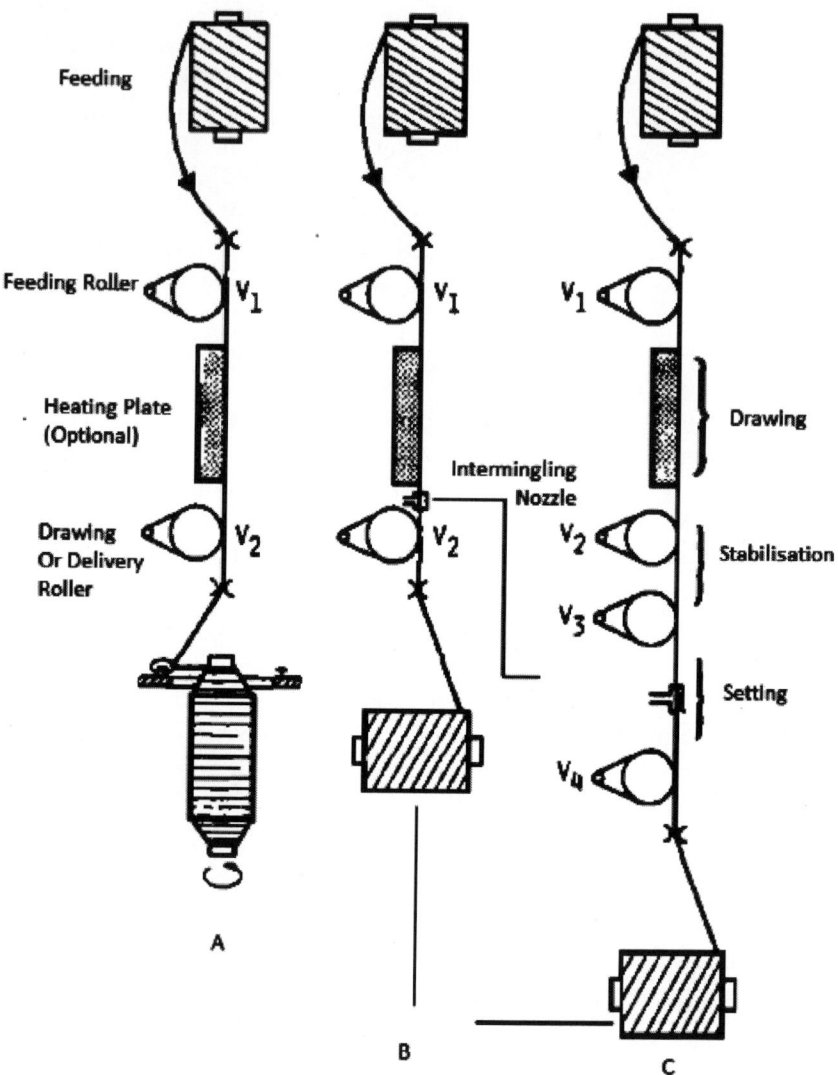

winding speed (up to 2000 m/min), the production of packages with higher weight (10–15 kg) and, from the quality point of view, of a yarn with more uniform properties thanks to a more accurate control of the variations in the winding tensions (the winding frame with spindle winding can cause tension peaks). Yarns wound on spindle present a twist which binds together and protects the filaments; on the contrary the filaments of yarns wound on bobbins are parallel so that, to make up for this deficiency, an intermingling device (a nozzle with intermittent flow of compressed air) placed before the winding device can be envisaged. For some applications as technical uses, additional cylinders (some of which heated) are positioned after the main drawing zone, for the scope of stabilizing the yarn and of fixing a prearranged thermal retraction (Fig. C).

5.1.4   Structural composition of polyester

Polyester fibres may be considered to be composed of crystalline, oriented semi crystalline and non crystalline (amorphous) regions. Stabilization distances between atoms in neighbouring molecules are usually van der Waals contact distances. The cohesion of PET chains is a result of van der Waals interactions, caused by induced dipole interaction and dispersion forces among the chains. The aromatic character leads to chain stiffness. The unusually high melting point of PET (compared to aliphatic polyesters) is not the result of any unusual intermolecular forces, but is attributed to ester linkages. The interactive forces create inflexible tight packing among macromolecules, but the limited flexibility in the macromolecule is mainly due to the ethylene group. A number of basic structural models are required to represent the different states of the fibre: amorphous (no orientation) after extrusion, amorphous (no orientation) after cold drawing, crystalline orientation after thermal treatment and after hot drawing, stretching and annealing.

5.1.5   Physical properties

PET is strong, resistant to stretching and shrinking, resistant to most chemicals, quick drying, crisp and resilient when wet or dry, wrinkle resistant, mildew resistant, abrasion resistant. They are able to retain heat-set pleats and creases and can be easily washed.

5.1.6   Mechanical properties

In polyester, an increase of molecular weight increases the tensile properties, modulus, and elongation. High tenacity filament and staple fibre curve (A and B) have very high breaking (curve E),strengths and moduli, but relatively low elongations. Partially oriented yarn (POY) and spun filament yarns, exhibit low strength but very high elongation.

Given below a typical stress–strain curve of different polyesters

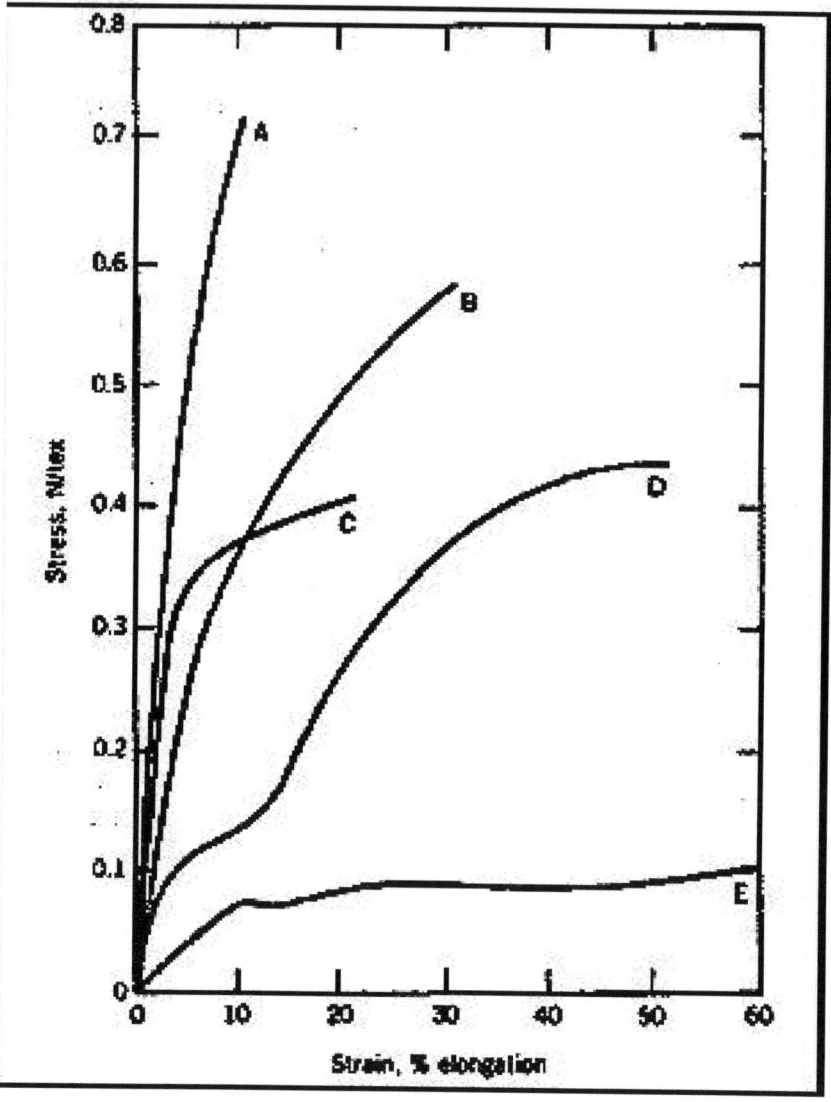

When exposing PET fibre to repeated compression (for example, repeated bending), so-called kink bands start to form, finally resulting in breakage of the kink band into a crack. If relaxation of stress and strain in the oriented fibre is allowed to occur through shrinkage during fibre manufacture, then shrinkage at the textile processing stage is reduced and initial modulus is lowered. It recovers well from stretch, compression, bending and shear because of its relatively high initial modulus.

Property	Filament yarn		Staple and tow	
	Regular tenacity	High tenacity	Regular tenacity	High tenacity
Breaking tenacity (Ntex)	0.35–0.5	0.62–0.85	0.35–0.47	0.48–0.61
Breaking elongation	24–50	10–20	35–60	17–40
Elastic recovery at 5% elongation %	88–93	90	75–85	75–85
Initial modulus (Ntex)	6.6–8.8	10.2–10.6	2.2–3.5	4.0–4.9
Specific gravity	1.38	1.39	1.38	1.38
Moisture regain (%)	0.4	0.4	0.4	0.4
Melting temperature (°C)	258–263	258–263	258–263	258–263

Reducing the molecular weight, which affects the abrasion resistance; flex life and breaking strength, results in a decrease in pilling tendency of PET fibre. The tighter the packing of molecular chains, the stiffer and more mechanically resistant the fibre is. Crimp stability of the fibre can be improved with an increase in heating temperature crimp compression of the fibre can be decreased by increasing draw ratio when the fibre is produced.

5.1.7   Chemical properties

Polyester fibres has good resistance to weak mineral acids, even at boiling temperature, and to most strong acids at room temperature. Hydrolysis is highly dependent on temperature. Thus conventional PET fibres soaked in water at 70 °C for several weeks do not show a measurable loss in strength, but after one week at 100 °C, the strength is reduced by approximately 20%. Polyesters are highly sensitive to bases such as sodium hydroxide and methylamine, which serve as catalysts in the hydrolysis reaction. Alkaline attack is sometimes used to modify the fabric aesthetics during the finishing process. The porous structures produced on the fibre surface by this technique contribute to higher wettability and better wear properties. Polyester displays excellent resistance to oxidising agents, such as conventional textile bleaches, and is resistant to cleaning solvents and surfactants. Concentrated solutions of benzoic acid and o-phenylphenol have a swelling effect. PET is both hydrophobic and oleophilic. The hydrophobic nature imparts water repellency and rapid drying. But because of the oleophilic property, removal of oil stains is difficult. Polyester fibres have a low moisture regain of around 0.4%, which contributes to good electrical insulating properties even at high temperatures. The tensile properties of the wet fibre are similar to those of dry fibre. The low moisture content, however, can lead to static problems that affect fabric processing and soiling.

5.1.8   Optical properties

PET has optical characteristics of many thermoplastics, providing bright, shiny effects desirable for some end uses, such as silk-like apparel. Recently developed polyester microfibre with a linear density of less than 1.0 denier per filament (dpf), achieves the feel and luster of natural silk.

5.1.9   Changing normal dyeing properties of polyester

Because of its rigid structure, well-developed crystallinity and lack of reactive dye sites, PET absorbs very little dye in conventional dye systems. Polyester fibres are therefore dyed almost exclusively with disperse dyes. Polymerising a third monomer, such as dimethyl ester, has successfully produced a cationic dyeable polyester fibre into the macro-molecular chain. The third monomer makes the structure of cationic dyeable polyester less compact than that of normal PET fibres. The disturbed structure is good for the penetration of dyes into the fibre. The disadvantage of adding a third monomer is the decrease of the tensile strength.

5.1.9.1   Low temperature (40 °C) dyeing polyester

This method employs a disperse dye in a micro emulsion of a small proportion of alkyl halogen and phosphoglyceride. The main advantage of this method is low temperature processing. The environmental problem that is produced by using toxic carriers.

5.1.9.2   Plasma technique

Spun bond PET nonwoven webs have been treated by (SO_2+O_2) plasma and (N_2+H_2+He) plasma. This treatment show that spun bond PET nonwovens web can be coloured by conventional water-soluble acid dyes. Plasma techniques open new avenues for colouring PET fabrics and are sure to be more evident in the colouring of polyester fibres in the future.

5.1.10   Other properties

Polyester fibres display good resistance to sunlight but long-term degradation appears to be initiated by ultraviolet radiation. Although PET is flammable, the fabric usually melts and drops away instead of spreading the flame. Polyester has good oxidative and thermal resistance and the resistance of polyester fibres to mildew, aging and abrasion is excellent. Molds, mildew and fungus may grow on some of the lubricants or finishes, but do not attack the fibre.

5.1.11   Applications

Because of their many desirable qualities, polyester fibres and fabrics have many uses.

Polyester is often used in outerwear because of its high tenacity and durability. It is a strong fibre and consequently can withstand strong and repetitive movements. Its hydrophobic property makes it ideal for garments and jackets that are to be used in wet or damp environments –coating the fabric with a water-resistant finish intensifies this effect. Polyester is used in pants, shirts, suits, and bed sheets either by itself or as a blend, because of its wrinkle-resistant property and its ability to retain its shape. Since these garments are frequently worn and washed, its stain-resistance and durability are also desirable.

Since polyester can be moulded into almost any shape, certain insulating properties can be built-in to the fibre. One method is to create hollow fibres. This process traps air inside the fibre which is then heated by the body. The warm air stays inside and helps warm the body in cool weather. A second method is to use crimped polyester in a fibrefill product. The crimp helps to keep in warm air.

5.1.12   Different variations of basic polyesters

Different fibres can be created by doing one or more of the following:

5.1.12.1   Adding additives (delustrant dulling agent, dyes)

Additives have extended the use of PET fibres into areas where the original commodity products had deficiencies, in, for example, soil-resistance, static protection, brightness, dyeability, etc. The techniques for the feeding of additives (dyestuffs, dulling agents TiO_2, polymer stabilisers) can be schematised as follows:

- addition of additives in solid state or in form of masterchips to the granules of the basic polymer during the feeding phase of the extruder; metering is carried out with a volumetric or gravimetric system, mixing takes place inside the extruder and usually continues in the melted state in a subsequent mixer.

- system for the injection of additives in melted or liquid state into the flow of melted polymer; the additive in melted state is obtained by masterchips treated in a separate (secondary) extruder, while the additive in liquid state is prepared inside a tank.

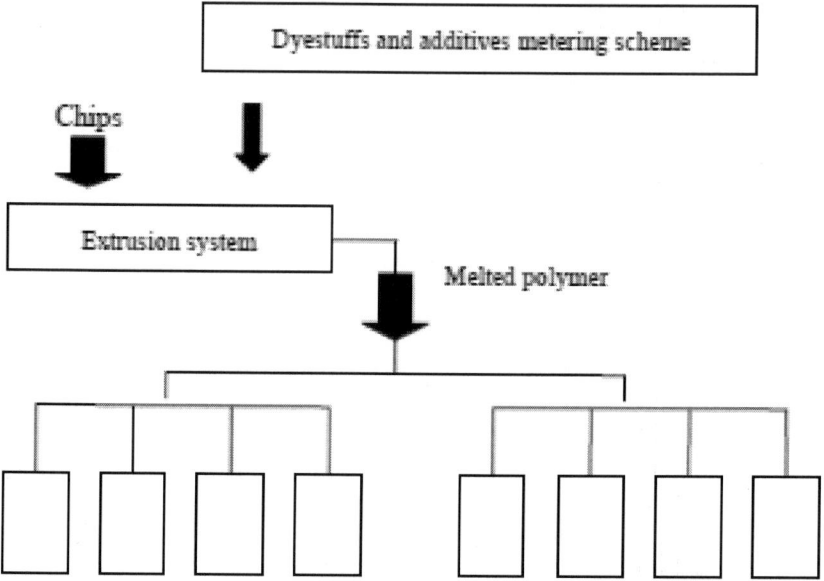

Metering is ensured by metering pumps and the injection points are placed on the main extruder or on the main spinning line, or even right before spinning pumps. The technique of additive feeding near the single spinning positions is increasingly widespread as, besides ensuring flexibility to the plant, it reduces the waste caused by frequent lot changing due to small productions and improves the quality of the fibre, owing to its reduced stay time in melted state (thermal stress).

5.1.12.2 *Mass colouration (spin dyeing)*

Normally polyester is dyed at higher temperature where by the disperse dye enter the amorphous portions of the fibre and gets entangle there when the material is cooled. The same process can be done during spinning process whereby the finely distributed dye pigments, which can withstand the thermal and chemical demand of the spinning process are added in the spinning solution/melted polymer and spun and on cooling after the fibre formation it gets dyed. Generally used for mass production, the dyeing process which is also called Jet Dyeing is used for upholstery fabrics, carpets, automobile articles and sometimes for menswear, work clothings, etc. There are various methods used for this:

1. Continuous process: dye is introduced to the already formed polymer (solid or molten state). Dye in powder or granule form is continuously distributed using dosage pumps and then mixed with polymer granules or chips, which supply the melting equipment (extruder or fixed melting furnace), or directly with molten polymer. Process is simple and flexible, but results in great variations in bright colours. Dyes (liquid or in paste form) are dispersed in binders (polymers of different molecular weights or polyethylene glycols of high molecular weight) and are injected into the molten polymer using additional extruder, which directly supplies the main extruder, in which polymer granules are melted.

2. Discontinuous process for polyester: in a unit of production for the manufacture of polyester granules, the polyester chips are dried in a vacuum for several hours at 140–180 °C in rotary driers or tumblers and at the end of this process specially selected dyes are added; they are fixed on polyester granules via thermo-diffusion. Suitable process for the manufacture of small quantities of spin-dyed fibres/threads.

3. Discontinuous process for polyamide: Polymer granules are simultaneously washed and dyed in the aqueous medium. After several hours drying at 180 °C, it is melted under the same conditions as colourless granules. Granule dyeing is carried out in a machine with bath circulation. It is indeed difficult to determine the cost difference between spin-dyeing and conventional dyeing even with only 10% accuracy; but it is estimated that the spin-dyeing of polyamide and polyester is 20–30% cheaper than dyeing according to normal processes. In addition, spin-dyeing has ecological advantages as well as simplicity and precision of technology. However, this type of dyeing cannot react to fashion trends in the short term and is therefore chiefly used for standard articles.

5.1.12.3 Changing cross sectional shape of the fibre

The simplest and most common shape is a circle, but by changing the shape of the spinneret, square, oval and bean-shaped fibres can be formed. As the polyester was introduced in the market it was accepted very well but slowly the negative characteristics of the fibres has surfaced. Thus the acceptance was reduced continuously. Researchers and manufacturers continuously looked at the methods to regain the acceptability. Comparison was always done with the natural fibres like cotton, silk, wool, etc. One of the apparent differences was the cross sectional shape of these fibres – cotton was dog 'bone' shaped, silk was triangular, wool is irregular. Some qualities are partially attributed these shapes and trials were taken to change a melt-spun fibre from circular cross-sectional shape to non-circular cross-sectional (NCCS) shape by varying the shape of spinneret orifice, provided the melt viscosity is high enough so that surface tension does not cause the filament to resume a circular shape to change cross-sectional shape of these fibres. Since the spinnerette holes were extremely small (0.015 in.) it was an engineering task to change the shape of the holes. Introduction of laser etching technique made it possible. The holes of the spinnerettes need not be the same as the required cross section and various trials were done to find the spinnerette shape for a required shape. For e.g., a hole shaped like a T gave trilobal filaments. Multilobed yarn cross sections (trilobal and octalobal) can give quite different appearances. Trilobal is glittery as the incident light reflects off the fibre surface, while octalobal gives an opaque matte effect, as the light is effectively absorbed by multiple reflections from the many acute angles. Sharp-edged filaments have the prized rustle and high frictional characteristics of pure silk, where it is called "scroop." Flat rectangular filaments give fabrics an unpleasant "slimy" handle. Gradually, these principles were applied to commercial yarns, and many filament yarns for the apparel and carpet markets now use NCCS fibres. One can even create a hollow fibre. In some applications it is desirable to generate increased bulk without adding weight, e.g. for insulation or padding. A solution is to make fibres that are larger diameter and stiffer, but with internal voids to reduce weight. By extruding hollow fibres, weight reductions of 20% or greater can be achieved – a considerable advantage. Foaming agents in the polymer (e.g. dissolved CO_2)have also been used to generate microporous, lightweight fibres. The different shapes affect the hand and strength of the fibre.

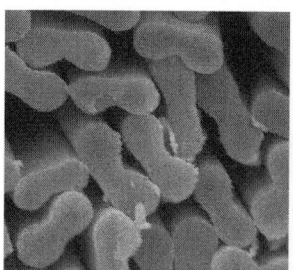  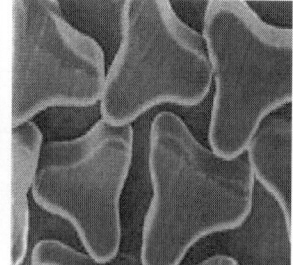

'Dogbone' CS   Multilobal CS Trilobal CS

· 1. Drawing out the fibre to five times its original length is normal, but polyester can be stretched even further. Drawing it out more than normal may also affect the strength, elasticity and dye-ability.

2. *Adding dye stuffs*

In its natural state, polyester is a slightly transparent off-white. Adding dye stuffs at the manufacturing stage can create brilliant colours like electric blue and atomic red.

5.1.12.4 Hollow fibres:

Hollow fibres are a type of core–sheath heterofil in which the core is composed of air. They are usually made in the form of crimped staple fibre and spun from a modified staple spinning pack. With advances in melt-spinner design, more complex geometries, are available. One patent describes a multiple-core circular polyester staple fibre with seven cores. Hollow fibres in the form of filament yarn have specialised uses in medical devices, but the largest volume market for hollow fibre is staple fibrefill for pillows, duvets, quilts and thermal outerwear. The desired quality here is "loft" and it is better if the fibre is light and bulky. The most desired quality is thermal insulation. Hollow polyester fibres are well suited to this end use: the air entrapped in the hollow cores adds significantly to their insulating properties; more hollow air cores increase this effect. The fibres are frequently crimped by a stuffer-box process or made as bicomponent hollow fibres, which develop spiral crimp on drawing. Such filler fibres are frequently treated with a permanent polysiloxane finish that makes them slippery, so that they slide easily over each other and resist clumping up, which reduces their insulating effect. In this form, they compete with goose down as a thermally insulating filling material. For end-uses like fabric interlinings, it is desirable to stabilise a filled hollow fibre structure by incorporating an additional thermally bondable bicomponent fibre.

5.1.12.5 Texturised fibre or crimping

When the fibre is drawn out, it is long and smooth. There are many deficiencies for this flat yarn. Crimping can give the fibre more texture and bulk, elasticity, soft touch, 'natural' aspect, hygroscopicity and can increase its insulation properties, as well as easier processing. Another

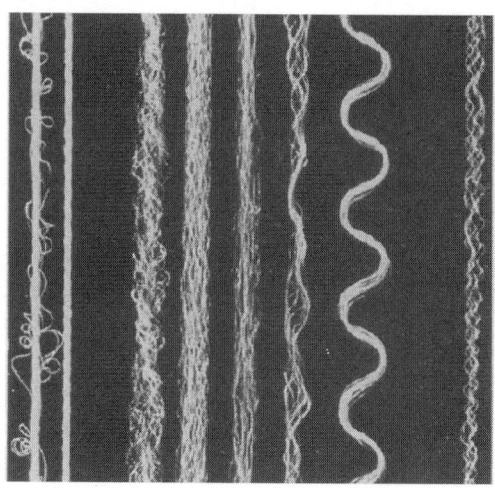

Some Texturised yarns

word for this is texturising. Flat yarn, whose structure can be compared to iron wires, gains properties which make it similar to a traditional cotton or woollen yarn. A textured yarn can be defined as "a yarn characterised by effective or latent crimps, waves, loops of single filaments which, after suitable treatment, can originate properties as bulkiness and/or elasticity".

Many methods are followed by manufacturers from the time texturised yarn was made available in the market, of which some methods are more popular and practiced than the other process, which are explained in short here.

Following scheme shows a general classification of the various processes.

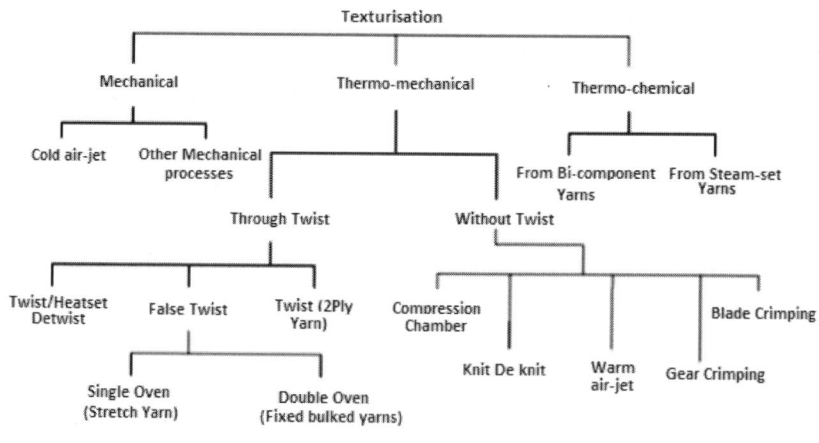

Various texturising process

Textured yarns can be grouped into two according to their manufacturing processes.

With the twist-texturing process following yarns can be obtained:

- *Set yarns (bulked yarns)*: This yarn will have only low crimp elongation and low crimp contraction; in the case of false-twist process, they are produced with two ovens and are named FTF (False-twist Fixed) yarns.

- *Highly elastic yarns (stretch yarns)*:They are characterised by high crimp elongation (extensibility) and high crimp contraction; in the case of false-twist process, they are produced by means of an oven and are named FT (False-twist) yarns.

5.1.12.5.1   Mechanical processes

5.1.12.5.2   Air-jet method

By this method a jet of a pressurised air directed through a nozzle with a certain angle and turbulence level against a yarn going through the nozzle; the yarn is overfed in order to create a compression which has an intensity defined by the ratio between feed and delivery speeds and which affects the typology of the loops generated in the yarn. In the process first the filaments is opened and then a whirl is created on the opened yarn. Then loops are created due the over feeding of the yarn which tend to come out of the yarn core. Finally the entangling and setting of the crimped wavy filaments takes place. Initially process speeds of 50 m/min required 20 m³/h of compressed air (which involves high energy consumption), while nowadays speeds of 500–600 m/min are attained with consumptions of compressed air (8–12 bar) limited to 5–6 m³/h.

 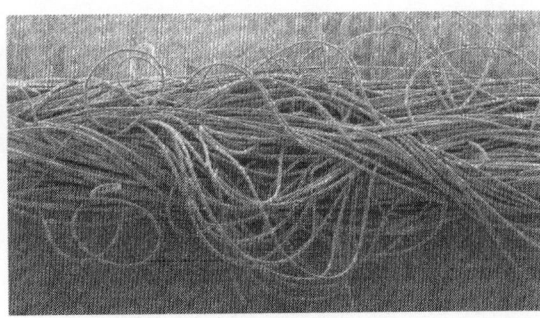

Air jet texturizing nozzle   Air jet texturized yarn (Enlarged)

5.1.12.5.3   Thermo-mechanical process
Thermo mechanical process without twist

This is based on bending, curving, waving systems which produce yarns with different bulkiness/elasticity properties, but without any tendency to rotate on their own axis and not on twist deformation principle.

There are different methods

1. Mechanical compression chamber method

In this method the yarn is forced, by 2 feeding rollers, into a heated crimping box and is submitted to a compression which generates planar bending points (saw-teeth type) on the yarn.

The pressure inside the chamber is regulated by a device which permits the delivery of the yarn at a constant pressure.

2. Air-jet compression chamber method

Air-jet compression method is a further development of the mechanical compression chamber method. Yarn compression is made by aerodynamic system with jets of heated compresses air or steam. The warm gaseous stream under pressure enters into a lateral opening of the box, flows into the yarn guiding channel, pushes the yarn into the expansion box where it is brought to a softening temperature (plasticization) and is compressed; the result on the yarn is a three-dimensionally shaped bending. The exhausted air gets out of leaks placed at the extremity of the box and the yarn is pushed out of the box. To permanently set the texturised yarn, it is quickly cooled on perforated drums with air suction.

3. Knit-de-knit method

This discontinuous method uses loop formation during knitting in texturisation. The knitting is done on single-feed small diameter circular machine; The knit fabric is thermo set by hot air or steam and then the yarn is unraveled and wound on cones at a speed up to 500 m/min. This method gives a wavy crimp of same shape and size as the stitch repeated throughout the length of the yarn. Used for specific purposes.

Two other methods named Blade and Gear methods are not much in practice now.

Thermo mechanical process with twist

One of the oldest method based on the thermosetting of a yarn after twisting and subsequent de-twisting but undergone many changes to solve non permanency of the textured effect. Presently it involves three discontinuous processes:

- Twisting on a twisting machine in 1 or 2 stages with a high number of turns (twists between 2500 and 4500 t/m, i.e. inversely proportional to the linear mass of the yarn).

- Twist setting on bobbin by thermal treatment, usually in autoclave with saturated steam and vacuum-steam cycles (130 °C temp. for nylon).

- De-twisting on a twisting machine, in a first stage zeroing the existing twist and successively applying a further slight twist in order to improve yarn stability and regularity.

As a result of the thermoplastic deformation caused by twisting, the single filaments composing the processed yarn show in the relaxed state a special/helicoidal waviness which has same direction as the imparted twist (the applied slight final twists are in fact aimed at reducing the yarn trend to rotate on its own axis). This process resulted into yarns with excellent elasticity (for use in hosiery), but some end-uses, in particular polyester for knitwear, a stabilisation of elasticity was necessary; to this purpose a further thermosetting stage was applied on the yarn, which was wound on bobbins with a present tension (stabilised yarns or "set" yarns).

This method is not very popular due to the numerous operations involved, to its low productivity and to consequently high production costs.

5.1.12.5.4   False twist method

If we imagine that a twist imparted on a yarn which is stationary, one can see that both sides of the twisting element impart twist in the opposite direction and when the twist force is removed the twists get de twisted and the twist is set to zero. Or in other words the opposite twists get de twisted and comes back to original state. In case of a moving yarn, the yarn coming out of the twisting element will have same configuration as the fed yarn.

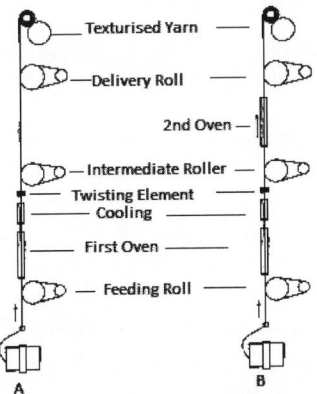

False twisting type texturisingmachine (schematic) A – with one oven, B – with two ovens

Now, suppose a yarn in motion is continuously twisted by a twisting element and brought into contact with a heating medium (oven) and is successively cooled (cooling zone), the filaments composing the yarn undergo a thermoplastic deformation of twisting and permanent nature, which is absolutely similar to that of previously described discontinuous process. If we wish to stabilise yarn elasticity, we shall need to place a second oven after the twisting element.

There are many types of twisting elements. Some are described below:

Hollow spindle type

It is a hollow tube with a pin inside which is fixed crosswise which can turn on bearings. Yarn is wound on this pin through the hollow spindle with a full rotation of 360°. The spindle is rotated around its own axis entails the insertion into the yarn of a false-twist which moves in length direction along the yarn. The present machines with magnetic bearing can insert up to 900,000 turns per minutes which accounts to a yarn speed of around 200 m/min.

Friction disc type

The system consists of a series of disks mounted on three spindles, which have their axes spaced out in such a way, that the disks result partially super imposed; the running yarn gets twisted by its contact with the disks, as it performs a helicoidal evolution around an imaginary cylinder generated by the superimposition of the disks. The false-twist aggregate is composed of a disk package placed on a three-axial system, in which the first package at yarn entrance and the last one at yarn delivery simply have the function of positioning the yarn (guiding disks) without originating any friction, while the inside disks exert a torsional friction force on the yarn (working disks). The effectiveness of the system depends on aggregate configuration, on geometry and material composing the disks, on spindle gauge and on number of twists applied. Particularly important is the type of material used for the disks: it must ensure on one hand a good grip and on the other a good abrasion resistance.

5.1.12.5.5   Thermo-chemical processes

There are mainly two types of thermo-chemical process. In one process bi component yarns of either side by side (s/s) or a skin-care (s/c) configuration with different thermal shrinkage is given thermal treatment which gives a three dimensional crimp. In the other process uses bi retractable yarns composed of two groups of yarns with differential thermal properties are made to undergo thermal treatment and texturisation is affected. This method is not much used for polyester.

5.1.12.5.6 Draw-texturisation

Once spinning machine with higher speeds (to 4000–5000 m/min) were introduced, a partially drawn yarn (POY) became available, giving a substantial contribution to the expansion of the false-twist texturisation technology. In fact this yarn has excellent extensibility during texturisation, thus permitting to complete the drawing on the texturing machine (drawing-texturisation). The drawing operation (R= 1.2–1.7) can be carried out by means of rollers or drawing shafts during the first stage before texturisation (sequential process) or directly in the texturing zone itself (simultaneous process). This last system is at present the only one used both for PES.

5.1.12.6 Antistatic and anti-soiling fibres

Static behaviour of the polyester due to low moisture regain was another reason for the non-acceptance of polyester in the markets. Synthetic fibres in general, and polyester in particular, are hydrophobic materials—polyester has a moisture regain of 0.4% at 60% RH. Polyester fibres are difficult to wet and rapidly build up static electrical charges by friction because as water effectively leaks away, voltage is produced. It is possible to build up potentials as high as 50 kV by rubbing a polyester fabric. Static charges also lead to attraction of dust and dirt. To avoid these problems, the moisture uptake of the polyester should be increased by combining it with hydrophilic materials that are wash-fast. One additive that has been used repeatedly is polyethylene glycol (PEG), a stable, functional, highly hydrophilic, water soluble, and humectants polymer:

$$HO-CH_2CH_2O[CH_2CH_2O]_n-H$$

An addition of PEG with polyester was possible since it does not drastically affect the PET properties but greatly improve the fibre moisture uptake, but at the expense of severe reduction in the light stability of dyed fibres. Other processes used a PET=PEG block copolymer in aqueous dispersion that was padded and baked onto the fibre as a textile finish. This relied oncocrystallisation of the PET segments with the polymer to make the treatment wash-fast. The most satisfactory technique is probably to make a bicomponent fibre with a thin coating of a PET=PEG copolymer on a PET core in a core–sheath configuration. This does not affect fibre properties and minimises the light fastness issue.

It is also possible to add PEG at the last stages of polymerisation, so that much of the additive remains agglomerated in a separate phase. In this case, subsequent scouring of the fibres removes much of the soluble PEG, leaving microscopic voids in the fibres. These voids can act as reservoirs for moisture, and will also decrease light transmission and increase the reflectance properties of the fibres.

5.1.12.7   Cationically dyeable polyester

Cationic dyeable polyester fibre is prepared by incorporating an additive containing sulphonic acid groups in the polyester chain during the polymerisation process. Many additives are recommended in the literature such as 5-sulphoisophthalic acid, m-or p-phenyl sulphonic acid, 2-naphthol 8-sulphonic acid, hydroquinone sulphonic acid, sulphonated pyrrole, alkylbenzene sulphonic acid, 5-sodium sulphoisophthalic acid, etc. However, its polymerisation technique is quite tactful and requires special skill.

Dyeing mechanism (D+dyestuff cation, X–anion, ∿PES components)

Advantages

1. The low temperature dyeing can be used to dye blends with elastomers, which would likely suffers from thermal degradation while dyeing at 130 °C with normal polyester.

2. Excellent colour fastness properties.

3. Two-colour effects can be produced on fibre blends of cationic dyeable polyester and normal polyester.

4. The depth of the shade against normal polyester will remain same.

5. Has less pilling as compared with normal dyeable polyester fibre.

6. All other properties of the normal polyester will be retained.

7. Cationic dyeable polyester is useful in minimising the problem of frosting in polyester/cellulosic blends, dyed in contrast shades. Frosting occurs because of the difference in abrasion resistance of polyester and cellulosic fibres. The cellulosic fibres have much lower abrasion resistance than polyester. Hence, during wear, part of the cellulosic portion is removed, leading to patchy dyed appearance at that particular portion. In order to minimise frosting, cationic dyeable polyester is used in combination with regular dyeable polyester and cellulosic fibre. Thus, a blend containing 33% CDP, 33% PET and 34% cotton is prepared. By dyeing disperse dyeable polyester to

lightest shade, cationic dyeable polyester to deepest shade and cotton portion to an intermediate shade the frosting can be minimised.

Cationically dyeable PE versus normal polyester

It contains $-SO_3Na$ group in polymer. Therefore, it is dyeable with cationic dyes. It contains higher DEG (di-ethylene glycol) units. Therefore, it dyes darker with disperse dyes (as well as with cationic dyes) due to lesser crystalline regions in yarn. They have lower intrinsic viscosity but melt viscosity is not so low than being expected from the intrinsic viscosity, based on normal polymer's Melt Viscosity *vis* Intrinsic Viscosity relationships. The number of TiO_2 agglomerates is more and size of TiO_2 agglomerates are larger. The polymer is less crystalline and more amorphous. Their weight reduction possibilities with hot caustic soda solution are higher. Cationically dyeable PE will have lower tenacity of yarns and its wash and wear and crease resistance is poorer but crimp permanency is lesser.

5.1.12.8   Others

As for improvements in dyeing properties, modified polyester has been introduced that contains ionic sites to facilitate ionic dye attachment as well as the use of a copolymer that lowers the compact structure of polyester. To increase the aesthetic quality of textured yarns, high refractive index inorganic particles have been incorporated into fibres and silk-like polyester fibre have been developed. However, although these modifications enhanced the lustre of polyester, softness in handle was lacking. During the 1970s and 1980s, the 'touch' of polyester fibre was enhanced; alkali de weighting treatments were used to make the fibre more delicate and a wrinkle finish imparted it to an appearance similar to that of silk. In addition, there have been many other activities in fibre development, such as polymer modification, fibre blending, surface treatment, the mixture of various fibre cross sections, special spinneret design and fine denier spinning for polyester and silk-like fibre developments.

A flame-retardant polyester fibre contains, e.g. 0.6% of phosphorus as a co-monomer which is polymerised in the spinning mass. The essential difference of a fibre modified in this way, in comparison to a conventional polyester fibre, is its self-extinguishing property. But it should be noted that, the post processes like pretreatment, dyeing, finishing, etc. has to be designed in such a way as not to affect the effectiveness of the retardant present in the fibre so as to retain the fire retardant property in the finished product.

5.2   Processing of polyester materials

5.2.1   Polyester process chart

Given below the general process finishing chart for polyester (woven and knitted).

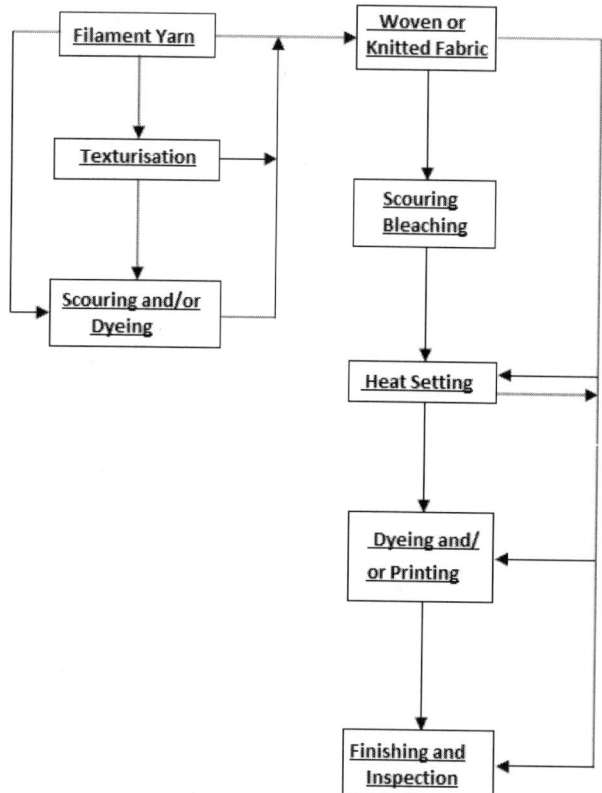

5.2.2   Heatsetting

Only by heat-setting do fabrics made of partially crystalline synthetic fibres acquire the dimensional stability, crease resistance and resilience desired in use. Heat-setting is usually indispensable for ensuring satisfactory behaviour of the material during other finishing processes and is itself, therefore, one of the most important finishing processes employed with materials containing polyester fibres or their mixtures with other fibres.

Stresses are set up within the fibres during their manufacture, e.g., during the spinning process and subsequent stretching, owing to the re-formation and re-orientation of the crystalline regions. In addition, further stresses are introduced as a result of the new shape imparted to the fibre during subsequent processing such as yarn spinning, weaving or knitting, and these attempts to assert themselves when the opportunity arises. These latent stresses lead to shrinkage during subsequent wet and dry heat treatment, e.g., during washing, dyeing, drying, ironing or pleating. Piece goods which have not been heat-set also tend to form folds and creases, and possess a poor handle. If sufficient heat energy is introduced into such a fibre or group of fibres so that some inter molecular linkages are broken and re-formed, stresses of the type described can be relieved. Usually, the material is in a tension-free state during the heat treatment so that the groups of fibres can shrink. After heat treatment has been completed, the newly acquired state should be "frozen-in" as quickly as possible so that it can be maintained indefinitely. Properly set polyester fibres possess very little residual tendency to shrink.

Mainly woven or knitted piece goods are heat-set, but single or double yarns are also treated in this way. The latter are heat-set or steamed to fix the twist. Yarns which have a tendency to shrink are, if necessary, pre-shrunk in hot liquor or by steaming before being dyed in the form of cross wound packages. Special effects can be obtained by heat-setting under special conditions, as for instance, in the manufacture of high bulk yarns, texturised yarns, safety belts and sewing threads. Polyester fibre-containing woven or knitted fabrics are mostly heat-set on a pin-stenter.

The treatment can be considered as an equivalent to permanent press finishing of cotton fabrics. Heat-setting, is normally carried out before dyeing. Pre-setting makes it possible to employ the very high temperatures required for obtaining dimensional stability and crease resistance without having to consider the fastness to sublimation of the dyes. At the same time, pre-setting also has a favourable effect on dyeing behaviour and, because of the acquired shape and dimensional stability, on the running properties. Setting can be combined with other processes like thermosoling, optical whitener treatment, etc. also. With polyester in particular, post-setting as a final finish can produce particularly high dimensional stability or a given type of handle.

Setting has to be done at a higher temperature and after the heat treatment the material has to be immediately cooled to retain the setting as otherwise any creases or folds formed before cooling can be undesirably set on the goods. There are chances of free shrinkage of the fabric, especially where the material is stretched and set. Setting can be done by treatment with hot air, heat treatment in presence of water (hydro setting), or high temperature steaming.

5.2.2.1   Hot air setting

Woven and knitted synthetic materials are usually heat-set on a pin stenter. Hot air is used as the heat transfer medium. The release of latent tensions in the material by heating leads to considerable shrinkage, the extent of which depends on the kind and type of fibre and the weave structure. The kind and extent of these changes in shape must be known so as to obtain the optimum results on the material by appropriate adjustment of the stenter. The cloth "relaxed" by hot-air treatment must be cooled down as quickly as possible to effect stabilisation. Usually a cooling zone is sufficient. Sometimes the heated fabric is led over a cooling drum. Polyester fabrics are almost exclusively set with hot air. Depending on the source of the fibre, weave of the fabric, and the kind of machinery employed, the temperature ranges between 150 and 220 °C, the time being 20–30 s; or with simultaneous fixation of the dye, as long as 60 s. The lower temperature range is used more especially with knit goods made of texturised filament yarns so as to preserve the crimp effect. The lower temperature region is enough to obtain dimensional stability of the goods, whereas stabilisation against crease formation is only obtained in the upper range.

Hot air is usually employed and is directed from above and below by jets onto the material. A controlled lengthwise and width wise shrinkage is possible on this machine; the width of the frame and the overfeed can be adapted to the shrinkage to be expected. To ensure a good flow of hot air between the selvedge and the pin chain, and to avoid an impression of the pin-bed on the edge of the material, it is recommended that use be made of hook shaped pins with a thickened base and "Quenching" of the material is effected in a cooling zone by blowing on cold air.

Setting of the material can be achieved without controlled width wise shrinkage by passing it through a heated chamber of the hot-flue type. The material can also be treated on a roller setting machine without temples. However, weft wise shrinkage is somewhat diminished by fluting the rollers. The fabric is led over heated metal cylinders (contact heat-setting) in such a way that alternating sides are subjected to heating.

5.2.2.2   Hydrosetting

Though hot air is mostly used for heat-setting, hydro setting is of some importance, especially with polyamide materials, on account of the fuller and softer handle obtained. In certain cases, texturised polyester goods are also hydroset. Hydro setting is effected with water at 125–135 °C in autoclaves. The hot liquor is allowed to flow for 20–30 min through the goods from within outwards. Hydro setting can be combined with dyeing or optical brightening of the material. In hydro setting done on beam dyeing machines, it should be noted that the control of width wise shrinkage is minimal. Setting in rope dyeing machines is not recommended as the creases, rope marks, etc. will get set. It is a general rule that the material has to be set in the flat form unless a particular effect (e.g., pleating) is required by setting.

Hydrosetting of woven materials has to be done on flat form without creases, as any creases formed during the setting operations remain as permanent creases unless it is again set at higher temperature after removing the creases. Hence usually this operation is done on beam dyeing machine or may be in HT jiggers. But in case knitted goods n Jet dyeing machines, the running folds are displaced so quickly that they can hardly become set and hence can be used. Indeed, with many types of material, it is possible to wash, dye and set them in these machines.

5.2.2.3   Steam setting

The effects obtained with steam-setting lie between those of hot air and hydro setting, depending on whether saturated steam (similar to hydro setting) or super heated steam (similar to hot-air setting) is used. In order to achieve the same temperature throughout when steaming large rolls of cloth, it is recommended that, after steaming for 2–3 min at 130–132 °C, the steam be stopped, vacuum applied, and steam again pressed through for some minutes at 130–132 °C. This procedure is repeated several times to remove air as far as possible. After this, polyester materials are steamed for 20–30 min at 2.8–3 bar (1.8–2 atm gauge), this corresponding to a temperature of 130–132 °C. After steaming, cooling proceeds for 5 min in vacuum to set the goods. Fabrics can also be set with superheated steam on the pin-stenter. The steam is blown in via a heat exchanger and has the same effect as hot air. Compared with setting with hot air, the fixation time needed for superheated steam is about 25% less because of the higher rate of heating up the material. In addition, the danger of yellowing is much less with steam-setting. But this type of setting is rarely practiced.

The most universally practiced method is hot air – pin stenter setting which is also more versatile and easy method. Heat setting may be done before or after scouring depending upon the extent of soiling. Soiled goods must be scoured before heat setting. Polyester filament yarn are headset by steaming for 15 min at temperature 5–10 °C higher than dyeing temperature, to prevent shrinkage during dyeing.

The heat-setting of polyester woven or knitted fabrics has the effect of:

(a) Imparting better shape and dimensional stability, i.e., shrink resistance to both warp and weft or stability of loops;

(b) Imparting a better resilience;

(c) Modifying the handle;

(d) Modifying the dyeing properties; and

(e) Reducing the tendency to pilling.

The material becomes stiffer with increase in setting temperature and time, but this effect can be removed by a subsequent wet treatment. Since all the heat-setting effects mentioned are modified by slight differences in treatment, it is essential to maintain uniformity in the heat-setting process. Polyester fibres are set within a few seconds at around 200 °C, but it must be remembered that some time is required to bring the materials to the setting temperature. Damp material dries at different rates at different places so that it would be unevenly set under high temperature conditions. Efforts should therefore be made to ensure that a material is dry before being subjected to heat-setting.

Woven and knitted fabrics which are to be coated, laminated or used as interlinings must be pre-set and pre-shrunk with particular care so that their shape remains unaltered. The shrinkage which occurs during wet finishing and heat-setting must be allowed for in calculating the width of the material on the loom or knitting machine. Working under excessive tension to prevent shrinkage, or a subsequent stretching of the material in order to achieve a prescribed width, can only be done at the expense of quality.

5.2.2.4   The point at which the material should be heat-set

In principle, polyester material can be heat-set either before or after dyeing. With the thermo sol dyeing process, the fibre is heat-set at the same time that the dye is fixed.

In some cases it may well be necessary to heat-set a given article more than once in order to ensure a correct finish or a desired effect.

5.2.2.5   Presetting

If heat-setting precedes dyeing, the material is first scoured although this necessitates intermediate drying. Though the heat-setting of unscoured material is more economic, there is the danger that impurities and spinning preparations will be "burnt in". It has already been mentioned that the dyeing properties of polyester fibres change after a heat treatment. Polyester fibres which have been heat-set with hot air at 160–180 °C are dyed noticeably weaker by many disperse dyes than are non-set fibres or those which have been set at an even higher temperature. If the article permits, it is advisable for presetting to be carried out at a temperature of at least 190 °C in order to obtain a good colour yield.

It must be pointed out that fluctuations in pre-setting can lead to unlevel dyeing. The tendency to reveal differences in setting is strongest when the subsequent dyeing is carried out at the boil in the presence of a dyeing accelerator (Carrier). Fluctuations in heat-setting are not so marked when the HT process is used, and are least apparent after the Thermosol process. Usually, however, it is possible to suppress the effects of fluctuations in setting by a suitable selection of dyes.

Polyester fabrics are normally present if it is intended to subject them to a wet treatment at an elevated temperature in rope form (winch, jet, soft flow) or in batch form (in a beam dyeing machine). While finishing the material in rope form gives rise to the danger of considerable crease and fold formation, the processing of a non-set material on a beam often produces a moire effect on account of shrinkage in the hot liquor, and this also results in a very much diminished liquor circulation. Polyester knitwear, texturized material, generally needs to be pre-set before being subjected to HT dyeing on a beam at 120–130 °C only if tension differences due to knitting have to be eliminated. If dyeing is to be carried out on the jigger, it is not absolutely essential to pre-set the material but it is advisable, since it eliminates any tensions within it that could lead to faults during processing and to a non-uniform appearance of the material. Heat-setting of the material before dyeing actually has the advantage that no consideration need be given to thermal stability when selecting the dyes. There are, however, exceptions in those cases where garments have subsequently to be subjected to intensive heat treatment, e.g., pleating or resin finishing.

General guide for heat setting conditions

Type of fabric	Temperature of hot air(°C)	Time (s)
Polyester staple	220–225	20–30
Polyester continuous filament	210–215	30
Polyester fibre/cotton	195–205	30
Polyester fibre/viscose	180–190	30
Polyester/wool	170–190	30
Polyester staple	190–200	30
Polyester continuous filament	180–190	30
Polyester fibre/cotton	185–190	30
Polyester fibre/viscose	170–180	30
Polyester/wool	170–180	30

5.2.3   Pretreatment of polyester

5.2.3.1   Scouring

Since the polyester as such is a strong yarn sizing, etc. are not used much. But winding oils, lubricants and in case of knitted materials, knitting oils, etc. maybe available on the yarn/fabric. Hence the goods should be scoured to remove extraneous impurities added during the manufacturing process. The usual method for this is to treat the goods in alkaline detergent solution. The concentration of the chemicals and severity of the treatment depends upon the extent of oiling. For heavily soiled goods stronger alkali like sodium carbonate should be used. Lightening and any other coloration is removed by adding sodium hydrosulfite and caustic soda to the scouring bath.

Recipe

Additions	Unit	Light-moderate soiling	Heavy soiling	Tinted or any coloured material
Scouring and dispersing agent	g/l	0.5–1.0	0.5–1.0	1
Trisodium phosphate	g/l	1*	1*	
Tetra sodium pyrophosphate	g/l	2*	2*	
Ammonia conc.	ml/l		2	2–3
Soda ash	g/l		2	2
Caustic soda 36 °Be	g/l			2
Sodium hydrosulphite	g/l			1–5
Grease solvent	g/l		2–5	
Temperature	°C	50–70	80–90	80–95
Time	Min	15–30	60–90	90

*Either one to be used.

After scouring the material,it is well rinsed in cold water till free from alkali. If caustic soda or soda ash is used in scouring and the goods are to be dyed at high temperature, it is suggested to neutralise them by treating in a bath containing 1–2 ml acetic acid (30%) per litre.

5.2.3.2   Bleaching

Usually, polyester fibres possess very little inherent colour when supplied. If, however, the degree of whiteness is insufficient, the fibre can be chemically bleached or optically brightened. Where the degree of whiteness required is very high, both treatments can be applied.

Only sodium chlorite is suitable for bleaching polyester fibre, other bleaching agents not being sufficiently effective.

Recipe

Quantity	Unit	Additions
1–2	g/l	Sodium chlorite
0.8–1.8	g/l	Stabiliser for chlorite
1–2	ml/l	Formic acid 85% (pH 3–3.5)

The goods are entered at 40–50 °C into the liquor set with the above ingredients, the temperature then being raised to the boil within 20–30 min. Bleaching proceeds at the boil for 30–45 min. Warm and cold rinses are then given. The well-known precautions that apply to the use of sodium chlorite (chlorite-resistant machinery, good ventilation and the like) should be borne in mind.

5.2.3.3   Optical brightening

In some cases, polyester fibres are already brightened during the fibre spinning process, e.g., as with some cotton-trade types. Polyester fibres can also be brightened during finishing by using the light-fast optical whitening products available in the markets. They are usually stilbene based disperse dyes and they can be applied either by the exhaustion or thermosol process. When applied by the exhaust method, as with dyeing, either the high temperature process,or the carrier process at the boil In order to obtain particularly brilliant dyeings, either of the optical brightening agents can be applied together with disperse dyes from one bath.

5.2.3.3.1   High temperature process

Recipe

Quantity	Unit	Additions
0.05–0.3	%	Suitable optical brightener
0.5–1	ml/l	Acetic acid 30% (pH 5–6 or as recommended)
0.5–1	g/l	Dispersing agent

Treat for 1–1.5 h at 120–135 °C and then rinse.

5.2.3.3.2   Pad thermosol (Bake) method

Polyester fibres can be optically brightened merely by padding them at room temperature with suspensions of suitable water insoluble brighteners and drying them at high temperature, a technique which differs basically from standard procedure. No fluorescent white is produced on polyester fibres at normal drying temperatures of around 100 °C, but it can be developed by giving the dried fibres a subsequent heat treatment. This is the principle of the pad-bake method, which is most popular way of brightening polyester, since this fibre is less prone to heat discoloration than other synthetic fibres. Polyester fabrics are padded at 30 °C with an aqueous dispersion of optical brightening agent and baked atleast 130–190 °C (as per the brightening agent used). The brightening effect only develops during this treatment. When applied by the thermosol method, yield white effects of almost the same intensity and fastness as those produced by the much longer batch wise exhaustion method, as long as the brighteners are fully developed on the fibre. This can easily be checked by placing a hot iron on the treated fabric. If no increase in whiteness is apparent, the temperature and the length of the heat treatment are correct. The thermosol method is ideal for optically brightening 100% polyester piece goods and piece goods composed mainly of polyester.

5.2.4   Dyeing of polyester

5.2.4.1   Disperse dyes

Disperse dyes are colorants with low water solubility that, in their disperse colloidal form, are suitable for dyeing and printing hydrophobic fibres and fabrics. Disperse dyes are unusual in that they are dispersed, rather than dissolved, in the dye liquor, where they exist as finely divided particles of c. 1 μm cross-section. Forerunners of the disperse dyes were the ionamine dyes of British Dyestuffs Corp.; these were N-methane sulfonic acids of amino azo or amino anthraquinone dyes that released the N-methane sulfonic acid group in the dyeing process and, thereby, precipitated as disperse dyes on the acetate fibres. The understanding of this mechanism in 1923 initiated the development of genuine disperse dyes. British Celanese and British Dyestuffs Corp. were the first companies to introduce these dyes into the market for colouring acetate fibres.

From 1924 to 1930, products of other companies appeared in the market, initially as pastes; later, when the materials could be dried successfully without interfering with their dispersibility, they were also marketed as powders. Furthermore, new dyeing processes necessitated the development of special disperse dyes. For instance, dyes characterised by special ease of sublimation are preferred for transfer printing. The demand for new fastness properties such as thermo-migration fastness and automotive light fastness also led to new dyes, as has the ongoing pressure on market prices.

Models for the dyeing of polyester fibres with disperse dyes have been developed. When the dye is applied from aqueous medium, it is adsorbed from the molecularly dispersed aqueous solution onto the fibre surface and then diffuses into the interior of the fibre. The following parameters determine the rate of dyeing and, to some extent, the levelling properties: (1) the dissolution rate during the transition from the dispersed crystalline state of the dye into the molecularly dispersed phase, and (2) the diffusion rate at the fibre surface and, especially, in the interior of the fibre. The rates of both processes vary with temperature.

Differences in geometry and polarity of the dye molecules can lead to wide variations in these finishing or dye-specific properties and can have a marked effect on the absorption characteristics of all dyes, irrespective of whether single component or combination dyeing processes are used. For instance, uneven dyeing may occur when an unequal distribution of particle size results in insufficient dispersion stability and hence crystal growth and precipitation at the substrate surface.

Molecules of disperse dyes contain no charged groups; they are non-ionic.

Chemical constitution

Industrially applied disperse dyes are based on numerous chromophore systems.

Almost 60% of all disperse dyes belong to azoic class, out of which only monazo dyes contribute for nearly 50%; anthraquinoids around 25% while rest from other miscellaneous classes, viz. methine (3%), styryl (3%), acrylenebenzimidazol (3%), ouinonapthalones (3%), naphthalimide (1%), napthoquinone–imine (1%) and nitro disperse dyes (1%).

5.2.4.1.1   Azo dyes

Azo dyes are currently employed to create almost the entire range of shades; anthraquinone derivatives are used for red, violet, blue and turquoise. The remaining dye classes are used mainly to produce yellow shades.

Azo dyes represent the largest group of disperse dyes for two reasons: (1) the ease with which an extraordinary number of molecular combinations can be generated by varying the diazo and coupling components and (2) the relatively simple process by which the dyes can be produced. With this class of dyes, manufacturers can respond much more easily to customers' and end users' requests for special shades and fastness characteristics. Azo dyes are defined as compounds containing at least one azo group attached to sp^2-hybridized carbon atoms, such as benzene, naphthalene, thiazole and thiophene. Under normal conditions, azo dyes exist in the more stable planar trans-form so that the carbon nitrogen bond angle is approximately 120 °C.

$$R-N=N-R$$

Most important disperse azo dyes are divided into mono- and disazo types; then each of these classes is subdivided according to the diazo and the coupling components. The diazo component is further subdivided where appropriate into aromatic and hetero aromatic amines. Azo dyes cover a whole gamut of colours, from yellow to blue-green hues, by varying the intermediates especially when heterocyclic diazo components are coupled to amino benzene couplers substituted with powerful electron donating groups, giving bright blue-green colours. Although some deficiencies of azo dyes are not easily overcome, such as duller shades, lower fastness to light and breakdown into carcinogenic amines derived from the cleavage of the azo linkage, and have somewhat restricted their extensive usage against anthraquinone dyes, the cost effectiveness undoubtedly compensates for the drawbacks mentioned above.

5.2.4.1.1.1   Monoazo dyes

About 50% of all disperse dyes are monoazo dyes, which thus represent the largest single group. Relatively simple syntheses enable a range of shades from greenish yellow to cyan to be produced with this chromophore system. Of all the disperse azo dyes, this class has the greatest economic importance. Commercial products are most often represented by structure, in which 4-nitroaniline and its substituted derivatives constitute the diazo component.

$$O_2N-\underset{Y}{\overset{X}{\bigcirc}}-N=N-\underset{A}{\overset{B}{\bigcirc}}-N\overset{R1}{\underset{R2}{}}$$

$$Z-\underset{Y}{\overset{X}{\bigcirc}}-N=N-\underset{R1}{\overset{R2}{\bigcirc}}-N\overset{R3}{\underset{R4}{}}$$

X,Y = H, Cl, Br, CN, NO$_2$, CH$_3$SO$_2$
Z = –NO$_2$
R1 = H, CH$_3$, Cl, NHCOR
R2 = H, CH$_3$O, C$_2$H$_5$O, Cl
R3, R4 = H, CN, OH, OCOR, COOR, aryl (R=alkyl)

As a typical donor–acceptor chromogen, the electron-accepting substituents, X, Y and Z and the electron donating substituents R1 and R2 are favourably sited to create visible colours from yellow to red, and more recently, blue (see below). In general,

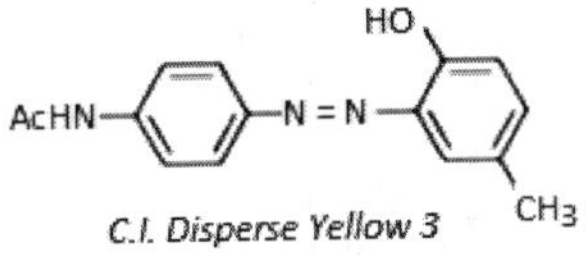

C.I. Disperse Yellow 3

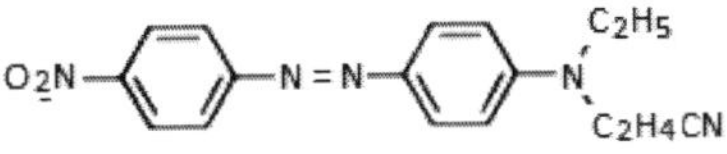

C.I. Disperse Orange 25

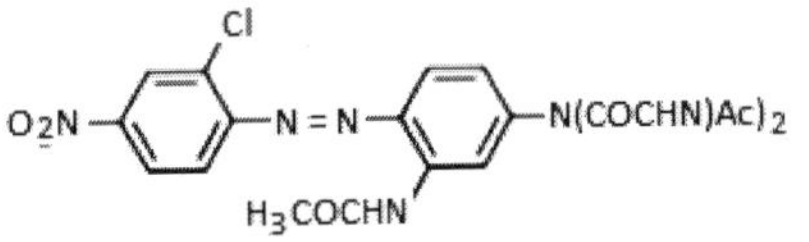

C.I. Disperse Red 167

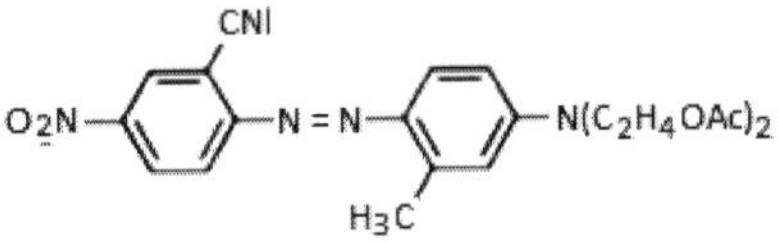

C.I. Disperse Violet 33

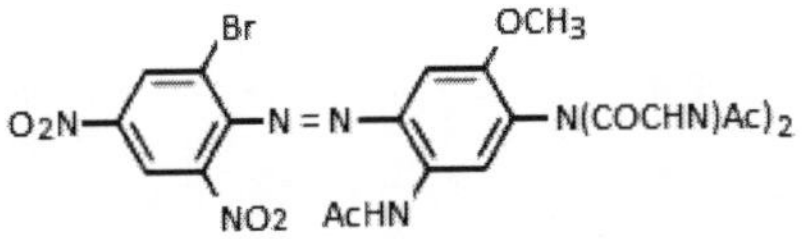

C.I. Disperse Blue 79

amino benzenes are easily diazotized by normal diazotization reagents and the resultant diazonium salts are comparatively stable so that high yields of dyes of good purity are obtained.

5.2.4.1.1.2   Disazo dyes

About 10% of all disperse dyes are disazo compounds. Even the simplest hydroxy disazo dyes, such as 4-aminoazobenzene coupled to phenol and 4-aminoazobenzene coupled to o-cresol, have a good affinity for polyester fibres and yield lightfast reddish yellow hues. However, these shades are frequently less bright than those obtained with monoazo dyes. The introduction of an alkoxy group into the central benzene ring causes a distinct shift toward orange. A similar bathochromic effect is obtained by replacing this benzene with naphthalene. Substitution of the first benzene nucleus by electron acceptors also causes bathochromic shifts. Disazo dyes of this type such as 6are frequently incorporated as components of black mixtures.

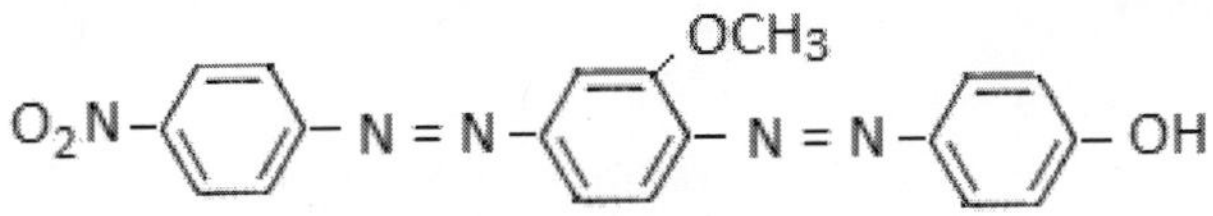

Disperse Orange 29, C.I.20677

5.2.4.1.2   Anthraquinone dyes

Anthraquinone disperse dyes were among the early 'acetate' dyes and have made an important contribution to the violet and blue shade range. They produce bright dyeings of excellent light fastness and cause no dye stability problems during dyeing. From a historical point of view, anthraquinone dyes are the oldest to mimic a natural chromogen. As seen from the Colour Index Classification, anthraquinone systems of basic formula are second only to the azo chromogen for the manufacture of disperse dyes.

Anthraquinone systems basic formula ; 9,10-anthraquinone

The brilliant red, blue and turquoise anthraquinone dyes have major industrial significance. The most important red shades are produced by alkyl or aryl ethers of 4-amino-1,3-dihydroxyanthraquinone. Anthrquinone dyes includes, for example C.I. Disperse Red 60 which acts as a basic colour for trichromic combinations in pale shades and is one of the best dyes available for heat transfer printing, C.I. Disperse Red 15 which can be used for the colouration of

C.I. Disperse Red 60   C.I. Disperse Red 15   C.I. Disperse Violet 26

cellulose acetate, although of limited colour range and of moderate fastness. In the anthraquinone group, when more hydrophobic substituents are introduced to enhance the affinity for polyester fibre, various shades can be produced. This dyes like C.I. Disperse Violet 26 is a widely used dye in the dyeing industry.

Other chromophores

As explained earlier there are other chromophores which contributes smaller part of the range, adopted to cater to some special characteristic or shade of the dye.

5.2.4.1.3   Quinophthalone dyes

Currently, 3'-hydroxyquinophthalone is employed extensively in a number of product lines used to dye synthetic fibres in greenish yellow hues with good light fastness and generally sufficient sublimation fastness. Suitable substitution in the phthalic acid residue or the quinoline nucleus may improve thermosetting fastness.

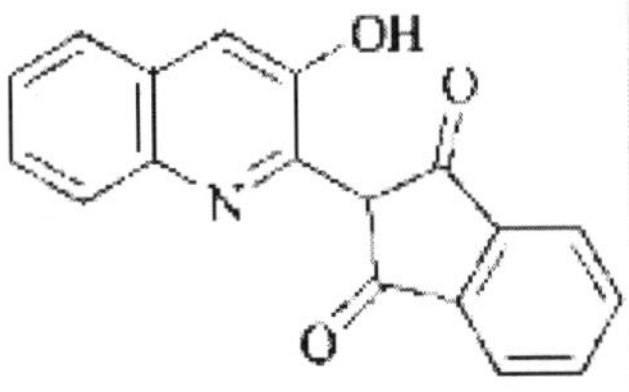

C.I. Disperse Yellow 54

5.2.4.1.4   Methine Dyes

The condensation products of 4-dialkylaminobenzaldehydes with cyanoacetic esters have long been used to dye acetate fibres. Brilliant greenish yellow dyes with excellent lightfastness are obtained on polyester fibres with the corresponding condensation products of malonodinitrile. The sublimation fastness of this dye type can be improved by introducing suitable substituents into the alkyl residue of the amino group or by doubling the molecular size, e.g., C.I. Disperse Yellow 99.

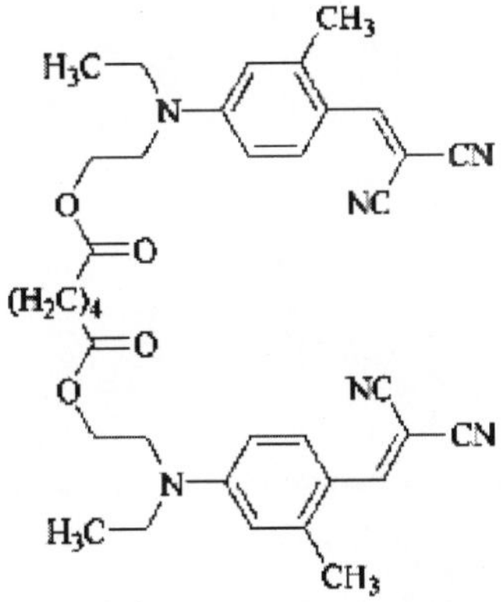

Disperse yellow 99

5.2.4.1.5   Naphthalimide dyes

Derivatives of 4-aminonaphthalimide were used initially to dye acetate fibres. For polyesters, condensation products of 1,8-naphthalenedicarboxylic acid (e.g., below) or 1,4,5,8-naphthalenetetracarboxylic acid with 1,2-diaminobenzenes are used.

5.2.4.1.6   Indigo dyes

Whereas indigo itself is not suitable for exhaustion dyeing of polyester, thio indigo gives fast brilliant red shades in pale to medium depth

5.2.4.1.7   Nitro dyes

2-Nitrodiphenylamines are readily obtained by condensation of derivatives of 2-nitrochlorobenzene with suitable aromatic amines. Because of their accessibility and good light fastness, these dyes became very important for dyeing cellulose acetate and, more recently, have gained a solid position as disperse dyes for polyester fibres. Examples are C.I. Disperse Yellow 42.

5.2.4.1.8   Commercially available dyes

Disperse dyes are currently used to dye cellulose 2.5-acetate, cellulose tri acetate, synthetic polyamides, and to a lesser degree, polyacrylonitrile and polypropylene. Their major application is clearly for dyeing polyesters.

Chemical synthesis produces dyes of varying particle size. When the dyes are applied in this form, uneven and spotty dyeing results, and the dyeing process may be slow and frequently accompanied by incomplete absorption. To assure high yield, good reproducibility, and faultless dyeing and printing in commercial use, especially when densely woven fabric or wound material is involved, the dye must be applied as a fine dispersion that is stable under the process conditions.

Most important requirement of a dye is that it should not change its structure during application. Some dyes during synthesis itself, this form is produced for other dyes after the manufacture the dyes are treated by heating the dye in an aqueous suspension in the presence of non-ionic detergents. The eventual dispersion process consists of grinding the moist dye, generally in a ball, bead, or sand mill, in the presence of dispersing agents (like lignin sulfates (sulfite cellulose liquors); condensation products of naphthalene, sulfuric acid, and formaldehyde; condensation products of m- and o-cresol, formaldehyde, and 2-hydroxynaphthalene-6-sulfonic acid; or mixtures of these products). Since the dyeing rate depends on particle size, the aim is to reduce the particle size to <1 m. A narrow particle-size distribution is important to minimise recrystallisation during storage and application. Difficulties in the production of disperse dyes may be encountered, for example, in grinding or drying dyes with low melting points. Therefore, only dyes that have a melting point >120 °C, and preferably >140 °C, should be used. Disperse dyes are sold either as aqueous pastes obtained by grinding or as powders obtained by gentle drying of the paste. Powders are generally produced by spray drying.

5.2.4.1.9   Disperse dye classification according to application

Disperse dyes are classified into major four groups A, B, C and D and an additional group without any suffix with respect to dyeing polyester. The dyes performance gradually improves from A to D with respect to the criteria of critical dyeing temperature, migration, build-up of shade, rate of diffusion and energy required for dye transfer from bath to fibre, etc., the worst performing will be in A group and the best performing will be in D group.

(i) 'No suffix' dye – it is not suitable for polyester, but suitable for acetate and nylon.

(ii) Group A dye – it has poor sublimation fastness (~Grade 2), suitable for acetate and nylon but may be used for polyester.

(iii) Group B dye – it shows moderate sublimation fastness (~Grade 2–3), is suitable for acetate and nylon. Well-levelled shades on polyester, coverage on 'configurational barre' is excellent and so is invariably applied on textured polyester.

(iv) Group C dye – it has good sublimation fastness (~Grade 3–4), is suitable for all methods to dye polyester, e.g. carrier, HTHP, thermosol and other methods, where maximum sublimation fastness is not required.

(v) Group D dye – it has maximum sublimation fastness (~Grade 5). Exclusively for dyeing of polyester in HTHP or thermosol technique but not in carrier method; few dyes produce wash fast shades on nylon.

The sublimation fastness grades shown are exclusively for polyester; the same for triacetate will be little better whereas for acetate the highest. Sublimation fastness can be improved by selecting a dye of higher molecular weight.

5.2.4.1.10   Principles of disperse dyeing

Solubility of disperse dyes is too feeble in water at room temperature (~0.2 mg/l) which is increased with heating of bath beyond 90 °C or addition of dispersing agent or both.

It is now generally accepted that disperse dyes are transferred from the solid phase to the fibre as individual molecules. Although disperse dyes have only a low solubility in water, they are highly substantive in the dissolved state. Thus, the solid phase serves as a reservoir to maintain a saturated solution.

A disperse dye is dyed in the presence of a dispersing agent either added during the dyeing operation or some present in the dye itself. This dispersing agent plays a crucial part in the dyeing process. As disperse dyes are specifically applied at higher temperature, the dye when added, remains present in bath in two basic forms: (i) a very little part remains in completely soluble form and (ii) the rest in finely dispersed insoluble form; the later remains predominant and the ratio is being determined by solubility of dye in water at that temperature. Thus the dispersing agent its dual character results in the formation of micells above critical, but low, concentration. The hydrophobic tails of the dispersing agent molecules are inside the micelle which, as a consequence, is able to solubilise the disperse dye molecules, so conferring a higher apparent solubility on the dye. When fibre is entered in dyebath, dissolved dyes molecules slide past the narrow pores present in fibre the dye from the micelle is transferred and get attached with physical forces causing reduction in share of dissolved dye in bath which forces dispersed insoluble dye particles to break up and go into solution to maintain share of soluble dye in bath. As micelles empty their dye, they re-from and dissolve more dye from the solid particles. The size of the dispersed dye particles is in the region of approx. 0.5–1 μm. Because of their hydrophobic character, disperse dyes have an affinity for the likewise hydrophobic synthetic fibres. The dye particles are able to penetrate the fibre during dyeing in a molecularly dispersed state and are held in the fibre in this form. It has been proved that the disperse dye inside the fibre is in monomolecular state. At the end of the dyeing process, the dye that has been absorbed by the fibre is in a state of dynamic equilibrium with the dye that remains in the bath, and the fraction of the latter that is in aqueous solution must be present in the same state of aggregation as the dye in the fibre. Hence it can be inferred that the transfer of the dye to the fibre takes place from a monomolecular aqueous solution, the concentration of which is maintained during the first phase of the dyeing process by the progressive dissolution of solid dye from the particles in dispersion in the bath. The dyeing process can be schematically shown as follows:

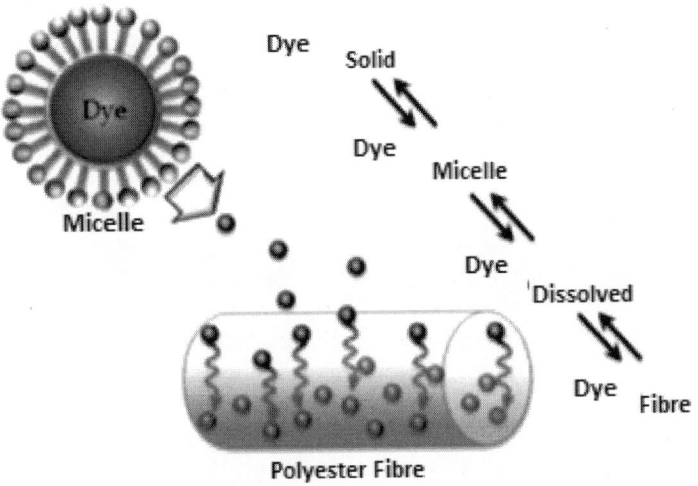

The whole process can be summarised into five steps:

    a. Dispersion of the dye in the bath.

    b. Some of the dyes dissolve in the water of the dyebath in the presence of the dispersing agents.

    c. Molecules of dye are transferred from solution to the surface of the fibre.

    d. The solution in the dyebath is replenished by the dissolution of more solid material from the dispersion.

    e. The adsorbed dye diffuses monomolecularly into the fibre.

       Though dispersing agents enhance solubility of dye for efficient dyeing and facilitate diffusion inside fibre, increase in temperature breaks down dye clusters improving solubility of dye and opens up fibre structure facilitating diffusion of dye inside, known as 'rate determining step'. The overall rate of dyeing is controlled by combined effect of temperature and type of dispersing agent used.

After the dye molecules have been adsorbed on the fibre surface it diffuses into the interior of the fibre by a relatively simple mechanism, which appears to obey Fick's equation i.e., the rate of diffusion of dye through unit area (transverse to the direction of diffusion) at any point in the fibre is directly proportional to the concentration gradient of the dye at that point. As would be expected, the amount of dye taken up by polyester fibres from a bath of constant concentration is found to be proportional to the square root of the dyeing time, until a saturation value is approached. Very similar results are observed during the earlier stages of the process in dyebaths of normal composition and concentrations, such as are employed in commercial 'exhaust dyeing' processes. It is found that the rate of dyeing is quite independent of the concentration of the dyebath, practically up to the point at which equilibrium is established. For dyeings carried out at a constant temperature, a plot of the instantaneous fractional 'dye uptake' $(C/C\infty)$ against time of dyeing gives a steeply-rising asymptotic curve, which appears to fit a law based on the hyperbola or, possibly, on the hyperbolic tangent. In case of the other fibres than polyester even though they can be dyed with disperse dyes their rate of dyeings are different. Polyester fibres dye very slowly at temperatures much below 100 °C. Cellulose secondary acetate can be dyed with disperses dyes readily over approximately 1 h at 80 °C. It is a blessing that it can be dyed this way as higher temperatures has to be avoided as otherwise acetate groups on the cellulosic fibre can be hydrolysed to hydroxyl groups, which can spoil the surface of the fibres and reduce their substantivity towards the disperse dyes. Cellulose triacetate is more difficult to penetrate with disperse dyes because of its more compact molecular structure, but it can be dyed at the boil. Nylon fibres can be dyed under conditions similar to those used for cellulose acetate fibres. In case of acrylic fibres, the presence of anionic groups such as $-SO_3H$ and $-COOH$ permit only pale shades to be obtained under normal conditions with disperse dyes.

Since the majority of synthetic fibres do not possess any ionic groups, they cannot be dyed by an ionic mechanism as is the case with e.g. acid, metal-complex or cationic dyes. These synthetic fibres can only be dyed with non-ionic dyes (= disperse dyes) which are practically insoluble in cold water due to the absence of any solubilising groups in the molecule. Disperse dyes are applied in the form of fine aqueous dispersions (disperse dyeing systems). The size of the dispersed dye particles is in the region of approx. 0.5–1 mm. Because of their hydrophobic character, disperse dyes have an affinity for the likewise hydrophobic synthetic fibres. The dye particles are able to penetrate the fibre during dyeing in a molecularly dispersed state and are held in the fibre in this form. Disperse dyes are produced with various chemical structures. The two main classes are the azo and anthraquinone dyes. Azo disperse dyes contain relatively unstable azo linkages in the molecule so that the sensitivity to reduction of these dyes must be taken into consideration. Anthraquinone disperse dyes have more stable structures and a smaller molecular size which gives them greater mobility. The solubility of disperse dyes in cold water is very limited (only a few mg/l). Solubility rises rapidly with increasing temperature, however, and some dyes can achieve a solubility of 100 mg/l at dyeing temperature.

5.2.4.1.11   *Factors affecting disperse dyeing*

5.2.4.1.11.1   Dispersing agent

The role of the dispersing agent was discussed under the principles of disperse dyeing. The dye molecule as available in the commercial form is in the clustered molecule form and agitation of the bath is in adequate to make into a dispersed form. It is essential to have a dispersing agent to make the dye in a dispersed form. But this dispersing agent does more work than a mere dispersing agent in the dyeing process of disperse dye. Dispersing agent imposes negative charge on dye to improve its aqueous solubility and inhibits formation of dye cluster through repulsion among negatively charged dye molecules; retains dye in consistent dispersion throughout dyeing by enclosing single dye particle with a protective film, as shown below, where 'D' denotes disperse dye and surrounding '−' ions are dispersing agents.

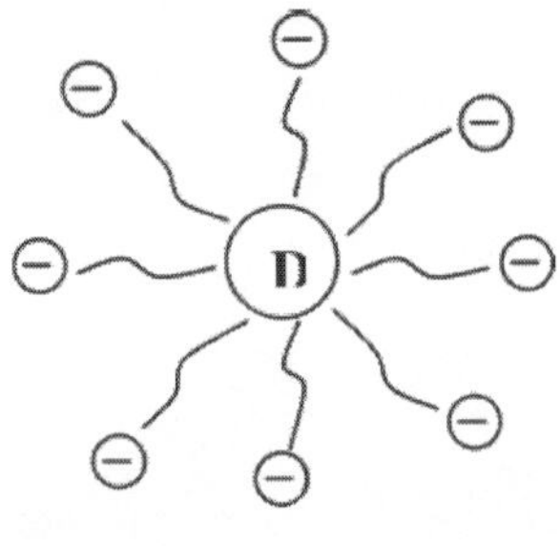

The zeta potential of commonly used disperse dyes range from −30 to − 80 mV. Dispersing agent, if in excess, enhances aqueous solubility of dye, reduces exhaustion and develops poor rubbing fastness. These are invariably mixed with disperse dye during grinding to keep control over particle size and do not allow to form cluster. While in bath, dispersing agents being highly soluble in water, surround a disperse dye molecule, keep it in dispersed state and release it slowly to get solubilised in bath for its acceptance by the fibre. Each disperse dye possesses varying solubility pattern at higher temperature and also in presence of dispersing agents due to difference in hydrophilicity of it which in turn is based on hydrophilic groups in its structure. Indeed, optimum concentration of dispersing agent required for each dye is different.

In a disperse dyeing bath, during the dye dispersion and dyeing, including the three successive phases of heating up, maintaining dyeing temperature and subsequent cooling the dye molecules can exist in various particle sizes and transform from on form to another (see below). These can exert critical effects on rate of dyeing, final degree of sorption and levelness.

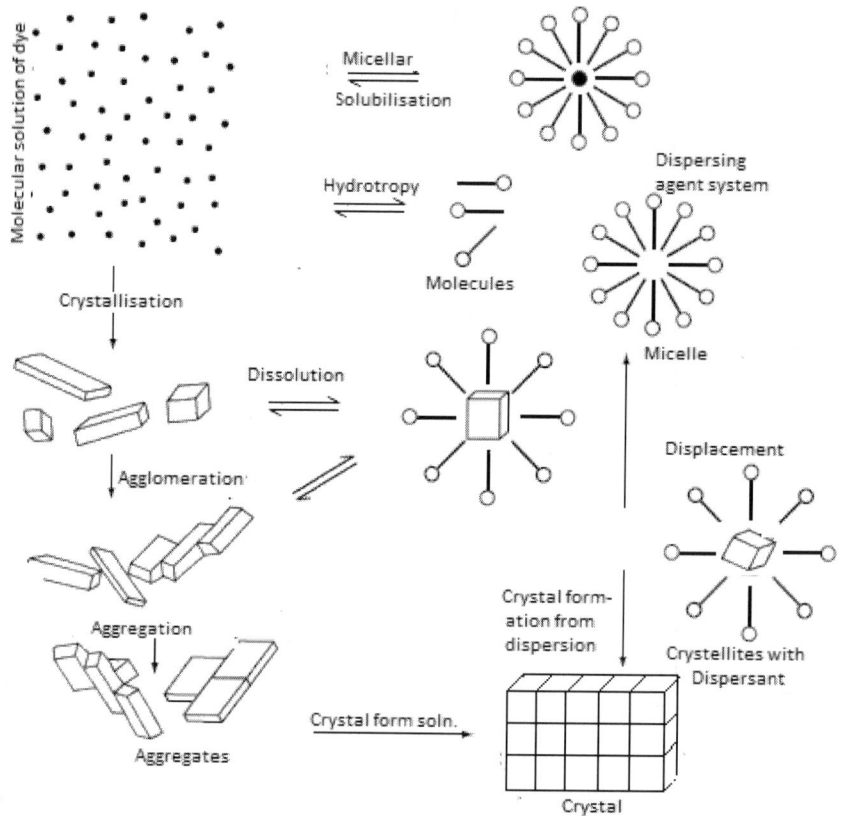

Disperse dyebath model (with dispersant and other auxiliaries)

As dispersion of dye is achieved through interaction between dyes and dispersing agent, concentration of the latter is based on amount of dye and do not have any relation with liquor ratio. Concentration of dispersing agent can further be reduced with increase in dyeing time to get a better levelled shade:

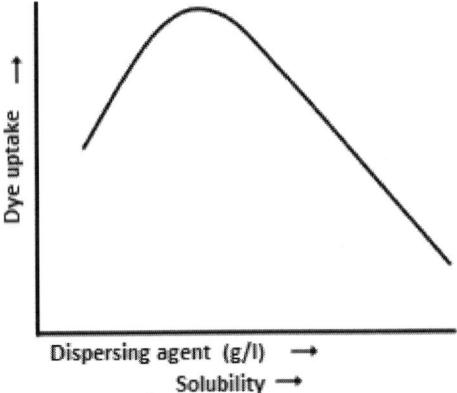

Influence of dispersing agent on solubility and uptake of disperse dye.

Dispersing agents belong to mainly formaldehyde condensates of either of naphthalene sulphonic acid, cresol, 1-naphthol 6-sulphonic acids, fatty alcohol-ethylene oxide condensate, alkyl aryl sulphonates or lignin sulphonates; the condensate of naphthalene sulphonic acid is the most popular one. Lignin sulphonates brings about chemical reduction of some azo disperse dyes and so mild oxidizing agent is used to protect dyes (especially for polyester–wool blends) e.g. m-nitrobenzene sulphonates. While anionic dispersing agents increase solubility up to limited extent, other surfactants increase solubility considerably. Dispersing agents must have cloud point at least 5–10 °C higher over dyeing temperature to avoid precipitation of dye during dyeing.

But there are some disadvantages in using dispersing agents. Powder brand dispersing agents, which is generally used in dye manufacturing, cause dustiness. In the processing it may not be a big problem since the dye powders supplied to the dyer have been treated already to render them essentially non-dusting. However, dustiness is a problem during dye manufacture. During dyeing the dispersing agents are hardly absorbed or used up by the fibres, hence they remain in the exhaust dyebath and are discharged to effluent. This can cause environmental problem since its inadequate and slow biodegradability.

5.2.4.1.11.2   Crystal form of the disperse dyes

The possible different crystal forms (e.g. α, β, δ, γ and ε) of disperse dyes have been suggested to influence the saturation values achieved on polyester, due to differences in the vapour pressure and solubility of the different forms of dye in both water and fibre.

5.2.4.1.11.3   Particle size in dispersion of the dyes

The aqueous solubility of disperse dye particles in a dispersion increases with decreasing particle size. Thus an increase in the severity of milling that accompanies a reduction in the particle size of the dye enhances the solubility and adsorption of disperse dye.

5.2.4.1.11.4   Substantivity of disperse dyes

The substantivity of disperse dyes towards polyester fibres is one of the most critical factors in determining dyeing behaviour and there have been many studies carried out to evaluate the substantivity of disperse dyes towards hydrophobic fibres, including polyester, in order to select suitable dyes. These attempts include the 'Solubility Parameter Concept' and 'IOR (Inorganic/Organic) values'.

5.2.4.1.11.5   Temperature

The adsorption and diffusion of disperse dyes on polyester are greatly influenced by temperature as an increase in temperature increases the mobility of the polymer chains in the amorphous regions of the fibre. Polyester fibres dye very slowly at temperatures much below 100 °C. At 85 °C, the temperature normally used for the dyeing of secondary cellulose acetate, it was found that polyester fibre dyed at rates between 700 and 1000 times slower than those measured for the same dyes on secondary cellulose acetate and nylon. When, however, sufficient time was allowed for the polyester fibre to come nearly to equilibrium with the dye bath, it was found that it had taken up approximately the same amount of dye as had secondary acetate and in most cases about twice as much as nylon. For the dye-hydrophobic fibre system, the affinity of the disperse dye decreases with increase in dyeing temperature while the saturation value of the dye in the fibre increases with increasing temperature.

Normally dyeing temperature of any dyes on a particular substrate is best at which the swelling and opening of the substrate and the exhaustion of the dye is the highest. In the case of disperse dyes on nylon and acrylic, Nylon 66 and acrylic can be dyed at boil due to adequate opening of fibre structure at this temperature. Nylon 66 has T_g around 71 °C and beyond that opening is rapid offering adequate space to all disperse dyes at boil itself. Acrylic has T_g around 80 °C and opening completes at around 85 °C providing maximum opening just below boil; heating up beyond boil causes fibre degradation. In contrast, polyester has T_g at 80–85 °C, but opening is too slow to affect any such opening at boil. Little opening may cause production of light shades with poor wash fastness. Dyeing temperature varies from dye to dye; the larger the structure of dye, more opening is needed necessitating a higher temperature of dyeing. Relative molecular weight of some disperse dyes are given below:

Structural Formula of the dye	Relative molecular weight
O_2N—〈〉—$N=N$—〈〉—NH_2	242
NO_2 〈〉—$N=N$— (quinoline OH, O, N–CH_3 structure)	324
O_2N—〈〉—$N=N$—〈〉(CH_3)—$N\langle CH_2CH_2CH_3 / CH_2CH_2CH_3$	344
O_2N—〈〉—$N=N$—〈〉—$N\langle C_2H_4CN / C_2H_5$	347
OH O NH_2 (anthraquinone with Br), NH_2 O OH	349
O_2N—〈〉(Cl)—$N=N$—〈〉—$N\langle C_2H_4CN / C_2H_5$	358
OH O NH_2 (anthraquinone), NH_2 O OH —OCH$_3$	376
O_2N—〈〉—$N=N$—〈〉—$N\langle C_2H_4CN / (CH_2)_2COOCH_3$	416
O_2N—〈〉(Cl, Cl)—$N=N$—〈〉—$N\langle C_2H_4CN / (CH_2)_2COOCH_3$	451
O_2N—〈〉(CN)—$N=N$—〈〉(CH_3)—$N(CH_2CH_2OCOCH_3)_2$	453

However, all disperse dyes exhaust well at or below 117 °C and a little higher temperature is maintained to compensate heat loss and better migration of dye molecules for levelled shades. In tropical areas and specifically in chilling winter, production of deep and levelled shades with better wash fastness is difficult to achieve due to condensation of steam during its long passage from boiler to the dyeing machine; the situation worsens if insulation of supply pipe is not proper.

Heat setting changes the morphology of the polyester fibres. When fabrics of polyester are heat set in air under conditions of free shrinkage, the dye exhaustion first decreases and then increase with increasing setting temperature. The minimum exhaustion occurs after seeing at around 160–190 °C. If applied tension prevents fabric shrinkage during heat setting, the dye uptake/temperature profile is similar to that under conditions of free shrinkage, but with higher uptake values

5.2.4.1.11.6   Fibre fineness

Much attention has been given recently to dyeing micro fibres. In a broad sense, especially in Europe, the term microfibre means fine fibres of less than 1.0 denier. However, in South Korea and Japan, where fine-fibre technology is more advanced, fine fibres of 0.04–0.4 denier class are generally used in this filament area. A useful preliminary relationship between the percentages of dye on weight of goods (C1, C2) needed to achieve a particular depth of shade on polyester fibres of two different fineness (D1, D2):

$$C1/C2 = (D1/D2)^{1/2}$$

According to this equation, it takes much more dye quantity, to dye the micro fibre to the same apparent depth as the regular fibre. Therefore, such marked denier difference can affect dyers in a number of ways.

5.2.4.1.11.7   Levelling agent

Some people may confuse between dispersing agents and levelling agents. An efficient dispersing agent ensures dispersion stability by maintaining dye particles at constant size and minimal solubility. Hence a true dispersing agent do not support or enhance levelling. Hence levelling agents are used to promote uniform distribution of dye throughout fibre structure by reducing the strike rate. They are mainly anionic and non-ionic surfactants which tend to solubilise the dye much more effectively. Some anionic levelling agents are able to promote dispersion stability but non-ionic types have a destabilising effect by breaking the film of dispersing agent protecting the dye particles in the dyebath when the dyebath temperature increases, hence great care is therefore required in their selection. As the dispersing effect is reduced and high shear rates in jet dyeing machines and additives such as electrolytes, fibre lubricants or sizes, as well as oligomers from the fibre, can contribute to this effect further. This can lead to precipitation and crystal formation and grow in size whilst retaining their original form, or they may undergo a transformation from the original thermodynamically metastable form to a more stable but less soluble form. Such crystals may not form until the incompletely exhausted dyebath is cooled after the dyeing process; this problem may be avoided by blowing off the dye liquor at 125–130 °C. Hence anionic levelling agent appears a better levelling agent in a dyebath where the dye promote level dyeing by control of exhaustion during the heating phase of dyeing.

But non-ionic agents tend to solubilise the dye much more effectively and thus contribute to level dyeing both by a retarding effect and through the promotion of migration. Consequently they are generally more powerful levelling agents than anionic products although their effects are much more dye-specific. But non-ionic levelling agents are not without disadvantages. A major problem with non-ionic agents arises from their inverse solubility. Thus an agent with a low cloud point may increase dye precipitation, although once again the effect is dyespecific. It is generally accepted that a nonylphenol with a low degree of ethoxylation, having a cloud point of about 40 °C, should not be used as a disperse dye levelling agent. A non-ionic levelling agents (e.g., $CH_3(CH_2)_{15}(OCH_2CH_2)_{17}OH$) with cloud point of about 105 °C, can be used for a dyeing temperature upto boil but it is not suitable for high temperature dyeing. Selected products should have cloud point beyond dyeing temperature otherwise may get precipitated in bath. Cloud point of levelling agents may also be enhanced with addition of anionic surface active agents for example a mixture of the fatty acid ethoxylate with 7–10% sodium dodecylbenzenesulphonate has a cloud point of about 150 °C.

$$CH_3(CH_2)_7CH=CH(CH_2)_7 - CO - (OCH_2CH_2)_{14}$$

Fatty acid ethoxylate

Levelling agents enhance solubility of dye and excess use reduce dye uptake. Some non-ionic levelling agents absorb dye causing loss in colour yield necessitating application of anionic products. Foam produced due to agitation in bath is suppressed by applying defoaming or deaerating agents. Nonionic surfactants can be beneficial in minimising the redeposition of the sparingly soluble polyester oligomers that are released from polyester fibres during high-temperature dyeing.

Along with non-ionic surfactants of low cloud pint hybrid agents of the ethoxylated anionic type can be mixed, for example ethoxylated phosphate type (see below) to avoid adverse effect of its low cloud point. This has many advantages like it is sufficiently anionic to avoid the problems with non-ionic agents, has additional effect of protection from hard water (due to polyphospahtes),

$$R-(OCH_2CH_2)n \left[O-\overset{\overset{O}{\|}}{\underset{\underset{O-M}{|}}{P}}-O \right]_x M$$

R = hydrphobe
n = 10-20
x = 1-3
M = H, alkali metal, or organic base

maintain effective stability in high concentrations of electrolyte, pH control by varying the nature of M and fully effective at pH 4–5, the most useful pH range for application of disperse dyes.

Another possibility of an effective levelling agent is a mixture of ethoxylates with aliphaticesters. This combination exerts a retarding effect on many dyes during heating up to about 100–110 °C, especially if the dyes are present in low concentration. At higher temperatures this retarding effect is increasingly offset by the accelerating effect of the aliphatic esters. This temperature-dependent interaction is said to improve the compatibility of combinations of dyes applied with this system.

There was anew idea put forward by some researchers to use a dispersing levelling agents, which can work as a dispersing agent and levelling agents. Ethoxylated multi-ester compounds (so-called 'oligo-soaps') are useful in this context. The thermal stability of a multi-ester is much greater because this macromolecular structure resists thermal agitation to a much greater degree. This maintains more stable dye dispersion at high temperature under conditions of high shear. With advantages like solubilisation of dyes at lower temperature, at lower temperature the strike rates are much less, soliublisation of oligomer and any acrylic size left on the fabric, much lower foaming, etc.

5.2.4.1.11.8   Reducing agents

The presence of any reducing agents or reducing atmosphere can break down the azo disperse dye to yellow or colourless amines. Chances of the reducing agents coming into dye baths are size residues, cellulosic fibres, cysteine in wool, or certain dispersing agents (e.g. pyrocatechol groups in lignin sulphonates and sulphite-containing naphthalene sulphonic acid condensation products).

Hence it is always advisable to add a mild oxidising agent in the bath to neutralise any reducing atmosphere during dyeing.

5.2.4.1.11.9   Carrier

Carriers, when added to a disperse dye bath, interferes dispersion pattern of dye, physical characteristics of fibre, accelerate rate of dyeing to cause better dye up take at lower temperature. Chemically, these belong to hydrocarbons or substituted hydrocarbons, phenols, amides, alcohols, etc. Carriers for polyester fibres are special auxiliaries which increase the amount of disperse dye taken up by the fibres over a normal period of dyeing at 100 °C. They are organic, usually aromatic, compounds of various types. Apart from the effective substance, the majority of commercial carriers also contain emulsifiers which ensure satisfactory emulsification or dispersion in the dye liquor, the effective substance being itself sparingly soluble in water.

The following graph gives the dye take up relative to the solubility parameter of the added carrier.

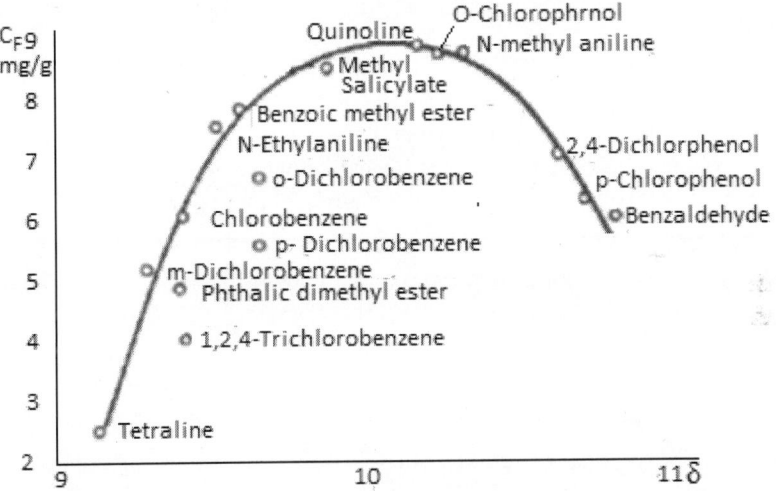

Dye take up relative to the solubility parameter of the added carrier

Few popular carriers are monochloro-o-phenylphenol, o-phenyl phenol, diphenyl, monomethylnaphthalene, halogenated benzenes, methyl salicylate, butyl benzoate, etc., and are highly toxic. Several theories were put forward to emphasise mechanism of carrier action, e.g. increased swelling of fibre, increased water inhibition, transport theory, increased solubility of dye in bath, film theory, liquid fibre theory, loosening of fibre structure (lowering

T_g), lubricity of fibre molecule and increased sites accessibility; exact mode of working of a carrier has not been established so far.

It is believed that these swell up fibre structure longitudinally, act as lubricant, cause adsorption and diffusion of dye rapidly at lower temperature. A high temperature method may be carried out in presence of carriers to get levelled results. Selection of a carrier for a specific dyeing process is based on its toxicity, biodegradability, availability, cost, ability to show synergistic action on rate of dyeing, compatibility in bath, ease in removal, non-volatile nature, no adverse effect on light fastness and change in handle, non-irritate to skin, uniformity in absorption by the fibre and if insoluble in water, should have good emulsion stability; not a single carrier fulfils all these characteristics. Efficiency of a carrier is calculated based on its acceleration factor (a) which is defined as:

$$a = \frac{\text{amount of dye taken up by fibre in the presence of carrier}}{\text{amount of dye taken up by fibre in absence of carrier}} = q_w/q_0 \leq 1$$

Carriers are of two types: water soluble and insoluble. Water soluble type includes benzoic acid, salicylic acid and phenol; higher dose is required and are not used in dyeing. Insoluble types are mostly preferred, e.g. diphenol, benzene, naphthalene, o and p-phenyl phenol, ochlorophenol, mono/di and trichlorobenzene, toluene, etc., which works better at 1–5 g/l but are highly toxic. Tumescal OPE (ICI) is a popular carrier on these grounds.

Carriers may have some characters which is preferable and some characters which is not preferable bay a processor. In practice, the demands placed on a carrier are various and sometimes very great. Probably an 'ideal carrier' should have the following characteristics:

1. High efficiency at the smallest possible concentration, and minimum dependence on the constitution of the disperse dye.

2. Rapid emulsification and good stability of the carrier emulsion under dyeing conditions. There is a danger of carrier spots resulting from the breaking down of the emulsion if the stability is inadequate.

3. Low volatility in steam. With carriers with high volatility there is the danger that, with some types of dyeing machine, the volatilised substances may condense on cooler parts, this resulting in drops falling on the material to form spots which then dye more strongly. Apart from this, a further addition is necessary with volatile carriers where the dyeing time is lengthy.

4. The odour should not be unpleasant otherwise the products can only be employed in closed dyeing machines.

5. It should have little effect on shrinkage of the fibre. This is important with cross wound packages.

6. The uptake of disperse dyes by blend fibres (especially wool) should not be increased.

7. It should be readily removable from the material.

8. The light fastness of the dyeing must not be impaired.

Actually there is no one carrier with all these preferable characteristics and one is compelled to make concessions on one side or the other, depending on the circumstances. If, at a given temperature, disperse dyes are dyed in the presence of increasing amounts of a dyeing accelerator, the colour yield rises up to a given concentration of carrier, whereupon it falls again as this is further increased. The optimum effect is dependent both on the carrier as well as on the dye employed. It is therefore not possible to replace a carrier by one with a different constitution without making an appropriate correction in the dyeing recipe. At temperatures of over 100 °C, optimum efficiency is obtained with smaller amounts of carrier.

Though carriers do accelerate the penetration of dyes, it is also possible to strip the dye from the fibre to a greater or lesser extent by means of carriers, especially when the concentration of the latter exceeds the optimum for dyeing. This effect can be put to good use in levelling out or in partially stripping dyeings which are too dark. However, too high a concentration of carrier must be avoided, otherwise the fibre will be damaged; this applies particularly to lengthy treatment times, especially at temperatures of over 100 °C.

Emulsification of carrier

The carrier is used in the emulsified form in dyeing. Hence the carriers must be carefully emulsified before use. Only the milky-white homogeneous emulsion may be added to the dyebath. The emulsification procedure depends on the carrier. Normally, depending on the carrier, it is mixed with 3–5 times the amount of water at 20–100 °C while stirring vigorously; this liquor is then brought up to the boil by an open steam pipe. With large quantities, it is advisable to employ a high-speed stirrer (1000–1500 r.p.m.). The resulting emulsion is intended for immediate use and should therefore be made up just before adding to the dye liquor.

Quantity of carrier requirement

The quantity of carrier required for a dyeing depends on the carrier and the dyeing temperature at which it is used and should be generally followed as per manufacturers advice. As an approximation the following table can be used.

Dyeing temperature	Jigger	Yarn dyeing	Winch	Soft flow/Jet
MLR	5–1	10:1–20:1	20:1–50:1	8:1–10:1
Boil	5–15	1.5–6	2–8	5–15
103–108 °C		0.5–4		
110–120 °C		0.5–2		0.3–2

The above figures are for 100% polyester. For blends containing upto 50% polyester the same amount of carrier is used as in 100% polyester goods. However, if the polyester component is less than 50% in a mixture, the amount of carrier can be reduced by 20–30%.

Disadvantages of carrier dyeing:

1. Residual carrier in the dyebath contributes to effluent pollution and may be environmentally harmful.
2. Carrier that is volatilised during dyeing or subsequent heat setting becomes an atmospheric contaminant.
3. Residual carrier in the fibre can be a health hazard, as well as causing an unpleasant odour on heating or during storage.
4. Some carriers, such as o-phenylphenol, tend to lower the light fastness of many dyes if carrier residues remain in the dyed fibre.
5. Some carrier, left on the dyed fabric may give obnoxious odour.
6. Many carriers are harmful to aquatic life and not biodegradable. Biphenyl is relatively non-toxic to river life but is not readily biodegradable; methylnaphthalene, also of low toxicity, is moderately biodegradable. COD and BOD of many carriers are high (see table below), hence environmentally harmful.

BOD and COD of some of the carriers used for polyester dyeing

Carrier type	COD (mg/l)	BOD (mg/l)
o-Phenylphenol	1000–2000	200–800
N-Alkylphthalimide	1000–2100	100–200
Arylcarbonate ester	900–1900	700–800
Methyl cresotinate	800–1700	200–800
Dichlorobenzene	500–1000	0
Trichlorobenzene	300–1000	0

5.2.4.1.11.10   Calcium and magnesium and heavy metal salts

Heavy metal ions can make metal–dye complexes with the disperse dyes and give shade changes and dyeing properties. Calcium and magnesium salts can cause shade changes on anthraquinone red dyes towards violet after drying. Water hardness does not have much effect on the colour and depth of shade of disperse dyes. Even hardness ratings of up to 50° (English) produce only slight differences in shade as compared with those of dyeings made in distilled water. Nevertheless, it is advisable to use softened water in order to eliminate all possible deleterious effects due to hardness, e.g., incrustations on the fibre or impairment of dispersion stability.

Heavy metal salts, especially those of copper and iron, can have a much more marked effect on disperse dyes. However, troubles of this nature arise comparatively rarely. When they do so, however, it is recommended that dyeing be carried out in the presence of a sequestering agent (e.g., Trilon B). When dyeing with Brilliant Red BEL or with Brilliant Violet 4REL it is preferable to add 0.25–0.5 g/l sequestering agent to the dyebath on principle, in order to prevent the formation of dark blue spots on the fibre due to reaction with iron salts. Comparative dyeings have shown that this addition of sequestering agent to the dye liquor produces no marked change in the shade of other disperse dyes if the alkaline reaction of the product is counteracted by the addition of a little more acetic acid so that dyeing proceeds within the usual pH range of 5–6.

5.2.4.1.11.11   pH

Dyebath pH exerts a marked influence on the efficacy of lignosulphonate dispersing agents,since this factor determines the degree of dissociation of phenolic and carboxylic acid groups,influencing the extent to which they are able to interact with the dye molecule. The situation is somewhat different with amine salts of lignosulphonates, since these differ from the sodium salts in their degree of ionisation. For example, when the pH is lowered from pH 7 to 4, more amino groups are protonated, thus increasing the proportion of ionised sulphonate groups present. The solubility of the dispersant is increased and so the adsorption of agent by the disperse dye particles becomes more difficult. However, the overall situation is complex because lignin derivatives vary in their content of phenolic and carboxylic acid groups and these exhibits a range of pK_a values according to their location within the macromolecule. It is normally best to dye within a pH range of 5–6, since the depth of shade and the colour obtained with some disperse dyes are unfavourably affected in a strongly acid, neutral or alkaline region. However, the change in the pH towards alkaline can hydrolyse some of the disperse dyes, especially the substituents in the dye molecule. Hydrolysis can split the substituents from the dye molecule resulting in change of shades, fastness and dyeing properties:

$$O_2N-\langle\ \rangle-N=N-\langle\ \rangle-N\!\!\begin{array}{c}CH_2CH_2OCOCH_3\\ CH_2CH_2CN\end{array} \xrightarrow[130°C]{pH\ 6.5} O_2N-\langle\ \rangle-N=N-\langle\ \rangle-N\!\!\begin{array}{c}CH_2CH_2OH\\ CH_2CH_2CN\end{array} + CH_3COOH$$

Acetic acid is especially important for dyeing polyester, to ensure that no trace of alkali is present during dyeing. Alkali degrades polyester.

The most commonly used acid for maintaining the pH of dyeing is acetic acid, because the danger of over-acidification is then avoided. However, formic acid, sulphuric acid, ammonium salts or suitable buffer mixtures can also be used to attain the desired pH value.

5.2.4.1.12   Dyeing with disperse dyes

Disperse dyes can be dyed from aqueous medium or solvent medium.

Dyeing from aqueous medium

A lot of studies and papers has been published regarding the kinetics and thermodynamics of disperse dyeing on polyester. Even though the disperse dye is insoluble in water, it has been proved that the dye has to dissolve in aqueous media to be adsorbed on the surface of the fibre and subsequent migration into the fibre. The dissolution of the disperse dye in the medium depends on so many factors like: temperature, dispersing agent, particle size of the dye, crystal modification and crystal growth, melting behaviour, type of dispersion, presence of electrolyte etc. Even though the solubility of the disperse dye is only in milligrams per liter but it increases many folds in the presence of dispersing agent and temperature. The state of the dye in the fibre is often compared to a solid solution. Thus, the thermodynamic dyeing equilibrium of a disperse dye between water and fibre follows the Nernst distribution law, i.e., $C_F/C_L = K$ at infinite time. At constant temperature, the distribution coefficient K [i.e., the ratio of the concentration of dye dissolved in the fibre (C_F) and in the liquor (C_L) at equilibrium] is constant. At normal dyeing temperatures the rate of dissolution of dispersed dyes in the dye liquor is assumed to be high, so that CL is constant as long as dissolved dye is present.

Actual dyeing takes place after the diffusion step. Hence the diffusion into the fibre determines rate of dyeing. As the diffusion continues more and more dye is dissolved in the aqueous medium and the dyeing moves forward. The amount of dye d that diffuses into the fibre within certain time t follows Fick's first law:

$$\acute{d} = \frac{D\left(C_1 \ \square C_F\right)}{\hbar}$$

The diffusion coefficient D is 10^{-10} to 10^{-12} cm^2/s and can be increased only by an increase in temperature. The diffusion path h is predetermined by fibre geometry; the concentration gradient $C_L \cdot C_F$, by C_L.

During the course of dyeing many changes takes place. The course of dyeing is the resultant of many factors. During the dyeing as explained earlier the dye is adsorbed on the surface of the fibre. The fibre contains varying proportions of crystalline and amorphous portions in their structure. Crystalline and amorphous portions in a fibre depends on the degree of drawing and the heat setting. Diffusion takes place only in the amorphous portions above the glass transition temperature. During dyeing, the amorphous portion can change as a function of stress and shrinkage. In addition, the dyeing medium (water) and the dye already diffused into the fibre change the fibre structure and thus the dyeability. Second, in the dyebath, the dye goes through a large number of stages, from solid particle to a single molecule, which interacts with each other. These include (1) Dissolving in the liquor and possibly melting, (2) Adsorption of dispersing agent and association to dispersing agent, (3) Incorporation into dispersing agent micelles, (4) Diffusion through the laminar liquor layer on the fibre surface and adsorption to the fibre, (5) Agglomeration, aggregation, and crystal growth in the dyebath, (6) Mutual influencing of dyes, isomorphism of different dyes (formation of mixed crystals with different solution behaviour than pure components) and (7) chemical changes, e.g., reductive destruction.

Third, the levelness of dyeing, that is, the basic requirement of colouring, is endangered by differences in temperature and concentration. In practice, such differences are unavoidable because of irregularities in the velocity and direction of the liquor flow at various positions in the dyeing aggregate and gradual depletion of dye in the liquor during passage through the dyebath. Hence, a constant supply of dye at the fibre surface is not guaranteed, and diffusion no longer determines the rate of dyeing. Despite many careful studies of model systems and individual processes, which permit a semi quantitative interpretation of individual steps under standardised conditions, dyeing results still cannot be mathematically predicted unambiguously. Thus, an industrial dyeing process must be supported by dyeing experiments. Recipes are drawn up on the basis of trial dyeings with the help of colour measurements.

Thus the dyeing process is always designed to be controlled by temperature and levelling agents which can help in controlling fast- and slowly diffusing dyes, especially when used together. The levelling is further enhanced by adequate migration capacity of the dyes, especially at high temperature. The dyeing process can be also controlled with dyeing accelerants (carriers). Carriers are aromatic compounds that exert a swelling effect on PES fibres and a dissolving effect on disperse dyes. They increase the rate of absorption and diffusion into the fibre and improve level dyeing.

Dyeing Methods

    I. Batch wise exhaust methods: these involve exhaustion of the dispersed dye from the dyebath on to the fibre.

    II. Continuous methods: in this case, the dye is applied to the fibre mechanically after which it diffuses into the fibre during a thermal treatment.

5.2.4.1.12.1   Batch wise dyeing

Dyeing is carried out from relatively long, aqueous liquor in which the dye is taken up slowly by the fibre. The dye is free to migrate in this discontinuous procedure, and, provided a sufficiently prolonged dyeing time is given, equilibrium between dye in the liquor and in the fibre is established, the point of equilibrium depending on the dyeing conditions. Though it is fundamentally possible to dye from solvents or water/solvent mixtures, such solutions have hitherto not found any practical application.

Batch wise exhaust method is subdivided into:

    1. At atmospheric temperature using carrier: this method is almost obsolete now due to the environmental problems and since it is difficult to dye dark shades and the resultant shades are of poor overall fastness.

    2. At higher temperature and pressure or HTHP dyeing: This method are dyed in specialised machines since these are dyed at higher temperature or pressure or both.

5.2.4.1.12.2   Carrier dyeing

It was found that certain organic chemicals such as phenols, amines and hydrocarbons when added to the dyebath in soluble form or as emulsion accelerates the adsorption of disperse dyes on polyester at or below 100 °C. Even though the exact action of carrier in the dyebath and dyeing is not clear numerous theories exist concerning the mode of action, of which the following are the most likely:

1. Carriers act as dye solvents: a strengthening of the dye occurs at the fibre/liquor phase interface.
2. Swelling action of the carrier: loosening of the van der Waals forces or hydrogen bonds between the fibre chains.
3. Lubricating action by the carrier on the dye and the fibre.
4. Vein or hole formation in the fibre structure (supporting the formation of free volume).

Other theories are also put forward to emphasise mechanism of carrier action, e.g. increased water inhibition, transport theory, increased solubility of dye in bath, film theory, liquid fibre theory, loosening of fibre structure (lowering T_g) and increased sites accessibility; exact mode of working of a carrier has not been established.

Examples of carriers:

1,2,4-Trichlorobenzene,

Phthalic dimethyl ester,

m-Dichlorobenzene,

p-Dichlorobenzene,

o-Dichlorobenzene,

N-Ethyl aniline,

Benzoic methyl ester,

Methyl salicylate,

Quinoline,

o-Chlorophenol

2,4-dichlorophenol

Benzaldehyde

Diphenyl

Monochloro-o-phenylphenol,

o-Phenyl phenol,

Monomethylnaphthalene,

Butyl benzoate, etc.,

most of them are toxic and banned.

X-ray investigations showed that even with water treatment at 100 °C, a noticeable rise in the degree of crystallinity occurs. In the presence of carriers, this increase iseven more significant. It was also possible to establish structural improvements in the crystalline areas. The increase in the degree of crystallinity appears to take place, to a considerable extent, at the expense of the originally present mesomorphic areas which were poorly or moderately organised. The remaining amorphous areas thus become more accessible, and the dye uptake capacity increases.

From infra-red spectroscopic measurements made on polyester, it can be concluded that the pre-diffusing carrier is absorbed by the material, but above all (insofar as it is constitutionally capable of this) it is relatively strongly bonded to the polyester molecule through hydrogen bonds. This is apparent from the OH valency variation band, which displaces itself after the longer frequency range. The pre-diffusing dye displaces a more or less large proportion of the bound carrier. A requirement for this is that the dye itself likewise carries groups which, possibly more strongly than the carrier, are capable of hydrogen bonds or dipole interaction with the polyester. The dye molecule will penetrate deeper (possibly also faster) into the material the less the dye is held up on its diffusion path by polar centres or by dye/polyester hydrogen bond interaction. The pre-diffusing carrier has the task of blocking the bond ready centres of the polyester molecule to a certain extent. The dye is thus not hindered on its diffusion path, or only slightly; it penetrates faster and more deeply. At the same time, it makes way more quickly for post-diffusing dye molecules; the dye uptake is increased. The following requirements are made of carriers: cheap, effective at low concentrations, easy and safe to use, low odour, colourless, neither toxic nor skin-irritant, no effect on fastnesses and nuances of the dyeing, not sublimable out of the liquor bath, not very soluble in water and with good emulsifying properties, not too strongly bound to fibres ecologically justifiable.

Main groups of carriers:

1. Hydrophobic swelling agents, e.g. benzene, chlorobenzene, diphenyl, o- and p-phenyl-phenol, mcresol: due to their good affinity to fibre, these products have excellent swelling action and can be effectively applied even in small amounts. A disadvantage in most of these substances is their toxicity and carcinogenicity, as is the poor lasting properties of the emulsion (forms a cream after boiling for a longer period).

2. Hydrophilic swelling agents, e.g. benzoic acid and its esters, benzyl alcohol: these substances are usually only effective when larger amounts are used, but have the advantage of being less toxic and, as a consequence of their simple emulsifying properties, they do not lead to drop formation even in open equipment.

For even distribution in the liquor bath, carriers mostly contain various added emulsifiers, and very often also other additives as well as solvents, in order to improve the emulsification behaviour, including storage stability.

Carrier type	Active matter	Emulsifier	Auxiliary solvent and other additives
Chlorobenzene	70–80%	10–20%	10–20%
Diphenyl type A	85%	15%	
Diphenyl type B	30–60%	10–20%	60–30%
Ester carriers	40–70%	10–20%	40–20%

Examples of formulations of carriers

Many of the carriers are toxic, odour producing and many of them are banned due to the toxicity, difficulty in removing form the dyed material and odour. In most cases a part of the carrier is absorbed by the fibre itself and removal from fibre is difficult

Table: Removability of carrier from the fibre

Thermosol conditions	Triclorobenzene	o-Phenyl phenol	Carboxylic acid ester
120 °C, 30 s	2%	2%	2%
120 °C, 300 s	40%	7%	15%
160 °C, 30 s	65%	32%	25%
160 °C, 120 s	96%	80%	60%
180 °C, 30 s	900%	70%	55%
180 C, 120 s	98%	98%	95%

% Removability of carrier from polyester during further processes at 120 °C, 160 °C,180 °C

Selection of a carrier for a specific dyeing process is based on its toxicity, biodegradability, availability, cost, ability to show synergistic action on rate of dyeing, compatibility in bath, ease of removal, non-volatile nature, no adverse effect on light fastness and change in handle, non-irritate to skin, uniformity in absorption by the fibre and if insoluble in water, should have good emulsion stability; not a single carrier fulfils all these characteristics. Efficiency of a carrier is calculated based on its acceleration factor (a) which is defined as:

$$\frac{\text{amount of dye taken up by the fibre in the presence of the carrier}}{\text{amount of dye taken up by the fibre in absence of the carrier}} = \frac{q_*}{q_0} \leq 1$$

Carriers are of two types: water soluble and insoluble. Water soluble type includes benzoic acid, salicylic acid and phenol; higher dose is required and are not used in dyeing. Insoluble types are mostly preferred, e.g. diphenol, benzene, naphthalene, o and p-phenyl phenol, ochlorophenol, mono/di and trichlorobenzene, toluene, etc., which works better at 1–5 g/l but are highly toxic. Tumescal OPE (earlier ICI) is a popular carrier on these grounds.

The benefits of carriers were overwhelming in the early days of polyester dyeing because polyester fabrics could be dyed in unpressurised becks with no more dyeing problems than with direct dyes. However, carrier dyeing has steadily declined since the development of suitable machines for dyeing polyester under pressure at temperature around 130 °C. Carriers are still used in some garment and small commission dyehouses where high temperature pressurised dyeing machines are not available.

The dyeing can be carried on any machine like winch, Jigger, jet dyeing soft flow, etc. The only advantage of this method of dyeing is that the dyeing does not need any specialised machines. In the first method, dyebath is set at a pH~5.5–6.5 with acetic acid or formic acid, carrier is added and the textile is treated in this bath up to 60 °C followed by addition of dye. The bath is further heated up at boil and dyeing is continued for 1 h followed by soaping and washing. The method is restricted for production of only light shades as fibre opening as well as solubilisation of dye are not just enough for diffusion of dye inside causing poor wash fastness. The adsorbed carriers must be removed from the polyester after dyeing, usually by hot-air drying the goods from 150 to 180 °C. Residual carriers can adversely affect light fastness if left on the goods.

Carriers are, however, technically outdated and ecologically harmful. For this reason, the carrier dyeing technique is recommended only when dyeing temperatures of 130 °C are impossible (e.g., with PESwool blends).

5.2.4.1.12.3   Theory of high temperature exhaust dyeing

Synthetic fibres can be exhaust dyed at higher temperature without the help of carriers (carriers may be added in smaller amount as levelling agent, but not as a dyeing assistant).

Disperse dye show very low solubility below 90 °C. As the temperature is increased the solubility of the dyestuff is also increased and the opening of the fibre. This principle is used in the development of the HTHP dyeing.

Equilibrium is reached only when the subsidiary equilibria are achieved through a balance in all the dyeing steps, viz.

(i) dye dispersed in bath ↔ dye dissolved in bath,

(ii) dye dissolved in bath ↔ dye adsorbed on fibre,

(iii) dye adsorbed on fibre ↔ dye diffused in situ fibre.

However, the increase in the solubility and the adsorption has to be controlled so that a level dyeing should take place. Hence normally 1 °C/min is normally followed upto 117 °C the temperature at which all existing disperse dyes complete exhaustion due to sufficient opening in structure, seems to be right enough for levelled shades. But normally in practical dyeing another increase by 10–12 °C is adopted in commercial application for uniform distribution and ease in migration of dye to compensate heat loss during dyeing, shown by the absorption curve below.

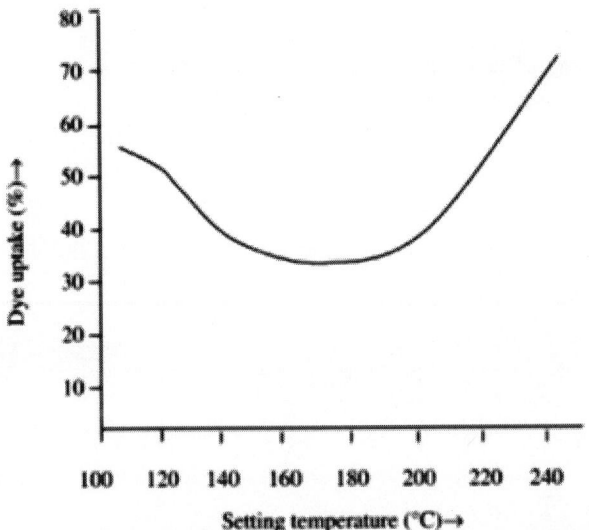

Effect of heat setting temperature on disperse dye uptake of polyester

Dye uptake decreases initially with increase in heat setting temperature of polyester; remains approx. same from 160 to 200 °C, again remarkably increased. If heat setting is carried out beyond 220 °C, uneven dyeing may occur due to improper setting. Setting temperature must be within 180–200 °C as explained in the below graph.

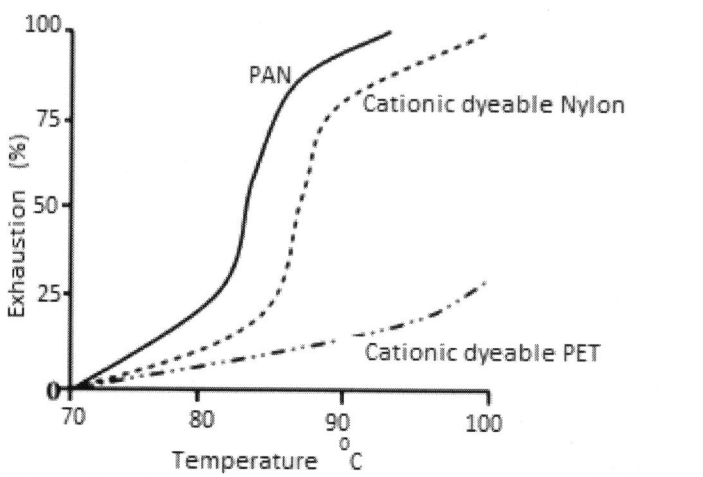

Dyeability of acrylic, cationic dyeable nylon and cationic dyeable polyester.

Because of the slow diffusion of disperse dyes at boiling temperature, PES fibres are dyed whenever possible at 125–135 °C under pressure (HT dyeing). To develop a course of dyeing that is as quick and efficient as possible, studies have been performed on optimising the temperature time program, using dyes together that have similar exhaustion properties and adapting the circulation of the liquor or goods to the type of material to be dyed. High temperature dyeing is the most widespread method of batch coloration. The temperatures (ca. 130 °C) require pressurised equipment and impart increased diffusion of the dyestuff (and therefore increased rate of dyeing) by reducing cohesion between polymer chains and increasing the kinetic energy of the dye molecules. A typical exhaust dyeing application sequence for polyester is shown in below figure, showing the three main phases of the process including the heating or adsorption phase, the high temperature or diffusion phase, and the clearing phase.

Adsorption phase

The heating or adsorption phase is the most critical in determining the levelness of the dyed fibre and it is essential therefore that the heating rate is appropriate to allow controlled adsorption of the dye. Although in the dyeing of polyester, levelling can occur through migration at top dyeing temperature, in rapid dyeing cycles the time at top temperature is minimal and it is even more critical to ensure dye is applied in a uniform manner during the adsorption phase.

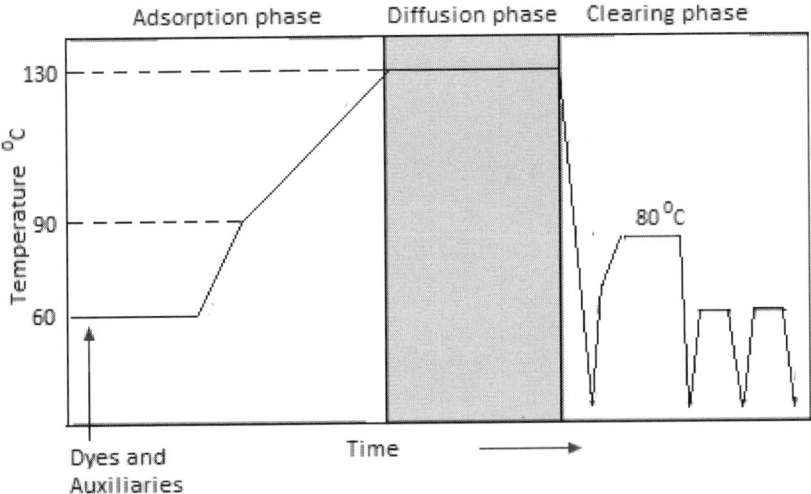

Phases in Exhaust dyeing of Polyester

The adsorption behaviour is strongly influenced by a number of factors, the most important of which are concentration of dye, temperature gradient, fibre type and auxiliary system. The rate of exhaustion of a disperse dye by polyester is controlled by the rate at which the temperature is raised. At some temperature between 80 and 120 °C the dyeing rate for that dye reaches a maximum. The temperature range over which the dyeing rate is at this maximum is known as the 'critical dyeing temperature' (CDT). Slow-diffusing high-energy dyes have a high CDT, whereas more rapidly-diffusing dyes have a lower CDT. Specific values of CDT depend on the rate of temperature rise, dye concentration, liquor flow rate, liquor ratio and the substrate to be dyed. Rapid-dyeing procedures depend on adding the disperse dyes at a temperature just below the CDT and then raising the temperature slowly in the vicinity of the CDT to ensure that the exhaustion rate that just permits level dyeing is not exceeded. The temperature is then raised from just above the CDT to the top dyeing temperature at the maximum rate.

Diffusion phase

The dyeing of polyester is often described as a diffusion-controlled process. This is because of the diffusion phases shown in Fig. 16, including convective transfer through the liquor adsorption and molecular diffusion into the fibre, is the rate-determining step. Where the time needed for the adsorption phase is largely influenced by the machine conditions, in the diffusion phase the time required at top temperature is directly related to the diffusion characteristics of individual dyes and dyeing depth, but generally, the standard time is 10–20 min for dyeing up to pale shades, 20–30 min for medium shades and 30–35 min for deep shades.

In the high temperature phase of the dyeing process, another important property of dyes is migration, or their tendency to level out. This phenomenon can be very important, particularly if dyes have been adsorbed in a non-uniform manner, perhaps due to in adequate liquor circulation or too rapid a heating rate. While the migration properties of disperse dyes may become a key factor if dyes are applied unevenly during the adsorption phase, the key parameter in the diffusion phase is the diffusion rate of the disperse dye.

Clearing phase

Because disperse dyes have such limited solubility in water, some particulate disperse dye may still be occluded on fibre surfaces after the dyeing phase is complete. If not removed, this surface contamination can undermine the brightness of shade as well as the wash, sublimation and crock fastness results. Commonly, the dyed polyester is cleared of surface-deposited dye as well as auxiliaries (e.g. carriers, surfactants) by means of treatment with detergent or reductive or oxidative treatments, in order to secure optimum fastness of the dyeing and also to improve the brightness of shade.

The usual treatment carried out, especially in heavy depth, is reduction-clearing, where the dyed fibre is treated in a strong reducing bath, usually made up of sodium dithionite and caustic soda. A treatment for 20 min at approximately 70–80 °C, is often sufficient to clear the fibre surface, but the ease of removal varies from chromophore to chromophore and dye to dye. This treatment acts to destroy loose azo disperse dye through chemical reduction of the azo link. Anthraquinone disperse dyes are not fully destroyed by such a treatment but a degree of removal of surface dye is achieved through temporary solubilisation of the disperse dye to the alkali-leuco form.

Reaction during the reduction clearing process

Research indicates that the polyester dyer will typically reduction-clear in the range of 30–50% of production shades and sometimes an even higher proportion in blend dyeing.

Minimising the need for reduction-clearing can lead to substantial productivity improvements and water and chemical savings as well as a reduction on the effluent load and should be a key objective in implementation of a rapid dyeing approach for polyester.

Disperse dyes are insoluble in water up to 90 °C, beyond which becomes soluble which increases with increase in temperature of bath, also opening of fibre (polyester) starts at 85–90 °C and increases with increase in temperature of bath. To control diffusion of dye for a levelled shade, the temperature is controlled from 90 to 120 °C, beyond which the bath is heated up to 130±2 °C for more opening required to effective dye migration inside to produce levelled shades. The method is very popular due to (i) efficient levelling as well as covering of yarn irregularities with superior fastness through effective diffusion of dye and (ii) process efficiency is higher due to better exhaustion of bath. No auxiliary are required except little acid and dispersing agent.

Level dyeing of a shade is highly dependent on the control of temperature. It should be noted that up to 100 °C, as high as 10% of the bath is exhausted, 80% exhaustion takes place during 100–120 °C and the rest 10% between 120 and 130 °C. Neither the initial nor the final absorption stages affect levelling and so such control over heating of bath is necessary, but an efficient control is needed during heating in the range of 100–120 °C.

After dyeing a reduction clearing is done as described below. In the cases of pale and medium-depth dyeings or for those dyes which cannot withstand reduction-clearing, the dye fibre can be given an alkaline scour; residues of anthraquinone dyes which may remain after reduction-clearing, may be removed using an oxidative treatment.

In favourable cases, the entire dyeing cycle can thus be shortened to less than 60 min. The same goal can be approached by using complex dye mixtures. HT dyeing can be made still more economical by using specially designed dyeing equipment to reduce the liquor ratio. The evenness of dye absorption is improved by dyeing in the presence of a levelling agent. Carriers can also be used as HT levelling agents.

5.2.4.1.12.4   Practical HT Dyeing

The HT dyeing process has its own advantages against carrier dyeing. Now a days carrier dyeing is almost obsolete and in batch wise dyeing HT dyeing is the most followed process. The advantages of HT dyeing are: The dyeing times are frequently shorter, the polyester fibres are better penetrated, there is more compensation for fluctuations in setting and stretching, sometimes a higher colour yield, and often somewhat better fastness properties. Furthermore, possible sources of fault which can exist when dyeing with carriers, e.g., the formation of spots due to the condensation of the carrier, breakdown of the carrier emulsion, and so on, do not arise. Apart from all this, there is no need to take special precautions to remove completely residual carrier from the fibre material where inadequate removal would mean impairment of light fastness properties.

Quantity	Unit	Bath additions
x	%	Disperse dye
0.5–1.0	ml/l	Acetic acid 30% to pH 5–6
0.5–1.0	g/l	Dispersing agent (e.g., Setamol WS)

Procedure:

The acetic acid and dispersing agent are first added to the dye liquor, and the dye dispersion is then passed through a sieve into the liquor when this is at 50–60 °C. The material is run for a few minutes at 50–60 °C to ensure uniform distribution of the dye, whereupon the temperature of the liquor is raised within 30–40 min to 120–135 °C (normally 125–130 °C) at a corresponding pressure of about 3–3.5 kg/cm², dyeing proceeding at this temperature for 1–2 h depending on the depth of shade required. The pH value of the dyebath should lie between 5 and 6 during the whole of the dyeing process.

After the dyeing is over it is preferred to discharge the bath at 130±2 °C to drain out liquefied low molecular weight oligomers which otherwise get precipitated on fibre surface during cooling below 120 °C and causes harsh feeling as well as deteriorates rubbing fastness of dyeings. The material is rinsed and given a (reduction) clear after completion of dyeing.

Notes:

1. When dyeing pale shades, it may well be advisable to commence dyeing at a lower temperature (30–40 °C) and then to raise the temperature more slowly.

2. The amount of dispersing agent is increased as the liquor is made shorter and the dye concentration lower.

3. Precautions must be taken to ensure that, where no automatically operating exhaust equipment is available, the removal of air from the system is adequate during the whole of the dyeing process. It is important to achieve the pressure inside the close system required to maintain the required HT. The static pressure must be maintained at least so high that the pump operates at its maximum efficiency (no cavitation) at the dyeing temperature employed.

After clearing

The dyeing of polyester fibres with disperse dyes is almost invariably followed by a clearing process in order to remove residues of dye and auxiliaries (including carrier). The following improvements are achieved by after clearing:

(a) The fastness properties are better, e.g., rubbing, washing, sublimation and light fastness properties;

(b) The colour is often rendered clearer, this being important for brilliant shades; and

(c) In some cases further processing of the fibre material is made easier, for instance, when spinning dyed polyester slubbing.

After clearing may consist of a simple scouring process. Usually, however, a reduction clear is given, the method normally depending on the type and quantity of dye and auxiliary employed, the type of fibre or the stage to which it has been processed, and on the dyeing technique used. The safest procedure for removing the last traces of carrier is to give a hot, alkaline scour followed by hot-air treatment of the material.

A washing/scouring may be given as follows:

Quantity	Unit	Bath additions
0.5–1.0	g/l	Scouring agent
0.5–1.0	g/l	Soda ash

Treat for 15–20 min at 70–80 °C before rinsing.

Reduction clearing

In the most of cases, especially medium to dark shades it is customary to give a reduction clear, by which the best possible fastness properties are obtained for a given dyeing. Where permissible, a reduction clear is also given with polyester fibre blends after the polyester component has been dyed.

Quantity	Unit	Bath additions
3–5	ml/l	Caustic soda 38 °Be
2–3	g/l	Sodium hydrosulphite
1–2	g/l	Dispersing agent or a scouring agent

The temperature of the reduction liquor is rapidly increased to 70–80 °C, or the hydro sulphite is not given until the liquor is hot. The dyed material is then treated at this temperature for 15–30 min, after which it is rinsed warm and cold and then acidified with acetic acid.

Notes:

1. It is useful, especially with deep dyeings, to add a little hydrosulphite (0.5–1 g/l) to the first rinsing bath to prevent any premature re-oxidation of anthraquinonoid dye residues.

2. With a liquor ratio shorter than 20:1, it may well be necessary to go beyond the quantities given for the hydrosulphite and alkali. Again, a further addition of hydrosulphite and caustic soda may be required when working with open vessels, particularly when the material is outside the liquor for part of the treatment time.

5.2.4.1.12.5   Continuous dyeing

Continuous methods are subdivided into:

1. Thermosol process: the dye is applied to the fibre mechanically with water acting as a transport medium. The water is then removed by drying (approx. 1 min at 200–220 °C). The fibre becomes thermoplastic in this temperature range which facilitates diffusion of dye into the fibre.

2. Steam fixation: Pressure steaming at 130 °C gives relatively good fixation whilst superheated steam, which finds wide application in the fixation of printed disperse dyes, is only used on a small scale in continuous dyeing.

5.2.4.1.12.5.1   Thermosol process

Thermosol dyeing process is important continuous process for dyeing polyester and polyester/cellulose fibre mixture with disperse dyes, which is used mainly for woven and knitted materials. This method has some advantages that could avoid a batch wise process (a conventional dyeing method), which is regarded as time-consuming and tedious, and would speed up the dyeing process. This is an important continuous process for dyeing polyester and polyester/cellulose fibre mixtures with disperse dyes, which is used mainly for fabrics and knitted materials.

Continuous thermosol dyeing of polyester fibre and polyester/cotton blends with disperse dyes came into its own in the early 1950s as textile mills began using more of these fabrics in place of 100% cotton. It provides a means of attaching the dye without exhausting colour onto the fabric and at one-third the cost. In this process, which applies the dye by a padder, the fabric is gradually dried and heated to temperatures of 400–430 °F for 90 s, depending upon the energy needs of the disperse dye in use.

It is still particularly popular for a number of applications such as apparel, workwear, home furnishings and industrial fabrics. Over the years techniques have been perfected for optimising its use and avoiding pitfalls that can still be encountered. Here are recommendations that should help achieve good success with the process.

5.2.4.1.12.5.2   The principle of thermosol process

The principle of dyeing is in forcing the dye to get solubilised in the solid fibre by partial loosening of the intra molecular bonds by heat when the internal structure opens up and allow the dye to diffuse into the interior of the fibre and when the materials is cooled to the room temperature the dye molecules are trapped inside and not allowed to come out as the pores become smaller. There are problems in dyeing 100% polyester by this method since the fibre is highly hydrophobic it is difficult to apply dye on the fibre by padding before the thermosoling process. Another problem is the migration of the dye due to heat which is also attributed to the hydrophobic nature of the fibre. Whereas the method is recommended for dyeing polyester cotton etc., where the dye is absorbed by cotton portion of the blend and at thermosoling temperature the dye is transferred and diffused into the polyester fibre from the cotton. Cotton can be further dyed separately as required. If the fabric is prepared well and the procedures are followed correctly a level dyeing is possible only care must be taken to avoid migration mainly.

Due to hydrophobic nature of polyester, dye concentration is kept on higher side. Liquid disperse dyes provide excellent result due to high rate of dispersion and diffusion; these diffuse more rapidly into fibre structure with saving of energy and short time of fixation. Dyes from B and C classes possessing moderate sublimation and wash fastness properties may be selected for levelled shades; dyes form D class produce excellent fast shades but are not recommended on grounds of having higher sublimation temperature and excess of dye is required to produce desired depth due to poor diffusion co-efficient. However, a reduction clearing definitely improves wash fastness and the latter can be verified by dipping dyed sample in dimethyl formamide at room temperature – stripping of colourin solvent confirms presence of dye on surface. Introduction of a migration inhibitor in padding liquor like sodium alginate or carboxymethyl cellulose arrests migration of dye and produces levelled shades.

First dye, thickening agent or migration inhibitor and possibly a hydrotropic agent (urea, thiourea) and an acidifier are padded. The coating of dye dispersion on fabric is very sensitive to touch and so the padded material be instantly passed through drying zone without any contact with guide rolls or other accessories. The best way is to place infra-red heaters just above the last nip for mild drying without migration of dye followed by drying in float drier at 90–100 °C.Finally, the dry fabric is then heated in air, to a temperature in the range of 190–220 °C for 1–2 min.

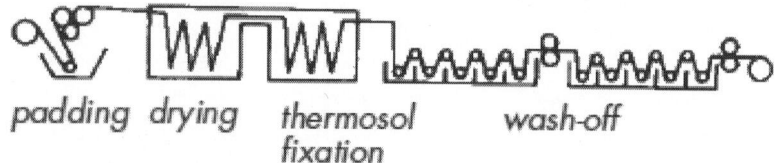

padding drying   thermosol   wash-off
                   fixation

Process flow of thermosol dyeing

In hot air, as the fabric approaches the maximum temperature, the disperse dyes begin to sublime and the polyester fibres absorb their vapours. Dyes of lower molar mass tend to sublime more readily, but they also suffer from low fastness and poor resistance to heat treatments. Dyes of higher molar mass have better fastness properties but are more difficult to apply. After thermo fixation, scouring or even a reduction-clearing treatment is necessary to remove any dye remaining on the fibre surfaces. PET seatbelt webbing is typically dyed with disperse dyes using thermosol dyeing processes in which the webbing is dipped continuously into a dye solution and passed through a hot chamber (ca. 220 °C) for approximately 2–3 min. The dyestuffs can penetrate the molecular chains of the fibres during their exposure to the hot chamber.

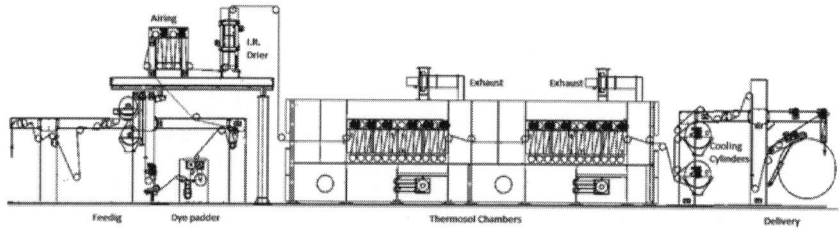

A two chamber thermosoling unit

Increase in draw ratio during melt spinning promotes orientation and decreases rate of diffusion; even heat setting at 180–210 °C causes surface melting of the polymer with less dye uptake. Uncontrolled drawing in filaments produce barré; stains on fabric must be removed prior to dyeing to produce levelled shades. Sometimes, a ring dyeing effect is produced due to insufficient diffusion of dye which may be rectified by exposing dyeings at 180–210 °C for 30–45 s. Non-ionic surfactants, when applied as dyeing accelerants, promote rate of dye dissolution in auxiliary melt, thereby increasing the overall rate of thermosol dyeing. At high surfactant concentrations, dye retention in surfactant may adversely affect dye fixation.

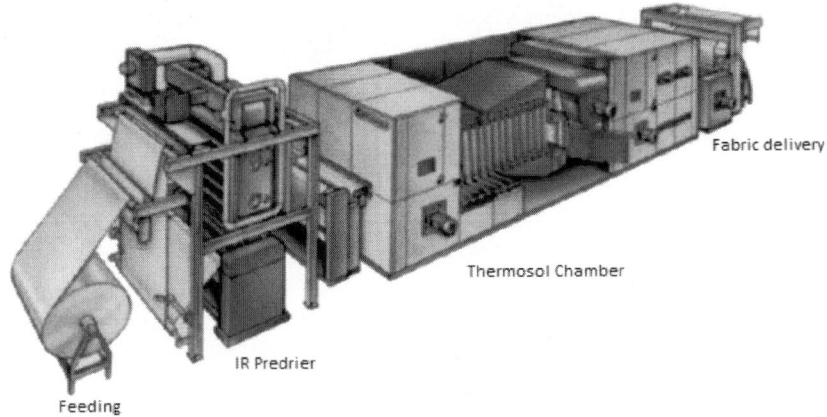

Feeding

IR Predrier

Thermosol Chamber

Fabric delivery

Thermosol Dyeing Plant

Diffusion of the dye from the surface to the fibre interior takes place during subsequent thermosol fixation in the vapour phase. The thermosol fixation process is also suitable for triacetate fibres. It is only suitable for polyacrylnitrile fibre when certain cationic dyes are used.

During thermosol fixation, disperse dyes are set on polyester fibre at a temperature of 200–225 °C which is applied for 30–60 s. The setting speed, the degree of setting and the penetration of dye into the fibre depend above all on the following factors:

- chemical constitution and finish of the dyes,
- dye combination,
- concentration of the dyes and fibre surface,
- type of polyester fibre,
- ratio of polyester/cellulose fibre,
- fabric structure,
- auxiliary additives in the pad bath,
- temperature of the fabric and thermosol fixation time.

The fibre material is specified before dyeing; dyes, auxiliaries, thermosol fixation temperature and time can be selected. The temperature of the fabric and the thermosol fixation time are not necessarily dictated by the temperature of the heat medium and the time spent in or on this medium. They are dependent on the heating speed which in its turn depends on the fabric structure, residual moisture and type of heat transfer.

Fluctuations in the temperature of the fabric and the thermosol fixation time cannot be completely eliminated at reasonable expense under operating conditions. Therefore, disperse dyes are preselected which result in level dyeings over a temperature range which is as wide as possible. Temperature dependency of dye fixation is documented by so called thermosol curves on which either the degree of fixing or the dye strength is recorded against the temperature. The degree of fixing indicates the quantity of fixed dye in relation to the padded dye quantity. The relative dye strength corresponds to the visual or colour measurement determined dye strength in relation to the maximum achieved dye strength. Both dimensions are usually quoted as a percentage. Migration during intermediate drying continues to cause problems in practice. The migration of dyes is not only the cause of two-sidedness and listing; strongly migrating dyes also lead to a non-uniform appearance, double colour effects (dichroism) and colour deviations in the sewing sample. Thermosol dyeing on a polyester mix fabric can have a worse appearance than an exhaust dyeing due to inadequate dye penetration. Optimum machining prerequisites, drying conditions, the migration inhibitors used and also disperse dyes with a minimum tendency to migrate can have a positive impact on the appearance of a thermosol-dyed product. The migration capacity of a thermosol dyemixture can be simulated in a laboratory test.

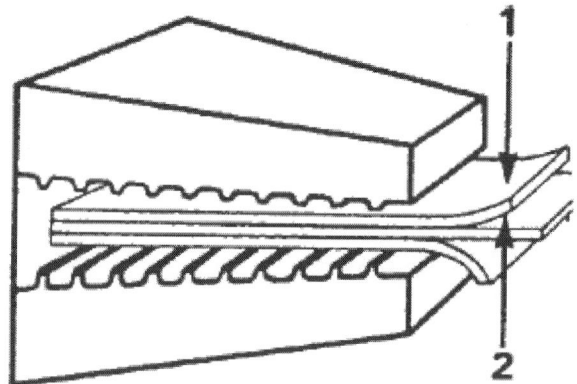

Migration during intermediate drying – effect of the dispergator system on the migration code using the example of Foron marine blue S-2GL (Sandoz).

Fabric: PES/CO 65/35 merc.; padding liquor: 32 g/l Foron marine blue S-2GL, 20 g/l Sanapol AM liquid (Sandoz); dry/thermosol: 2 m, 150 °C/min, 215 ° reductive cleaned, PES skeleton. 1 = fabric upper surface; 2 = fabric lower surface

5.2.4.1.12.5.3   *Practical dyeing process*

A practical dyeing process can be divided up into the following operations:

1. Padding the dye on to the material;
2. Drying the padded material;
3. Fixing the dye in the fibre (the Thermosol process);
4. After treating the material, or cross-dyeing the other component in the case of mixtures.

Fundamentally, three things are necessary for obtaining satisfactory results with the thermosol process:

(a) very carefully and uniformly pretreated materials,

(b) specially selected dyes and textile auxiliaries, and

(c) a uniformly operating thermosol dyeing plant with units that are well synchronised with one another.

Essentially, a thermosol dyeing range consists of:

the padding machine,

the drying plant, and

the thermo fixation (thermosol) installation.

Two- or multi-bowl padding machines are employed, usually the latter. Drying is best divided up into pre- and residual-drying processes. "Infrared-driers" are often used for predrying, while hot-flues, float-on-air driers or cylinder driers are employed for residual drying. Thermo fixation, or the thermosol treatment, is carried out on stenter frames, hot-flues with supplementary heating (heat chambers), float-on-air driers or even cylinder driers.

The heat energy required for drying, and especially for the thermosol process, can be applied as:

(a) hot air,

(b) dry-superheated steam,

(c) contact heat (solid metal surfaces, liquid metals, glass beads and so on), and

(d) infra-red radiation.

Various heat transfer arrangements can be combined together, for example, heated metal cylinders blown with hot air, or infra-red heating combined with hot air. Small thermosol dyeing installations can be obtained for working out recipes in the laboratory. When suitably adjusted as regards the passage of the material, these machines can also be used for the thermosol dyeing of belts, tapes, tow, slubbing and yarns on the commercial scale.

5.2.4.1.12.5.4   Pretreatment of the material for thermosol process***

Since the polyester materials are basically hydrophobic the pretreatment process should ensure that the material wets out equally overall. Any difference in wett ability causes uneven uptake of the dye during padding, and hence a skittery and unlevel dyeing. The surface of the fabric must be smooth and free of soiling. Thick knots and projecting ends of fibres must be carefully removed, since they are pressed in during padding and cause localised paler dyeings. Materials which contain a large number of burls or knops are not suitable for thermosol dyeing.

Materials can be prescoured with suitable scouring agent (e.g., Kierlon B – BASF). If spotting has to be done with solvent containing products to clear localised soiling care should be taken to avoid strong rubbing and the spotted material should not be allowed to dry with solvent and immediately taken for scouring. The solvent may have their affinity for the dye modified apart from the possibility of leaving a halo around the spotted area, the places treated with solvent.

Polyester (mixture) materials do not normally need to be present if the thermosol process is to be carried out on a stenter, since the material is actually set during the process, appropriate adjustments to the machine and feed of the material being made. Presetting is employed only when tensions are present in the material which might upset the regular progress of the material through the machine, or if irregular lengthwise and breadthwise shrinkage is to be avoided, since such could occur during the thermosol process in the heating chamber or on the heated cylinders.

When polyester (mixture) fabrics are to be dyed by the thermosol process, singeing can be carried out as a pretreatment. The resulting melted ends of the fibres do not dye to a different shade in the thermosol process as indeed they would do from aqueous liquor. The most economic procedure is to singe the loom-state material. However, it is sometimes preferable to desize and scour before singeing, since the fibre ends are more exposed and the burning in of impurities is avoided. Singeing must be uniform over the whole width of the material (staggered arrangement of the burners). Non-uniform singeing of the material leads to similar differences in affinity of the polyester fibre for the dye to those caused by non-uniform setting. It is advantageous to make two passages at a high speed with a small flame.

5.2.4.1.12.5.5   Padding

5.2.4.1.12.5.6   Auxiliaries used in pad liquor and their functions

Padding auxiliaries are almost always added to the padding liquor and thus have the following functions:

(a) maintenance of the finely dispersed state of the dye (stabilisation of the padding liquor);

(b) improvement of the adhesion of the dye to the fibre;

(c) prevention of dye migration during drying;

(d) prevention of soiling of machine parts;

(e) promotion of the transfer of disperse dye from the natural fibre to the neighbouring polyester fibre when thermosol treating mixture fabrics;

(f) acceleration of dye fixation and increase of the colour yield.

Whatever auxiliaries are used, these should have the quality of easily washing off after dyeing.

Normally, a rapid wetting out agent and an anti migrating agent along with a pH buffer or acetic or formic acid to maintain the required acidity (pH 5–6) are the auxiliaries used in padding. Wetting agent is more important in the case of blends like polyester/wool.

The padder employed should ensure virtually water-insoluble, finely dispersed dyes uniformly applied on to the water-repellent, non-absorbent, smooth-surfaced polyester fibres, and in such a manner that they do not migrate under the conditions of drying. This problem is much less in the case of blends polyester fibre/natural fibre fabrics are somewhat more absorbent but there still remains the problem that the polyester component, which usually comprises the greater proportion of the mixture, and for which the padded disperse dye is intended, has little ability to pick up liquor. Efforts must be made therefore to carry out padding under the most favourable mechanical conditions, and to improve the adhesion of the dye to the polyester fibre by the addition of suitable auxiliaries. In the majority of cases, it is an advantage in padding to have a long immersion path so as to saturate the material as much as possible with the liquor. Normally, the material is given two immersions and passed twice through the nips, especially when good penetration of the goods and a satisfactory levelness of padding is to be obtained on densely woven blend fabrics, these comprising a very large part of all polyester materials submitted for thermosol dyeing. Preference is therefore usually given to a 3-bowl padding machineor a five (horizontal bowls forming a trough) rather than to a 2-bowl padding machine (see under coupled pad batch dyeing of reactive dyes section). The hardness of the bowls should be 60–70° Shore, and should, as far as possible, be the same for all bowls. With structured fabrics, the raised portions of the material are usually squeezed off more than the rest of the fabric so that they are dyed paler. The use of softer bowls can be of advantage in such a case. Padding is normally carried out at room temperature, though 40 °C is sometimes employed, especially where the material is difficult to wet out, as, for instance, with polyester/linen blends. As a rule, raising the padding liquor temperature above 20–30 °C for any length of time impairs the stability of the dye dispersion. The dye must be brought on to the material absolutely uniformly (edge to edge)during padding. It is practically impossible to adjust any irregularities in this respect later on.

5.2.4.1.12.5.7   Pre- and post-drying

The function of a predrier is to reduce the moisture content of the fabric to prevent the dye migration during drying. Usually, an infra-red predrier is used (explained in detail under E – control dyeing in reactive dyeing section In Dyeing of Textile Substrates I – Cotton by the same auothor). The moisture loss should amount to about half the initially padded amount of 30–40% in the case of polyester (filament) piece goods, i.e., there should be a residual moisture content of 15–20%, and one of about 30% with polyester/cellulosic fibre fabrics (initial pick-up 60%). Moisture measuring instruments are not only able to indicate the extent of predrying, but can also be used as regulators. After leaving the predrying shaft, the material is finally dried either in a hot-flue, float-on-air drier, or on a cylinder drying machine.

To avoid the depositions on the initial roller are eased by using Teflon-coated or electro-polished rollers. It is important, both in predrying and final drying, that the heat should be applied uniformly all over the material, and that no folds/creases should develop. No thermo fixation (diffusion of dye into the fibre) should occur during the drying stage. Normally, drying is carried out over 100–140 °C, *the actual temperature depending on the material and the drying system.*

5.2.4.1.12.5.8   Thermosoling

Thermosoling can be done as a continuous operation after drying or may be treated as a separate operation. Since thermosoling units in continuous operation can frequently operate at a higher speed than the stenters so often used for the thermosol process. When treated as a separate process enables the material to be inspected after padding and drying. If faults, such as spots of colour, spotting caused by droplets, running folds and so on, are detected at this stage, the goods can still be washed out. It is hardly possible to deal with such irregularities after the thermosol process. If stenter is used for thermosoling it should preferably be equipped with at least three heating zones. If fewer heating zones are available, apart from the necessarily slow through put of the goods it is not possible always to ensure uniform treatment. This is mainly due to the fact that the pin chains require a certain time to warm up on entering the heating zones, and the conditions for a uniform temperature over the whole width of the material are not attained until this is achieved. In order to avoid marking off on the pin bed and to obtain in a better flow of air between the material and the chain, it is best to have hook-shaped pins, or pins with a thickened base, or the like. This prevents the material lying directly on the pin bed itself. Clip chains are of little use here, since they may mark the material. The greater the number of fixing zones, the greater the rate of through put possible. A thermosol treatment with hot air usually requires 30–60 s at 190–220 °C, this depending on the type of material, depth of shade and the individual dye. Normally, the conditions are 60 s at 200 °C.

Apart from appropriately designed stenters, machines without a constant-width device are employed, e.g., heat chambers, special thermosoling units, etc. are used. These give a greater rate of production, but, when using such a method, even with present material, the fact must be taken into consideration that a constant width of material cannot be guaranteed. However, it is possible to adjust the width to a certain extent by means of a short after setting treatment on a stenter, (about 20 s at 200 °C). After completion of the thermosol treatment, the goods should pass through a cooling zone and should never be rolled up or plaited down hot.

Guideline recipe

Quantity	Unit	Bath additions
x	g/l	Selected disperse dyes
15–20	g/l	Padding asst. (migration inhibitor)
2	g/l	Dispersing agent
y	ml/l	Acetic acid to pH 5–6

The auxiliaries are dissolved by pouring on a little hot water. The disperse dyes are dispersed or alternatively diluted with a little water. The products are poured separately through a sieve into the mixing vessel. The mixture is then made up to the required volume with cold water and homogenized by stirring.

Padding is usually carried out at 20–30 °C.

Hot flue (drying unit) drying temperature should be around 120 °C but not over 140 °C is usually employed for drying. Depending on the type of material and the dyes used, the thermosol process is applied for

30–60 s at 190–220 °C.

(on the average 60 s at 200 °C)

using hot-air. When employing contact heat on heated cylinders (rarely used now) the thermosol treatment takes 10–20 s at 215 °C.

5.2.4.1.12.5.9   Reduction clearing

Reduction clearing can be conducted on a jigger, winch, open-width scouring machine, or liquor circulating machine.

Jigger

Quantity	Unit	Bath additions
8–10	ml/l	Caustic soda 38 °Be (32.5%)
3–4	g/l	Sodium hydrosulphite
2–4	g/l	Dispersing agent
1–2	g/l	Scouring agent (e.g. Kierlon B-BASF)

2–4 ends at about 70 °C; rinse and acidify.

Winch and circulating machines

Quantity	Unit	Bath additions
3–5	ml/l	Caustic soda 38 °Be (32.5%)
1–3	g/l	Sodium hydrosulphite
0.5–1	g/l	Dispersing agent
0.3–0.5	g/l	Scouring agent (e.g. Kierlon B-BASF)

Treat for 20–30 min at about 70 °C; rinse and acidify.

Guideline recipe for a OW scouring machine (6 Compartments):

1st compartment –rinse cold;

2nd, 3rd, and 4th compartment –reduction clear;

5th compartment –rinse warm;

6th compartment –acidify cold.

Quantity	Unit	Bath additions
8–10	ml/l	Caustic soda 38 °Be (32.5%)
4–5	g/l	Sodium hydrosulphite
1–2	g/l	Dispersing agent
0.5–1.0	g/l	Scouring agent (e.g. Kierlon B-BASF)

Hydrosulphite check should be made during reduction clearing. Should more be necessary, 1 l caustic soda 38 °Be should be added with each kilogram hydrosulphite.

5.2.4.1.12.5.10 Problems of thermosol dyeings and solutions

Poor fabric preparation: This is probably the single biggest cause of problems. Be sure to follow recommended preparation procedures. Skimping on preparation usually ends up costing dyers twice the amount they saved in terms of light fastness deficiencies, excessive crocking, colour bleeding, unlevel dyeings and side/center/side shading.

Improper dye weighing and makeup: Disperse dye mixes must be properly stirred and strained. Specking can occur if dye has been stored for a while and dries out on the side of the drum or if it is not stirred properly before weighing. Dye that is not properly protected can also grow moles which cause its molecular size to change. Reliable dye manufacturers incorporate a herbicide in their products to control this problem. The water used in dye mix makeup should not be hotter than120 °F, and the mixer should be run slowly at all times. Never subject disperse colours to live steam which can cause dispersion break-down or low cloud points. This in turn can cause specks, unlevel dyeing or dye separation from other colours in the bath.

Mixing of dyes from different suppliers: It is important to use dyes from the same supplier in the colour combinations to help assure that the dispersion systems are the same and that they will be compatible. The plant control laboratory should make quality checks to verify compatibility and take speck tests of colour combinations.

Excessive amounts of salt and anti migrants: Excessive amounts of these agents used to control migration tend to reverse the appearance and migration of the fabric as well as build up on the machinery. Amounts of each in the range of 15 g/l are normally sufficient. In the dyeing of 100 percent polyester, add a thickener with anti migrant to insure even pickup and good appearance. With the use of Disperse Blue 60 or Reactive Turquoise, common salt in the presence of static electricity from fabric will sometimes precipitate the dye in the padder. If this happens, find a means of discharging the static before the dye enters the padder or do not use salt.

Adverse effect of acetic acid: If acetic acid is used as the pH control, it can create sodium acetate in the presence of highly alkaline fabrics which tends to reduce the yield of some disperse colours. The use of MSP instead of acetic acid for pH control is recommended. The pH of the thermosol dye batch should be in the 5.5 range to obtain the best yield of disperse colours and should remain constant during the run. A pH below that range can cause problems with the dispersions, anti migrants and wetting agents: above it the dye yields may deteriorate or the shade may become duller due to scorching of the cotton fabric.

Improper dye selection: Low energy disperse dyes usually have limited light fastness and poor sublimation when thermosoling and can contaminate the oven. Selection of disperse dyes in the medium to high energy level range give the best results in the continuous thermosol process.

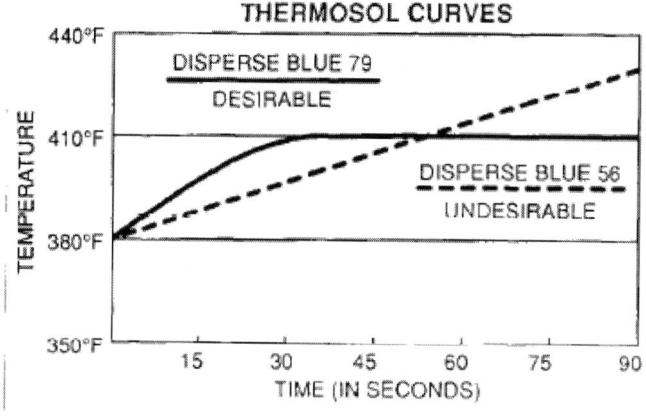

In thermosol dyeing the best results are obtained by using medium to high energy dyes

The disperse dyes can be divide into three groups:

Group I –Low energy dyes includes dyes which can be thermosoled at as low a temperature as 190 °C (±5 °C approx.). This temperature is employed, for instance, where a considerable degree of fibre protection is necessary. However, the temperature must not fall below185 °C, even in the case of fluctuations.

In Group II – Medium energy dyes. The dyestuffs are employed at around 200 °C (±5 °C approx.). This is the temperature range which is normally chosen for the thermosol process in any case.

Group III contains dyes which give the best results at about 215 °C (±5 °C approx.).

Generally medium and high energy dyes are used for thermosoling. Avoid mixing two groups even though the temperature ranges given with the three groups should not be considered as comprising rigid limits for the use of the dyes. The manufacturers can give guideline of which are the most suited dyes for thermosol process considering energy, temperature of dyeing, heat stability and sublimation fastness criteria.

Improper clearing of stains: A common mistake is to assume that a sulphur or a vat procedure will provide even removal of disperse colour stains which is a necessary part of the process. The keys to preventing re-deposits after the removal procedure are water flow, temperature and use of the proper dispersant. If the surfactants used in the removal of unfixed dye or disperse stain are not free-rinsing, they will carry colour to the final drying cans and cause drying of the unfixed colour on the surface of the fabric. This, in turn, can lead to poor cold water bleed and poor light fastness, even if initial crock test results are acceptable.

Wear of wash boxes and nips: Do not overlook the maintenance of wash boxes and nips which will wear over time. This is especially important in dyeing deep shades.

Common problems: Two common problems encountered by some in continuous thermosol dyeing with disperse colours are dye spots and side center shading. The following are suggested practices for avoiding them or eliminating them once they occur. To avoid dye spots, stir the dye well before using and strain through a micron strainer. Dilute the dye with warm water before straining and be sure the strainer is clean.

To avoid side center shading, follow these practices. Maintain an even chamber temperature within 5 degrees from side-center-side. Avoid uneven predryer burning or heating from one side to the other. Select the proper energy-level disperse colours and make sure the nips on the washer are even. In addition, do not use too much salt or anti migrant (30 g/l is usually the maximum for each), keep damp fabric from coming to the machine or entering the oven, and do not place a fan or have a direct air flow on one side of the fabric only.

5.2.4.1.12.6   Special dyeing processes

5.2.4.1.12.6.1   HT steaming

This takes place in continuous HT loop steamer. Hot steam above 100 °C is the heat transfer medium. Specially suitable for fixing polyester or triacetate when printing with disperse dyes. It is possible to work tension-free so they are especially suited for processing tension-sensitive knitwear. Operated at temperatures between 160 and 180 °C for 6–8 min. Higher fixation temperature and dwell time require careful selection of dyestuff (above-average fastness to sublimation); as a rule, printing paste contains a fixation accelerator. They are also used, with suitable dyestuff choice and fibre type, with saturated steam (100 °C) so these steamers have a temperature range between 100 and 180 °C.

Very light shades are sometimes dyed without the use of pressure and without a carrier. Only those disperse dyes that diffuse sufficiently quickly into PES fibres even at 100 °C are suited to this purpose (e.g., *C.I. Disperse Yellow 7; Orange 1, 33; Red 4, 11, 60; Blue 56, 81*). The dyeing equipment must be tightly closed, and top heating in vats and jiggers is advantageous. Dyeing time is extended to several hours. Since, in a process of this type, a portion of the dye adheres to the fibre surface without being fixed, dyeing must be followed by a reductive clearing step.

5.2.4.1.12.6.2   Dyeing from foam

For saving energy, attempts were made to entirely dispense with an aqueous dye liquor and to apply disperse dyes to PES articles from a foam medium. Problems can arise with regard to the levelness of dyeing. When considering the energy balance of the foam dyeing technique, it is important that although water is saved in the dyeing process, considerable amounts of water are required e.g., for pretreatment and after cleaning.

5.2.4.1.12.6.3   Air as transport medium

A clear economic advantage is observed when air is used as the transport medium for the textile material (e.g. Then Air Jet). Here, efforts are made to prevent foam formation. The small amount of aqueous dye liquor present carries dyes and auxiliaries.

5.2.4.1.12.6.4   Dyeing in alkaline medium

Oligomeric fibre components migrate out of the fibre during dyeing and can cause problems due to formation of agglomerates with dyes and deposition on the textile material or in the dyeing equipment. The cyclic trimer of ethylene terephthalate is especially harmful because of its low water solubility (<0.5 mg/lat 100 °C). In order to reduce the number of problems with oligomers, polyester lining fabric for example can be dyed at pH 9.5 in accordance with the HMK (Hoechst/Mitsubishi/Kaney) process with selected disperse dyes on a beam dyeing machine (but also suitable for jet and tops dyeing). In this process, oligomers leaving the fibre are hydrolysed to water-soluble components before their deposition. To achieve this the pH is adjusted to ca. 10.5 with NaOH at the start of dyeing; it then decreases gradually to about 7.5. Alkali-stable dyes must be used. Many blue azo dyes are unsuitable for this process. As many azo dyes have a tendency, depending on pH, temperature and reduction conditions, to deteriorate in boiling, anthraquinone dyes in particular are used as more stable types for this purpose. A range of Dianix (anthraquinone) and Samaron (azo) dyes perform for the HMK process.

The reductive alkaline after treatment can be omitted.

5.2.4.1.12.6.5   Dyeing from organic solvents

Chlorinated hydrocarbons, in particular, have been recommended as a medium for continuous or batch dyeing of acetate and PES fibres. For continuous dyeing, dye application from either chlorinated hydrocarbons or aqueous liquor with fixation in solvent vapour is possible. This method is dis favoured for ecological and toxicological reasons.

5.2.4.1.12.6.6   Dyeing from supercritical carbon dioxide

According to studies performed in the early 1990s, dyeing from supercritical CO_2 is possible because of its adequate solubilizing power for disperse dyes [128]. Temperatures up to 150 °C and pressures near 30 MPa are used. Short dyeing times, good levelness, and low disposal costs are expected. The CO_2 is recovered. The equipment is naturally very complex. Dyeing to a predetermined shade is difficult.

5.2.4.1.12.6.7   Dyeing via the gas phase

Dyeing with dye vapour in the vacuum or in a gas stream as the medium has also been investigated. Although advantageous from an energy standpoint, the problems encountered on a practical scale have not.

5.2.4.1.12.6.8   Continuous dye fixation with microwaves

The moist fabric padded with dye liquor is irradiated with microwaves in the presence of vapour. The addition of urea or carrier is recommended for rapid diffusion of the dyes.

5.2.4.1.12.6.9   Fastness properties of the disperse dyed materials

Wash fastness

In general terms the wash fastness of disperse dyes are satisfactory. When a fastness test is done as per ISO 105 C03 wash test, which simulated domestic washing with soap and examined staining on polyester and cotton or wool. Nearly all disperse dyes gave very good to excellent results. However, from the 1970s onwards, achieving satisfactory fastness to washing became more problematic as the result of a customer-driven rise in standards being implemented in the guise of more severe testing protocols which requires using adjacent nylon, many dyes has shown problems which showed a greater tendency to stain than wool or cotton. The high temperatures lead to thermo migration where dye molecules tend to move from the core of the fibre towards the surface, and thus counteract the reduction clearing and lower the wash fastness.

When fastness was done by ISO 105 C06 etc. using detergents which simulated domestic laundering, detergent, containing much less surfactant, is able to remove surface dye, but is not as effective at solubilising it and keeping it in solubilised form in the bath like soap, promoting staining of the adjacent fabric which affected the staining ratings. This has rendered many previously satisfactory dyes unsuitable and made dye selection far more important in this respect. An extreme requirement is that for polyester microfibres: the high surface area to mass ratio of these materials means that higher than normal depths of dyeing are needed for satisfactory shades. These problems are overcome by using dyes incorporating easily washable groups into dye structures. Thus, the modified dyes containing diesters, thiophenes, benzodifuranones, phthalimides and fluorosulfonyl group have been designed to improve the wash fastness.

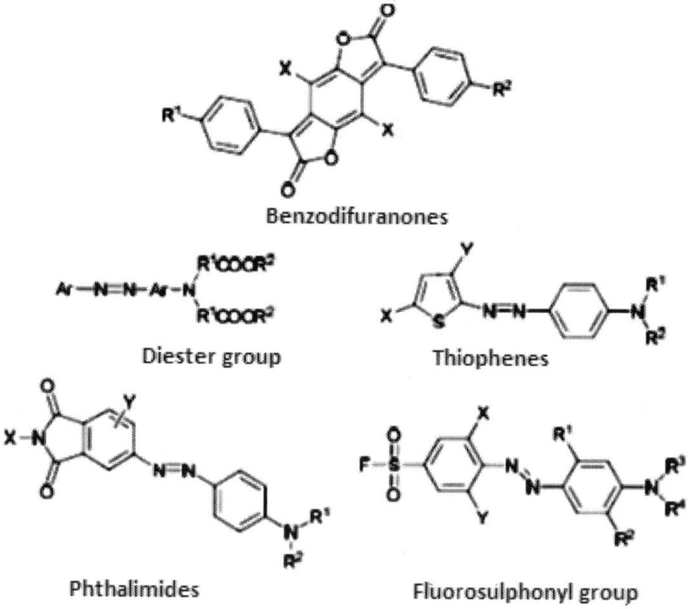

Benzodifuranones

Diester group

Thiophenes

Phthalimides

Fluorosulphonyl group

Disperse dyes with high wash fastness

The most often patented dyes with enhanced wash fastness and build-up properties were based on heterocyclic amines such as 2-aminothiophene, and 2-aminopyrazole, and were used separately or in combination.

Perspiration fastness

Disperse-dyed polyester generally performs well because its structure is still close-packed at body temperature (the level at which the test samples are maintained). Perspiration fastness tests have been designed to simulate the action of various other wet agencies including perspiration. Much effort has been expended on trying to find ways to mimic human sweat, which is a complex mixture containing salt and amino acids. Fresh perspiration is slightly acidic, but goes alkaline with bacterial action, hence the development of two test, one at pH 5.5, the other at pH 8. Disperse dyes gives good fastness ratings in both cases.

Gas fading of disperse dyes

Polyester, triacetate and acetate fibres dyed with anthraquinoid disperse dyes possessing primary or secondary $-NH_2$ groups in structure show fading effect on exposure to N_2O gases or other fumes. This happens when either primary $-NH_2$ groups are diazotised or secondary $-NH_2$ groups are converted to nitroso compounds. The problem can be solved by treating man made with chemicals possessing affinity for fibre having general formula $R1R2N(CH_2)_n N R1R2$ during or before dyeing or by introducing electronegative groups in place of $-NH_2$ groups. Azo disperse dyes mostly remain stable against gas fumes except a blue dyes.

5.2.4.1.12.7   Exhaust methods

Mainly, there are four methods available for dyeing polyester goods.

    a) Dyeing at boil (98–100 °C) with carrier.

    b) Dyeing at 100–110 °C with less carrier.

    c) Dyeing at high temperature (120–130 °C) in closed machines.

    d) Thermosol dyeing.

Exhaust dyeing with disperse dyes proceeds through the following stages: Stage A: movement of individual dye molecules at the molecular level due to thermal influences; Stage B: dye molecules are captured in the adsorption zone of the polyester fibre surface; Stage C: transport or migration of adsorbed dye molecules into the fibre interior. Stages A and C represent the bottlenecks of the dyeing process. Dye solubility plays a decisive role here and can be improved by chemical additions and rising temperature. Suitable chemicals for this purpose are, e.g. carriers.

5.2.4.1.12.7.1   *Dyeing at boil (98–100 °C) with carrier:*

i. Open Becks

Mostly these methods are of academic interest only because of the environmental problems and restrictions of use of the carriers.

Carriers

Carriers for polyester fibres are special auxiliaries which increase the amount of disperse dye taken up by the fibres over a normal period of dyeing at 100 °C. They are organic, usually aromatic, compounds of various types. Apart from the effective substance, the majority of commercial carriers also contain emulsifiers which ensure satisfactory emulsification or dispersion in the dye liquor, the effective substance being itself sparingly soluble in water.

There are quite a number of theories on the way in which carriers act. It will, apparently, be a very difficult thing indeed to find a simple concept which is applicable to all variations in the type of carrier, dye, fibre and method of processing, especially as the various mechanisms involved tend to overlap one another. It is assumed that the absorbed carrier opens up the internal structure of the water repellent polyester fibre so that the mono molecular disperse dye can penetrate more rapidly into the interior of the fibre. This hypothesis is supported by the fact that a polyester fibre, after pretreatment with a carrier and subsequent thorough extraction of the carrier from the fibre, can be dyed in a normal manner without the presence of a carrier. However, this explanation does not exclude the possibility of other modes of action,depending on the type of carrier and the constitution of the dyestuff. Although, according to modern concepts, the designation "carrier" is no longer quite appropriate, it has become well established. "Dyeing accelerator" is another frequently employed term.

In practice, the demands placed on a carrier are various and sometimes very great. Amongst others, the "ideal carrier" should fulfill the following requirements:

1. High efficiency at the smallest possible concentration, and minimum dependence on the constitution of the disperse dye.

2. Rapid emulsification and good stability of the carrier emulsion under dyeing conditions.

There is a danger of carrier spots resulting from the breaking down of the emulsion if the stability is inadequate.

3. Low volatility in steam. With carriers with high volatility there is the danger that, with some types of dyeing machine, the volatilised substances may condense on cooler parts, this resulting in drops

falling on the material to form spots which then dye more strongly. Apart from this, a further addition is necessary with volatile carriers where the dyeing time is lengthy.

4. The odour should not be unpleasant otherwise the products can only be employed in closed dyeing machines.

5. It should have little effect on shrinkage of the fibre. This is important with cross wound packages.

6. The uptake of disperse dyes by blend fibres (especially wool) should not be increased.

7. It should be readily removable from the material.

8. The light fastness of the dyeing must not be impaired.

However, there is no carrier which fulfils all these requirements in an ideal manner, and one is compelled to make concessions on one side or the other, depending on the circumstances.

If, at a given temperature, disperse dyes are dyed in the presence of increasing amounts of a dyeing accelerator, the colour yield rises up to a given concentration of carrier, whereupon it falls again as this is further increased. The optimum effect is dependent both on the carrier as well as on the dye employed. It is therefore not possible to replace a carrier by one with a different constitution without making an appropriate correction in the dyeing recipe. At temperatures of over 100 °C, optimum efficiency is obtained with smaller amounts of carrier.

Though carriers do accelerate the penetration of dyes, it is also possible to strip the dye from the fibre to a greater or lesser extent by means of carriers, especially when the concentration of the latter exceeds the optimum for dyeing. This effect can be put to good use in levelling out or in partially stripping dyeings which are too dark. However, too high a concentration of carrier must be avoided, otherwise the fibre will be damaged; this applies particularly to lengthy treatment times, especially at temperatures of over 100 °C.

Procedure

The dye bath is set at 50 °C with

Additions	Unit	Quantity
Dispersing agent	g/l	0.5–1
Acetic acid (30%) pH 5.5–6*	g/l	4.8
Disperse dyes	g/l	x

*pH as per dye manufacturers recommendations.

The requisite quantity of properly dispersed dye is and the scoured goods are entered and the temperature is slowly raised to boil within 30 min. After 5 min suitable carrier emulsion 3–5 g/l parts is gradually added. The dye bath should be kept in between pH 6 and 7. This can be done by adding acetic acid. Dyeing is continued for about 1–1.5 h. The goods are then rinsed thoroughly and removed for after treatment.

Jigger dyeing:

Recipe

Quantity	Unit	Additions
0.5	ml/l	Acetic acid 30%
0.5–1	g/l	Dispersing agent
5–10	g/l	Carrier (emulsified)

The liquor is made with acetic acid and dispersing agent and it should have a weakly acid reaction (pH 5–6).The temperature of the dye bath is again raised to boil, if required. Jigger is then closed and goods are run for two ends 6–8% of suitable carrier is added over next two ends. The pH of the bath should be in between 6 and 7. The requisite quantity of dyestuff properly dispersed is added and goods are worked for 2 h or 10–12 ends at boil. The dyed goods are then rinsed and carried forward for after treatment.

Notes:

In dyeing dark shades including blacks and navy blues it is advisable to add the dyestuff in several portions within the first half hour.

It is unnecessary to make any further addition of carrier in shading or if the dyeing time is prolonged.

5.2.4.1.12.7.2   *The combined high-temperature/carrier dyeing process*

This combined dyeing procedure is always employed when it is desired to make use of the advantages of dyeing over 100 °C but, for certain reasons, namely, the maintenance of the fibre properties, the dyeing temperature cannot be raised as high as that normally employed in an HT process. This applies, for example, when dyeing polyester/wool mixtures where the temperatures often range over 103–108 °C will affect the wool. Apart from other machines, low-pressure dyeing machines are used for hank yarns and cross wound packages in which polyester fibres and their blends with other fibres can be dyed at around 105 °C. Compared with machines which are suitable only for dyeing at temperatures up to the boil, the low-pressure dyeing machines possess the advantage that the dyeing time is shortened and that smaller amounts of carrier are required. The temperature range of 110–120 °C can also be employed for dyeing polyester fibres when it is not desired to subject these to higher temperatures. A carrier may be added even when dyeing at temperatures above 120 °C. This is the case, for instance, when it is desired to use dyes with very high sublimation fastness, but for which high dyeing temperatures of 130–135 °C are normally required in order to attain full depth of shade within the usual dyeing time. It sometimes happens that this comparatively high temperature is undesirable in view of its unfavourable effect on the fibre, for instance on the resilience of high-bulk (texturized) polyester fibres, the migration of oligomers out of the polyester fibre, and so on; it may also be that such high temperatures cannot be attained within the works for various reasons such as type of equipment, available steam pressure, and the like. In such cases, a dyeing can be effected at, say, 120–125 °C by using fairly small amounts of carrier. Fibre shrinkage must also be considered when deciding whether dyeing should be carried out at an elevated temperature without the addition of a carrier, or whether a lower temperature with carrier should be employed. The shrinking effect of carriers, depending on the type of carrier, increases to a greater or lesser extent with rise in temperature. On the other hand, it must also be remembered that considerable shrinkage can occur at an elevated temperature even in the absence of a carrier. When dyeing packages which are very much inclined to shrink, it may be preferable to dye at, say, 110–115 °C in the presence of a carrier with little shrinking effect, e.g., Palanil Carrier A or AN, than without a carrier at a higher temperature. The higher the dyeing temperature, the lower the amounts of carrier usually needed. "The Carrier Dyeing Process", with the exception that the required higher temperature is employed instead of working at boil.

5.2.4.1.12.7.3   Cares to be taken in using carriers

Some carriers contain anionic emulsifiers. Normally, these carriers are compatible with all other anionic and non-ionic auxiliaries present in a dye bath. However, cationic agents, excepting can, under certain circumstances, affect the stability of the emulsion and cause precipitation. This must also be remembered if cationic softening agents or "retarders" are to be included in the dye bath. Phenolic carriers, may form insoluble adducts with oxyethylation products under unsuitable conditions in an acid, ammonium salt-containing liquor, and this can give rise to the material being spotted with dye.

Carriers which are volatile (Phenolic type) or ones which are having volatile solvents should be avoided. Inhalation of the vapours of the dye liquor must be avoided at all costs. If there is a possibility of carrier vapours escaping into the dye house, provision must definitely be made for extractor fans or ventilators to ensure efficient change of air.

Carriers usually induce higher shrinkage in dyeing, which shrinks in hot water. If special measures are not taken to allow for pronounced shrinkage when dyeing yarn in a package machine, or piece goods in a beam dyeing machine, major difficulties may arise, such as unsatisfactory levelness and penetration, together with moire effects on fabrics. In case of some carrier the effect is strikingly low, so that the difficulties just described can be avoided by employing these products.

Experience has shown that residual carrier remaining on the fibre can very much impair the light fastness of the dyeing. It is therefore essential to give a thorough hot rinse and to ensure careful after- or intermediate-clearing of the material, as is also customary for removing excess dye. Of course, residual carrier can best be removed from the material by hot drying or subsequent heat-setting.

The procedure of using of using a particular carrier has to be followed strictly as per the manufacturers recommendations.

5.2.4.1.12.7.4   After treatment

An after-wash improves the fastness properties of the dyed goods. For pure polyester material a reducing bath containing

Quantity	Unit	Additions
5–8	ml/l	Caustic soda 38 °Be
3–4	g/l	Hydrosulphite
1	g/l	Wetting agent (alkaline stable)

The after treatment is carried out at 70–80 °C, and 4 ends are normally adequate.

Notes:

The carrier may be completely removed as the residual carrier may affect the light fastness of the dyeing.

5.2.4.1.12.8   High temperature exhaust dyeing

In simple words the pores on the polyester fibres are comparatively small compared to the size of the disperse dyes and hence the diffusion disperse dyes at boiling temperature. Hence methods have to be adopted by which the pores are made larger to help the dye into the fibres. Hence polyester fibres are dyed whenever possible at 125–135 °C under pressure (HT dyeing). In favourable cases, the entire dyeing cycle can thus be shortened to less than 60 min. HT dyeing can be made still more economical by using specially designed dyeing equipment to reduce the liquor ratio. The evenness of dye absorption is improved by dyeing in the presence of a levelling agent.

The dyeing of polyester at temperatures above 100 °C has been established as having a number of advantages over dyeing in conventional machinery which may be summarised as follows:

1. The time of dyeing is reduced. This is achieved by the more rapid rate of penetration into the fibre by the dyestuff when dyeing at 120–130 °C.

2. No 'carrier' is required.

3. The fastness to light and wet treatments of the resultant dyeing is usually higher. This is achieved by deeper dyestuff penetration into the fibre at the higher dyeing temperatures.

The use of carriers which may have a deleterious effect upon the light fastness properties is avoided.

4. Superior exhaustion of the dye bath is obtained and heavier shades can therefore be produced.

5. Levelling and coverage of yarn irregularities are greatly improved.

Dyeing is usually carried out in closed liquor circulating machines. The goods are present and well batched.

There are various machines available to dye polyester at high temperature by exhaust method.

5.2.4.1.12.8.1   HT jiggers

Jigger is a very versatile machine for dyeing as it can process the material in a very low liquor ratio and the material is given treatment in open width and in single layer, which is not possible in many machines. In the atmospheric jiggers disperse dyes cannot be applied as the dyeing process cannot be done above boil. But HT jiggers are developed for this purpose since the whole unit is enclosed in an air tight autoclave whereby the temperature can be raised around 130–150 °C which is sufficient for dyeing disperse dyes on polyester.

HT jigger

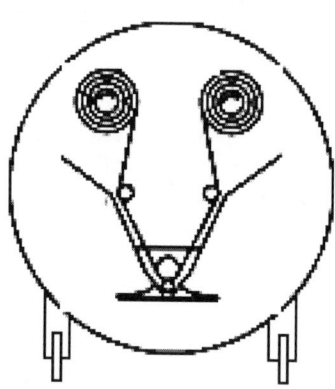

HT jigger (schematic diagram)

Dye bath is prepared as per following recipe

Quantity	Unit	Additions
0.5–1.0	g/l	Dispersing agent
1.0–2.0	g/l	Sequestering agent
0–0.5	g/l	Carrier or levelling agent
1	g/l	Acetic acid for pH 4.5–6 (or as per dyes manufacturers requirement)

The material is wet out in this bath in two ends and the well dispersed disperse dye is added in two ends at room temperature.

Heating is started and the temperature is continuously raised to 130 °C. Check the pH intermittently and correct if necessary.

Run at this temperature for 30–45 min and finish the dyeing in even ends.

Rinsing, reduction clear, soaping, etc. are followed.

5.2.4.1.12.8.2   Jet dyeing and soft flows

The outstanding features of these machines are their rapid and almost tensionless transport of the fabric and their vigorous liquor circulation. Compared with winch-becks the danger of running folds becoming set is very much reduced on account of the quicker shift of the position of the folds, the lesser lengthwise elongation and the uniformity of the temperature. Other advantages of jet-dyeing machines include the relatively short liquor ratio and the shorter heating-up and dyeing times. Jet-dyeing machines have been developed in which only part of rope of material is immersed in the liquor at any given time, and others which are entirely filled with liquor.

Quantity	Unit	Additions
0.5–1.0	g/l	Dispersing agent
1.0–2.0	g/l	Sequestering agent
0–0.5	g/l	Carrier or levelling agent
1	g/l	Acetic acid for pH 4.5–6 (or as per dyes manufacturers requirement)

In jet-dyeing machines, the rate of heating-up can be increased up to 4 °C/min. In addition, it is pointed out that, on completing dyeing in winch becks and jet-dyeing machines, the liquor should be allowed to cool off slowly (1 °C/min), so that running folds are not set. The rate of cooling varies from fabric to fabric. With sensitive, and especially smooth, knit goods and closely woven fabrics, cooling must proceed slowly down to 50 °C. Too rapid cooling can lead to setting of folds, even with well-present goods, and these folds can hardly be eliminated afterwards. A reduction clearing is given, particularly with medium and deep shades.

Auxiliaries

A slip agent or lubricating agent that is employed to prevent the setting of running folds when the goods are dyed in rope form. This product makes it easier for the running material to slip over itself, so that folds remaining localised at one place for only short time and, hence, are not fixed in the material. The use of a cationic antistatic agent is recommended in the last rinsing bath to improve the handle and to impart an antistatic effect.

Notes:

1. In circulating liquor machine dyeing it has been found that the best fastness to rubbing is only obtained if the bath is dropped at temperatures above 90 °C/195 °F. Cooling the bath below this temperature may lead to recrystallisation of the dye still in the bath which is then retained by the dye goods.

2. In particular un levelness and faulty penetration likely to arise in both high temperature and carrier assisted dyeing can be effectively overcome by an addition of 0.25–0.5 g/l non-ionic levelling agent.

Excessively high concentrations impair the dye yield.

5.2.4.1.12.8.3   HT beam dyeing machine

Apart from the thermosol process, dyeing on the HT beam dyeing machine is the most common method for polyester fibre blend fabrics. HT beam-dyeing machines are often used only for the dyeing of woven and smooth knit goods that have little structural pattern.

It is imperative that the goods to be dyed should be heat set at, say, 200 °C for 30 s in hot air, as otherwise they would shrink considerably under HT conditions, and the roll of cloth would become so hard that the dye liquor could no longer penetrate. Furthermore, shrinkage of the material produces a moire effect and impressions of the seams on the fabric. The washed, dried and set material is wound on dry, low tension, onto the beam by means of a hyperbolic situated at the exit of the stenter and under feed roll. It is also particularly important to know, how much the material will shrink across its width during dyeing, since the covering of the beam perforations must be accordingly allowed for from the start.

In most cases, the heat-set goods are wound onto the beam in a wet state, and it is advisable to employ specially constructed batching devices which ensure a uniform rolling on of the material with only a moderate tension. It is an advantage to lead the goods through a boiling water-bath immediately before rolling onto the beam, in order to eliminate residual shrinkage which is usually still present, even after the material has been well set. A highly practicable idea is to arrange for the hot water treatment to be given on a padding machine which is linked up with the winding-on device, thus making one operation out of two. The thickness of the wound layers on the beam, and hence the length of the batch, is dependent on the diameter of the beam, the internal diameter of the machine, and the type of material.

The following data serve as a guide:

Length of the batch in meters = Diameter of the beam (mm) ×factor

The GSM of the material and the factor is given below as a guideline

Fabric (woven or knit) GSM	Factor
120	4–5
250	2–3
400	1–2

During dyeing, the liquor circulates only in an outer direction. The pressure difference should be as low as possible in order to maintain the bulk and structure of the goods, and should be of the order of 0.5 kg/cm^2.

Apart from the tension applied on batching the material, and the thickness of the roll, there is another factor which requires very close attention, namely, how far the material should extend beyond the perforations in the beam. It must be remembered that the liquor will seek the way of least resistance. It is therefore necessary to ensure that the resistance should be the same at all points on the roll of fabric, including the edges. The overlap beyond the perforations must accordingly be more extensive, the greater the thickness of the roll and the denser the material; the frequently very high density of the selvedges must not be overlooked here. A consequence of too little overlap is that there is insufficient flow of the dye liquor towards the outermost corner of the material, so that the fabric is dyed paler there (see figure below)

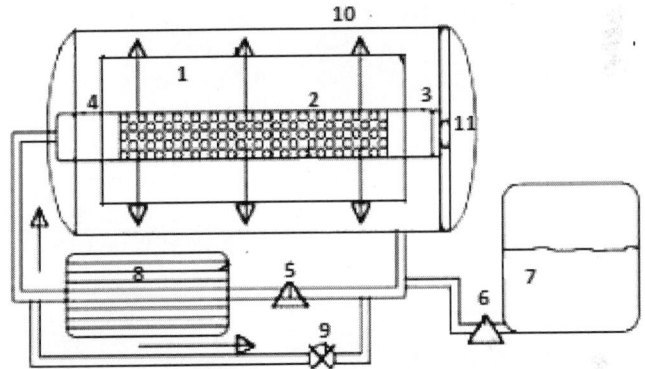

Beam dyeing (schematic diagram): 1. Fabric; 2. Perforated beam; 3. Sleeves; 4. Covers; 5. Circulation pump; 6. Addition pump; 7. Addition tank; 8. Heat exchanger, 9. Bypass valve; 10. Autoclave; 11. Door

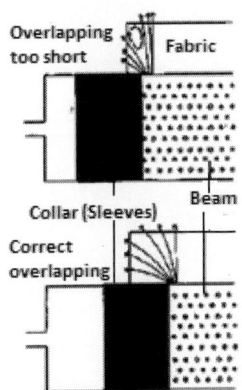

Sleeve and overlapping positions

To what extent the selvedges should overlap in individual cases can often only be determined by prior tests. With dense materials, good results are usually obtained when the overlap is of the same order as the thickness of the roll. With open-weave goods, on the other hand, an overlap of 3–5 cm is often quite sufficient, even when the roll is fairly thick. Naturally, it is assumed here that there is no significant residual shrinkage in the material, either lengthwise or across the width. On dyeing, the apparatus is first slowly filled with water (cold, or up to a maximum of 50 °C without circulation, in order to drive out the air from the package. Circulation should not begin until this stage has been completed, and then it should be in an outward direction. The liquor is heated simultaneously, and it is set with acetic acid, auxiliaries and dyes.

After dyeing, rinsing proceeds with overflow where possible. The outlet valve is not fully opened, and an amount of fresh water equivalent to that of the liquor draining off is allowed to flow in, while the circulation is maintained. With deep shades, it is an advantage to add 1 g/l hydrosulphite to the rinsing liquor which is allowed to circulate for 3–5 min to effect reduction clearing. Rinsing, with overflow, proceeds until the rinsing water runs clear. After this, oxidation is carried out by adding 0.5 cc/l hydrogen peroxide 30% to the rinsing liquor, and circulation proceeds for 5–10 min at 40–50 °C, whereupon 0.5–1 g/l detergent is added, the liquor heated to about 90 °C, and the goods soaped for 10–15 min. As far as possible, circulation must not be interrupted as from the commencement of dyeing up to, and including, oxidation.

Alternatively the reduction clearing and washing can be done after unloading in open width form.

5.2.4.1.12.8.4   HT/HP steam fixation

Not a very commonly used method. But if the facility is available with the processor this method can be tried. Satisfactory results are not achieved by steaming in saturated steam at 100 °C. Pressure steaming at 130 °C gives relatively good fixation whilst superheated steam, which finds wide application in the fixation of printed disperse dyes, is only used on a small scale in continuous dyeing.

5.3   Dyeing of polyester in alkaline bath

Dyeing of polyester is predominantly done in acid bath. But many disperse dyes give good results over a wide pH range say, pH 2–9. Some will only give satisfactory results over a narrower acidic range (pH 2–6) and a few require careful control to within pH 4–5.5. Both the groups and other dyes give good results at pH 5 and hence the pH has been considered as standard for disperse dyeing. An addition of acetic acid for closed systems and soft water is sufficient to maintain the required pH or a buffered system is preferred. EDTA is widely used to counteract the effects of metallic impurities, which not only affect the hue and fastness of a few susceptible dyes but may also catalyse dye reduction and promote deterioration of dispersion properties, as described above.

Since there are disperse dyes which can dye in alkaline bath, one can think of dyeing in alkaline conditions at pH 9.0–9.5 which has certain advantages especially dyeing blends. Main advantages being economically more profitable, dyeing can be done directly after preparation processes including bleaching, merce rising of polyester/cotton and caustic weight reduction of polyester, after reduction clearing after dyeing, etc. In most cases the dyeing can be started without complete neutralizing but only adjusting pH to 9–9.5 by either washing or slight acidification. Other advantages include improved handle of the substrate, more effective solubilisation and removal of oligomer, less frequent and easier cleaning of machinery and possible avoidance of reduction clearing. There are possibilities of simultaneous application of disperse and reactive dyes to polyester/cellulosic blends.

Alkaline bath dyeing of polyester is not without obstacles even though theoretically it is possible. It is slightly difficult to have a buffer to keep the pH stable at pH 9–9.5 even at higher temperatures. A second problem is the disperse dyes has to be carefully selected so that it remains stable in alkaline conditions at high temperature. Another problem is the oligomer formation during dyeing which may be hydrolysed and form –COOH group at high temperature which can upset the pH of the bath. But a mixture of auxiliaries or one auxiliary which can function as a buffer, a stabiliser for dyes, assist in the dissolution of oligomer, chelate metal ions can work in dyeing polyester in alkaline bath.

Many buffer systems have been suggested like, amino acid derivatives applied in combination with an alkali –N,N-bis (hydroxyethyl) glycine (see below) in combination with

$$HOOC - CH_2 - N \begin{array}{c} CH_2CH_2OH \\ CH_2CH_2OH \end{array}$$

N,N-bis(hydroxyethyl)glycine

sodium hydroxide, N,N-dimethylglycine, N-methylalanine, a mixture of a phosphonate and a polycarbonic acid, N-methylglycine and along with NaOH or Na_2CO_3, $NaHCO_3$, Borax, etc.

Alkaline HT Dyeing

After a careful selection of such dyes one can dye polyester in HT process in alkaline bath of about 9.5 pH. Machines like Beam Dyeing, Jet and soft flow dyeing can be used for this purpose. Azo based disperse dyes may get deteriorate under reducing conditions, which is possible in a higher temperature conditions and hence may be avoided. Since anthraquinone dyes does not get affected under reduction atmosphere and hence is more used for this purpose.

5.4   Highlight fastness disperse dyes for special applications

Fastness properties of normal dyes supplied by manufacturers suffice the requirements of dyeing dress materials. But there are some requirements where very high light fastness requirements are necessary, for example, auto interiors. Although a close relationship does exist between the chemical structure of a dye and its light fastness rating, it is important to remember that other factors are of relevance, the most critical of which are:

1. The inherent photo stability of the dye molecule; in general, the chromophoric nucleus is the most important element in determining the light fastness of a dye, but nuclear substituents may alter the fastness significantly.

2. The concentration of the dye within the fibre; usually the fastness of a dyed fibre increases with increasing dye concentrations.

3. The nature of the fibre in which the dye is dispersed; different fibres contain different chemical groups and these substituents can have a significant effect on the light fastness rating of a dye on a given fibre.

4. The wavelength distribution of the incident radiation; not all the absorption are equally effective in starting a fading process.

5. The composition of the atmosphere; the moisture content of the atmosphere can have a marked effect on the fading rates of certain dyes.

Disperse dyes show highest light fastness on polyester, better than di-, tri-acetate and then nylon in that order. It has been observed that 4-aminoazobenzene, β-cyanoethyl, o-acylation groups has shown increasing light fastness on polyester. Oxidation and reduction reactions are known as the two most important pathways for the fading of dyes. It is assumed that photo-oxidative processes are responsible for fading on non-protein substrates, such as polyester and acetate, as shown

below, but it is not well proved. But it has been proved that increased fading in the presence of oxygen. Direct evidence has however been obtained of an oxidative pathway, the dealkylation of 4-N,N-dialkylaminoazobenzenes on polyester and nylon in which dye-sensitised singlet oxygen attacks the terminal nitrogen lone pair electrons of the dye in the ground state. While the dealkylated products will not differ radically in colour, the creation of carbonyl species and peroxides may cause destruction of the azo group, leading to more significant fading. Anthraquinone dyes containing electron-donating substituents are more susceptible to fading than those containing groups, such as –NHCOPh, which contribute towards better light fastness. Further incorporation of less basic substituents, notably heterocyclic moieties, into β-positions of 1,4-diaminoanthraquinone leads to bright turquoise blue dyes, for example C.I.Disperse Blue 60 and Blue 87 that were originally developed by DuPont and BASF in 1955and 1963, respectively (see below). The superior light fastness of these dyes is attributed to the

C.I. Disperse Blue 60   C.I. Disperse Blue 87   C.I. Disperse Blue 56

Dyes with improved (Higher) Light Fastness

electron-withdrawing effect exerted by the imide groups. These two bright dyes are still preferred for the coloration of polyester in intrinsic turquoise shades and in a bright green by the addition of an appropriate yellow dye.

Researches has shown that photochemical decomposition of azo anthraquinone dyes is closely related to the wavelength of UV radiation which differs with the substrates –maximum photodegradation takes place at 230 nm on cellulose acetate, 259 nm on triacetate and 316 nm on polyester.

The primary mechanism of fading of the latter azo dye was a reductive reaction, whereas decomposition of the anthraquinone derivative Red 60 was essentially an oxidative process. An electron-withdrawing 4-nitro or especially 3-nitro group in dyes based on azobenzene derivatives on polyester, enhances light fastness. An electron-donating 3'-methoxysubstituent in the opposite ring boosts these effects further. The 2'-methoxy-5'-methylsubstitution pattern also reinforces the favourable influence of the 3- or 4-nitro group but markedly lowers the rating of 2-nitroazobenzene. All three nitroazobenzenes are adversely affected by a 2'-hydroxy-5'-methyl arrangement.

Higher light fastness of different substituents on azobenzene dyed on polyester

Azobenzene substituents	Light fastness
3-Nitro-3'-methoxy	7–8
3-Nitro-2'-methoxy-5'-methyl	7–8
4-Nitro-2'-methoxy-5'-methyl	7–8
3-Nitro	7
4-Nitro-3'-methoxy	7
4-Nitro	6-7
3-Nitro-2'-hydroxy-5'-methyl	6-7

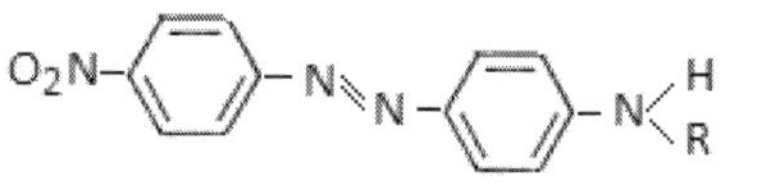

Azobenzene derivatives R = Ph - C.I.
Disperse Orange 1 R = H - C.I. Orange 3

Substitution and corresponding improvement in light fastness is substrate specific on nylon, light fastness in the p-nitrophenylazo series (3.157; X = NO₂) is marginally improved by electron donation (methyl, methoxy) but markedly lowered by electron withdrawal (chloro and especially cyano or nitro) in the ortho position but on polyester, electron-withdrawing groups oriented ortho (chloro, cyano) or para (nitro, acetyl, methylsulphonyl or diethylaminosulphonyl) to the azo linkage improve the fastness but electron-donating substituents (methyl, methoxy) lower the light fastness ratings (See below table)

Changes in light fastness by same substitutions on aminoazobenzene structure (RHS) on nylon and polyester

Substitution on the structure shown on RHS	Resultant Dye (C.I. Disperse)	Light Fastness on Polyester	Light Fastness on Nylon
X=NO₂, Y=CN, Z=H	Red 73	6-7	1-2
X = NO₂, Y =Z = H	Orange 25	5	4
X = NO₂, Y = CH₃, Z = H		4-5	4-5
X = Y = NO₂, Z = H		2	1

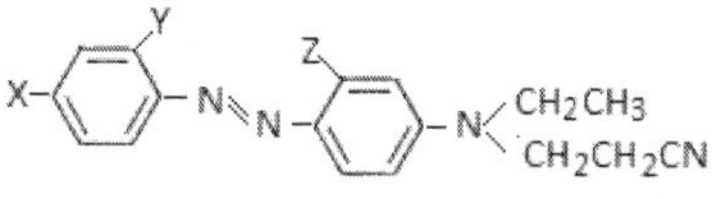

Oxanilide has been found as a photostabilising groups, which on substitution on monoazo and nitro diphenylamine structures has given dyes with very high LF which can be used for automotive trims dyeings.

Similarly, by substituting different groups, one can design dyes with better wet fastness or heat fastness as required for special requirements.

Chapter 6
Processing of micro polyester

Normal polyester once accepted whole heartedly by the whole world as it was introduced, as later on it was slowly lost its attractiveness due to many disadvantages in wear properties like, hydrophobicity and related characteristics like static current, thermal properties, attraction of dirt, comfort, handle, etc. Due to this during the last few years, the comfort, fuller handle of the natural fibres has scored heavily over the wash and wears properties of the synthetic fibres. After the introduction of polyester micro and super fibres these disadvantages were solved to some extent. The advantages of this product include excellent durability and dimensional stability, in addition to a soft, full hand not normally associated with synthetics. Microfibers are designed for a soft handle and silky appearance with good air permeability and exceptional drape.

6.1   What is microfibre?

The fineness of a synthetic yarn is usually described in 'decitex' or 'denier'.

Decitex = the weight in grammes of 10,000 metres of yarn

Denier = the weight in grammes of 9,000 metres of yarn

Hence a 167 decitex yarn can also be described as a 150 denier yarn.

Note – for simplicity, only the decitex description will be used here.

Decitex per filament (dtexpf)

It is described as the decitex of a yarn divided by the number of filaments within the yarn.

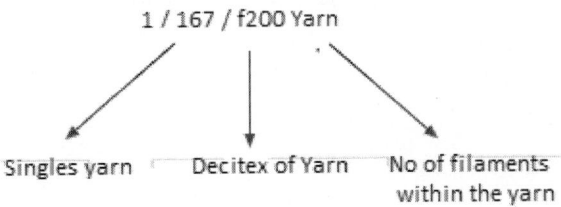

1 / 167 / f200 Yarn

Singles yarn   Decitex of Yarn   No of filaments within the yarn

Thus this yarn has dtexpf = 167/200 = 0.835.

Any fibre of approximately 1 dtex or 1 denier or less is generally defined as a microfibre. In other words, microfibres are termed as fibres having fineness of less than 0.0001 denier. The microfibres have many differences when compared to the regular denier fibres (typically 2–3 dtex), even though there is no difference in the chemical structure. Fibre producers all over the world have been continuously trying for finer and still finer fibres for almost 20–25 years. The most popular definitions of microfibres and super microfibres are,fibre containing fibrils of

Coarse fibres: >7.0 dtex,

Fine fibres: 7.0–2.4 dtex (normal fibres),

Very fine fibres: 2.4–1.0 dtex,

Microfibres: <0.3–0.99 decitex/denier per filament,

Super microfibre:<0.5 decitex/denier per filament.

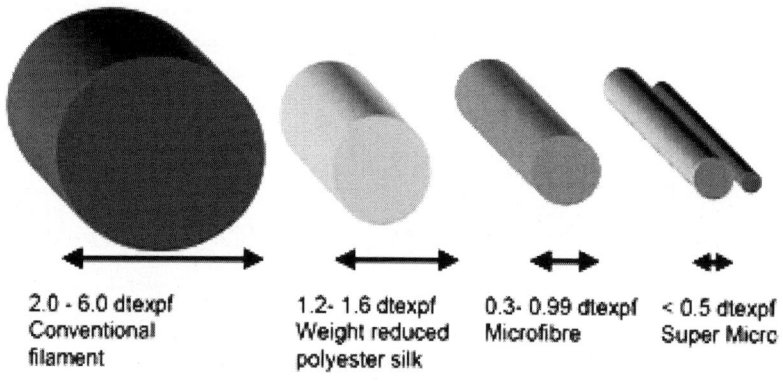

2.0 - 6.0 dtexpf
Conventional
filament

1.2- 1.6 dtexpf
Weight reduced
polyester silk

0.3- 0.99 dtexpf
Microfibre

< 0.5 dtexpf
Super Micro

Comparison of polyester fibre sizes

Fabric constructions based on microfilament synthetic yarns are commonly referred to as:

Microfibres in Europe and the USA and Shin Gosen in Japan (meaning New Generationsynthetics) fabrics from these sources use a combination of conventional, fine and/or superfine filaments, filaments of widely differing shrinkage characteristics, very high and low twist yarnstextured, air-textured and flat filament yarns, variations in cross-sectional shape, etc., to produce a very wide range of high density materials which have novel characteristics and properties which demand very specialised processing.

Effect fabrics available to date include:
- imitation silks,
- peach skins,

- suedes (light to heavyweight),
- air and moisture permeable waterproof fabrics,
- staple blends and yarn mixtures with viscose, cotton and wool.

Popular outlets for these rapidly growing effect fabrics include apparel, sportswear, leisurewear (by virtue of very soft, silk-like handle, and improved moisture absorbency wicking and evaporation given by microfibers), furnishings and upholstery fabrics, lens wipes, etc.

The microfibre in the blend serves to improve the handle, wear properties and care properties of natural fibres without impairing their functionality. Emerising effects are particularly fashionable in men's fashion. As dyeing chiefly takes place after emerising, a mild "wash-out" effect corresponding to the trend in fashion results from this. Light, flowing qualities are to the fore in women's fashion. The greater surface area of microfibres requires a greater addition of lubricants in spinning, etc. These lubricants must be removed in pretreatment. The very high surface area associated with closely packed microfiber filaments means that a greater proportion of incident light is reflected when compared with larger diameter filaments. This results in a visually lighter appearance and necessitates higher percentages of dye needing to be applied to achieve the same relative visual strength as on coarser fabrics. More dye is required for any given shade – wet fastness will be lower on microfibres than on conventional fibres in the same shade.

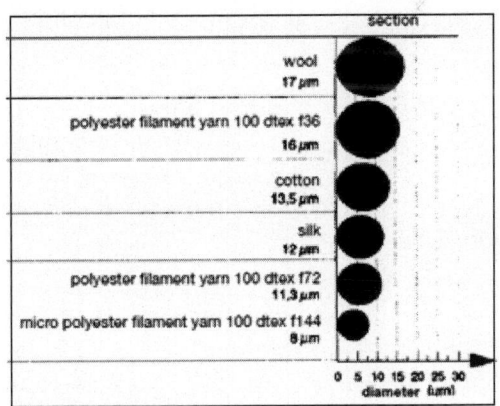

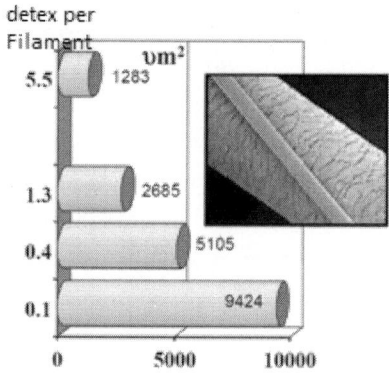

Surface Area of 1 m of 100 dtex Yarn

6.2   Properties of microfibres

A fabric made of microfibre will contain four times more filaments as compared to conventional fabric. It will therefore have more surface area and bulk, thereby giving distinguishing properties such as soft, full handle with excellent drapability and comfort. This is in addition to the inherent properties of synthetic fibres like easy-care and dimensional stability. The tightly woven fabrics made of polyester microfibres, give good water proofing and wind resistant effects, much better than those obtained with coated or laminated or finished conventional fabrics. These tightly woven fabrics impede water droplets from penetrating but allow water vapours to permeate,resulting in increased comfort, especially for the sportswear.

Microfibre fabrics are particularly suitable for sports clothing, for which functionality is relevant and four primary functions are demanded:

1. Good weather proofing, e.g. wind-resistance, is achieved using dense, but microporous fabrics; yarns with a large specific fibre surface area increase density and covering power, but at the same time produce a high porosity in the fabric, which even without coating achieves amazingly good serviceability.

2. Good thermal insulation, heat constancy and wearing comfort are determined by fine fibrillness, light weight, low moisture absorption, no swelling of fibres, good capillary moisture vapour transmission, softness and smoothness. Air pocket, material thickness and low moisture absorption play a key role in the effectiveness of insulation.

3. Good water vapour transmission is achieved using light, shape retaining, double flat knits made of texturised yarns with good

capillary water transport capability on the inside and yarns made of water absorbing and water transporting fibres on the outside.

4. Optimum elasticity through selection of the correct yarn type and material design.

6.3   Methods of manufacturing microfibers

6.3.1   Conjugate technology

Developed by Kanebo Ltd. of Japan and involves spinning of bicomponent filaments comprising of a nylon 6 matrix with polyester fibrilles. Following the spinning and weaving stages, the fabric is subjected to a solvent swelling treatment. The poor cohesion in solvent of the polyester fibrilles and the nylon matrix causes the individual wedge-shaped polyester segments to move outwards. Each polyester segment then acts as an individual super-microfibre filament.

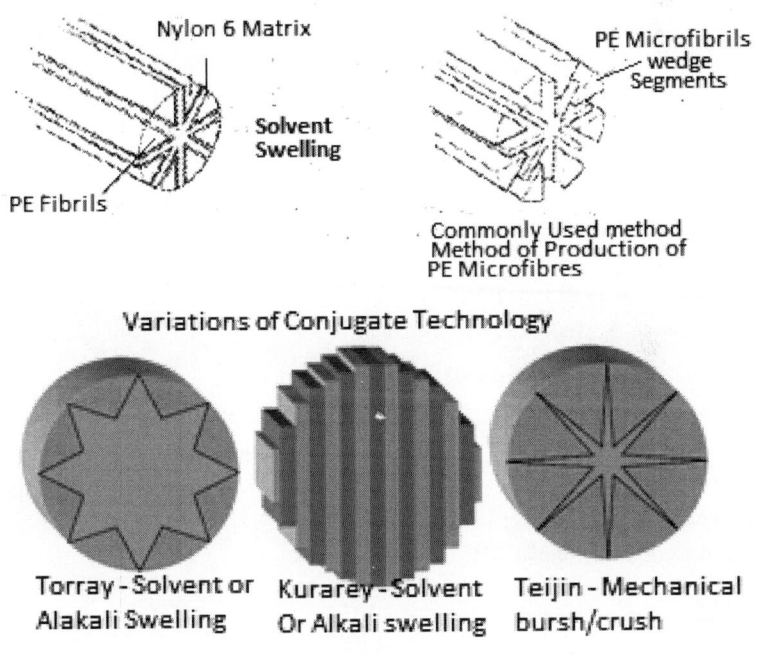

Conjugate technology

6.3.2   Sea island technology

Another method involves the spinning of islands of polyester in a sea of polystyrene followed

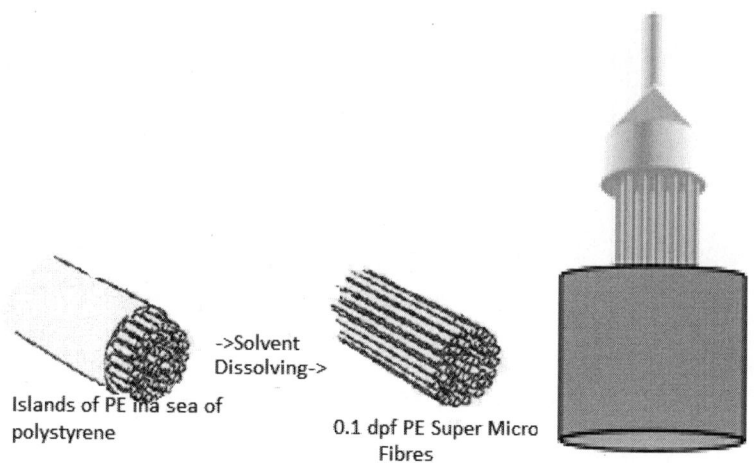

->Solvent
Dissolving->

Islands of PE in a sea of
polystyrene

0.1 dpf PE Super Micro
Fibres

Sea island technology   Directmelt spinning

by spinning and weaving steps, the polystyrene is dissolved using chlorinated hydrocarbon solvent to leave polyester super microfibre filaments.

6.3.3   Direct melt spinning

Single component filaments are extruded through spinnerettes using conventional melt spinning technology, it is possible to produce microfibers down to 0.4 dtexpf. Developments by Asahi Kasei and Unitika of Japan have enabled spun fibrils of 0.1 dtexpf.

6.4   Applications of microfibres

6.4.1   Imitation of silk

Microfibrils can make imitation silks. But to produce a synthetic with all the qualities of silk is next to impossible. However with new technological innovations and probably blending a near to silk materials are made with some additional properties.

6.4.2   Peach skins and suedes

Since these fibres are fine there are possibilities of making peach skin fabrics using microfibers. Thus a wide range of fabric from delicate peach skin finishes through to heavy suedes, similar in handle to chamois leather can be produced using micro-fibres and super micro-fibres. Combining yarns of widely differing shrinkage characteristics, combining filaments of differing dtexpf and differing cross sectional shape in special effect blends are used to produce such fabrics. The combined effect of differential shrinkage filaments is to form a 'micropile' during relaxation and shrinkage, which takes place during fabric processing.

6.4.3 Worsted effects

The combined effect of multi false twisting and air texturising techniques.in warp and weft blends produces interlaced or so called structured fabrics.

6.4.4 Blends with viscose

High twist microfibres in both polyester and nylon are being used in blends with viscose to imitate theunique drape and handle of 100% viscose to impart considerable reduced shrinkage, anti-crease characteristics, tensile strength anddurability.

6.4.5 Moisture vapour permeable waterproof effects

Breathable but shower proof effects for casual outwear can be produced by using high density fibres. The fabrics are either fluorocarbon finished and or alternatively polyurethane coated from a solvent such as DMF or MEK, which is subsequently flashed off. The evaporation of the solvent leaves micropores in the surface coating, which allows breathability whilst still fully retaining the water proofing effects.

6.4.6 Blends with cotton

The aesthetics, which the microfibre offers like, softness of handle, warmth and moisture absorbency has a considerable impact on polyester cotton blends. For high quality shirtings, PES microfibre staple isbeing used and similar development is expected in knit goods particularly in apparel sector. Wiping cloths (Lens wiping),the combined effect of ultrafine filaments, very high surface area and the capillary effect of conjugates which imparts moisture absorbency enables a wide range of wiping clothes for lenses of cameras and spectacles.

6.4.7 Knit goods

Napped knit goods are another range of use for microfibres, which are used both in the leisure and sports sector due to their comfort and their good wearing properties. These fabrics are normally only napped after the dyeing process. While the requirements aim for a pleasant and soft, voluminous handle for leisure fabrics, water repellent properties are also demanded for sports items. This effect is normally produced using fluorine chemicals, as normal silicones result in an oily and less voluminous handle on these items. Easier care results as an additional effect.

6.5   Wet processing of micropolyester

Whilst many of the recommendations for wet processing of microfibres will vary depending upon components, construction and end use, there are important factors in each of the three stages of wet processing, preparation, dyeing and finishing, which apply generally to microfibre- based fabrics.

6.5.1   Processing flow charts

Most widely used

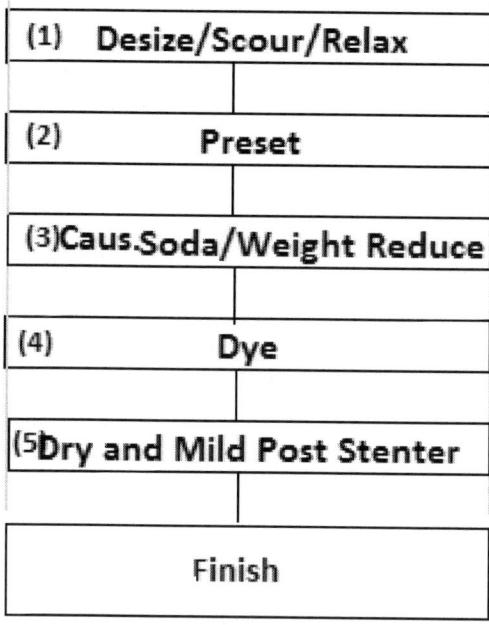

Alternatives

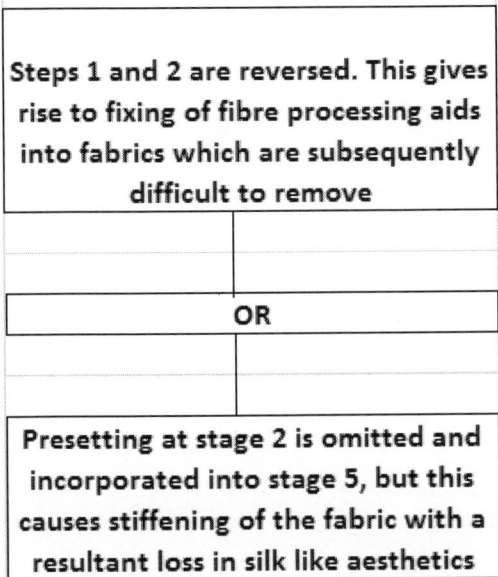

Wet processing flow chart for light weight silk and rayons

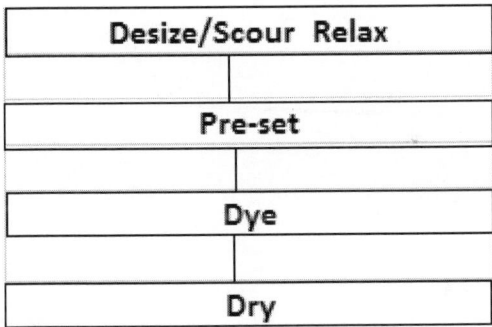

Wet processing flow chart for woven blends with cotton or viscose

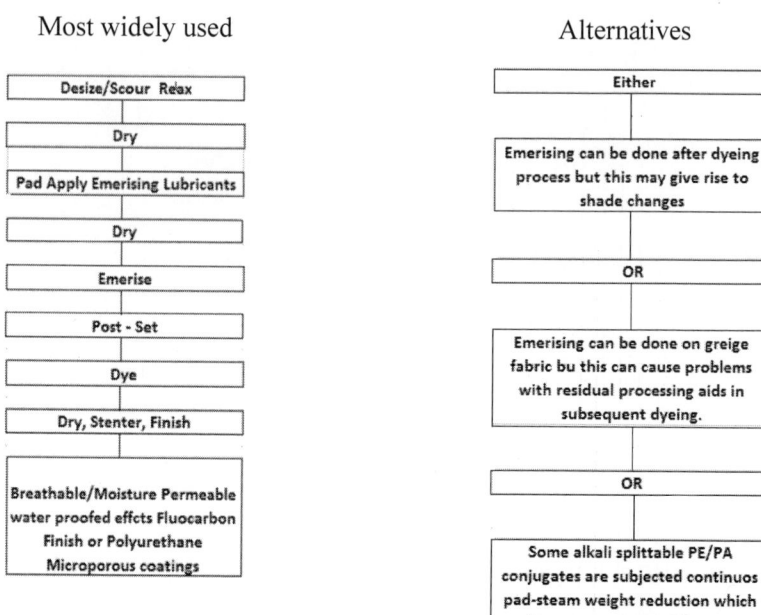

Wet processing flow chart for peach skins and suedes

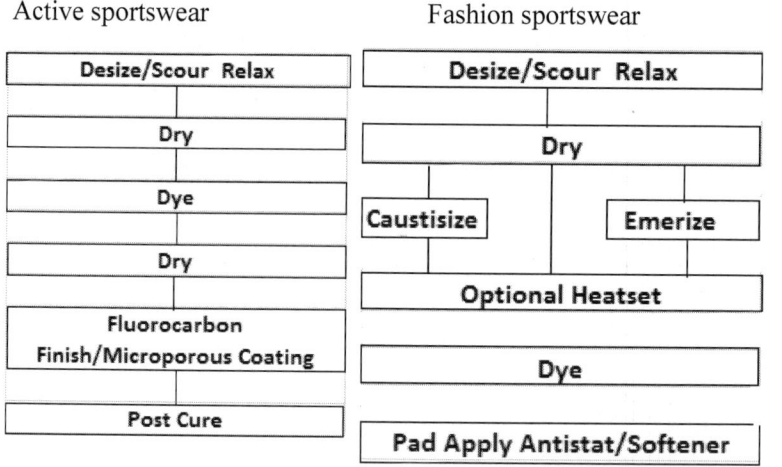

Wet processing flow chart for woven sportswear

6.5.2   Heat setting

Heat setting can be done in the grey state or scoured state. If there is no adverse effect on the removal of the unwanted size, the setting in grey state is preferred since it will reduce the chances of crease formations in the further processes. However the setting procedure and conditions are to be followed as per fibre manufacturer's recommendations. Pre-setting conditions should be as advised by the fibre manufacturer. Many microfibre based fabrics dyed without pre-setting to take account of such factors as high shrinkage, which are controlled at the preparation or dyeing.

6.5.3   Preparation

It was understood from the explanations of the manufacturing methods of the microfibers that lot of spinning aids are used during manufacturing. The preparation has to be designed to remove all these production aids. Also it should be seen that an even preparation with removal of high melting waxes on the fibre happens during this stage with relaxation to avoid any crow feet marks and unlevelness in dyeing.

Relaxation of the fabric must be under minimum tension to allow the controlled shrinkage during scouring. Creasing and 'crow feet marks' will result if insufficient care is taken at this stage. Shrinkage relaxation of microfibres starts at lower temperatures than conventional polyester filaments. Controlled shrinkage with relaxation is a must in case of microfiber preparation especially in case of knitted goods.

As well as acrylate or polyester-based sizes, large amounts of oils and waxes are frequently applied to aid fibre processing and help protect delicate polyester micro fibre fibrilles, because of the very large surface area of micro fibre-based fabrics, 2–5 times more fibre processing aids must be applied compared with conventional polyester based fabrics. To avoid problems with levelling and coverage of the disperse dyes applied in dyeing, the size and fibre processing aids must be efficiently removed. These sizes may be identified and specific solvent or desizing methods can be followed ensuring complete removal of these sizes. It is recommended to carry out pretreatment on machines operating with the lowest possible tension which will facilitate subsequent wet finishing processes.

Woven polyester fabrics will often contain very high twist, high shrinkage yarns incorporated into the fabric. Microfibre based fabrics have a high bulk density and a particularly large surface area. An efficient preparation stage thus becomes vital to optimise removal of fibre processing aids whilst minimising creasing and maintaining the desired aesthetics.

6.5.4   Identification and methods to remove specific sizes

Identification

Generally polyester, polyacrylate or polyvinyl acetate sizes are used. They can be identified as follows:

Test	Observation	Inference
Dip a sample of grey fabric in solution of 0.5% C.I. Basic Red 22. Wet out well, take out and leave for 10 s in the air and rinse thoroughly in cold water	Light or dark red	Polyester, polyacrylate, polyvinyl acetate size present
Wash stained fabric with dichloromethane	Stain removed	Polyester size
Wash stained fabric with methanol	Stain removed	Polyacrylate size

After identification following recipes can be used in each case for their removal.

Sizing material	Desizing bath conditions
Polyester	Adjust the pH to 7–9 with 0.2–0.5 g/l soda ashorwith 1.5–2.0 g/l disodium hydrogen phosphate
Polyacrylate size	Adjust pH 10–11with 1.0–3.0 g/l soda ashorwith 1.5–2 g/l trisodium phosphate
Polyvinyl acetate	Adjust pH to 14 with 2.0–5.0 g/l caustic soda

At pH 7–9, polyester size is removed as a dispersion with the aid of auxiliaries, whereas at pH 14, it is saponified to produce the soluble monomer. The high concentration of caustic soda is essential to ensure complete saponification of size.

Preparation process of microfiber fabrics usually follows 3 stages:
There are number of options when polyester microfibre based woven fabrics are to be prepared for dyeing, which includes batch-wise or continuous.

6.5.5   Desize – scour – relax

A prewashing of the textile fabrics is required, as the warp ends are also sized in the field of microfibres. They are washed without tension in wide washing machines. In special cases(e.g. for severely contaminated materials), the fabric should be treated and/or washed twice in accordancewith the dwell process. After treatment, the material should be tested for remaining size residues. If size is remained a post-washed before dyeing on the jet machine using 1.5–2.0 ml/l of caustic soda liquor (38 °Bé), afterwards rinsed hot and cold and neutralised. Waxes have a high melting point and inefficient removal at the preparation stage can cause subsequent unevenness and patches problems at the dyeing stage. The preparation process should be very gentle to avoid any breakage of the fibres or creasing but should be efficient to remove the size and oils added on to the fibres to facilitate the manufacturing process and help protect the delicate polyester microfibrils.It is known that for the same volume, the surface area of a fibre inversely proportional to the radius. Thus surface area of the same weight of the fibre, the fibre surface will be much higher compared to the normal polyester and hence calls for 2–5 times addition of the processing chemicals especially preparation chemicals. To minimise creasing and maintain desired aesthetics it is necessary to control shrinkage of microfiber filaments during scouring stage. During relaxation, it is important to maintain minimum tension. Controlled shrinkage under relaxed conditions is vital during preparation stage. Shrinkage commences at lower wet processing temperatures when fabrics contain microfibers.

Scouring recipe

Additions	Alkaline normal	Alkaline jet/soft flow	Neutral	Neutral jet/soft flow
Suitable scouring agent (g/l)	0.5–2		0.5–2	
Or Low forming scouring agent (g/l)	0.5–2	0.5–2	0.5–2	0.5–2
Defoamer (g/l)	0.2–0.1		0.2–0.1	
Lubricating agent	1.5–2	1.5–2	1.5–2	1.5–2
Soda ash	2–3	2–3		

Load the machine and treat for 30–45 min at 60–70 °C:

Hot wash,

Cold wash,

Neutralise (if necessary).

Notes:

1. Use higher quantities of the products while operating with low M: L ratio.

2. Use neutral recipes with PVA sizes.

6.5.6   Bleaching

Set the bath with following recipe:

Quantity	Unit	Additions
1–3	g/l	Sodium chlorite

1–3	g/l	Sodium nitrate
1–2	g/l	Formic acid (adjust pH to 4)
1–3	g/l	Suitable scouting agent

And treat for around 60 min at 90 °C,

Hot wash,

Neutralise with a mild alkali at 90 °C,

And antichlor with 2 g/l sodiumbisulphite at 60–70°C for 20 min.

6.5.7   Weight reduction

Even though microfibers are produced to get closer feel to silk, etc. but still it is missing drape and aesthetics. Several efforts to make the fibre closer to silk, etc. have been tried. The weight reduction is most effective among them. Since the basic fibre is polyester the same weight reduction which is described elsewhere in this book for polyester is basically applicable to microfibers also. In the caustic weight reduction, the polyester is hydrolysed to produce water soluble, sodium terephthalate and ethylene glycol. In weight reduction treatment time, temperature, MLR, purity and strength of caustic soda, MLR, other auxiliaries, etc. controls the final finish. Addition of cationic agents like quaternary ammonium compounds can be used to accelerate the rate of weight reduction, but care should be taken to avoid yellowing, deterioration of tensile strength, removal of cationic residues.

After the weight reduction step, the woven fabric has large amounts of decomposition products likeoligomers, sodium terephthalate and ethylene glycol adhering to its surface, which are to be efficientlyremoved. A soaping bath at 90–95 °C is recommended before dyeing.

6.5.8   Dyeing process

Following factors should be considered in designing a dyeing process of polyester microfibers:

1. Disperse dyes exhaust on microfibres at a much faster rate than on conventional polyester.

2. Fabrics based on microfibres exhibit much greater shrinkage as the temperature of dye bath rises.

3. Exhaustion of dyes and shrinkage of yarns occurs at lower temperatures when compared with conventional polyester based yarns.

4. The more highly stretched finer sections of the filaments require more dye for a specific depth of shade, around 30–60% depending on the fineness.

5. In dyeings of microfibres, the light fastness of disperse dyes is partially degraded in comparison with normal textile polyester fibres so that a dye selection is required. If a setting process is carried out after dyeing, besides light fastness the dry heat setting fastness is also important. So the light fastness in particular may fail less depending on the depth of shade, material and finishing; a pH of the dyeing < 6 produces deterioration in polyamide microfibres. The same applies for the wet fastnesses, chiefly caused by the increased dye demand and (in polyester fibres) the thermo migration.

6. Because of the enlarged fibre surface area, it produces an increased absorption and a greater exhaustion rate of the dyes at the start of the dyeing process, from which levelling problems result.

7. In addition, the changed optics and the enlarged surface area result in an increased dye demand, which is 23 times the demand of classical fibres in microfibers made from polyester and twice the demand in microfibers made from polyamide (see figure below).

Typical fibre titre (ET) = 3 dtex

Cross sectional shape

Round, shiny   Round, dull   Profiled, shiny   Profiled, dull

Dye cons.   100   appx. 120   appx. 130   appx. 140

Effect of the fibre section and the dulling on the dye consumption on polyester microfibres

After desizing and washing, the material is dried at 140–150 °C. The free shrinking when dyeingthe unfixed cloth attains the optimum density of the fabric. It is dyed on jet dyeing systems at high fabric speeds, which should be at least 250 m/min depending on machine type.

Since, for the same standard depth on normal PES and microfibre higher amount of dye is required on microfibre, therefore exhaustion starts earlier in microfibres, hence the following precautions to be taken

1. Chemicals and dyes to be added at, at least 10–20 °C lower temperature than normal PES – dyeing temperature.

2. As the exhaustion on microfibre takes place faster at lower temperatures you need to do the following. (a) Metered addition of dyes if facility exists or addition in four portions at 2–3 min intervals. (b) Slow heating rate from 40 °C upto 90–100 °C.

3. Addition of a dispersing agent in excess of normal additions, as more dye is required to produce the same depth on microfiber (in some cases as high as 300% of normal PES) and the dispersion is required to be stable.

4. Addition of a strong levelling agent, which starts retarding even at lower temperatures.

5. Addition of lubricant is a must, as microfibre fabrics are prone to creasing.

It should be noted that the exhaustion behaviour for a particular standard depth is not affected to a large extent by the fibre fineness. For the same amount of dye on fibres with differing fibre fineness, it is seen that the exhaustion on the finer denier fibre takes place at 10–20 °C lower temperatures than the normal and the strike rate is also much faster than the coarser denier. Finer denier fibres permit higher saturation under standard dyeing condition of 130 °C for 60 min. The migration behaviour of disperse dyes, on microfibre is more dye specific than fibre specific. The sublimation, wet fastness and light fastness

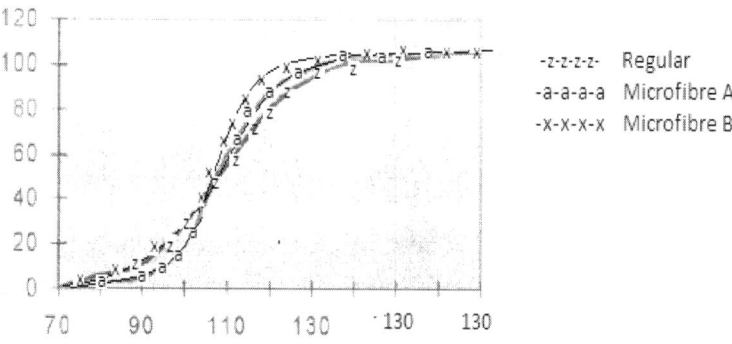

-z-z-z-	Regular
-a-a-a-a	Microfibre A
-x-x-x-x	Microfibre B

The migration of disperse dyes on microfiber is not fibre specific properties for the same visual depth on microfibers may appreciably decrease due to the fact that microfibres are dyed with higher percentage of dye for the same visual/standard depth. On fabric made from a blend of micro fibre and normal

PES fibre, solidity of shade becomes difficult and at top temperature during hold times the dye tends to migrate from microfibre to the normal PES fibre.

6.5.8.1   Exhaust dyeing

Dyeing is done as per polyester dyeing procedure except for the gradient for raising the temperature.

Quantity	Unit	Additions
2	g/l	Levelling agent
1.5	g/l	Lubricating agent
1–2	g/l	Formic acid (adjust pH to 4.5–5)

Bath is set with usual polyester dyeing auxiliaries, but higher quantities as explained earlier. Start the bath at 40°C add all the chemicals (A), run for 10 min. Divide the disperse dyes in 4 equal portions. Add the first quarter of the dye (B)at same 40 °C and run for another 5min. Raise the temperature to 90 °C at 0.5–1°C per min gradient. Add the balance 3 portions of the dye at 90 °C at 2–3 min intervals. (C–E) From 90 °C, one can raise the temperature at 1.5–2 °C per minute upto 130 °C or as per critical temperature zone and rate of rise recommended for that dye combination. Hold at 130 °C for 20–30 min more than used for normal PES.

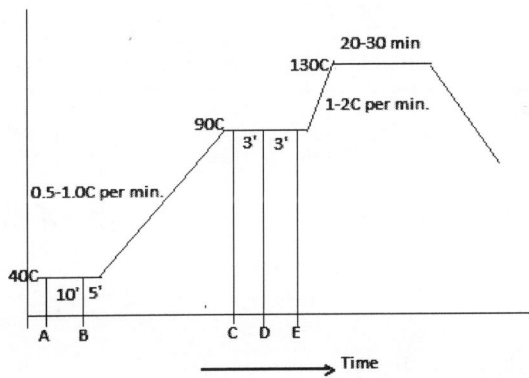

For a blend of normal PES and microfibre fabric, similar precautions to be taken as stated above. However, as at lower temperatures the distributions of dye on microfibre will be more than the normal fibre to reduce the difference in shade between normal and microfibre, the top temperature and hold times should be reduced or else dye will transfer from microfiber to normal fibre and solidity of dyeing will be affected.

6.5.8.2   Thermosol dyeing

Even though the exhaust dyeing is most practiced method of dyeing considering the penetration of the dyes and the handle of the finished product, thermosol dyeing taking enough precautions also can be considered especially for light and medium shades for level dyeing . Even though normal PE can be dyed at 200 °C, in case of microfibers it is often found that hardening of the fibre takes place. Hence the thermosol dyeing at 180 °C is more preferred. It has been found that sufficient penetration of dyes even for darker shades takes place at this temperature. However, it is advised to take enough trials before venture in to this method of dyeing as there are different types of microfibers are available in the market.

6.5.8.3   Finishing

Natural handle and wear comfort are the main qualities in using microfibres in shirts and blouses. Possible surface arrangements are sanding for "peach skin" or "angel's skin" and reduction of pilling and shaping for stylish effects. The finish should serve to emphasise the natural character. Silicon elastomer micro emulsions are particularly suited to attaining a soft, dry and coolhandle character. As shirts and blouses are manufactured in almost all possible blends, the final finishing should be matched with the blend component.

6.5.8.3.1   **Microfibre effects**

The key effects achievable with microfibres can be categorised as general and specific effects.

6.5.8.3.1.1   General effects

Softness – from the low profile effect of the fibres compared to regular 100% polyester.

Absorbency - from the capillary action of fine fibres compared to regular 100% polyester.

Dimensional stability – like normal polyester *but unlike cotton.*

As microfibre in polyester/cotton; compared to regular polyester/cotton.

Greater durability and softness – compared to regular polyester/cotton.

Better absorbency -use of less cotton for the same absorbency compared to regular polyester/cotton.

6.5.8.3.1.2   Special effects

More rapid cooling through moisture transport. Excellent anti-creasing and stability properties.

Special surface effects, such as peach or suedeing.Special feel like "Micro powder" or "Moist Touch". Water barrier effects while permeable to vapour. These effects lead to improved comfort and desirability and texture effects that are durable and dimensionally stable.

Finishing is always based on the purpose for which the fibres are used. Most popular areas where micro fibres are used include:

1. Fashionable apparel and outerwear: in fashion sector, natural fibres and viscose have been extensively used for washed out and emerised effects. Natural fibres, and particularly viscose, have been popular because of the associated smooth and soft handle, but potential has been severely restricted because of dimensional stability and creasing problems. Microfibres play a vital role in overcoming these difficulties. Application of silicone elastomers by padding or exhaust techniques followed by stentering at 170–180 °C is a popular method of enhancing softness of handle.

2. Functional sportswear: the high density polyester microfibre based woven fabrics comprising of flat filament or textured yarns are an ideal basis for water proof effects, and performance is considerably better than similar constructions based on cotton or PES/cotton. Microfibres with a fluorocarbon finish, even without coating, can achieve a resistance to water penetration equivalent to a water column of 350–700 mm, which is generally sufficient for ordinary rainwear or skiwear. Their water vapour permeability and associated wearer comfort is appreciably better than coated or laminated fabrics.

3. Surface effect finishes: the formation of micropile and microloop surfaces, by using differential shrinkage yarns, results in formation of peach skins and suedes. Before emerising, it is better to pre-lubricate the fabric with a non-permanent softener and this is best done on fully scoured and relaxed fabric, as it ensures better visual appearance after dyeing closer control of the emerising operation.

6.5.8.3.1.3   Peach skin, emerised finishing

Fine-fibril polyester yarns and microfibers are emerised with 400/500/600 grain sizes. Stillfiner grain sizes, around 600–800, are not normally used because they really "polish" rather than emerise. Details of grain size in Cylinder roller, Lattice roller and the direction of rollers are given below:

Data	Cylinder roller	Lattice roller
Grain size for fine-fibril polyester yarns and microfibres	400/500/600	80–120

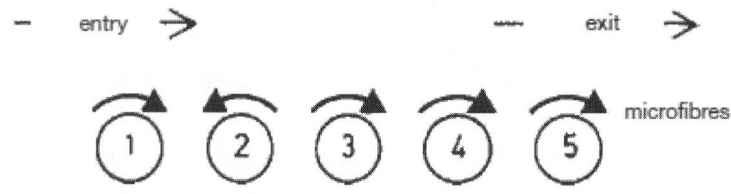

It is usual to give uniformly higher fabric tension is employed with micro fibres and fine-fibril yarns, allowing them to make only light contact with the emerising rollers (tangential contact).

6.5.8.3.1.4   Water repellent, hydrophobic finishing

Hydrophobic finishing may be carried out after dyeing and calendaring (as required). A post shrinking on stenter also may be carried out. The final finishing should be carried out so in such a way that the required testing is passed (say, >500 mm water gauge after 5th wash). Sports clothing and rain wear are to a large extent manufactured from 100% micro fibres (polyester or polyamide). Properties such as water vapour transmission, water impermeability, wind impermeability, but also soft handle and electrostatic properties are important. Fluorine chemicals are used for the water repellent finishing, as normal silicones can reduce water impermeability via channel formation due to their gliding effect.

6.5.8.3.1.5   Soft finish

Fabric for normal wear, usually made by blending with cotton, demands more softness, wear comfort, easy care and functionality than wind and water impermeability, which is required for sportswear. Surface treatments such as sanding or emerising are the most wide spread processes. With regard to the finishing operations, these operations differ only negligibly from those of sportswear. The sanding is normally carried out after dyeing, as the dyeing process may impair the effect on the cellulose part. Products, which do not have an influence on the water-repellent effect or can be washed out easily, should therefore be used as raising finishes.

Chapter 7

Processing of cationically dyeable and texturised polyester

Normally dyeing procedures of cationic dyeable polyester fibres are similar to those of normal polyester fibres, but quite different dyeing conditions and special precautions are required. Depending upon the end use requirements cationic dyeable polyester fibres can be used alone or in form of blend yarn with normal polyester or cellulosic fibres.

Cationic dyeable polyester is the modification of normal polyester, which makes it dyeable with cationic dyes at lower temperature developing bright colours with good all round fastness properties. The main advantages of it are:

- Low temperature dyeing can be used to dye blends with elastomers, which would likely suffers non thermal degradation while dyeing at 130 °C with normal polyester.
- Excellent colour fastness properties.
- Two colour effects can be produced on fibre blends of cationic dyeable polyester and normal polyester.
- The depth of the shade against normal polyester will remain same.
- All other properties of the normal polyester will be retained

7.1 Mechanism of converting normal polyester to cationic dyeable polyester

To convert the normal polyester to cationic dyeable polyester a stronger anionic group such as $-SO_3Na$ groups are introduced into the polymer. $-SO_3Na$ group is introduced in the ester interchange through charging 5 – sodium sulpha dimethyl isophthalate.

7.2 Properties of cationically dyeable polyester vs. normal polyester

- Because of the presence of higher DEG (di-ethylene glycol) units it is dyed darker with disperse dyes (as well as with cationic dyes).

- Polymer is less crystalline and more amorphous.
- Lower tenacity.
- Wash-and-wear property or crease resistance is poorer due to lower crystallinity. Crimp permanency is lesser.

Pretreatment

All these fabrics should be first given a pretreatment to remove spin finishes, dirt and impurities. The scouring treatment is given in mild alkaline condition using

Quantity	Unit	Additions
1–2	g/l	Suitable scouring agent
x	g/l	Soda ash to pH 8–9

At 70 °C for 20 min.

7.3   Dyeing

7.3.1   Dyeing at boil using a carrier

Recipe

Quantity	Unit	Additions
x	g/l	Cationic dyes
1–4	ml/l	Carrier
6	g/l	Glauber's salt calc.
1	g/l	Sodium acetate
0.5	g/l	Levelling agent

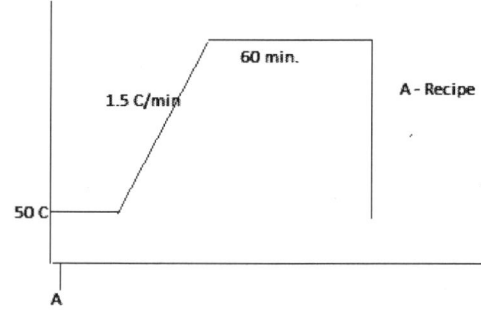

Procedure

Set the dye bath as mentioned in the recipe.

Start dyeing at 50 °C and raise the temperature to boil at the rate of 1.5 °C/min and hold for 60 min.

During dyeing pH should be maintained at 4.5.

Drain and give hot wash at 60 °C with suitable detergent (about 1 g/l) for 10 min.

Notes:

1. To prevent fibre hydrolysis, it is recommended to add Glauber's salt to the dye bath. The recommended concentration of salt as per the dyeing temperature is as follows:

Concentration (g/l)	Dyeing temp. (°C)
3	Boil (100)
4	106
6	120

2. Any non-ionic carrier can be used.

3. Anionic carriers are only suitable along with an antiprecipitant in the dye bath. Pre-trial is essential.

4. When dyeing blend in one bath with basic, acid and disperse or direct dyes (e.g. in fibre blends with wool, polyester or cellulosic fibre), precipitation of the dye may occur. This can be avoided by using an anti-precipitant. It also improves the reserve.

5. Pale to medium shade can be dyed using one bath two stage method. This also applies to dyeing with disperse and basic dyes.

7.3.2   Dyeing at 100–105 °C

Set the bath with the following at room temperature

Quantity	Unit	Bath additions
x	%	Cationic dyestuff
1–4	ml/l	Carrier
6	g/l	Glauber's salt
1	g/l	Sodium acetate
0.5	g/l	Levelling agent

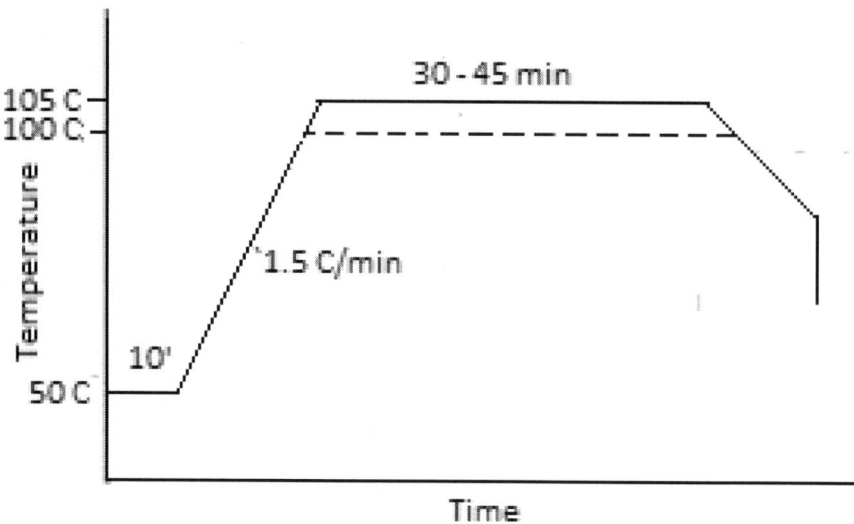

Procedure

Raise the temperature to 50 °C (alternatively one can start the bath at 50 °C) run for 10 min for homogenisation and even wetting and raise the temperature to 100–105 °C at 1.5 °C per min gradient. Dye at this temperature for 30–45 min, cool and drain.

It is not necessary to do a reduction clearing after dyeing; it is sufficient to wash at 60 °C with an ion active detergent.

While dyeing very deep shades it is recommended to remove unfixed dyestuff by washing with 1.0 g/l detergent at boil.

Refer notes in dyeing at boil. These notes apply to this dyeing also.

7.3.3   High temperature dyeing

Recipe

Quantity	Unit	Additions
x	g/l	Cationic dyes
1–2	ml/l	Carrier
6	g/l	Glauber's salt calc
1	g/l	Sodium acetate
0.5	g/l	Levelling agent

Procedure

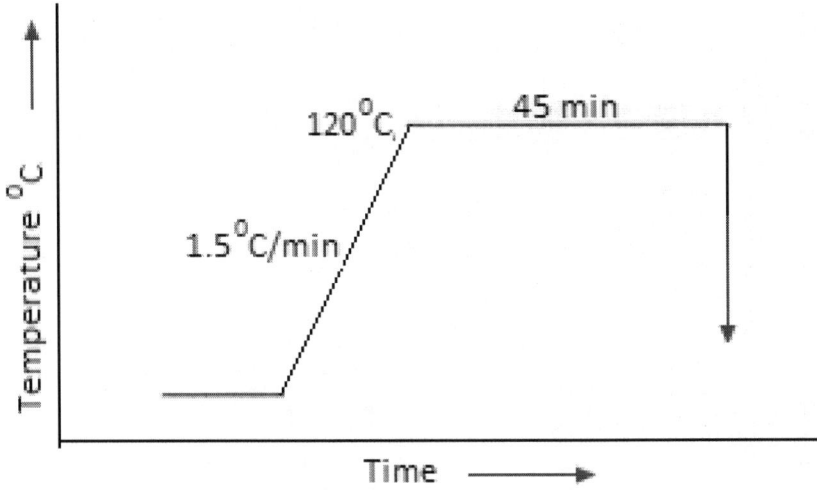

Set the dye bath as per recipe:

> Add the dye and auxiliaries at room temperature and run for 10 min.
>
> Raise the temperature to 120 °C at 2 °C per min.
>
> Hold at this temperature for 30–45 min.
>
> Cool and drain.

No reduction clear is necessary. For light to medium shades washing at 600 °C is sufficient whereas for dark shades washing at boil with suitable soaping agent is adviced.

Notes:

1. To prevent fibre hydrolysis, it is recommended to add Glauber's salt calc to the dye bath as follows:

   The recommended concentration of salt as per the dyeing temperature is as follows:

Quantity	Unit	Dyeing temperature
3	g/l	At boil
4	g/l	At 106 °C
6	g/l	At 120 °C

2. Any non-ionic carrier is used only when a levelling agent (antiprecipitant) is being used in the dye bath. Pre-trial is essential.

3. When dyeing blends in one bath with basic dyes and acid or direct dyes (e.g. in fibre blends with wool or cellulosic fibres), precipitation of the dye may occur. This can be avoided by using a suitable anti precipitant.

Pale to medium shades can be dyed by the one-bath/two stage method. This also applies to dyeing with disperse and basic dyes.

Stripping

Set the bath with the following keeping MLR 10:1 to 30:1.

Quantity	Unit	Additions
5	ml/l	Sodium hypochlorite
4	g/l	Sodium nitrate

Adjust pH to 4.0–4.5 with acetic acid. Enter goods at 75 °C, raise temperature to 95 °C within 20 min. Treat at 95 °C for 30 min. Cool down to 80 °C in 10 min. Rinse thoroughly in warm and cold water. Drain.

Treat with 2.5-g/l sodium bisulphate at 45 °C for 15 min. Rinse in warm and cold water. Drain.

About 90–95% of the dyeing can be stripped by this method.

7.3.4   One bath dyeing of cationic dyeable PE (CDPE)/ PE/viscose

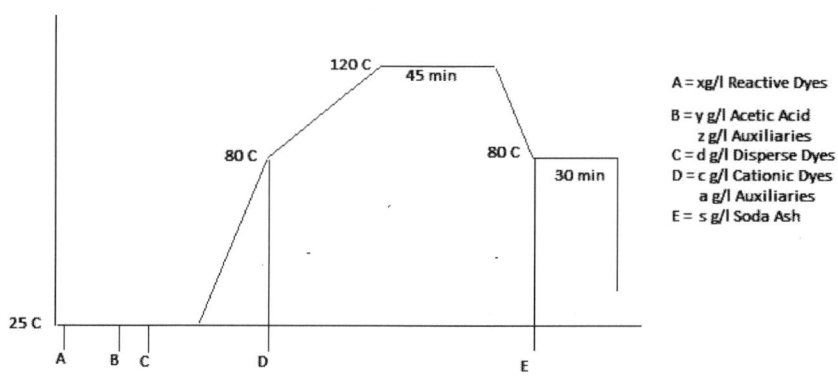

Procedure

Load the material, add required amount of common salt/Glauber's salt, run for 5 min.

Add the reactive dyes, run for 20 min at room temperature.

Add the acetic acid (50%of the quantity required for pH 4.5).

Levelling agent 1 –2 g/l.

Run for 5 min at room temperature and add pre dissolved disperse dyes.

Run for 5 min and raise temperature by 2 °C/min up to 80 °C.

Add cationic dye (dissolved with the help of acetic acid and wetting agent with hot water) and raise temperature to 120 °C by 1 °C/min and hold for 45 min to complete the dyeing of disperse and cationic dyes.

Cool down up to 80 °C and add required amount of soda ash (predissolved), run 30 min at 80 °C.

Drain and give hot wash at 60 °C for 10 min. Soap at 80 °C. and neutralise with acetic acid at room temperature.

7.3.5   Dyeing of CDPE/PE blends

It is possible to dye cationic dyeable polyester/polyester blends with

- Reverse effects,
- Shadow dyeing, and
- Two tone effects.

The combination of these two fibres is mainly used for outer wear. The fastness properties required for garment fabrics are obtained with the cationic dyes and disperse dyes and a careful selection of chemicals. Contrast and shadow effects can only be achieved if the normal polyester component is dyed pale with disperse dyes. Anionic modified polyester (basic dyeable) has a higher affinity to disperse dyes. It is therefore only possible to over dye the anionic modified polyester component with basic dye if the initial dyeing is pale.

Recipe

MLR 10:1	MLR 30:1	Unit	Additions
2	1	g/l	Dispersing agent
1	0.5	ml/l	Carrier
4	4	g/l	Glauber's salt
x	x	%	Modified cationic dye
y	y	%	Disperse dye

Procedure

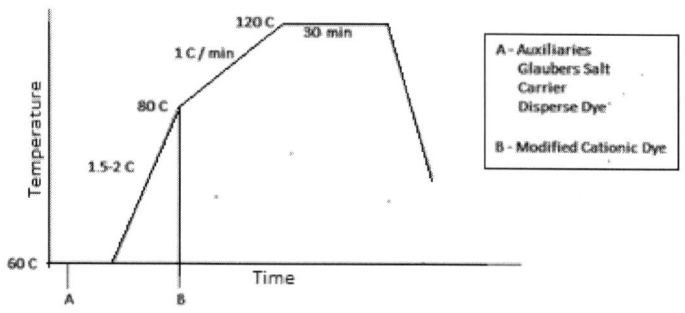

Follow the dyeing graph.

It is better to reduction clear the dyeing to remove the tinted basic dye on the normal PE to give a better contrast (in case of contrast shades) and improve the fastness properties.

For reduction clearing treat the dyeing with 1 ml/l detergent, 2 g/l sodium hydro sulphite and 1 ml/l ammonia 28% at 70 °C for 20 min.

Stripping of the dyeings

Set the dye bath with

Quantity	Unit	Additions
5	ml/l	Hypochlorite
4	g/l	Sodium nitrite
x	g/l	Acetic acid to adjust pH 4–4.5

Enter goods at 75 °C and raise temperature to 95 °C within 20 min.

Treat at 95 °C for 30 min, cool down to 80 °C in 10 min.

Rinse thoroughly in warm and cold water, drain.

Treat with 2.5 g/l sodium bisulphite at 45 °C for 15 min.
Rinse in warm and cold water and drain.

7.4   Processing of texturised polyester materials

Generally the main problem in processing and dyeing of texturised polyester fibres is unevenness or stripes of the dyeing's. This unevenness is, in many cases, attributable to differences in the internal structure of the fibres, that is, differences in crystallinity and orientation of the amorphous zones in the fibre, these determining the dye ability of polyester fibres. Structural differences of this kind can occur in texturised polyester fibres as a result of:

- fluctuations in the conditions during melt-spinning, e.g., differences in the purity and chain-length of the polymers, spinning temperature, stretching, and so on.

- fluctuations in the conditions during texturising above all, during setting, e.g., differences in twist, thread tension, temperature and the duration of heat treatment.

Normally the texturised polyester material in the grey form will not have the inherent elasticity of the fibre. The crimp, bulking, etc. has to be regained by suitable process. It is achieved by treating the material at a temperature of 60–100 °C in water without any tension. Normally this treatment is combined with a scouring operation (if no scouring is done this treatment has to be given separately or in a continuous operation this may be achieved by a steaming and then continuing the scouring operation). The crimping temperature chosen between 60 and 100 °C is governed by the kind of material and the extent of bulking required. In case of knitted material treated in jet there are chances of crease formation and hence the treatment may be given lower temperature with trials and the full bulking can be achieved while drying with hot air at 150 °C. The best results in bulking, are obtained by using winch becks. In open-width treatment it should be noted that the whole process has to be done with minimum tension on the material at a temperature 10–20 °C higher than the treatment temperature in winches.

During this washing, the goods are given the opportunity to develop the crimp. Loom-state cloths have internal tensions which are released by the simultaneous action of heat, water and mechanical movement and sometimes by steaming. The crimp effect, which has been temporarily lost because of elongation of the yarn during knitting, is regained.

The goods after the bulking operation should be set to retain permanent elasticity, bulkiness, etc., which can be done by heatsetting by the methods, explained earlier. This is mainly done with hot air or steam. A smooth fabric, i.e., without any raised pattern, can be set with water in an HT piece beam dyeing machines at temperatures of 120–130 °C. The majority of knitted or woven fabrics can be set in a stenter, bearing in mind the mechanical conditions imposed. Depending on the polyester material, the time required for fixation is from 15 to 30 s at a temperature between 160 and 190 °C. The material can be passed through a stenter with very little tension by arranging for an appropriate overfeeds and adjustment of the width. The processes just described, namely, bulking or scouring, followed by drying and heat-setting, usually precede the actual dyeing process. This order of operations holds true irrespective of whether dyeing is to be carried out on ordinary winch becks, HT winch becks, or piece beam dyeing machines. The best results, as regards handle, are obtained by this procedure. However, the amount of work involved is very costly, since a drying and heat-setting process comes in between the actual wet treatments.

7.4.1   Machines used for relaxation wash

As explained earlier the washing machines has to be selected taking into consideration whether woven or knitted and the structure to give almost zero tension wash. These days washing machines are available with least tensions, however we shall look at some of the options.

7.4.1.1   Winch becks

These are the most suitable machines for washing tubular knitwear, though they are also employed for slit-goods, as well as warp-knit and Raschel-knitgoods. The distance between the winch and the surface of the liquor should be kept as short as possible to prevent elongation of the knit goods. The liquor is heated to about 700 °C within 10 min, and the material is treated for 20–30 min at this temperature. It is particularly important that the liquor should be cooled slowly down to a temperature of below 60 °C in order to prevent fixation of running folds. However due to high MLR and low production these are not much used now.

7.4.1.2   Open width washing machine

The most versatile machines which is suitable for all type of fabrics and is widely used. To get a crimp and bulkiness achieved on winch becks one should make sure that the tension is minimum and preferably with load cell based tension adjustments and the washing temperature should be as high as possible, at, say, between 80–90 °C.

7.4.1.3   Perforated drum/conveyer washing machine

This is most suited for knitted material. Since the material sticks on the drum and moves along with the drum or on the conveyer practically no tension is applied on the fabric.

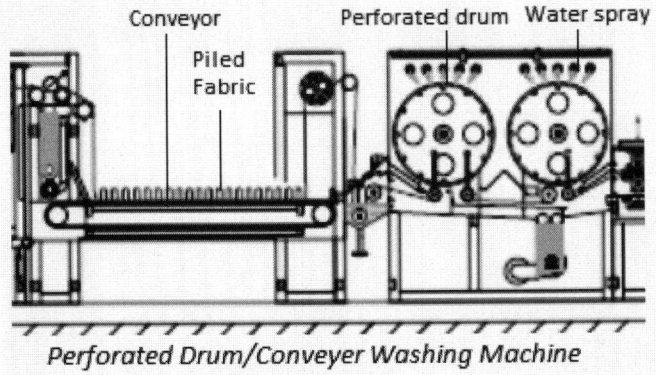

Perforated Drum/Conveyer Washing Machine

During the washing operation oils, dirts and other stains also has to be removed. One may use the following recipe in washing:

Quantity	Unit	Additions
0.5–1	g/l	Solvent based scouring agent
0.5–1	g/l	Soda ash

If there is iron stains one can have pre washing with

Quantity	Unit	Additions
1	g/l	Oxalic acid
0.5–1	g/l	Wetting agent or scouring agent

at 50 °C for 30 min.

Given below a summary of machines and process parameters:

Process	Woven	Knitted OW	Knitted tubular	Garments
Stabilising/ partial setting	Stenter 130–140 °C	Stenter 130–140 °C		Shaping machine
Development of crimp and washing	Winch 60–80 °C, OW washing M/c, perforated drum washing, jet 80–90 °C	Winch 60–80 °C, OW washing M/c, perforated drum washing, jet 80–90 °C	Winch beck, jet dyeing Mac 60–80 °C	Drum washing M/c 60–80 °C

Dewatering	Squeezing, suction	Squeezing, centrifuge	x	g/l
Drying	Stenter 100–105 °C, loop dryer 80–90 °C	Stenter 15–30 s 100–105 °C, Loop dryer 80–90 °C	x	Tumbler 60 °C
Heat setting	Stenter 15–30 s 150–180 °C, hydro setting during dyeing 120–130 °C, etc.	Stenter 15–30 s 150–180 °C, hydro setting during dyeing 120–130 °C, etc.	Hydro set during dyeing or slit and heat seats OW	Ironing press at 130–150 °C

Notes:

Since drying and heat setting after washing is expensive, these operations can be dispensed with only in special cases.

In case of knitted goods that have a pronounced relief structure, there are less chances for the folds to set, such goods can be dyed, immediately after being washed without setting.

When dyeing under HT conditions, setting with hot air can be replaced by hydro setting at 120–130 °C simultaneously with dyeing.

The hydro setting in beam dyeing during dyeing is not advised as there are chances of heavy shrinkage and the perforations of the drum may get exposed causing uneven liquor circulation and uneven dyeing

7.4.2   Dyeing

Dyeing can be done as in the case of normal polyester.

Suitability of various machines for set and unset goods in woven, knit, warp knit and garments are given below.

Recipe and procedures follow the normal polyester dyeing section.

Machines	Knitted tubular	Knitted OW	Warp knitted	Garments	Woven
Winch at boil	Partially suitable if not preset, chances of folds	Avoid not preset goods, suitable for heat set goods	Avoid not preset goods, suitable for heat set goods		No unset goods suitable for heat set goods

Machines	Knitted tubular	Knitted OW	Warp knitted	Garments	Woven
HT jets and soft flows	Partially suitable if not preset, chances of folds	Avoid not preset goods, suitable for heat set goods	Avoid not preset goods, suitable for heat set goods		No unset goods, suitable for heat set goods
HT beam dyeing machine	Not suitable	Avoid not preset goods, suitable for heat set goods	Avoid not preset goods, suitable for heat set goods		No unset goods, suitable for heat set goods
HT drum dyeing machine				No unset goods, suitable for preset goods	

Chapter 8

Dyeing of Nylon

8.1   General

Nylon is a generic name representing a family of synthetic poly condensate fibres made of a chain of repeating functional amide groups (–NH–CO–). Even though there are many types of polyamides (see table below) the most important types are Nylon 6 and Nylon 6.6. Polyamides were the first bulk produced synthetic fibres and was instantly accepted in the market. It was holding the first place as a synthetic fibre, but has now fallen to second place after the introduction of polyester in the market. The nylons are named according to the number of the carbon atom forming the monomer in question. The melting points of the polyamide-6 and -6.6 polymers are well-suited to fibre manufacture and application hence most important.

Polyamide types	Melting point (°C)	Moisture regain (%)	S.G (g/cc)	Softening point (°C)	Raw material
Polyamide 3	340	7.7			β-Lactum (βalanine)
Polyamide 4	256	7.7			γ-Butyrolactam
Polyamide 6	215	4.3		210	ε-Caprolactam
Polyamide 7	225	3		220	7-Amino-oenanthic acid (7-oenantholactam
Polyamide 8	200	2.9		180	8-Caprolactam
Polyamide 9	210	2.5		190	9 Amino-nonanoic acid
Polyamide 9.1					1.9-Namethenediamine and urea
Polyamide 11	190	1.2		175	11-Amino undecanoic acid
Polyamide 12	180	0.9			Lauryl lactam (12-amino dodecanoic acid)
Polyamide 6.6	255	3.8		236	1.6-Hexamethylene diamine and adipic acid
Polyamide 6.9					1.6-hexamethylene diamine and non adicarboxylic acid

Polyamide types	Melting point (°C)	Moisture regain (%)	S.G (g/cc)	Softening point (°C)	Raw material
Polyamide 6.10	214	2.6			1.6-Hexamethylene diamine and sebacic acid
Polyamide 6.12					1.6-Diamino hexamethylene and dodecane carboxylic acid
Polyamide 6.6/6					Adipichexamethylene diamine salts and e-caprolactam

Details of manufacture of this fibre are given in the book 'Textile substrates – Fibre, Yarn and Fabrics' by the same author. The raw material for the manufacture of Nylon 6 is caprolactam which can be manufactured by the following sequence starting from coal tar. The raw material for Nylon 6.6 is adipic hexa methylene diamine or hexa methylene diamine adipate– monomeric exchange product of adipic acid and 1,6-diamino hexane (hexa methylene diamine). The poly condensation is mainly depends on upon the temperature of the polymerisation. For example, 89% conversion can be achieved at 260 °C. The acid chain-stopper may be acetic acid, stearic acid or adipic acid, which at a particular stage of polymerisation reacts with the amino group, present at one end of the macro molecule and prevents further polymerisation. Hence, a desired viscosity of the molten polymer may be obtained. Once most of the volatile diamine has reacted, water is gradually removed by distillation. Finally, the molten mixture is heated to 280 °C under nitrogen to avoid oxidation. The final liquid polymer is extruded as a thick ribbon, cooled to solidify it, and cut into chips.

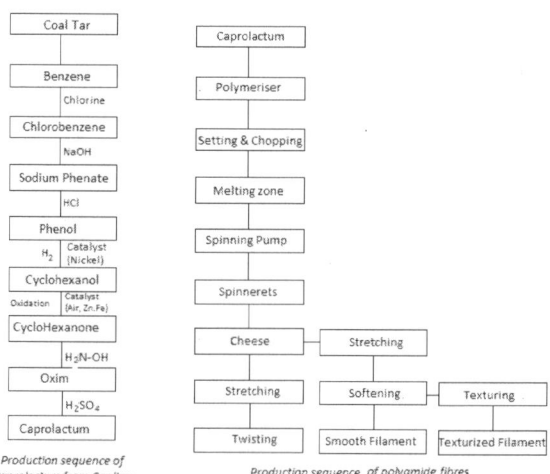

Production sequence of Caprolactum from Coaltar

Production sequence of polyamide fibres

The nylon chips are re-melted and the molten polymer extruded to form fine filaments. The melting point of nylon 6.6 is around 250 °C. The molten polymer coming out of the spinneret, i.e. the spinning jet is solidified in falling through the air, to a take-up machine which is about 6 m below the spinneret. Before reaching the mandrel on which the cheese is built up, the filaments are moistened by passing them through a steam chamber to ensure dimensional stability of the final package which may contain about a kilogram of filament.

The take-up machine operates at a very high speed of about 1000 m/min. About 45–50% RH and 20 °C temperature is maintained around take-up machine. The filament at this stage is not strong enough and of low dimensional stability. To improve this, it passes through a process called drawing. The nylon filaments are drawn by passing them around and between sets of rotating cylinders of increasing speed. A typical set is with upper set of nip rollers, running at 100 m/min, and a lower set of rollers running at 300–400 m/min, and hence stretch the yarn 3–4 times its original length before the stress begins to increase sharply. Drawing decreases the filament diameter. The elongation causes folded polymer chains to unfold and the chains to slide past each other, gradually increasing the degree of orientation of the polymer molecules along the filament axis. Drawing is carried out at room temperature (cold drawing). Finally, the filament yarn which is delivered from the lower nip rollers is twisted on up twister operating at a speed of 8000 rpm, and, is collected on the spindles. Twist imparted depends on the end use. Thus, hosiery yarn is imparted a twist of about 1182 tpm (30 tpi) while for weaving purpose a low twist of about 275.8 tpm (7 tpi) is considered sufficient. The residual caprolactam (about 3%) on the filament yarn is removed by washing with boiling or nearly boiling water, which also sets the twist in the yarn. The production sequence is given above.

Nylon 6 is also manufacture in a similar way as Nylon 6.6. Caprolactam is heated with a little water, a chain stopper such as acetic acid, and an acid catalyst. The water initiates the hydrolysis caprolactam to 6-aminohexanoic acid and then slowly removed by distillation. 6-Aminohexanoic acid reacts with caprolactam, opening the ring and generating a new amino end group. This amino group then undergoes the same addition reaction with more caprolactam to form polymer chain. The polymer chain thus formed will have an amino and carboxylate as end groups (see the formula below). This Nylon 6 can be directly extruded

$$nHN(CH_2)_5 CO + H_2O \rightarrow nH_2N(CH_2)_5 COOH$$

Opening of caprolactum forming amino caproic acid

$$nH_2N(CH_2)_5 COOH \rightarrow H\text{-}HN(CH_2)_5 CO\text{-}OH + nH_2O$$

Formation of Nylon 6 from aminocaproic acid (Condensation polymerisation)

$$nHN(CH_2)_5 CO \rightarrow H\text{-}HN(CH_2)_5 CO\text{-}OH$$

Formation of Nylon 6 directly from Caprolactum (Addition polymerisation) from the polymer or can be extruded into thick bands and cut into chips and spun into filament separately at a later stage. Any unreacted monomer or low molecular oligomers may be removed by water washing after extrusion.

The numbers of repeat units along the polymer chains of nylon 6.6 and nylon 6 are about 65 and 130, respectively, corresponding to molecular weights of about 15,000 g per mol. Even though a synthetic fibre, nylons are not totally hydrophobic like polyester, and their moisture regain is about 4% (nylon 6, 6.6). Since this low moisture regain does not allow the effective dissipation of electric charge and thus soiling takes place easily. The heat that is released with the hydration (wetting) of dry fibres with water (heat of hydration) is regarded as a measure of the hydrophilic property. Heat of hydration of polyamide is polyamide 31.8 kJ/g (wool 113.1 kJ/g, polyester 3.4 kJ/g).

Like wool, this fibre is also having weakly acidic carboxylic acid groups and weakly basic amino groups in them and the terminal amino groups can be of the level of 100 mmol/kg and the number of carboxylic acid group will be of about 90 mmol/kg. Since some of the terminal amino groups would have been reacted with the acetic acid chain blocker, the number of acid groups are more than the acid groups in nylon. In regular nylon it may be almost same (55 acid groups and 45 amino groups).

Polyamide fibres are used for woven and knitted goods used for blending with wool, cellulose, polyester and acrylic fibres. Woven polyamides are used for anoraks, rain wear wind breakers, ski pants and jackets, umbrellas, linings, tenting, awnings, rucksacks, suitcases, narrow fabrics, blankets, upholstery, etc. Whereas knitted materials are used as sports and bathing wear, stockings and tights, socks, underwear, bed linen, car upholstery, lace, nets, tulle, etc. Polyamides are also used as textile floor coverings – woven carpets, tufted carpets, needle felts, etc.

8.2   Preparation and bleaching

8.2.1   Scouring

Nylon can be scoured in mild alkaline bath with certain precautions. The scouring agents or detergents should be non-ionic as anionic detergents can block the dyeable $-NH_2$ sites by getting absorbed and consequently reducing the exhaustion of anionic dyes. Emulsified organic solvents may be added to help in the removal of oil stains on the material.

Guideline recipe for batch wise scouring

Machines – Jigger, winch, Soft flow, etc.

Quantity	Unit	Additions
2–3	g/l	Non-ionic detergent
1–2	g/l	Emulsifiable solvent
1–2	g/l	Soda ash

Treatment temperature: 70 °C.

Treatment time: 20–30 min.

Scouring also can be done on an open soaper.

Recipe

Quantity	Unit	Additions
2–3	g/l	Non-ionic detergent
1–2	g/l	Emulsifiable solvent
1–2	g/l	Sodium pyrophosphate

Treatment temperature: 90 °C.

Treatment time: 10–20 min.

Notes:

1. It is better to scour the fabric after heatsetting. There are two advantages – the scouring can be done at higher temperature say, boiling and better scouring and stain removal. Creasing, problem is reduced even in rope form scouring.

2. If scouring has to be done before heatsetting for some reasons the maximum temperature of treatment should not go above 60 °C and avoid rope form scouring, as the creases may set.

In both cases after scouring the material may be washed at room temperature and an acid treatment may be given to neutralise the alkali and remove any cationic finishes on the fibre.

8.2.2   Heatsetting

Heat setting is essential to preserve the best characteristics of polyamide fibre/ fabric and knitwear, namely, a high degree of elasticity and shrink resistance. The process can be carried out before or after bleaching or dyeing.

Woven fabric is usually set by hot air on a hot roll machine or on a pin stenter with the necessary equipment for setting. For knitwear the main methods are saturated steam or high temperature hydro fixation. These can be carried on a vacuum steamer, pressure steamer or hydro fixation machine.

Temperature and time for optimum fixation

Type of nylon	Saturated steam	Hot air	Hydrosetting
Nylon 66	130 °C	225 °C	135 °C
Nylon 6	130 °C	190 °C	135 °C
Time of setting	20 min	20–30 s	30 min

8.2.3   Bleaching

As the nylon fibre is supplied reasonably white it does not need bleaching. However, if it is yellowed during heatsetting or exceptional whiteness is required for full white, it may be bleached with sodium chlorite. (Both hydrogen peroxide and hypochlorite bleach are not recommended). Hypochlorite can affect the strength of the fibre due to chlorine combining with the secondary amino groups.

Recipe

Quantity	Unit	Additions
1-2	g/l	Sodium chlorite 80%
1-3	g/l	Sodium nitrite
0.2	g/l	Acid stable detergent
1-2	ml/l	Acetic acid or Formic acid 85% to pH 3.5 - 4

Treat for 1 h at 80–85 °C.

Notes:

1. Acetic acid is preferred to formic acid for adjusting the pH (3.5–4) because it has a buffering action on sodium chlorite solution in the required region.

2. The detergent helps in a combined scouring and bleaching effect and control the chlorines dioxide escaping to the atmosphere by creating form on the surface of the bath.

3. Sodium chlorite activator, if added in thebath, the bleaching can be done in a near neutral bath (pH 6–6.5) e.g. ethyl lactate (2–3 g/l) for short liquor and 1 g/l for long liquor bath or ammonium chloride.

Peracetic acid is a common bleaching agent for bleaching polyamide (and polyester). The bleaching is carried at a pH of 8.2. Alkaline pH is necessary to consume the acetic acid released during bleaching at the same time the pH has to be maintained. An effective buffer or automatic pH control device by continuous addition of an alkali into the bleaching bath. A pH higher than 8.2, say >9 can cause the peracetic acid break down at an incredibly high speed, resulting in an inadequate whitening effect and possible fibre damage.

Oxidative bleaching treatments are the most effective in overcoming the yellowness of grey-set nylon but analysis reveals that some amino end-groups are oxidised and dye substantivity is lowered. This effect significantly impairs the resistance of the nylon to photo degradation and lowers the light fastness of dyeings and prints produced on the bleached substrate. The more drastic is the oxidative bleaching treatment, the greater the sensitivity of the treated nylon to photo degradation.

Hence, in most instances nylon is bleached under reducing conditions using sodium dithionite and a suitable fluorescent brightener under acidic conditions at pH 4–5.

8.2.4   Optical whitening

Nylon is supplied sufficiently white. But if whiteness has to be improved after scouring or bleaching operation whiteness can be considerably increased by applying optical whitening.

Optical whitening is available in acid or acid milling type. Application can be done exhaust as given below:

Quantity	Unit	Additions
0.2–0.5	%	Acid or acid milling type OBA
2–4	%	Acetic acid 40% **Or**
1–2	%	Formic acid 85%
		pH adjusted to 3–5

Temperature of treatment is 80–90 °C for acid and 50–60 °C for acid milling type for about 20 min.

Optical whitener can be applied from a reductive bleach bath as follows:

Quantity	Unit	Additions
0.2–0.5	%	Reduction stable OBA
3	g/l	Sodium hydrosulphite

Temperature is 85–90 °C for 20 min.

8.2.5   Photo degradation of nylon

When nylon is in simultaneous contact with light, heat, air, water vapour, ozone and gas fumes photo degeneration takes place. Photo degradation can be controlled by:

1. Optimal purity of starting materials for polymerisation.
2. Minimal temperature of polymerisation, with exclusion of oxygen.
3. Inclusion of protective inorganic salts, particularly those of transition metals such as manganese (II).
4. Inclusion of protective organic compounds as antioxidants.
5. Exclusion of fluorescent brighteners from the polymer.
6. Increased average molecular mass of the polymer, increased content of amino end-groups, as in deep-dye nylon variants.
7. Minimal temperature of the polymer melts, with exclusion of oxygen.
8. Minimal concentration of delustrant, using the optimal grade of titanium dioxide.
9. Selection of suitable spinning lubricants.
10. Draw ratio as high as possible to ensure maximum crystallinity, consistent with other commercial requirements.
11. Storage of grey fabric away from direct sunlight and away from heating systems.
12. Minimal temperature of heat setting in the grey state, consistent with the attainment of adequate set.
13. Treatment in a stenter designed for steam injection.
14. Alkaline rather than neutral or acidic conditions of scouring.
15. Oxidative bleaching should be avoided, but fluorescent brighteners can be applied under reducing conditions.
16. Dyes of high fastness to light (6–7) are preferred, with particular care in selecting suitable trichromatic combinations.

17. Certain metal-complex and milling acid dyes exert a protective effect on nylon, whereas others may catalyse photo degradation.

18. Preferable to dye at or near pH 5, since more strongly acidic conditions can adversely affect antioxidants.

19. Inclusion of thioureaor hydroxylamine in the dyebath eliminates dissolved oxygen and minimises the risk of degradation in high-temperature dyeing.

8.3   Dyeing

Polyamide fibres take dye at relatively low temperatures. For this reason, energy-saving low temperature processes (e.g. at 55–75 °C) have been proposed. However, to optimise, the low energy costs must set against the higher time costs, since in the case of polyester, a temperature which has been reduced by 10 °C means that the necessary dyeing time is approximately doubled. The optimum maximum temperature, at which the overall costs (steam and time) are minimal, depends on the ratio of steam costs to time costs, as well as on the formula (rate of dyeing), the fibres (staining speed) and the heating rate (see figure below). The necessary dwell time in the case of this optimum maximum temperature is, by contrast, almost independent of the formula and fibres, and is essentially given by the cost ratio of steam/ time. If the quality of the goods permits,

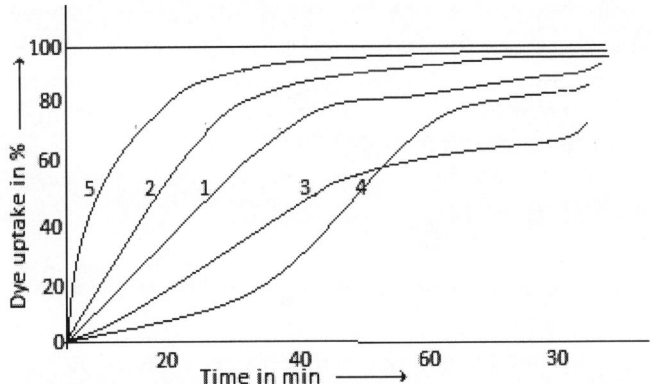

Absorption curves of various dyestuffs applied to polyamide whereby curve 1 is approximately optimum. Curves 5 and 2 represent a rapid dyeing process with the accompanying risk of unevenness. Cruves 3 and 4 guarentee the dyeing but require expensively long dyeing times. Curve 1 shows the optimum dyeing curve.

it is consequently more economical to dye different formulae at different maximum temperatures but the same dwell time than (as has been more common to date) to dye for different lengths of time at the same temperature.

Owing to their chemical and physical structure, nylons have an affinity for every dye class, although the most commonly used are acid, premetallised and disperse dyes.

8.3.1   Dyeing with disperse dyes

In case of disperse dyes the dyeing mechanism is little different than in the case of polyester. Disperse dyes are insoluble in water and hydrophobic, but readily colour nylon. Here the nylon acts as a solid solvent in which the disperse dye is soluble, so the nylon extracts the dyes from the aqueous bath almost as in the case of cellulose acetate and other synthetic fibres with disperse dyes. Their mode of attraction to the fibre is non-ionic and they are relatively insensitive to chemical and physical variations in the nylon and in pH. Small disperse dye molecules with relative molecular mass of around 400 show high rates of diffusion into nylon fibres and have good migration properties. Hydrogen bonds are often involved in dye/fibre attractions and can participate in dyeing mechanisms in many ways, especially in the case of synthetic fibres. In case of nylon, the bonding between disperse dyes (NH–, NO– groups) and polyamide happens. Although disperse dyes are quickly absorbed by nylon at temperatures up to the boil, unfortunately they are also desorbed during washing, leading to poor wash fastness. Washing normally takes place well above the T_g of wet nylon and disperse dyes can migrate freely within the non-crystalline regions of the nylon and desorb into aqueous solutions. Disperse dyes are used on nylon where wash fastness is not critical, for example in the pale shades on stockings and tights. The light fastness of disperse dyes is good and they are therefore used for dyeing nylon carpet yarns where washing fastness is not critical.

8.3.2   Dyeing with direct dyes

Direct dyes can be used to dye pale to light shades on nylon but it gives duller shades with poor wet fastness due to poor diffusion of large dye molecules. Hence used only for cheaper qualities. Direct dyes with more SO_3H – groups in them gives better fastness on nylon. Direct dyes are also dyed from an acidic bath as in the case of acid dyes. The fastness can be improved by suitable after treatments.

8.3.3   Dyeing with acid and metal complex dyes

8.3.3.1   Parameters influencing dyeing

8.3.3.1.1   Characteristics of fibres

Polyamide belongs to the poly condensate group of fibres, the most important of these being nylon 6 (Perlon type) and nylon 66 (Nylon type), which differ some what in their physical and dyeing properties.

The Perlon type is slightly less temperature-resistant, but easier to dye. Some what better fastness properties can be expected with the nylon type.

The following fibre properties and/or pretreatment processes are among the parameters that can influence the result.

Fibre surface/cross section (round, trilobal, etc.): The more highly structured and the larger the surface, the lower the colour strength and thus the lower the light fastness.

Fineness (dtex): The finer the fibres or fibrils, the lower the colour strength.

Delustring: The higher the degree of delustring, the lower is the colour strength and light fastness.

Texturing: Colour strength is lower on textured than on smooth yarns.

8.3.3.1.2   Dyeing temperature

The best dyeing temperatures for polyamide depend mainly on substrate form, dyeing equipment, type of fibre and required shade depth. Higher temperatures result in more rapid bath exhaustion, higher migration rates and better coverage of barrenness, but can harshen handle.

PA 6 textured to max. 96 °C,

PA 66 textured to max. 108 °C,

PA 6 non-textured to max. 120 °C,

PA 66 non-textured to max. 120 °C.

8.3.3.1.3   Dyeing pH

The optimum pH depends on type of dye, shade depth and type of fibre. A constant pH throughout dyeing is crucial to reproducibility. It is advisable to adjust the pH with a buffer.

Buffers for various pH for acid dyeing

Buffer chemicals	pH >	4	4.5	5	5.5	6	6.5	7
Sodium acetate or ammonium acetate	g/l		0.5	1	2	2	2–3	
Acetic acid	ml/l	1–1.5	0.5	0.4	0.3	–	–	–
Monosodium phosphate	m/l	–	–	–	–	2	1.5	0.5
Disodium phosphate	g/l	–	–	–	–	0.5	1	1.5

The pH depends closely on the process water, and the above figures should thus be regarded as guidelines. A rise in pH is to be expected on heating permutit water.

8.3.3.1.4   Migration

The movement of the dye in or on the substrate from points of higher to points of lower concentration is known as migration. Migration depends primarily on the constitution of the dye, but is also affected to a varying degree by the type of fibre and treatment conditions (pH, auxiliaries, temperature and time). Migration is an indication of the levelness of a dye.

8.3.3.1.5   Barrenness

Production-related physical and chemical differences in the fibre material and also stretching differences can result in barreness. Barreness can be covered to varying degrees by selecting suitable dyes and using appropriate dyeing auxiliaries.

8.3.3.1.6   Build-up

Build-up refers to the maximum shade depth achieved with a specific dye. Build-up is primarily determined by the chemical constitution of a dye. Fibre data, especially saturation limit of the fibre, also play a role.

8.3.3.1.7   Blocking

Mutual displacement of individual acid dyes from the fibre during dyeing is known as blocking. This effect depends mainly on dye constitution, but also on dye concentration, fibre saturation, auxiliaries, pH and temperature. Thus disulpho dyes on polyamide are blocked to a varying degree by monosulpho dyes. Dye selection plays an important role in preventing.

8.3.3.2   Dyeing with acid dyes

8.3.3.2.1   Theory of dyeing with acid dyes

Nylon can be dyed with anionic dyes from weakly acidic pH solution. The affinity of acid dyes for nylon depends on the electrostatic attraction between the negatively charged sulphonate groups in the dye and the amine end groups on the fibre that have become positively charged through dyeing under moderately acidic conditions. This process can be considered as a simple anion exchange process in which a counter ion such as acetate, associated with a protonated amino group in the nylon, is exchanged for a dye anion.

$$HOOC–Nylon–NH_2 + CH_3COOH \rightarrow HOOC–Nylon–NH_3{}^{+-}OOC–CH_3$$

$$HOOC–Nylon–NH_3{}^{+-}OOC–CH_3 + Dye \rightarrow HOOC–Nylon–NH_3{}^{+}Dye^- + CH_3COO^-$$

Since the reaction with the ammonium ion groups are limited the number of anionic dyes can be taken up by nylon are also limited hence dyeing a dark shade or black by this method is impossible.

From a dyeing point of view, imido (–CONHCO–) and imino (–NH–) groups represent the most important components of the protein fibres (wool, silk) and polyamide fibres:

PA6: $–CONH–(CH_2)_5–CONH–(CH_2)_5–CONH–(CH_2)_5–$

PA6.6: $–CONH–(CH_2)_6–CONH–(CH_2)_4–CONH–(CH_2)_6–$

Wool: –CONH–CHR1–CONH–CHR2–CONH–CHR3

Silk:$–CONH–CHR1–CONH–CHR2–CONH–CH_2–$

Acid dyes are also used for bright shades in light to medium shades with medium to poor fastness. Acid dyes are anionic dyes characterised by possessing substantivity for protein fibres such as wool and silk and polyamide (e.g. nylon 6 and nylon 6.6) fibres or any other that contain basic groups. Acid dyes are normally applied from an acid or neutral dye bath. As the size of the acid dye molecule generally increases the colour fastness to washing increases because of the increasing strength of the non polar forces of attraction for the fibre. The polyamide fibres have $–NH_2$ which gets protonated in the acidic dye baths and the group is transformed into $–NH_3{}^+$. The dye anion with $–SO_3{}^-$ group gets attracted to fibre and is adsorbed, forming an ionic linkage, a salt link, with the fibre dye site. Since the basic dye sites in the fibre is limited mono sulphonated acid dyes are more suitable than di-, tri- and tetra sulphonated dyes as against the direct dyes as the $–SO_3{}^+$ sites involve in the linkage formation here. When more sulphonyl groups are available in the dye more basic sites gets blocked and there by less dye molecules are able to form an ionic linkage. Even though the degree of sulphonation helps in dyestuff solubility, the colour build up is generally greatest with mono sulphonated dyes.

The relative molecular mass of the polymer is therefore important for the dye affinity of the fibres. The higher the relative molecular mass, the lower the number of amine end groups, but not all of these groups will be accessible to the dye molecules. It should be noted that the nylon 6 dyes more easily than other nylons even though the acid dyes are attracted to the amine end groups (AEG) in polyamide fibres. The amine end groups in the crystalline regions of the fibre are less accessible as acid dye receptor sites. The dye affinity depends therefore on the crystallinity of the fibres and this can be influenced by the rate of cooling of the extruded yarn after extrusion, the draw ratio, exposure to steam, the application of swelling chemicals such as phenol and the texturing conditions. The reason being nylon 6 has a more open physicochemical structure and a lower temperature compared with nylon 6.6. Thus, acid dyes diffuse more readily into nylon 6, but the colour fastness to washing of a similar dye on nylon 6.6 is generally superior because of the more compact fibre structure. The AEG can vary on nylon fibre due to various reasons temperature and/or tension differences during manufacture or processing which can cause dye ability variations with acid dyes. This problem often referred to as barré or barrenness, which is common in nylon dyeing.

The acid dye uptake on nylon depends on the pH of the dyebath. From a pH 7 to 4 the dye uptake slowly increases. At pH 2.7 the fibre attains maximum cationicity, which is called the 'isoelectric point' of nylon and from pH 4 to 2.7 the pH is more or less stable. If the pH is brought below 2.7 excess acid is absorbed because after all $-NH_2$ groups are protonated and as per the reaction shown above, formation of positive charges on the fibre increases as the dye bath pH becomes more and more acidic. As the protonated sites are increased more and more acid dyes are reacted and hence the higher dye uptake. As the temperature increased to boil the dye uptake happens much faster and dye fibre salt linkage takes place through the formation of ionic bond between acid end group of dye and protonated $-NH_2$ group of fibre.

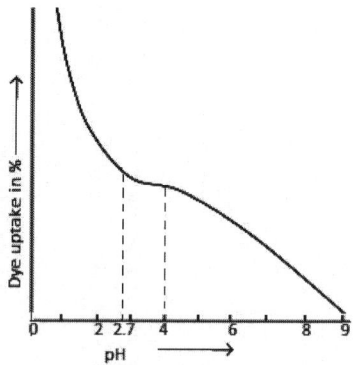

Graph showing pH of the dye bath and acid dye uptake by Nylon

But with an excess of acid and lower pH the acid further attacks the − CONH groups turning them also cationic and dyes are also absorbed on these sites and thus at very low pH of 2.7–1 the pickup is steeply increased. This results in over dyeing and degradation of the fibre also.

The Fibre
HOOC - Nylon - NH₃

First stage
HOOC - Nylon − NH₃⁺⁻OOC − CH₃ + ⁻O₃S - D ⟶ HOOC - Nylon − NH₃⁺ ⁻O₃S - D

Second stage (at lower pH)
HOOC - Nylon − NH₃⁺ ⁻O₃S - Dye + D - SO₃H + CH₃COO⁺ ⁻H ⟶ H⁺ ⁻OOC - HN - CO - Nylon - NH ⁺⁻O₃S - D
                                                                                H⁺⁻O₃S - D

For an even dyeing the −NH₂ groups should be slowly protonised and converted to dye fibre salt linkage while the −CONH sites are kept intact.

In a practical dyeing process for light shades the pH is kept on the higher side (towards neutral as all the −NH₂ groups do not have to be protonated and entered in to bondage with the dye. Whereas in case of deeper shades to achieve a levelled shades pH is kept at 4–4.5 so that most of the −NH₂ groups has been involved in the reaction and 80–85% dye has been exhausted. At this stage the pH can be further reduced to 2 safely to complete the exhaustion. The dyeing can be washed and after treated. Since the fastness of acid dyes on nylon is poor, especially nylon microfibres, it can be improved by after-treatments such as syntanning, that is, the adsorption of a sulphonated synthetic tanning agent that provides a physical barrier to desorption by blocking the fibre pore structure in the fibre surface regions and by providing electrostatic (ionic) repulsion to dye desorption.

The acid levelling dyes have smallest molecule and hence high rates of diffusion and there is little hindrance to the movement of the dye molecules in the fibre structure. The exhaustion rates are high, which can lead to the dye molecules attaching to the most accessible sites and hence poor evenness. In dyeing nylon it is usual to restrict the rate of dyeing initially by using a moderately acidic pH and then lowering the pH as the reaction proceeds. This will then give a level dyeing. The acid milling dyes, which have a larger molecular size and exhaust onto nylon in the pH range 3.0–5.0. They diffuse into the nylon more slowly and are not as good at covering physical and chemical variations in the fibre, but the wash fastness is higher. The usual dyeing technique is to start with the dyebath neutral and then to lower the pH.

Super milling acid are dyes with the largest molecular size among acid dyes and they are the most sensitive to chemical and physical variations in the fibre. They have a high affinity under neutral conditions and it is most difficult to obtain level dyeing with super milling dyes on nylons. But these dyes give the dyeing with best fastness properties among the three groups of acid dyes. Thus, when using this category of acid dyes it is worth using levelling agents during dyeing. Anionic levelling agents (e.g., Turkey red oil, calsoline oil HS with multiple –OSO$_3$ groups. Carboxylic soaps) enter the fibre first and interact with basic dye sites, restricting dye uptake. Amphoteric levelling agents, which contain both a positive and a negative charge, block the basic dye sites in the fibre but also complex with the acid dyes in the dye bath, slowing the rate at which the dyes exhaust onto the fibres. As the dye bath temperature is increased, the anionic (or amphoteric) levelling agent desorbs from the fibres, allowing the dye anions to diffuse and fix within the fibre. At higher temperature the complexes formed with acid dyes will also break down releasing the dye anion free for diffusion and fixing inside the fibre. Both these action together result in a levelling action by the thus gradual adsorption by the fibre as the dye bath temperature is increased.

The affinity of nylon for acid dyes can also be varied during polymerisation. When monobasic carboxylic acid is added to the monomers before polymerisation to block some amine end groups gives a light acid dyeable nylon. Where as incorporating copolymerisable compounds containing tertiary amine groups, such as N-(2-aminoethyl) piperazine deep acid dyeing nylons are produced. This compound is incorporated into the polymer chain by reaction of the primary and secondary amines to form amide groups. These additional basic dye-site groups are not end groups, but are equivalent from the dyeing point of view. It is also possible to increase the acid dye ability by adding a strong mono functional acid such asphenylphosphinic acid to the polymerisation. The acid, which does not cause degradative reactions, complexes with the amine groups to produce amine salt end groups that do not participate in the amine–carboxyl–amide equilibrium reaction, but are available during dyeing as sites for acid dyes, which displace the strong mono-functional acid from the salt groups.

8.3.3.2.2   Dyeing process

8.3.3.2.2.1   Exhaust dyeing on winch/beam/jet machines

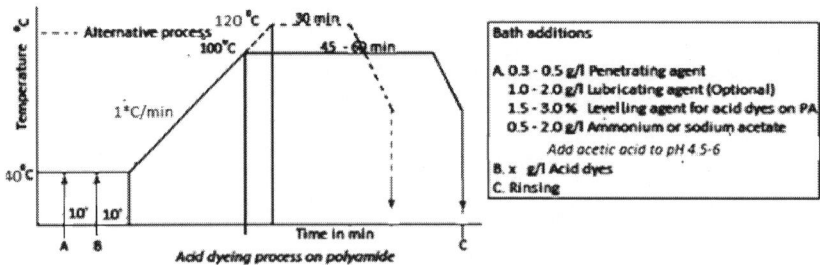

Acid dyeing process on polyamide

Notes:

1. Wet fastness properties are improved by after treatment with fixing agents.

2. When dyeing high concentration of dyes like blacks it is better to add 0.3% dispersing agents/emulsifiers and 2–4% acetic acid.

3. Usual pH of dyeing is 4.5–5.5.

4. Pale shades dyeing maybe done at higher pH i.e., near neutral.

8.3.3.2.2.2   Exhaust dyeing on jigs

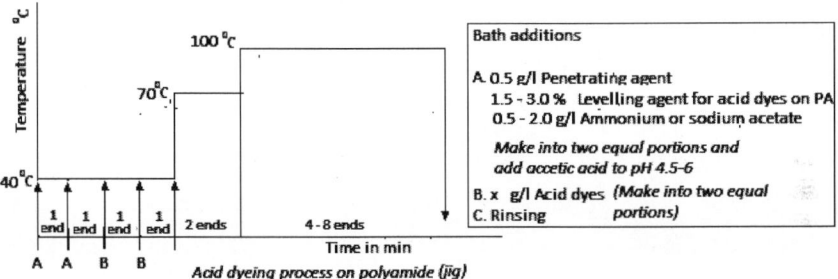

Acid dyeing process on polyamide (jig)

Notes:

1. Wet fastness properties are improved by after treatment with fixing agents.

2. When dyeing high concentration of dyes like blacks it is better to add 0.3% dispersing agents/emulsifiers and 2–4% acetic acid.

3. Usual pH of dyeing is 4.5–5.5.

4. Pale shades dyeing maybe done at higher pH i.e., near neutral.

5. Dyeing temperature: 90–92 °C rather than at the boil (less temperature drop between edge and centre).

8.3.3.3   Dyeing with metal complex dyes

Another class of dye which can dye nylon is metal-complex dyes. However, the use is limited as differences in the polyamide material cannot be levelled out in most cases and they form co-ordinate bonds with nylon. These dyes have good light and wet fastness properties but produce only dull shades.1:1 Metal complex dyes with monoazo compounds with sulphonic acid groups, which contain one chromium atom per dye molecule (chromium complexes) are chiefly dyed on polyamides. They have to be dyed from a strongly acid bath and show very good levelling power under these dyeing conditions. Good to very good light fastness as well as good wet fastnesses. Well suited to carbonized piece goods. Quantity of sulphuric acid can be reduced by specific auxiliary products. They are sensitive to metal-complexing agents. In case of 1:2 metal complex dyes they contain two tone components and a metal central atom with ionized, acid water-soluble groups or methyl sulphone, alkyl sulphone or sulphonamide groups. They are dyed onto polyamides (wool and silk) but with chances of uneven dying if not properly controlled. They tend form to streaky dyeing and the ability to cover affinity differences in the nylon is variable and blocking auxiliaries are normally used during dyeing Shades usually dull. These are large molecules and give excellent wash and light fastness in pale shades. This makes them particularly suitable for colouration of nylon car seat covers.

There are two types of metal complex dyes (1) 1:1 Metal complex dyes where one mole of dye is attached with one atom of metal and (2) 1:2 metal complex dye where two moles of dye are attached with one atom of metal. It is needless to say that the molecules of 1:2 type is much larger than 1:1 metal complex dyes.

8.3.3.3.1   Dyeing with 1:1 metal complex dyes

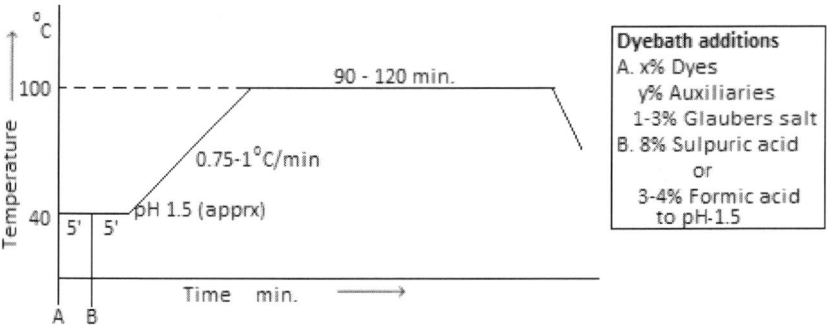

Dye bath is prepared with well dissolved dyes and the electrolyte (Glauber's salt or common salt 1–3%). Material is entered and worked for 5 min and the acid is added to adjust pH to approximately 1.5–2 (8% sulphuric acid or 3–4% formic acid). The temperature is slowly raised after homogenising the bath to boil and hold at this temperature for 1.5–2 h to complete the exhaustion and drop the bath. The dyeing is further soaped and washed.

8.3.3.3.2   Dyeing with 1:2 metal complex dyes

8.3.3.3.2.1   Exhaust dyeing on winch/beam/jet dyeing machines

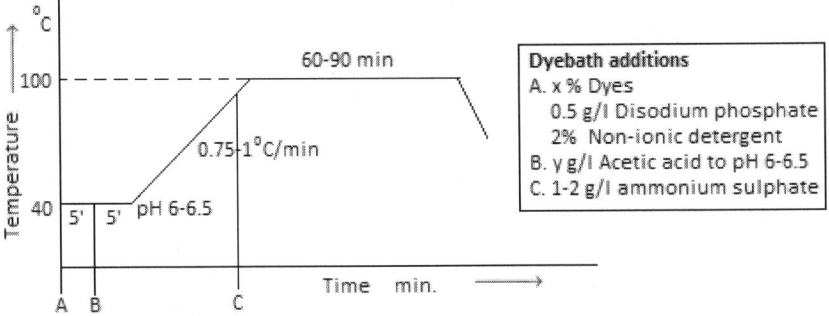

Set the dye bath is set at 40 °C with the dyes, 0.5% disodium phosphate, 2% non-ionic detergent, load the material and work for some time to homogenise the bath. Check the pH, and adjust the pH to 6–6.5 by adding acetic acid if necessary. After running for 5 min raise the temperature to boil at 1 °C per min reaching the boil an acid liberating agent like ammonium sulphate can be added and run at boil for 60–90 min or more to complete the exhaustion, as the migration of the dyes are slower due to the bigger size of the dye molecule. This may also cause poor coverage of the shades.

8.3.3.3.2.2   Exhaust dyeing on jigs

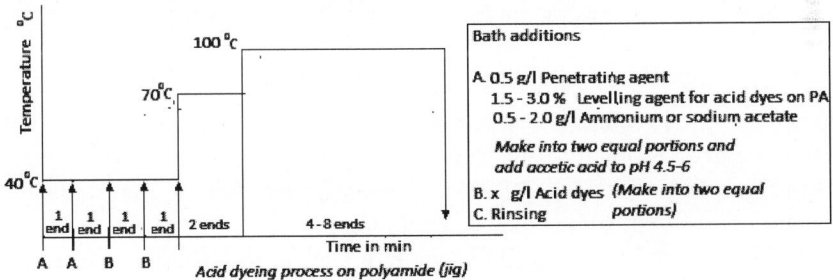

Acid dyeing process on polyamide (jig)

8.3.3.4   Other dyes

Reactive disperse dyes: dye able as disperse dyes; chemical bonding with the fibre takes place following an alkaline after treatment thereby giving good wet fastness properties. Another class suitable for polyamide dyeing is dispersed 1:2 metal-complex dyes. They do not contain any water-soluble groups and are applied as disperse dyes on polyamides. Good light and wet-fast, building up to the deepest tones. Tendency to streaky dyeing in brighter shades. Usually dull tones.

Chrome dyes: of interest only for black shades (applied by the so-called meta chrome process in which dye and mordant are applied simultaneously from a single bath).

Reactive dyes are also used for dyeing nylon for bright shades with good the light fastness and wet fastness but it varies appreciably on polyamide. There can be levelling problems due to affinity differences. Combination dyeings are only possible to a limited extent. The build-up is also limited.

8.3.3.5   The problem of barré or barriness

Barré is a special phenomenon more found in nylon dyeing, which can be caused due to various reasons. Barré is continuous visual stripy pattern parallel to yarn direction in knitted or woven fabrics. The reasons for barré may be, a physical, optical ordye differences in yarns or geometrical differences in fabric structure or poor coverage power of dye, chemical variations from changes in the AEG of nylon fibres which can give rise to dye ability variations with acid dyes, etc. Physical variations caused by temperature and/or tension differences in nylon fibres can similarly lead to differences in the uptake of disperse or 1:2 disperse pre metallised dyes. Geometrical and optical difference also can create barré effect. Such variations include irregular spacing of two consecutive yarns due to tension difference, mal adjustment in the production machine, difference in the physical or chemical characteristic of the subsequent yarns, difference in delustering of the yarns, cross sectional difference in the yarns, differences in denier of filaments (microfibres produce lighter shades), twist differences, fabric manufacturing faults, etc.

Barré also can be caused by the poor coverage properties of the dye involved, which is called orientation barré. Such barré can arise due to the difference in $-NH_2$ end groups which is the main groups reacts with the dye. If the degree of polymerisation differs in one yarn compared to the other or in one area of the yarn with respect to other the terminal $-NH_2$ group also will differ which in turn will change the depth of dyeing. It can also be caused by the difference in the merge numbers (different batches of production).

8.3.3.5.1   How to overcome barré problem

Physical variations can be minimised by dyeing at higher temperatures (e.g. up to 120 °C) with nylon 6.6, or prolonging the dyeing time. High-temperature dyeing is used with the larger 1:2 metal complex acid dyes in order to achieve better fibre levelling and fibre penetration, which leads to improved colour fastness to washing also. Using swelling agents like benzyl alcohol, n-propanol, which can help in the internal migration can also help the in controlling the barré problem. Another method is to use low molecular weight disperse dyes for dyeing. The dyes give better migration and dyeing takes place irrespective of the –NH2 groups and can cover the difference due to the number of –NH2 groups in the fibre. Barré problem also can be handled better by using equalising acid dyes which has more levelling and good migrational property and uniform reaction with terminal –NH2 groups, shades are bright but the disadvantage is that the dyeings are having poor wash fastness. Wash fastness can be improved by after treatments like syntan treatment and other suitable fixing agents. Levelling agents of the anionic type also can help. The action of these types of levelling agent is explained above. Cationic levelling agents also can help when anionic dyes are used, but it works in a different way. Cationic levelling agents form a complex with the anionic dyes making the molecular size bigger, thereby reducing strike rate. At higher temperature breaks and the anionic dyes are released which slowly migrates into the fibre thereby producing level shades. The disadvantage of using cationic levelling agents is that the material can yellow during storage which limits their usage.

8.4   Polyamide micro fibres

High comfort in wear and easy care make woven and knitted goods of polyamide microfibres popular for functional leisure and sportswear. Polyamide microfibres, which are supplied by numerous manufacturers, vary widely in fineness, degree of delustring, fibre profiles and surface modifications. Brand names include Supplex/Tactel (Du Pont), Meryl Micro (Rhône-Poulenc, Viscosuisse) and Siks Micro (Snia Fibres). Microfibres are polyamide filament yarns whose individual filament titer is less than 1 dtex. Fabrics are labelled "micro" if at least one system, warp or weft, consists of microfibres.

Nomenclature	Fibre fineness
Coarse	>7 dtex
Medium fine	7–2.4 dtex
Fine	2.4–1 dtex
Micro	1–0.3 dtex
Super fine	<0.3 dtex

Generally, light and wet fastness properties of polyamide microfibres are often considerably poorer (1/2–1 rating) than those of conventional polyamide fibres. Fastness depends closely on the type of fibre.

8.4.1   Process routes for different nylon microfibre fabric

(1)   Woven

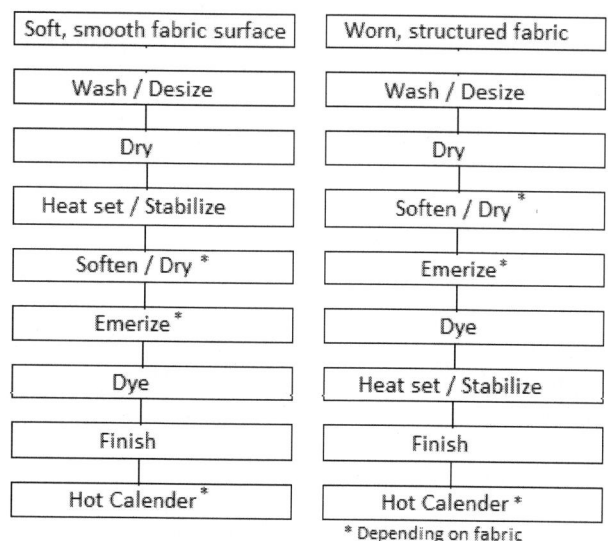

(2)   Knitted

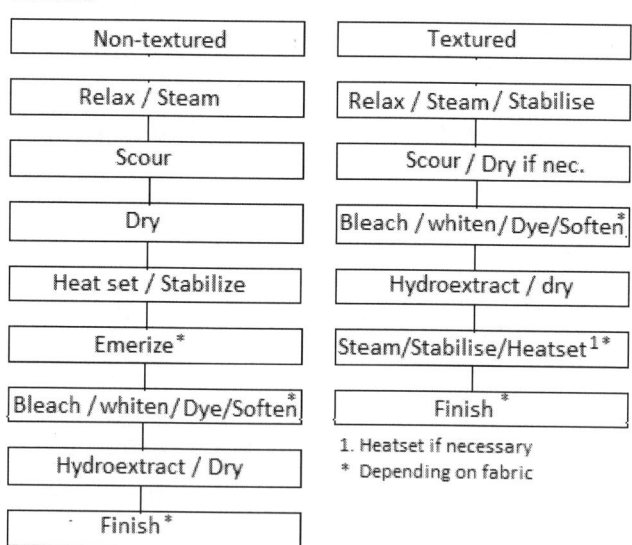

8.4.2   Pretreatment

8.4.2.1   Scouring/desizing/relaxation

Thorough scouring is essential for satisfactory dyeing of both knitted and woven goods. Spinning lubricants, antistats, stains and with woven fabrics sizes as well must be carefully removed from the goods. Sizes are either poly acrylic acids or those based on poly acrylic or modified poly acrylate.

Guideline recipes

Knitted fabric (batchwise)

Quantity	Unit	Additions
0.3–0.5	g/l	Penetrating cum antifoam agent
0.5–2	g/l	Detergent
0.5–1	g/l	Sequestering agent
0–2	g/l	Soda ash/trisodium phosphate/tetra sodium pyrophosphate

Continuous scouring and washing of woven fabric

Short-cycle washing machine – Open-width soaper, perforated drum and other systems

Box Overflow rinse, 80–85 °C.

Boxes 2/3: 1–3 g/l suitable detergent (e.g., Invadine NF – Ciba),

0.5–1 g/l Sequestering agent,

1–2 g/l Soda ash,

pH 10–11, 85–90 °C.
Feed auxiliaries and chemicals continuously and check pH.
Boxes 4/5:Rinse, 80–90 °C.
Box 6:Rinse, 40–50 °C (neutralise if necessary).
Scouring in jigs
1–2 Ends overflow rinse at 80–85 °C.
1–3 Ends in following bath.

Quantity	Unit	Additions
2–5	g/l	Detergent
0.5–1	g/l	Sequestering agent
2	g/l	Soda ash

At pH 10–11 and temperature 85–90 °C.

1–2 Ends hot rinse at 85–90 °C.
2 Ends warm rinse at 45–50 °C.
1–2 Ends rinse cold, unload.

8.4.2.2   Heat setting/stabilisation

If smooth goods are required, heat setting/stabilisation must be performed before dyeing, as otherwise creases, running marks and moiré effects can result, depending on dyeing equipment.

Fabrics with crash and crinkle effects are different. These are not given a preliminary heat treatment, but dried at 150–160 °C after dyeing, the effects achieved during dyeing thus being stabilised (Temperature of setting given earlier).

8.4.2.3   Emerising and softening

This mechanical dry finish improves comfort in wear, while at the same time giving the goods a more bulky, soft handle and a "natural" look. Emerising is usually performed after heat setting or before dyeing.

To achieve good, even emerising effects, it might be necessary to apply a suitable softener with antistat effects. Residues from emerising/softening can destroy the effects of a fluorocarbon finish.

Guideline process

Quantity	Unit	Additions
20	g/l	Softener
10	g/l	Antistatic agent (for emerising,non durable is sufficient)

Pad at a liquor pick-up of 45–60% and dry at 140–150 °C, heat set or stabilise if necessary 10–20 s at 170 °C.

Emerising usually depends on fabric or construction and the finish requirements. Usual grain size of emerising paper – 250–400 is used.

It is advisable to give a wash before dyeing in the same dyeing equipment with 1 g/l of a suitable dispersing agent (e.g., Ultravon LX – Ciba) with for 15–20 min at 60 °C for removing the fluff.

8.4.3   Dyeing

On account of their low titer, micro fibres have several special dyeing characteristics, which are discussed below. Micro fibre filament yarns require higher amounts of dye than conventional filament yarns to achieve the same shade depth. Fastness standards required of functional clothing are high and not always easy to achieve. Acid dyes are suitable for pale and medium shades. For good build-up and high fastness with medium and deep shades, it is better to use 1:2 metal complex acid dyes.

8.4.3.1   Acid dyeing

Guideline recipes and process

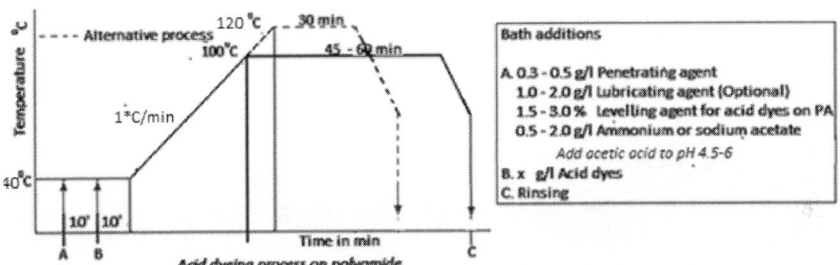

Acid dyeing process on polyamide

8.4.3.2   1:2 Metal complex dyeing

Guideline recipes and process on jet/overflow machines

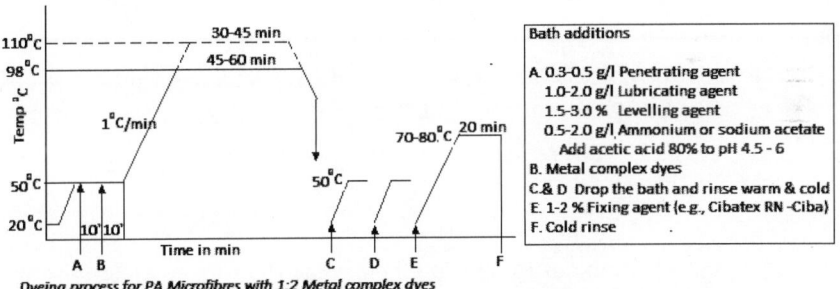

Dyeing process for PA Microfibres with 1:2 Metal complex dyes

Notes:

1. For black shades and very dark shades pH may be maintained at 4–4.5 with 1.5–3% acetic acid 80%.

2. Dyeing at above 100 °C in partially flooded machines can cause oxidation damage during dyeing of polyamide, resulting in a decrease of tensile strength. An addition of 1–2% fibre protectant (e.g., Ciba fast AO – Ciba) is recommended.

8.4.3.3   Guideline recipes and process for dyeing on jigs

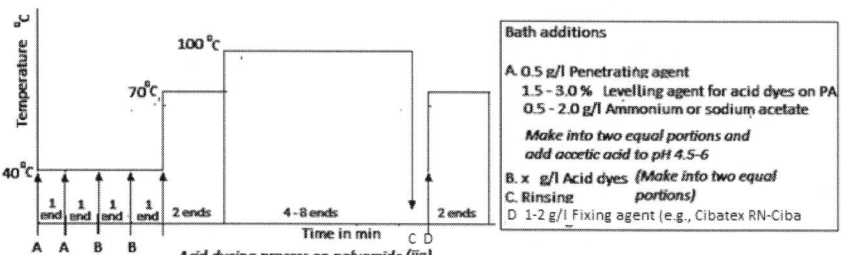

Acid dyeing process on polyamide (jig)

Notes:

1. For black shades and very dark shades pH may be maintained at 4–4.5 with 1.5–3% acetic acid 80%.

2. Dyeing temperature: 90–92 °C rather than at the boil (less temperature drop between edge and centre).

8.5   Mass Colouration

Mass coloration is a common process of dyeing synthetic fibres in the manufacturing stage which is much more widely used today for colouring manufactured fibres. The fibres that are coloured by the addition of pigments prior to extrusion are referred to as mass-dyed, dope-dyed, or solution or producer coloured. Dyes or pigments are incorporated into the polyamide polymer melt during fibre manufacture. The dyes that are being used pass into a dissolved phase on incorporation into the polymer melt or solution, but pigments remain as finely dispersed particles.

The dyes and pigments selected for mass colouration should have following specifications especially pigments: (a) Pigment particle size and particle size distribution, should be suitable both in the spinning mass and in the filament Fine pigment particles (≤ 1 μm) are required because coarse particles would interfere with the filterability of polymer solutions or melts and could impair the tensile strength of fibres, the diameter of which normally lies in the range 16–45 μm. (b) The dyes used should have enough solubility in the spinning mass (d) Good overall colour fastness properties covering the requirement of various customers since it is a mass colouration (c) The pigment and dyes should be able to withstand all the mass-processing conditions of the polymer. The pigments used must have good resistance to organic solvents and good heat stability for use in fibre manufacture. Following colourants are used for mass colouration of polyamides:

Dyes used for normal bath dyeing	Solvent dyes and pigments for mass colouration
Disperse	Aqueous dyes
Acid	Polymer soluble dyes
Metal complex	
Reactive	
	Pigment dispersion in polyamide

The pigments can be inorganic or organic, although inorganic compounds are added more frequently than organic owing to the higher heat stability. Inorganic pigments tend to be used for automotive applications such as upholstery and carpets where a high colour with a good rub and light fastness are fundamental requirements. Organic pigments are limited by the need for stability at the extrusion temperatures. Nylon 6 has a lower melting point than nylon 6,6 and hence more pigments are available for use. Normally the pigments are added in the form of a concentrate dispersed in polymer just before extrusion. Four main methods of incorporating colorants into manufactured fibres depend on the specific fibre production process. These are: (a) Batch process - Pigment preparations or solvent dyes may be used in the batch method in which the whole of the spinning mass is coloured, the colorant concentration being equal to that in the coloured filament after extrusion. (b) Injection process – In this process a master batch prepared which is a concentrated mixture of pigments and/or additives encapsulated during a heat process into a carrier polymer that is then cooled and cut into a granular shape. The master batch(concentrate) is continuously injected through metering into the spinning melt which is next extruded into fibres. The spinning mass is mixed properly mixed to a homogeneous mass to

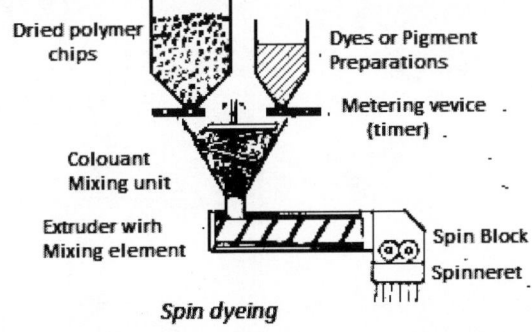

Spin dyeing

ensure even spun coloured filaments and fibres. Except for pigments to impart colour, the master batch could also contain other additives like, ultraviolet stabilisers, flame retardants, antistatic agents, antioxidants, antimicrobials, fibre lubricants necessary for the fibre performance requirements. (c) Chip blending – in chip blending polymer chips are homogeneously mixed with the colour concentrate prior to extrusion (d) Chip dyeing - Chip dyeing is a more specialised technique used with nylon 6 whereby the polymer chips are precoloured with polymer-soluble dyes and then melt spun.

Chapter 9
Dyeing of acrylics

9.1   General

Among natural fibres wool has its own importance, mainly as a winter wear. But the scarcity of the wool due to its animal origin and only limited places of production paved the way to find out synthetic equivalent of wool. Thus came the invention of the new fibre called acrylic or polyacrylonitrile. It was first produced by DuPont in 1941 and given the trade name as Orlon. However the bulk manufacture started only in 1950s. Other manufacturer's brand names are Acrilan, Cashmilon, Creslan, Dralon, Drytex, Beslon, Exlan, Toraylon, Vonnel, etc. All these fibres are basically polyacrylonitriles.

Acrylic has picked up well once it has come to the market as it was a light weight, soft and warm and wool like and much less expensive than woollen yarn. Another important advantage is acrylic is moth proof where as wool is not.

Acrylic fibres are two types – Non-ionic and Cationic. A prerequisite for their use in dyeing is that the dyes have affinity for the fibre. The affinity of a dye is mainly determined by its state of charge. It is therefore advisable to consider the state of charge as group characteristic.

The disperse dyes belong to the group of dyes which have no ionic charge. They are only sparingly soluble in water and thus require a special finish which ensures that they can be dispersed finely in water and that the dispersion can be maintained also under the conditions of dyeing. Disperse dyes have affinity for all synthetic fibres. Non-ionic dyes which can be reduced to water-soluble compounds are known as vat dyes.

9.2   Pretreatment

Acrylic fibres often contain, in addition to spinning and lubricating agents, contamination and oil. This contamination is liable to impair the levelness of shade dyed on the fibrous material. For this reason, it is usually advisable to precleanse the material before dyeing. This preliminary wash can be carried out in acid and also in weakly alkaline baths when the material is heavily contaminated. Following an alkaline bath, the fabric material should be rinsed, if necessary, with addition of acetic acid until it is free from alkali. The following recipes show how far the amounts of detergents should be modified to suit the liquor ratio.

9.2.1   Acid wash

Liquor ratio		Unit	Additions
10:1	40:1		
0.5–1	0.25–0.5	g/l	Non-ionic detergent (e.g. Nekanil LN)
2–3	2–3	ml/l	Acetic acid 30%

Approximately 30 min at 50–60 °C.

9.2.2   Weakly alkaline wash

Liquor ratio		Unit	Additions
10:1	40:1		
0.5–1	0.25–0.5	g/l	Nonionic detergent (e.g. Nekanil W/Nekanil 914)
0.5–1	0.5–1	ml/l	Ammonia 25% or
0.5	0.5	g/l	Soda ash

Approximately 30 minutes at 50–60 °C.

a.   Alternate recipe

This process is recommended for highly soiled goods.

Quantity	Unit	Bath additions
1–2	g/l	Non-ionic scouring agent/detergent
0.5–1	g/l	Trisodium phosphate

The material is treated at 50–w60 °C for 20–30 min. It is then rinsed thoroughly (first with water at 50–60 °C and then with cold water).

9.2.3   Alkaline wash

Liquor ratio		Unit	Additions
10:1	40:1		
1.5–2	1	g/l	Kierlonhighly conc
1–1.5	1	g/l	Nekanil 914
1–2	1–2	g/l	Soda ash

Approximately 30 min at 50–60 °C.

When the material has been stained with fat or oil, a suitable solvent (alkaline stable) can be added (say 0.5–2 g/l) along with the alkaline wash liquor.

9.2.4   Acidic scouring

This method of scouring is recommended for those acrylic fibres having anionic residues

Quantity	Unit	Bath additions
0.5–1	g/l	Non-ionic scouring agent/detergent
0.5–1	ml/l	Formic acid 85%

The material is treated at 50–60 °C for 20–30 min. It is then rinsed thoroughly (first with water at 50–60 C and then with cold water).

So long as the scouring agents are non-ionic, the kinds of scouring agents do not matter.

9.2.5   Bleaching with sodium chlorite

Since the original material of acrylics has a sufficiently pure whiteness, it is unnecessary to give bleaching before dyeing. Acrylic fibres which are to be used for white goods or goods dyed in brilliant shades must have a high degree of whiteness. Added whiteness may be obtained by a combined use of sodium chlorous acid ($NaClO_2$) and fluorescent dye (cationic and dispersing type). Depending on the type of fibre, however, they exhibit a certain amount of yellowing which often requires chemical bleach or even a combination bleach using an optical brightener.

Acrylic fibres and fabric can be bleached using sodium chlorite. A chlorite stable optical whitener (either cationic or non-ionic) can be used along with the chlorite bleach bath to obtain a full white material.

9.2.5.1   Effect of pH

As shown in figure below, the rate of decomposition of sodium chlorous acid is higher when pH islower, but whiteness remains unchanged at pH below 3.5. Accordingly, pH of the bath is set at about 3.5.

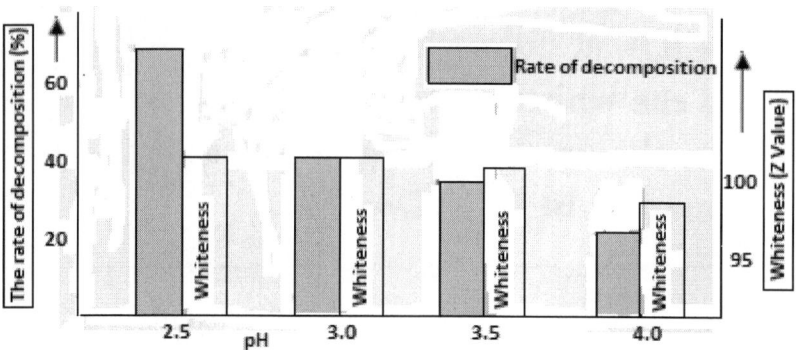

9.2.5.2   Effect of temperature

As figures below shows, the rate of decomposition of $NaClO_2$ increases proportionately to the increase of the temperature, but the whiteness of acrylic approaches a point of its maximum at 85 °C. This shows that it is sufficient if bath temperature is maintained at 85–90 °C.

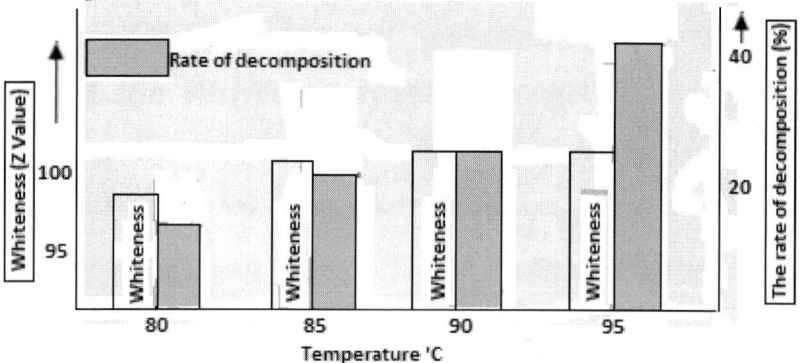

Recipe

Quantity	Unit	Additions
0.5–1.5	%	Optical whitener (chlorite stable)*
0.5–1.5	g/l	Sodium chlorite 80%
0.5–1	g/l	Chlorite stabiliser
1–2	ml/l	Formic acid (pH 3–3.5)

* A cationic or non-ionic OW or a combination of both can be used.

Start bleaching at 50 °C and raise the temperature to 70 °C and then heat to 90 °C at a 1 °C/min. Run at boil for 30–60 min. If procedure is not followed properly, especially heating from 70 °C to boil may result in an uneven whitening.

9.2.5.3   Dechlorination

Unless the dechlorination treatment is conducted after bleaching, light-fastness may sometimes be lowered. The general condition of dechlorination treatment is as follows:

Acid sodium sulfite or sodium thiosulphate 2–3 g/l.

Treating temperature, 60–70 °C.

Treatment time, 20 min.

Notes:

1) Levelling of uneven whitening: Treat the material with 2–3 g/l stabilised hydrosulphite (e.g., Rongolite C/Blankit IN) and 0.5 g/l suitable wetting agent for 30 min at the boil. After a short intermediate rinse, bleach with by heating the liquor within

Quantity	Unit	Bath additions
0.8–1	g/l	Sodium chlorite 80%
0.6–0.8	g/l	Sodium chlorite stabiliser
1–2	g/l	Formic acid to pH 3–3.5

approximately 30 min to the boil and treating for a further 30 min at this temperature. If the whiteness is reduced and it is not enough the material can be again treated with a small amounts of optical whitener and sodium chlorite. Otherwise give another bleaching treatment with

Quantity	Unit	Bath additions
0.5–1.5	g/l	Sodium chlorite 80%
0.5–1	g/l	Sodium chlorite stabiliser
1–2	ml/l	Formic acid to pH 3–3.5

For 30 min at boil.

9.2.5.4   Staple, slubbing and tows

Acrylic material in blends often uses dyed material since the second component dyeing is often more difficult. In this context dyeing of staple, slubbing and tows are more important.

1) Load the material into suitable cages or cylinders in the dyeing machine.

2) Preliminary scouring with non-ionic detergent.

Quantity	Unit	Description
0.25–1.0	g/l	Non-ionic detergent
0.5–1.0	g/l	Trisodium phosphate or tetrosodiumpyrophosphate

10–20 min at 50 °C.

3) Hot and cold rinse.

4) Dyeing at temperatures up to 106 °C, if suitable machinery is available.

5) Rinse.

6) Application of softening or antistatic treatments with suitable auxiliaries.

7) Hydro extract and dry at 80 °C or below (is sometimes done on the Fleissner).

Packages of spun staple wound relatively firm:

1) Preliminary scouring as for loose staple.

2) Dye on the package machine at up to 106 °C using cationic dyestuffs.

3) Rinse.

4) Softening and antistatic treatment with suitable auxiliaries.

5) Hydroextract and dry.

9.2.5.5   High-bulk yarns in hank form

1) Relaxation and preliminary scouring with

Quantity	Unit	Description
0.25–1.0	g/l	Non-ionic detergent
0.5–1.0	g/l	Trisodium phosphate or tetrosodium pyrophosphate

2) The goods being introduced at 98 °C. Scouring proceeds in a cooling bath (to about 50 °C),the maximum rate of cooling being approx. 1 °C per minute. It may be advisable to reduce the cooling rate still further over the region 70–98 °C, where the thermoplastic nature of the acrylic fibre makes it particularly sensitive to distortion.

3) Warm (about 50 °C) and cold rinse.

4) Dyeing: It is difficult to avoid unlevel results if the ordinary machine is used, in which the hanks are simply hung over a single set of rods. It is necessary to have a second set of rods through the bottom of the hanks to prevent the goods from floating during the dye process, as a result of liquor circulation and surface active content. The dyeing temperature is 98–106 °C, cooling of the exhausted dye bath being carried out at the temperature gradients described above, particular care being paid to the rate of cooling in the thermoplastic region.

5) Softening and antistatic treatment with suitable auxiliaries.

6) Hydroextract

7) Drying at temperatures up to 80 °C.

Remarks: It is feasible to relax the yarn in the grey state steaming for 15–20 min at 95–100 °C.

9.2.5.6   Knitted piece-goods made from high-bulk yarn

1) Relaxation and scouring in winch (enclosed) or jet

The cloth is introduced at 95–98 °C into the following liquor:

Quantity	Unit	Description
0.25–1.0	g/l	Non-ionic detergent
0.5–1.0	g/l	Trisodium phosphate or tetrosodiumpyrophosphate

After 5–10 min the liquor is cooled to 45–50 °C, paying special care to the rate of cooling in the thermoplastic region. A covered winch with heated cover or jet is indispensable, in order to ensure accurate control of cooling rate throughout the whole fabric. Streams of coolair can cause excessive creasing and fabric distortion.

2) Rinse.

3) Dyeing with cationic or selected disperse dyestuffs.

4) Softening, antistatic finishing auxiliaries.

5) Rinse and hydroextract.

6) Dry below 80 °C on a loop-dryer and possibly steaming. The relaxation process is omitted when dyeing knitted goods made from staple fibres. Instead, the preliminary scouring is done at 50 °C using the same additions.

9.2.5.7   Woven cloth made from acrylic staple

1) Preliminary scouring and stabilisation:The cloth is treated on the open-width washing machine without tension in the following liquor at 95 °C:

Quantity	Unit	Description
0.5	g/l	Non-ionic detergent
0.5–1.0	g/l	Trisodium phosphate or tetrosodiumpyrophosphate

The cloth is cooled rapidly with a jet of cold water at the exit to the washing machine and then further rinsed with cold water. In the absence of open-width facilities, the cloth can be washed at 50 °C on the winch although maximum dimensional stability is not obtained by this method.

2) Dyeing on an enclosed winch with slow cooling of theexhausted dyebath (heated winch cover).

3) Softening, antistatic finishing as usual.

4) Hydroextract.

5) Drying at 80 °C and possibly steaming.

9.2.5.8   Full white

The bleached acrylic material can be optically whitened by cationic OBA or even dispersed OBA. Cationic OBA is applied by exhaust method at 90–95 °C for around 60 min.

The pH is adjusted at 3 with acetic acid.

Bleaching and whitening can be done in one step following the recipe below:

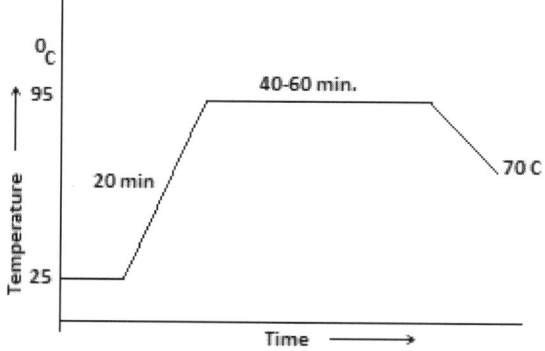

Quantity	Unit	Additions

1	g/l	Sodium Chlorite
0.8-1.0	%	OBA Cationic
x	g/l	Acetic acid to pH 3-4
0.5 - 1.0	g/l	Retarding agent (e.g., Levagal PAN)

Dechlorination with 2-3 g/l of acid sodium sulfite or sodium thiosulfate for 20min. at 60-70°C

9.2.5.9   Setting of PAN materials

Even though the set PAN fabric is not permanent as in the cases of setting in polyester,polyamide,etc., it is generally advised to do the setting of PAN fabrics. While with fabrics of polyester, polyamide,etc., it is possible to obtain a permanent setting effect which imparts good shape retention to the finished product, the setting effect obtained on acrylic fibres is almost completely lost under most dyeing conditions. This is caused by the fact that the acrylic fibre is plasticised as soon as the glass transition temperature is exceeded.

Despite this, fabrics of acrylic fibres are often "set". This is because certain fabric qualities which tend to form running creases or even breakage in which dyeing can be improved in this respect by a "presetting" operation. On the other hand, the purpose of presetting is also to ensure that the goods are shrunk completely. This is particularly important when the fabric is to be dyed on a piece beam in which case width wise or lengthwise change during dyeing is liable to cause trouble. A change in the shape under dyeing conditions is usually caused by inner tension in the fabric resulting from weaving which is relaxed during dyeing. Any changes in this respect during dyeing can be prevented by "presetting" the goods in a tensionless state and thus relaxing the latent stresses within the material before the latter is dyed.

Usually setting is done on stenters. The fabric should travel in such a manner that it can shrink freely. On leaving the stenter, no tension warp or weft wise should be exerted on the fabric and it should hang freely. To ensure this it is necessary to adjust the over feed and the width should be adjusted to suit the special shrinkage behaviour of the fabric being treated. The temperatureof the stenter is usually 170–190 °C and the treating time is 15–60 s.

9.2.6   Dyeing of acrylic fibre

9.2.6.1   Cationic dyes

Cationic dyes or basic dyes, as they are traditionally named were originally applied to many types of cellulosic and protein textiles and were notable for their brilliant shades. However, because of deficiencies in fastness, particularly in light-fastness, their application to these textiles gradually declined. Nowadays, cationic dyes are applied extensively to commercial acrylics, for which they have high substantivity. Acrylic fibres contain anionic groups, to which cationic dyes become strongly attached. However, the high substantivity suggests that non-ionic forces of attraction must also be significant. Cationic dyes are now marketed with good light- and wash-fastness on acrylics, yet still retaining brilliance of shade over a wide range of colours. Although these dyes are predominantly applied to acrylics, an affinity for them can also be imparted to polyamide and even polyester fibres by the introduction of monomers containing anionic groups, usually sulphonate groups.

Cationic dyes are ionised in aqueous media. The cation is coloured and generally contains an azo or anthraquinone group. Cationic dyes are often marketed as chlorides, though sometimes as oxalates and other salts cationic dyes are suitable for dyeing anionic fibres, e.g., wool and acrylic fibres.

Cationic dyes can be subdivided into two groups depending on the type of their positive charge.

9.2.6.1.1   Group I

These are basic cationic dyes where the positive charge is a part of the chromophoric system. The charge resonate within the dye structure and remains as part of chromophoric system. These dyes are of high tinctorial value produce brilliant bright shades but with comparatively low light fastness (cotton 1; Polyacrylonitrile 3) e.g. Malachite green. The chromophoreis TPM, i.e. triphenylmethane.

C.I. Basic Green 4

9.2.6.1.2   *Group II*

The first group is made up of dyes which have a point charge away from the chromophoric system such as the anthraquinonoid or the azo with a so-called external basic group (the positive charge is not a part of chromophoric system rather is positioned at some distance from chromophore). The advantage of this group is that they are extremely light fast but less bright and costly. The dye chrysoidinebelongs to this group. The positive charge of this dye is located at one single nitrogen atom.

Azoic chromophore (Chrysodine)

Anthraquinonoidchromophore

9.2.6.1.3   *Group IIa*

Cationic group exists in dye but in balance with halogen to start with, migrate for uniform distribution on PAN and the cationic group is generated during dyeing. They are also called migrating cationic dyes (see below).

The dye as such adsorbed on to the fibre as a disperse dye and then at higher temperature the halogen atom attached to the nitrogen atom separates and dye the molecule acquires a positive charge. The dye further acts like a cationic dye and migrates. With the dyes which belong to the second group, the charge is distributed through mesomery over the dye cation.

9.2.6.1.4   *Some commercially available cationic dyes:*

1. Azo dyes

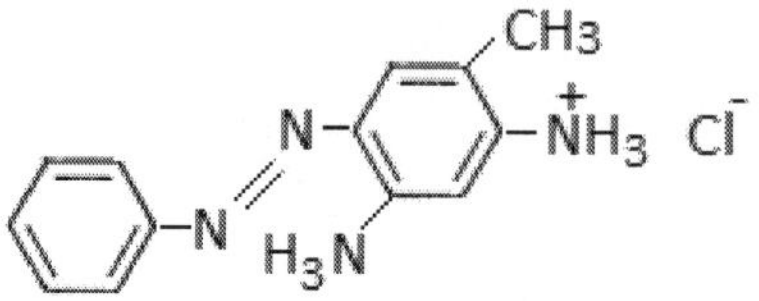

C.I.Basic Orange 2

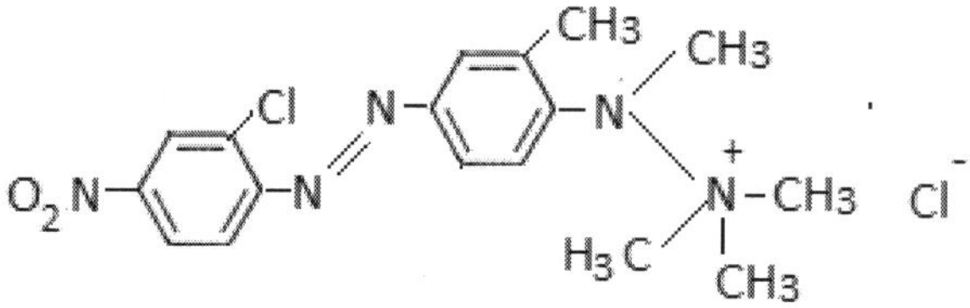

C.I.Basic Red 18

2. Anthraquinonoid

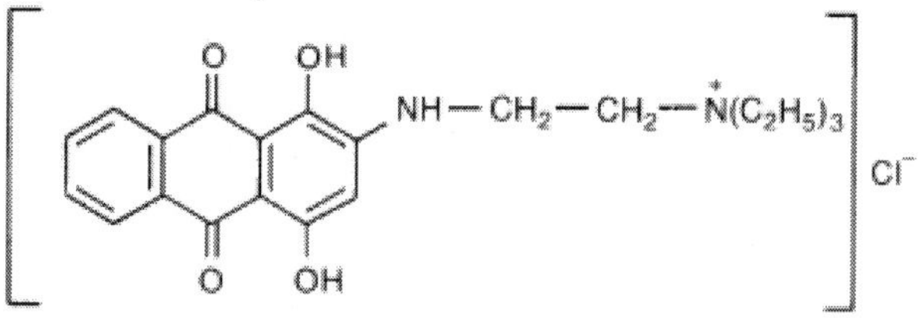

3. Diphenyl methane

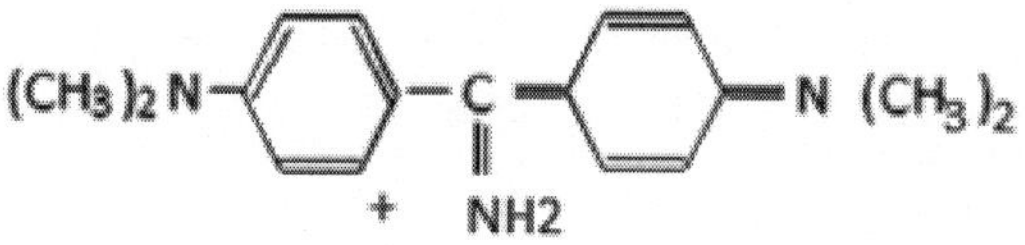

C.I Basic Yellow 2

4. Triphenyl methane

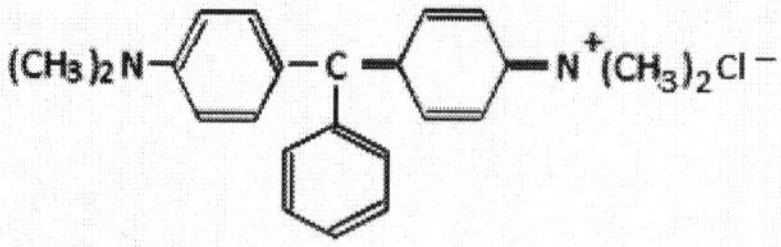

C.I Basic Green (Malachite Green 4)

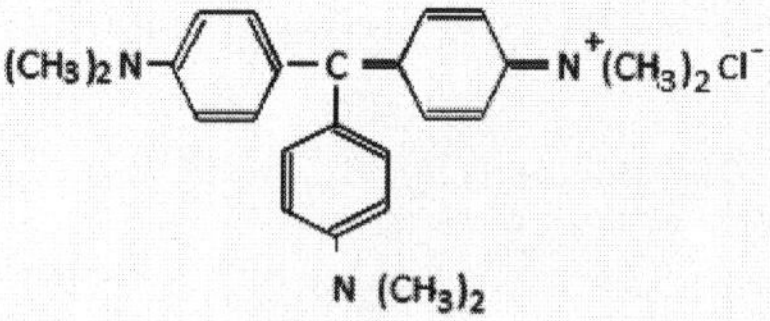

Crystal Violet

5. Methinedyes

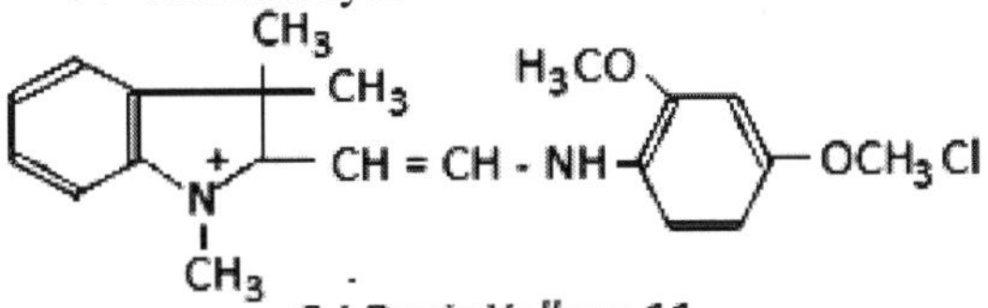

C.I Basic Yellow 11

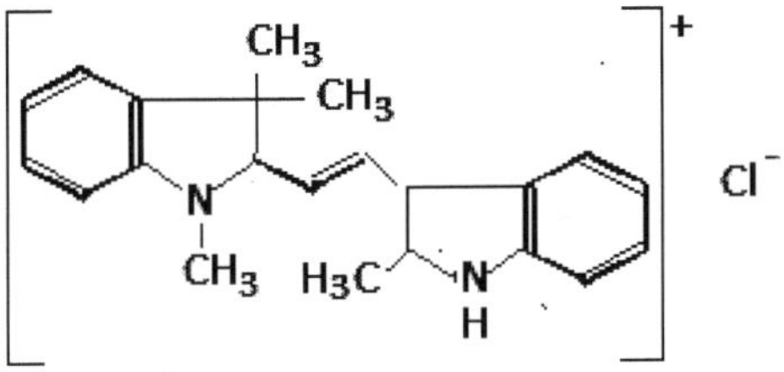

C.I.Basic Yellow 51

6. Acridine dyes

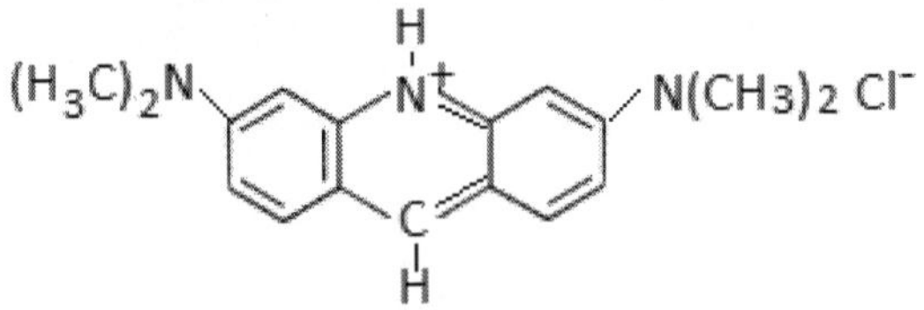

C.I Basic Orange 14 (Zinc chloride double salt)

7. Xanthene dyes

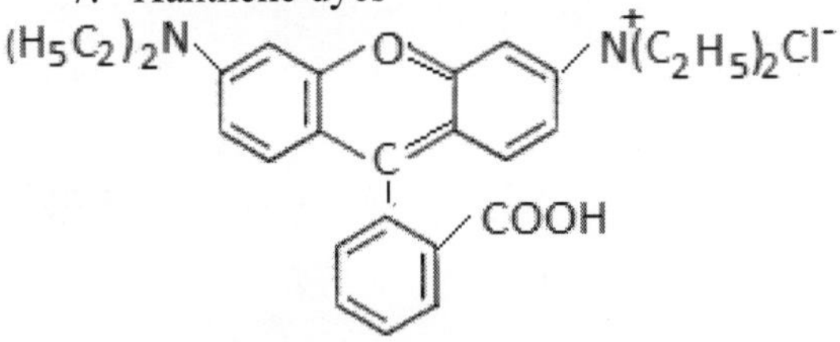

C.I. Basic Violet 10

8. Azine dyes

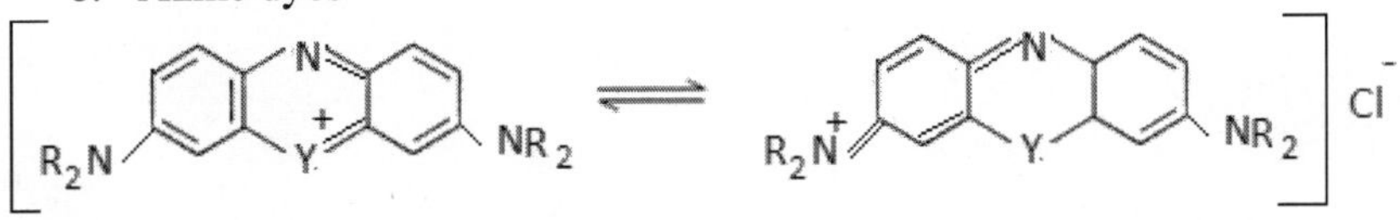

General formula of azine dyes (Y= -O-,-S-, -NR-: R = Alkyl, Aryl, Hydrogen

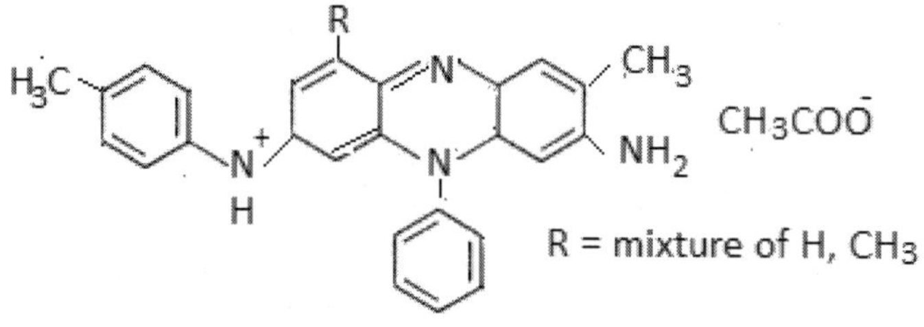

Mauveine - the first industrially manufactured dye

8a.   Oxazinedyes

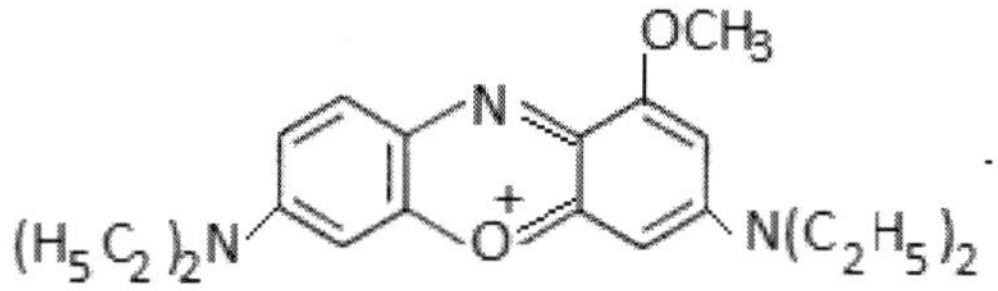

C.I. Basic Blue 4

8b.   Thiazine dyes

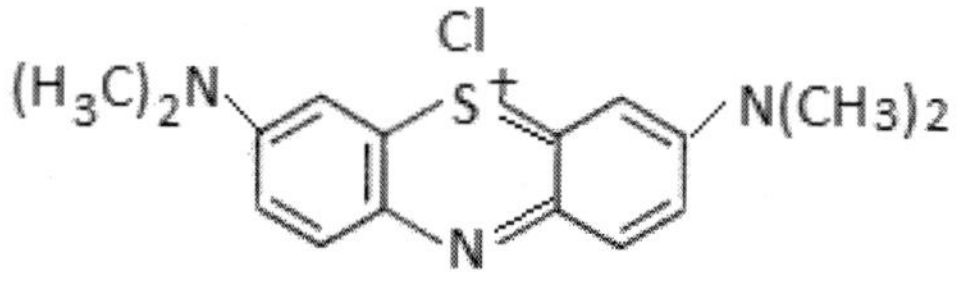

C.I. Basic Blue 9

9.2.6.1.5   *Mechanism of dyeing*

Like that in every other diffusion process, the time progress in dyeing acrylic fibres follows the Fick's law:

$$\frac{dn}{dt} = -F \cdot D \cdot \frac{dc}{dx} \qquad \text{Equation 1}$$

In this equation, the number of dye molecules in which diffuses within the time t through a defined area F becomes the greater, the larger the diffusion coefficient D and the greater the concentration gradient dc/dx. From the Fick's law the following Eq. (2) can be derived:

$$c_F = 2 \cdot F \cdot c_0 \cdot \sqrt{\frac{D \cdot t}{\pi}} \qquad \text{Equation 2}$$

In this equation c_F signifies the concentration of dye in the fibre after the time t, F the fibre surface available per fibre volume, c_o the dye concentration at the surface of the fibre and D the diffusion coefficient. Equation (2) applies only to a ring dyeing. It can no longer be used when the diffusion boundaries overlap in the fibre. This means that the Eq. (2) only gives a correct reproduction of a dyeing process at the start. To understand Eq. (2), it is necessary to split up the dyeing mechanism into two stages:

1. the adsorption of dye at the fibre surface,
2. the subsequent diffusion.

The two stages can be represented as follows:

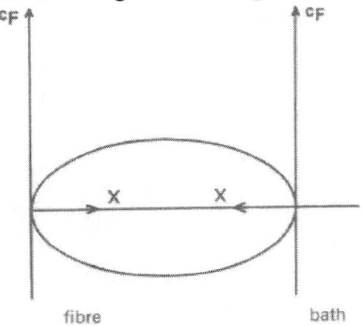

Schematic diagram to illustrate the dyeing mechanism

In the above figure, a circular cross section of the fibre is reproduced in perspective. The diameter shown is equivalent to the diffusion distance x. The two vertical tangents signify the boundary area between fibre and the dyebath. They simultaneously act as ordinates in which the concentration in the fibre c_F is marked off.

9.2.6.1.   Dye adsorption

The dyeing process starts when the dye approaches the fibre surface and is adsorbed there. This forms the surface concentration c_o contained in Eq. (2). The adsorption process rapidly and has therefore no bearing on the speed. The surface concentration c_o is determined by the bath concentration c_L and the affinity of the dye for the fibre. Both c_o and c_L are reproduced schematically in figure below. It is shown in the equation,

$$c_o = f(c_L)$$

whereco is a function of cL.

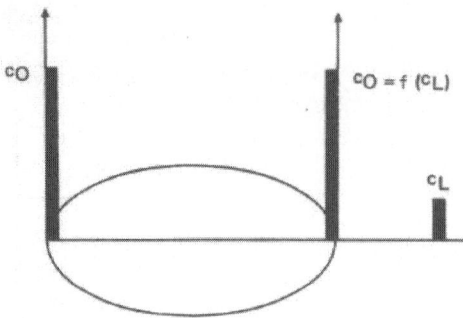

Dye adsorption and formation of a concentration gradient

The surface concentration results in a concentration gradient dc/dx which extends into the interior of the fibre. This concentration gradient is, according to Eq. (1), the activating force for the second stage of the dyeing mechanism, i.e., the diffusion.

9.2.6.1.5.2   Dye diffusion

Figure below shows schematically the diffusion of dye into the fibre as a function of time. Figure below illustrates the diffusion for the extreme case in which the dyes that diffuse from the surface of the fibre into the interior are constantly replaced by adsorption from the dye bath until the fibre is penetrated. This extreme case exists when the fibre is dyed in a practically infinite liquor ratio or if the amount of dye available is adequate for the fibre to reach its saturation point.

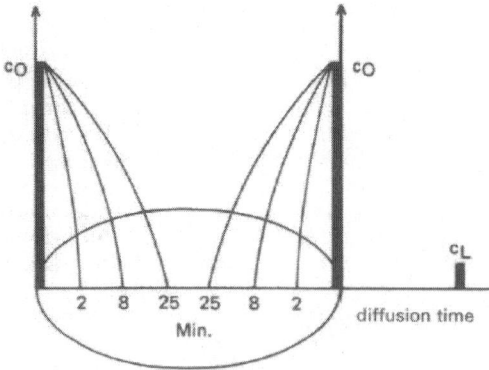

Diffusion process when the surface concentration remains constant

The progress of a practical dyeing in which the bath becomes exhausted and thus the surface concentration does not remain constant but is diminished is illustrated in below figure.

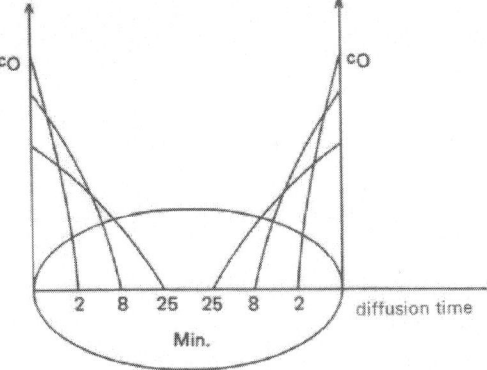

Diffusion process when the surface concentration decreases

Equation (2) which is reproduced in above two figures is one of the fundamental factors in the theory of dyeing acrylic fibres. It applies to the dyeing of acrylic fibres with disperse dyes and cationic dyes. Indications exist, however, that disperse and cationic dyes go onto acrylic fibres according to different rules. These differences become evident in equation c_o and bath concentration c_L.

9.2.6.1.6   Theory of cationic dyeing

The Nernst equation cannot be applied to the adsorption of cationic dyes to acrylic fibres. The reason for this is that the anionic groups at the surface of the fibre are responsible for the binding of dye and that the number of these acceptors is limited. The dye concentration at the surface of the fibre c_o therefore cannot increase linearly with the bath concentration c_L but approaches the limiting value asymptotically. A quantitative correlation between surface concentration c_o and bath concentration c_L is determined by the Langmuir equation,

$$c_o = \frac{K \cdot c_L}{1 + K \cdot c_L} \cdot S \quad \text{Equation 5}$$

in which K is the adsorption equilibrium constant and S the concentration of anionic groups in the fibre. Equation (5) shows that when a large amount of dye is offered (c_L) the number 1 in the denominator can be disregarded and c_o then approaches value S. The dependence of surface concentration c_o on the bath concentration c_L has the form of Langmuir isothermals. In below figure such a curve is given for the special case of adsorption of Red GL on Orion 42.

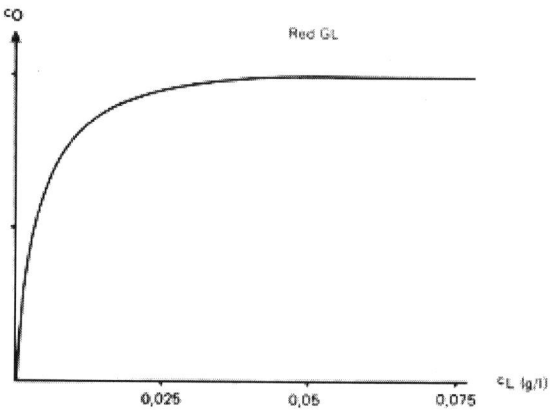

The surface concentration as function of the bath concentration in dyeing Orion 42 with Red GL

The above figure shows that, in this particular example, a dye concentration of only 0.03 g/l in the dyebath is adequate to saturate the surface of the fibre quantitatively.

This allows the conclusion that in dyeing acrylic fibres with cationic dyes the diffusion generally occurs from a saturated fibre surface and the condition of the limiting value

co = S   Equation (6)

is fulfilled at least at the beginning of the dyeing operation.

When Eq. (6) is incorporated into Eq. (2), Eq. (7) is obtained:

$$c_F = 2 \cdot F \cdot S \cdot \sqrt{\frac{D \cdot t}{\pi}}$$

Equation (7)

Equation (7) indicates that the rate at which the dye goes onto the fibre is not dependent on the dye affinity K and the concentration of dye in the bath c_L. According to Eq. (7), the rate at which the cationic dyes go onto the fibre is governed by the diffusion coefficient as the only dye constant.

Cationic dyes not only produce clear but also strong and brilliant shades with surprisingly high light-fastness on acrylic fibres. Even though dyeable with cationic dyes, it soon became obvious that it is far from easy in practice to obtain level dyeings in this manner. The fact that cationic dyes are taken up extraordinarily rapidly by acid modified acrylic which results in unlevel dyeings and, hence, unsalable goods.

9.2.6.1.6.1   Chemistry of dyeing

The mechanism of dyeing affinity of acrylic dyes by cationic dye is formed mainly by ion exchange reaction between dye cation and anionic sites in the fibre as shown below:

$$F\text{–}SO^{-3}M^+ + D^+ \rightleftharpoons F\text{–}SO^{-3}D^+ + M^+$$

Accordingly, when an acrylic fibre is dyed to the equilibrium with various degrees dye concentration it follows Langmuir type isothermal line. The saturated quantity of absorption will depend on the number of anionic sites in the fibre. (Thus for e.g., Cashmilon an acrylic fibre it

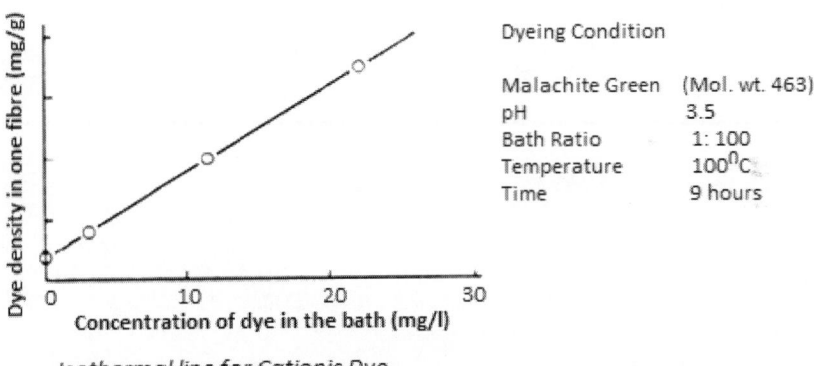

Dyeing Condition

Malachite Green   (Mol. wt. 463)
pH   3.5
Bath Ratio   1: 100
Temperature   100⁰C
Time   9 hours

Isothermal line for Cationic Dye

is 5.25×10^{-5} g/equivalent g. This corresponds to 2.43% owf of Malachite green of molecular weight is 463).

9.2.6.1.6.2   Thermal dependability

Owing to the fine fibre tissue similar to that of accrylic fibres and its hydrophobic property, acrylic fibre doesnot swell very easily, and at a temperature below 80 °C, dye molecules may not be diffused into the inner part of the fibre. But beyond the transition point which depends on temperature between 80 and 90 °C the property of movement of polymer segment will increase abruptly, and the absorption of dye will begin suddenly as shown in the figure below.

In practical dyeing, when dyeing is done at 100 °C, one has to hold at this temperature for a while after the dye has been absorbed, to allow internal diffusion of the dye, and if this is inadequate it can result in the so called 'ring dyeing' or thermal discolouration.

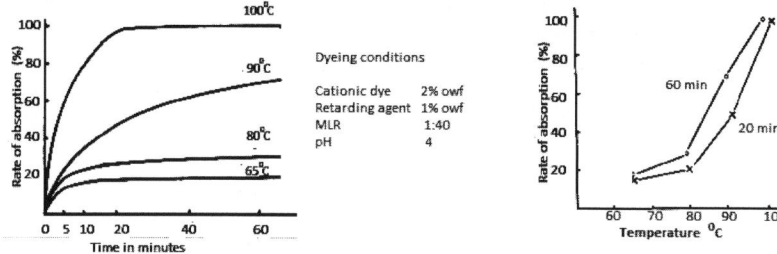

Graph showing thermal dependability

9.2.6.1.6.3   *Effect of dye concentration*

The graph below (a) shows dye concentration has not much related to dyeing speed when acrylic dye is dyed with a cationic dye. This is due to the fact that the saturation on the surface of fibres is quick and diffusion and the diffusion into the inner part of the fibre is exceptionally slow. Namely, only a small quantity of dye is enough to saturate the surface of fibres and any added quantity would not increase the dyeing speed.

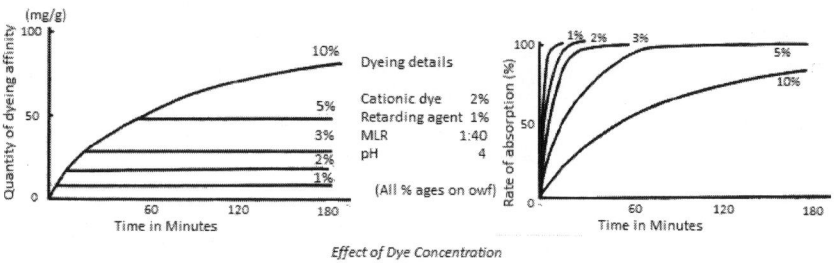

Effect of Dye Concentration

The graph above (b) shows the time-rate of absorption. The lower the concentration of dye is, sooner the absorption will be. However, care has to be taken in such a way that the rate of absorption may not be confused with the dyeing affinity.

9.2.6.1.6.4   *Effect of fineness of fibre (Denier)*

The dyeing speed is considered to be proportional to the surface area of the fibre. Accordingly, comparing fibres having same weight, the greater the denier is, the smaller the surface area will be, and the slower the dyeing speed becomes. We can generalise that provided the crystallinity and orientation remaining the same the dyeing speed is inversely proportional to the square root of the denier. The following graph shows dyeing of Malachite green (2.5%) on fibres of

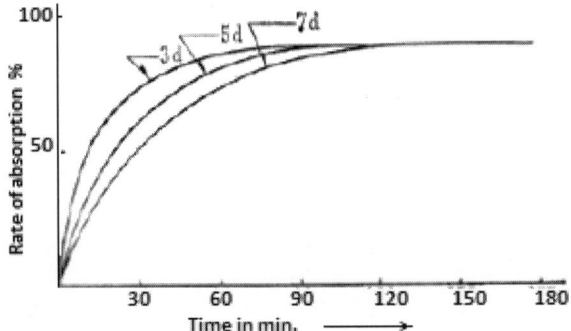

various deniers (3.5 and 7 D) at 100 °C, pH 3.5 and an MLR of 1:100. In the graph which ploted the rate of absorption against time one can see the rate of absorption is slower for higher deniers.

9.2.6.1.6.5   *Selection of dyes*

Leveling of basic dyestuffs can be a major problem which often can be traced partially to differing exhaustion rates of individual dyestuffs which may occur in combination shades. One method for measurement of exhaustion rates is the time of half-dyeing or the time in minutes required for the fibre to absorb half as much dye as will be absorbed if dyed to equilibrium. Half-dyeing times do not adequately describe the behaviour of basic dyestuffs in combinations where the individual dyestuffs could interfere with each other's exhaustion rate.

Following graph shows half-dyeing time of two dyes which behave differently. The dye A is having a larger quantity of the equlibrium dyeing affinity than that of the dye B, but they have the same half dyeing-time.

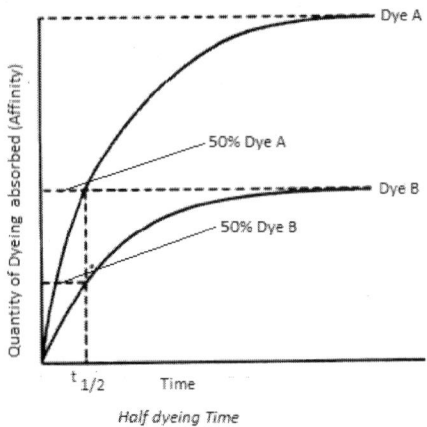

Half dyeing Time

A method has been developed for determining compatibility of K values. Dyeings of the basic dyestuffs, which are to be rated, are made in succession for a specified period of time with dyes having a known strike rate or K value. Five groups (K-1 through K-5) are used to classify dyes, with K-1 dyes being exhausted first when dyed in combination with dyes of any other group; dyes with a K-2 are taken up preferentially when combined with dyes of K-3 to K-5. For optimum leveling, the K values of basic dyes in a combination shade should be as close to each other as possible.

9.2.6.1.7   *Practical dyeing*

Generally, the cationic dyes on PAN have properties of bright shades and superior light-fastness, and can be dyed in dark, mild and light colours. They are dyed at pH 2–5 and decomposed at higher pH. While they have a strong bonding power with fibre and superior light-fastness, they are liable to cause an uneven dyeing. Accordingly, it is advisable to use retarders with them.

When cationic dyes are applied on PAN fibres there are two major characteristics of the dyes has to be taken into consideration.

1. Below the glass transition temperature (about 80 °C) exhaustion is very slight and then shoots up at temperatures only a little above this temperature.

2. Below 100 °C, very little or no diffusion of the dye takes place.

Because of these two characteristics of cationic dyes the rate of dyeing, and hence levelness, are very difficult to control. The degree of difficulty is maximum for readily dyeable fibres with a high glass-transition temperature. Another problem in cationic dyeing is the sensitivity of some basic dyes to alkaline hydrolysis, these dyes vary in their response to dyebath pH, again depending on fibre type. The pH must be controlled to within 4.0–5.5 in order to obtain reliable, reproducible results across the range of dyes and fibres. With proper dye selection these dyes can be dyed at lower temperatures. Thus during conventional dyeing auxiliaries may be added to(a) give the required pH, preferaably a buffer system, (b) control the rate of sorption in the critical temperature region and, as far as possible, to promote migration, generally a retarding agent.

Hence the dyeing method has to be designed in which special measures are taken, right from the start, to ensure that the dyes go onto the material evenly. In contrast to other classes of dyes, cationic dyes migrate very slowly. It requires a very long time to level out unevenly dyed shades and good results can only be obtained when temperatures above 100 °C are used. To avoid unlevelness from the start, several requirements must be met. In order to explain these n

Unlevelness occurs when a dye goes onto the various areas of fibrous material at different rates. Three factors are responsible for this difference in the rate at which the dyes go onto the fibre.

1. The first factor results from the characteristic of the fibre. Usually, this cannot be influenced and must be accepted. Thus, it is very difficult to obtain level dyed shades when differences exist in the degree of stretching or in the content of dye binding groups within the fibre. This applies in particular to dyeing continuous filaments. Unevenness is also often a result of differences in the degree of white caused, for instance, by differences in the concentration of optical brighteners in the fibre. Drying temperatures: The stability to hot air or contact heat of dyed and undyed acrylics can also be different, wet-spun fibres being usually more sensitive than the dry-spun types. But as a rule a drying temperature of 140 °F (60 °C) should not be exceeded for bleached or dyed high-bulk yarns in hanks and for yarns on wound packages. This precaution will prevent yellowing and noticeable changes of shade. Stenter temperatures should not exceed 212 °F (100 °C) for white fabrics and 248 °F (120 °C) for dyed fabrics.

2. Differences in the concentration of dye in the dyebath are the second cause for unevenness. These are liable to occur, for instance, in dyeing wound packages if the resistance of the textile material to liquor flow is not uniform. The dye concentration diminishes where the liquor flow is low while dye is replenished more rapidly where the liquor flow is high.

3. The third cause for unevenness is the temperature differences in the dyebath. Differences of only 4–5 °C result in doubling the rate at which the dyes go onto acrylic fibres or alternatively reduce it to a

half. If a temperature gradient exists within the dyeing liquor, e.g., when the bottom part of a hank has a temperature of 4 °C higher than the top, the lower part of the hank will take up twice as much dye per unit of time than the upper part. Thus the shade dyed on the hank becomes uneven.

In the below graph, the bath exhaustion was plotted on the ordinate. The values of the abscissa show that the dyebath is heated within 30 min from 80 to 100 °C and then kept at the boil for 60 min. A vertical line separates the exhaustion process in a heating phase and a boiling phase. One can consider that temperature differences occur only in the heating phase. When a

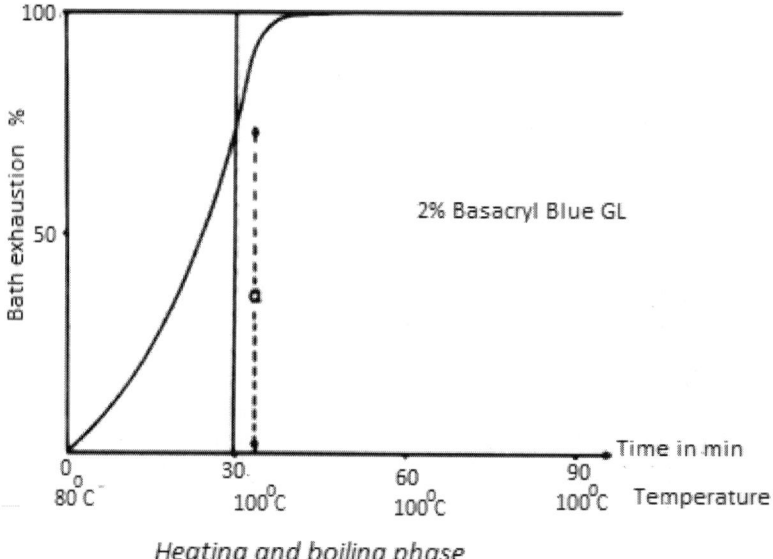

Heating and boiling phase

bath is heated, however, the liquid layers near the heaters warm up first. Thus, a temperature gradient occurs automatically and this extends from these liquid layers to the interior of thedyeing liquor. In dyeing, it must be ensured that these temperature differences remain as small as possible so that they cannot cause difficulties. This is done by circulating the liquor rapidly.

Another possibility of reducing the formation of temperature gradients is that the dyebath is heated slowly and uniformly. When heat is conducted towards the interior of the liquor more rapidly by the liquor circulation supplied by the heaters, the temperature gradient can be maintained so small that it has no effect on the result of a dyeing. In contrast to a liquor which is being heated, a boiling dyebath can be considered as a medium which has the same temperature throughout. Addition of heat to a boiling bath no longer increases the temperature but only converts water to steam.

When temperature differences occur only in the heating phase and thus cause unevenness, it is possible to consider the amount of dye which goes onto the fibre during the heating phase, designated "a" in above graph, as a quantitative value of the danger of unevenness. The larger the value of "a", i.e., the higher the reduction of dye in the dye bath during the heating phase, the lesser the chances for producing a level shade.

In below graph, the bath exhaustion curves have been plotted for a 1–6% dyeing of a cationic dye (e.g., Basacryl Blue GL). The curves show that "a" diminishes with increasing depth of shade. This explains the well-known effect that it is more difficult to dye pale shades than deep ones with good levelness.

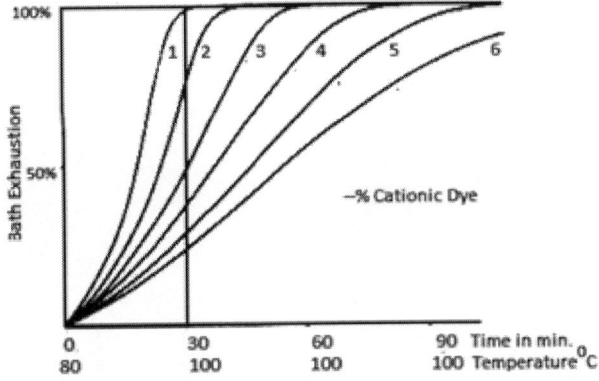

Bath exhaustion curves for various depths of shade

A way out of the difficulty was sought by using suitable dyeing auxiliaries in the form of surface-active cationic compounds which later became known collectively as "retarders". These retarding agents made it possible at last to obtain reproducible dyeings on acrylic fibres with cationic dyes. However, the retarders has two main disadvantages one it increases the cost of dyeing and second, due its blocking behaviour the retarder blocking can arise even when a retarder is added in only slight excess, and can prevent the required depth of shade being obtained at all in spite of prolonged dyeing.

Retarding agents for cationic dyes are of three types:

1. Cationic retarding agents: This type of retarding agent for example, is 4th class ammonium salt containing long chain alkyl group as shown in the following formula and may be considered to be a colourless cationic dyes.

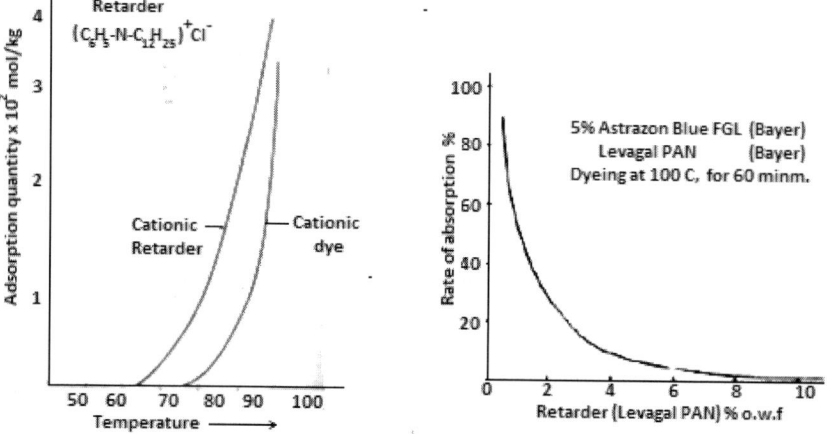

The cationic retarding agents, being ionic as cationic dyes, may compete with cationic dyes for acquiring the anionic dyeing seats. Accordingly, they have the effect of dispelling the fixed dye molecules, as shown in figures below there by retarding the adsorption of the cationic dye.

Though there is some difference depending upon the dye used, an abrupt absorption of dye begins when no retarder is used at about 80~85 °C, while, if a cationic retarder is used, the temperature at which the absorption of dye begins, is as high as about ९० °C. Consequently, it is possible to lessen the range wherein attention is to be paid to the control of the speed of temperature rise. In order to enable an effective competition for anionic seats in fibres, it is ideal to feed the retarder in dispersion above the temperature at which the absorption of dyes begins, but it troublesome in practice to do so, it is generally fed at the same time with dyes.

Quaternary ammonium compounds largely predominate as cationic retarding agents; their fundamental structure (see below) offers the possibility of varying up to four substituent groups around a quaternary nitrogen atom, and hence the variety of possible structures is enormous. A range of these compounds examined for their retarding effect in the application of basic dyes given in the below table:

Structures of some typical cationic retarding agents

$$R - \underset{\underset{R3}{|}}{\overset{\overset{R1}{|}}{N}}^{+} - R2 \ X^{-}$$

Typical formula

R	R1	R2	R3	Anion X
$C_{12}H_{25}$ (Dodecyl)	CH_3	CH_3	CH_3	Cl
Coco[1]	CH_3	CH_3	CH_3	Cl
$C_{16}H_{33}$ (Hexadecyl)	CH_3	CH_3	CH_3	Cl
$C_{18}H_{37}$ (Octadecyl)	CH_3	CH_3	CH_3	Cl
Tallow[2]	CH_3	CH_3	CH_3	Cl
Coco[1]	CH_3	CH_3CH_2	CH_3	CH_3SO_4
Coco[1]	CH_3	$C_6H_5CH_2$	CH_3	CH_3SO_4
Coco[1]	CH_3	CH_3	CH_3	Cl

1. Consisted of approximately 47% C_{12} and 18% C_{14} with lesser amounts of C_8, C_{10}, C_{16} and C_{18} hydrophobes.
2. Consisted of approximately 48% oleyl, 27% cetyl and 13% stearyl, with minor quantities of others.

The retarding agent for a dyeing is selected based on several factors like the rate and extent of sorption of the retarder compared with those of the dyes. The type of associated anion has only a minor effect on the properties of a cationic retarder. It is always better to select a retarder which has a compatibility value equal to or slightly lower than that of the dyes, so that itwill tend to be absorbed by the fibre either at the same rate as the dyes or somewhat more quickly. On the other hand, if the compatibility value of the retarder is significantly lower than that of the dyes,then there is a very real tendency for it to act as a blocking agent (with attendant problems),and if it is higher the efficiency of retarding effect is affected. The compatibility values of some quarternary ammonium compounds based on the above explained below:

Compatibility values of some retarding agents based on the above typical formula $R - \underset{R3}{\overset{R1}{N}} - R2 \ X^{-}$

R	R1	R2	R3	Compatibility values
$C_{14}H_{29}$	CH_3	$C_{14}H_{29}$	CH_3	1
$C_6H_5CH_2$	CH_3	$C_{14}H_{29}$	CH_3	2.5
CH_3	CH_3	CH_3	$C_{14}H_{29}$	3
$C_6H_5CH_2$	CH_3	$C_{7-9}H_{15-19}$	CH_3	5
$C_6H_5CH_2$	CH_3	$C_6H_5CH_2$	CH_3	>5.0

After the selection of the retarder is done, the concentration of the retarding agent in the dyebath has to be decided. Usually the theoretical retarder quantity will be given by the dye manufacturer. However, it depends on the number of anionic sites of the fibre, which again will depends on the fibre type. Generally it is assumed that level dyeing occurs when the number of cations in thesystem (retarder as well as dyes) is just enough to saturate the anionic sites in the fibre. It is understood that more retarder will be needed for fibres of high saturation value and for lower applied depths. Level dyeing also depends on many physical factors such as substrate form and machinery efficiency. Experience suggests that much less than the theoretical amount of retarder will often be adequate with a machine with good circulation capacity and temperature controls.

Cationic retarders may be classified into four categories:

1. Strongly cationic with a strong blocking effect,
2. Moderately cationic with a weak blocking effect,
3. Weakly cationic with no blocking effect,
4. Products with little or no retarding effect but giving some levelling.

For practical dyeing purposes types (3) and (4) are preferred, even though they may be costlier.

The agents retarding activity is required during the critical exhaustion phase as the temperature increases from about 80 °C to the boil. Thus, some cationic retarders have been designed to hydrolyse progressively in this temperature region, so reducing the retarding activity in the later stages of dyeing and safeguarding against blocking effects so that subsequent shading and redyeing are then less problematical. The dyeing method has to be adopted as per the dyes being used in dyeing according to the affinity of the retarder.

Other cationic retarders considered other than quarternary ammonium compounds are saturated alkylamines, dimethylcocoamine oxide, alkylpyridinium salts, imidazoles and imidazolinium

$$CH_3$$
$$|$$
$$RNH_2 \quad R = C_{10} \ C_{12} \ C_{14} \text{ and } C_{16} \qquad\qquad R - N \rightarrow O$$
$$|$$
$$CH_3$$

Saturated alkyl amines   *Dimethyl coco (R - mixture of oleyl, cetyl, stearyl groups) amine*

salts, alkyldiamines, alkylpolyamines, as well as sulphonium and phosphonium derivatives.

1.   *Anionic retarding agents*: Different from the cationic retarders, anionic retarders are not absorbed in acrylic fibers. The principal element of this type of retarder is such an anionic compound, for example, as formalin condensate of naphthenic acid, and in general is used together with non-ionic dispersing agents. This retarding agent

possess opposite electrical charge to that of dye and forms a complex through the combination of anionic compound and cationic dyes and thus, the equilibrium relation is achieved as shown below:

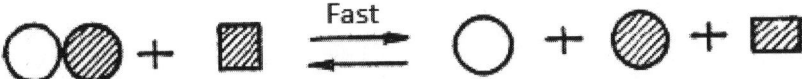

The complex increases molecular weight of dye, reduces its mobility and hence the strike rate. Precipitation of such a complex during dyeing is avoided by applying a non-ionic dispersing agent. Dye-retarder complex from different parts of bath strikes fibre surface at different times. With heating up of bath, the complex breaks, dye is released from complex and diffuses inside fibre. In many cases, in the early stages of the dyeing the retarding effect of the anionic product is relatively slight and for this reason it is necessary to heat the dye liquor more carefully and more slowly than when using cationic products.

The special features of anionic auxiliaries are:

1) that a level dyeing is easily attained as dyeing is effected in a bath with a definite low density;

2) that the shading is made easily, because the seats within fibres are not blocked by auxiliaries.

Anionic retarders act differently with different types of dyes, hence the compatibility has to be checked before using them. Sometimes different method of dyeing has to be employed. The disadvantages of the anionic retarding process are that:

1) the dyestuff and the dyeing auxiliaries complex precipitation, and;

2) the dyestuff in the bath is not completely absorbed, but 10–20% of the dye is retained inthe bath.

Undesirable precipitation of this complex, which is one of the drawbacks of the system, can be inhibited:

(a) by using excess anionic agent;

(b) by using an anionic agent that contains two or more sulphonate groups so that the resultant 1:1 complex retains solubility;

(c) by incorporating a non-ionic agent as an antiprecipitant.

The precipitation problem occurs with dyestuffs with additives (Complex) of low solubility. Special care is required for dyeing in a dark shade at low bath ratios. Generally, the use of the anionic retarder may be said to be an advantageous method for dyeing in light-medium c olour at high bath ratio.

Anionic retarders include sodium dinaphthylmethanesulphonates and polyethoxylatedalkylarylsulphates.

Advantages of anionic retarders:

1. When used in acrylic blends dyeing baths, it is compatible with anionic dyes and anionic dispersing agents in the dyeing bath corresponding to the other part in the blend.
2. Anionic retarders do not block the dyeing sites in the fibre.
3. They have no adverse effects on the bulkiness of certain bicomponent fibres.
4. They promote good migration of dyes.
5. They can be used as stripping agents to reduce the depth of colour in reprocessing.

Disadvantages of anionic retarders:

1. To prevent precipitation the quantity of anionic retarder should increase with increasing quantity of dye (the opposite of the situation with cationic retarders) and this conflicts with requirements for promoting exhaustion; hence exhaustion of dye when applying medium or heavy depths is poor.
2. The levelling action of the retarder is low during the exhaustion stage.
3. It is not possible to use the cationic softeners or other auxiliaries in the dyeing bath.
4. *Inorganic salts:* The action of retarding agents can be explained by similar considerations. If, for instance, a 0.8% dyeing with a cationic dye (e.g., Basacryl Blue GL) is dyed with increasing amounts of a retarding agent (e.g., Basacryl Salt G) the curves in the below Fig. (A)are obtained.

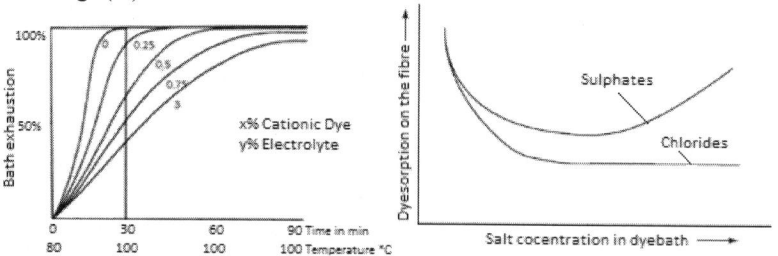

(A) Bath exhaustion curves for various amount of electrolyte (B)Dyes sorption for different electrolyte concentration

The retarding effect of electrolytes in the application of basic dyes to acrylic fibres increases with increasing concentration of salt up to a certain level.

Increasing the concentration beyond this point has no further effect on exhaustion with certain univalent anions, whilst with multivalent types there is an increase in dye sorption (Fig. (B), above). These results have led to the conclusion that ionic mechanisms alone do not entirely explain the complex interactions that occur between basic dyes and acrylic fibres.

These agents retard the rate at which dyes go onto the fibre and have the effect that rate is lowered and the exhaustion process occurs mainly in the boiling phase as is the case in changing over from pale to deep shades. Glauber's salt is very weak in the retarding effect compared with the cation retarder, but is effective when added 5–15% owf in the anhydrous form. However, its effect very much differs according the dyes employed. The larger the amount of retarder used, the less dye goes onto the fibre during the heating phase and the more readily level shades are obtained.

Electrolytes such as sodium chloride and sodium sulphate tend to retard dyeing through preferential adsorption and subsequent displacement by the dye of the more mobile sodium ions, although the effect is relatively weak even compared with the weaker cationic retarders. Nevertheless, the use of up to 10% owf sodium sulphate in combination with a cationic retarder may enable the amount of the latter to be reduced by up to 20–30%. The limitations of electrolytes, apart from this lower effectiveness, are that they reduce the final uptake of dye, their effectiveness decreases with increase in temperature and their effect is greatest with fibres containing weakly anionic groups such as carboxylate, rather than stronger ones such as sulphonate. Cationic softeners for acrylic fibres are sensitive to the presence of electrolytes, although sulphate-tolerant softeners may be used.

4. *Polymeric retarders:* Polymeric retarders are high molecular weight (1000-2000) polyquaternary ammonium compounds carrying cationic positive charges on structure and are attracted to the fibre surface. As the size of the molecule is big it cannot diffuse into the fibre and remain on the surface only. Thus the available surface for the actual dye to diffuse is reduced. This phenomena reduces the diffusion rate and thus retards the dyeing. They do not diffuse into the fibre but are strongly adsorbed at the fibre surface, reducing its anionic potential and hence donotassist migration. Even though they can retard much effectively with much less quantity than the quaternary agent but with some disadvantages.

Comparison of retarders

Property	Cationic retarders	Anionic retarders	Inorganic retarders

Principle of retardation	Competition over the anionic seats on the fibre	By producing a complex with the cation dye, the retarder decreases the concentrationof the free cation dye in the bath	The same as with the cation retarding process
Retarding effect	Good	Good	Lower
Exhaustion of the dye	Good	10–20% Retained	Good
Appearance after dyeing	Good	Hard	Normal
Correction	Difficult as the sites are blocked	Easy	Easy

9.2.6.1.7.1   Dyebath additives

9.2.6.1.7.1.1   Detergents and alkalis

Non-ionic detergents are used in scouring and soaping–mainly poly glycol ethers.

Wherever alkalis are required one may use phosphates like trisodium phosphates and tetrasodium pyrophosphate. Normally anionic detergents, especially ammonia, as alkali are not used for scouring polyacrylonitrile fibres, because they can give rise to spotty, uneven dyeings.

9.2.6.1.7.1.2 *Levelling agents*

There are different types of levelling agents working on different principles.

Levelling agent I

To get level shades on acrylics particularly in pale and medium colours, add a levelling agent to the dyebath. The levelling agent controls the rate of absorption of cationic dyestuffs and levels out the affinity of the components in dyestuff mixtures. During dyeing, the levelling agent is absorbed by the fibre in the same way as the cationic Acrylic dyestuff and occupies a position in the dyestuff-binding groups of the polyacrylonitrile molecule. The retarding effect of the levelling agent continues for quite some time after boiling temperature has been attained, so that any unevenness in dyeing caused by initial temperature variations in the dyeing can be corrected.

Levelling agent II

These types of levelling agents are cationic compound of medium affinity for acrylics. This product retards absorption of acrylic dyestuffs when dyeing begins. But the retarding effect of this levelling agent is only temporary once boiling temperature has been attained, because there is a progressive breakdown of this agent by saponification. Contrary to the effect produced by Levellingagent I, this prevents the often undesirable blocking of the dyestuff. Moreover, this levelling agent is used as a retarder to shade acrylic dyeings with cationic dyestuffs near the boiling point.

9.2.6.1.7.1.3   Migration agents as levelling agents

Control of the rate of dyestuff absorption is not the only means to get level dyeings. Migration agents can also be used to promote migration of cationic dyestuffs on the acrylic fibre. They have a low affinity for acrylics, in contrast to levelling agent. Hence the retarding action of migration agent is comparatively slight, and the dyestuff will not be blocked. It promotes migration of cationic dyes on acrylics and therefore helps you to get level shades. If necessary, it can also be used together with levelling agents.

Moreover, it will also be good for subsequent levelling of faulty dyeings, preferably at 221–226 °F (105–108 °C). Dyeing can be continued in the same bath, because there is no blocking effect.

9.2.6.1.7.2   Dyeing with selected disperse dyestuffs

The use of disperse dyestuffs presupposes fairly wide tolerance in steaming, pleating or decatizing fastness requirements.

Disperse dyes are commonly used for hand-knitting yarns, but for the most part only light shades are obtainable.

9.c.6.1   Dyeing' method

Recipe

Quantity	Unit	Bath additions
x	%	Selected disperse dyes
1–1.5	%	Acetic acid 30% pH 5–7
1	g/l	Monosodium phosphate
0.5	g/l	Disodium phosphate

1–2	%	Dispersing agent

The dye bath containing the disperse dyestuffs is heated to 50 °C and set with disperse agent and levelling agent if necessary, set pH 5–6 using acetic acid The material is entered and brought to the boil during 30–40 min, dyeing being continued for 1½–2 h at the boil. Careful cooling (max 1 °C per min) and rinsing to follow.

It is also possible to relax and dye in a single bath, if the high-bulk yarn is completely clean. The hanks are introduced into the stationary dye/chemical liquor at 85 °C. Relaxation takes about 5 min. Liquor circulation is commenced and the temperature raised to 98 °C in 20–40 min. Dyeing duration is 3/4–1 h. Liquor cooling and rinsing as already described.

Notes:

1.   Shade of some disperse dyes is affected by differences in the pH of the dye bath. Shades produced in an acid bath often change during steaming. This change results in a shade that is almost the same as that obtained in a neutral bath. The more acid the dyebath, the more marked the difference in shade. To avoid these changes in shade during subsequent steaming processes it is always advisable to dye in an almost neutral bath.

2.   Any shade correction during dyeing can be made by adding dye to the boiling dyebath with the steam shut off.

3.   Correction of uneven dyeing: Shades that have been dyed unevenly can be readily levelled out in boiling dyebath with the addition of levelling agent.

4.   Shades that have been dyed too deep can be partially stripped to about half their original depth with 2–3 g/l suitable levelling agent within approx. 30 min at 100 °C.

5.   Shades dyed with disperse dyes can be stripped in a boiling bath with

Quantity	Unit	Bath additions
1	g/l	Sodium chlorite 80%
1	g/l	Sodium chlorite stabiliser
2–3	g/l	Levelling agent
1–2	ml/l	Formic acid 85% (pH 3–3.5)

If the stripping is not adequate after boiling for about 30 min, it is advisable to add an additional

1 g/l sodium chlorite 80%,

and to boil for a further 30 min.

9.2.6.1.7.3   Dyeing polyacrylonitrile with cationic dyes

Cationic or basic dyes give generally fastness properties of the highest order. Acrylic fibre has a "fibre saturation value" is a characteristic constant for a particular fibre, indicating the limit of cationic dye uptake. After this limit generally lead to blocking phenomena, and dye baths do not exhaust fully.

9.2.6.1.7.3.1   *Dyeing acrylic without retarders*

Generally, acrylic material is dyed with retarders to contain the sudden strike rate of cationic dyes at higher temperatures. If dyeing without retarding agents, a complete controlled temperature increase can give a level dyeing for example: start the dyeing with dye and auxiliaries at 40 °C adjust the pH to 4–4.5 with acetic acid and raise the temperature to 70 °C at 1.5–2 °C/min. From 70 °C onwards the temperature is slowly raised(1 °C/min or lower) to 85 °C

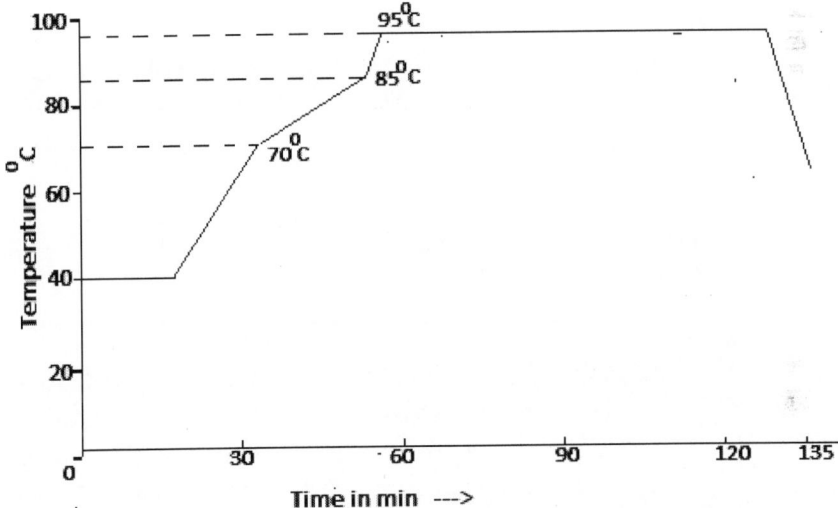

with proper mixing/agitation of the bath as the maximum diffusion takes place at this time (75% in 80–85 °C. If temperature is raised at 1 °C/min 75% diffusion takes place in 5 min and if it is raised at 0.5 °C/min the diffusion takes place in 10 min. Thus the temperature control is very important to achieve level dyeing. Temperature can be further raised to boil (95 °C) at 2 °C/min and hold at this temperature for 1–2 h for allowing the migration to complete.

Since the strike rate of cationic dyes on polyacrylonitrile is very high we have to use retarders for getting level dyeing. There are two types of retarding agents and they work in different principles.

1. *Cationic retarding agent:* These retarding agent exhausts onto the fibre in much the same way as a cationic dyestuff. It can compete with the dyestuffs for sites in the fibre. They exerts a good dyestuff retardation and with correctly selected dyestuffs permits good shade build-up. Provided the dosages carefully regulated, good dye bath exhaustion can be obtained. Amounts depend on the shade depth and on the uptake rate of the fibre.

Dye conc. (%)	Conc. of retarding agent (%)
Less than 0.2	5–4
0.2–1.0	4–3
1–2	3–2
2–3	2–1
3–4	1–0

For high affinity fibres and dyes with a rapid rate of exhaustion, higher retarder concentrations should preferably be used. The reduced amounts of retarder can beused in certain cases especially for packages, fabrics, knitted fabrics, etc. However,a preliminary trial is recommended.

2. *Anionic retarder:* When this retarder is added to the dye bath, it forms complexes with the cationic dyestuffs which break down during the dyeing process to release the cationic dyestuff. Agglomerations are prevented by including a non-ionic dispersing agent. Since the behaviour of the cationic dyestuffs towards anionic retarders is specific, it is essential to make very careful selection of dyestuffs.

3. *Glauber's salt as retarding agent:* Electrolytes such as Glauber's salt possess a certain retarding effect on cationic dyestuffs and they encourage migration within the fibre. The most suitable are those dyestuffs known to be sensitive to electrolytes, and possessing good levelling and migration characteristics. They are particularly suitable for light shades.

9.2.6.1.7.3.2 *Dyeing with anionic retarders*

The liquor is set at 50–60 °C with about

Quantity	Unit	Description
1–2	%	Anionic retarding agent
2–3	%	Non-ionic dispersing agent
		Acetic acid + Sodiumacetate to pH 4–4.5

Run the material for some time in the above bath and the dissolved dyestuff
is added and the

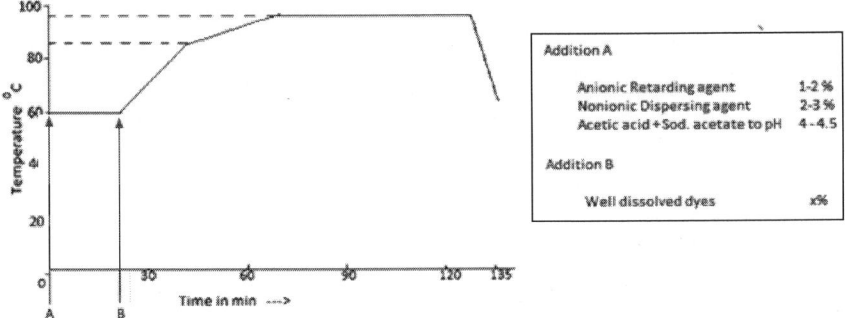

Addition A	
Anionic Retarding agent	1-2 %
Nonionic Dispersing agent	2-3 %
Acetic acid + Sod. acetate to pH	4 - 4.5
Addition B	
Well dissolved dyes	x%

temperature raised during 30 min to 85 °C and then during 1 h to the boil.
After 1–2 h at the boil the liquor is cooled slowly and the dyeing rinsed.

9.2.6.1.7.3.3   *General dyeing method*

Depending on depth of shade the dye bath is set with the following additions
at 50–60 °C.

Guideline recipes

Light to mediumshades	Dark shades	Unit	Dye bath additions
0–1	1.5–2	%	Sodium acetate
3–1.5	1.5–1.0	%	Acetic acid 60% to pH 4–5
0–10	10	%	Sodium sulphate calc.
1.5–4	0–2.5	%	Retarding agent
0.5–0.75	0.75–1.0	%	Non-ionic dispersing agent
0.5–0.75	0.75–1	%	Levelling agent

The goods are treated in the chemical bath for 5 min before adding the
dissolved cationic dyestuffs (through a filter cloth). After dyestuff addition
the temperature is raised to 80–85 °C during 10–15 min and then to the boil
during 45–60 min. This represents a heating rate of approx. 2 °C per minute
between 50 and 85 °C and approx. 1 °C per 3–4 min between 85 and 100 °C.
Duration of dyeing at the boil is 3/4–2 h. The dyebath is cooled to 45–50 °C
using the cooling rates already described. The goods are then rinsed.

Notes:

1. Cationic dyestuffs tend to exhaust rapidly above 90 °C and at 98–100 °C are completely fixed. Above 95 °C the dyeing rate increases by about 30% per degree centigrade temperature increase. In the temperature range 60–85 °C it is only possible to consider the dyestuff as being held adsorptive, at the fibre surface.

2. Some cationic dyestuffs are best applied at pH 4–5.5 from a buffered sodium acetate/acetic acid solution. Increased amounts of acetic acid can have a levelling effect by reducing the dyeing rate. This is particularly important when dyeing pale shades on acrylic fibre with a high dyeing rate. In such cases good results can be obtained by dyeing at pH 2 (sulphuric acid).

3. If the factory water is softened by ion-exchange, any sodium bicarbonate present should be converted into sodium sulphate by adding sulphuric acid. The pH value should be measured electrically and not by means of pH indicator papers.

4. Selection of dyes (Cationic) are very important in dyeing acrylic fibres. For light–medium tones it is much better to choose dyestuffs with low exhaustion rates and good migration and levelling properties, whereas for dark shades it is more economical to use products with more rapid dyeing properties.

5. It is feasible to raise the dyeing temperature to 105–108 °C in order to exhaust more quickly which is an important factor when dyeing dark shades. But dyeing at elevated temperatures brings with it the risk of excessive shrinkage and harsh handle.

6. If the acrylic fibre is to be dyed at temperatures between 100 and 106 °C, it is necessary to prolong the heating phase in order to allow enough time to heat the bath to the higher dyeing temperature. On the other hand, the boiling phase can be cut down because uniform penetration of the acrylic fibre is obtained more rapidly at the higher temperature. Dyeing at temperatures above 100 °C, therefore, usually offers a saving in time which is particularly noticeable when dark shades or when fibres with a high denier are being dyed. It must be considered, however, that dyeing acrylic fibres under hightemperature conditions is liable to result in deterioration of the handle and the nature of the goods. This applies in particular to dyeing at temperatures above 106 °C.

9.2.6.1.7.3.3.1   *Exhaust dyeing methods*

9.2.6.1.7.3.3.2 Dyeing of PAN piece goods on beam dyeing machine

If the fabric is dyed in piece beam dyeing machines, the beam is first covered with a loosely woven fabric before the wet or dry goods are wound on the beam. If possible, this looselywoven fabric should be of a material which is not stained by cationic dyes. Usually, fabrics of cotton or spun rayon are used. Generally speaking, about fivelayers are adequate to arrest the batched goods from slipping and to prevent the perforations of the beam from marking off on the goods. The fabric should be wound evenly and not too loosely. It must also be wound in such a manner that the perforations are completely covered by the material during dyeing. The usual method is to allow the fabric to project over the perforated part of the beam by an inch or so, thus allowing for any possible residual shrinkage of the material during the pretreatment. The distance between the selvedge and the perforation can usually be adjusted by metal sleeves. The optimum distance depends mainly on the resistance of the fabric to the liquor flow. When this is large the selvedge must project slightly further over the perforations than with goods which have low resistance.

Loosely woven goods through which the liquor travels more readily can be batched in larger yardages than densely woven goods. After the goods have been wound onto the beam it is covered tightly with an end cloth. This prevents, on the one hand, rising of the upper layers of the goods when the liquor is circulated. In addition, the end of the winding should not be sewn onto the acrylic fibre to ensure that the fabric is not damaged when the seam between theend cloth and the goods tears or distorts. Before batching the fabric onto the beam, the former should be shrunk by presetting. Shrinkage lengthwise and widthwise is liable to cause difficulties during dyeing. When the fabric shrinks widthwise, the perforations of the beam at the selvedges are liable to be uncovered. This, in turn, will cause the liquor to escape by way of the least resistance and the liquor throughput through the fabric will be reduced markedly. Like in dyeing loose stock, when channelling occurs, the shade dyed on the fabric will become uneven. Shrinkage lengthwise causes tension within the fabric batch and the pressure or abrasion areas between the individual layers of the fabric are liable to exhibit moire formation.

In dyeing acrylic fabrics on the beam, the liquor must always circulate from inside to outside. The liquor circulation should not be stopped throughout the finishing process. Piece beam dyeing offers the advantage that the fabric is held widthwise and therefore is not liable to form creases. On the other hand, the fabric often loses its full woolly nature and becomes flatter. With some fabric qualities this change during dyeing in piece beam dyeing machines cannot be accepted. In such cases it is necessary to dye the material in the winch.

9.2.6.1.7.3.3.3 Dyeing of PAN piece goods on winch dyeing machine

Winch dyeing should be considered in dyeing piece goods of acrylic fibres that they are liable to become plastic and distort under the conditions of dyeing. Everything must therefore be done to avoid tension warp and weft wise. The winch should be installed immediately above the surface of the bath to ensure that the air passage is as short as possible. To avoid the goods from slipping and thus forming glossy areas, it is better the winch should be covered with a suitable rough fabric.

Generally piece beam dyeing offers the advantage that the fabric is held widthwise and therefore is not liable to form creases. On the other hand, the fabric often loses its full woolly nature and becomes flatter. With some fabric qualities this change during dyeing in piece beam dyeing machines cannot be accepted. In such cases it is necessary to dye the material in the winch.

It is also advisable to power the guide rollers. Round guide rollers are often better than those having a profile. The reason for this seems to be that guide rollers which have a profile more readily drive out air entrapped in the fabric and thus accelerate formation of running creases.

It must be ensured that the winch turns at adequate speed. If the speed is too low, the fabric rope does not change in its positions and creases are liable to be fixed in the fabric. Furthermore, formation of running creases and breakage can be minimised by increasing the speed of the winch. Generally speaking, the winch should have a speed which ensures that the fabric is transported at a speed of 50–60 m/min.

When fabrics which tend to form running creases are to be dyed or winch becks which turn only slowly are to be used, it is advisable to sew the fabrics into a hose. The seams, however, should be so far apart that the liquor contained in the interior of the hose can be rapidly exchanged in order to obtain a uniform shade. In dyeing acrylic fibres, relatively small differences in temperature in the dyebath are liable to cause unevenness. It must therefore be ensuredthat the temperature is maintained constant over the entire width of the beck. This often cannot be ensured with open winch becks. Closed winches are far more suitable, some of which allow heating of the cover.

9.2.6.1.7.3.3.4   Dyeing of PAN goods with low affinity cationic dyes

Set the dyebath at 122–158 °F (50–70 °C) with the following additions, depending on the required depth of shade, in this sequence.

Pale and mediumshades	Heavy shades	Bath addition
0.5–1%	1.5–2%	Sodiumacetate
3–1.5%	1.5–1%	Acetic acid 60%
10%	10%	Glauber's salt
0.25–0.75%	0.5–1%	Dispersing agent
3–4%	1.5–0.75%	Levelling agent

9.2.6.1.7.3.3.5   Rapid dyeing method using cationic dyes

Dyeing of acrylics with cationic dyes can be cut down to minimum dyeing time if an elevated starting temperature is selected and the most suitable retarder addition is used (the quantity of retarder depends on the type of acrylic fibre which has to be dyed). This means you do not have to heat the dye bath to the dyeing temperature, which usually takes quite a long time, if you start right away at an elevated temperature. It will be necessary to know which acrylic fibre is used, and what are its

Dyeing rate (V) and

Saturation value of the fibre (SF),

Dyeing recipe.

From these the quantity of retarder and the starting temperature can be found. The quantity of dyestuff as given in the dyeing recipe must be expressed as a percentage in relation to the weight of the goods. Convert accordingly if the quantities are stated in a different way.

Retarder quantity

Fibre type V, S_F (From fibre manufacturers),

Dyestuff type,K (From dye manufacturers).

From these three values total concentration value is found out.

Dyestuff quantity percentage dyestuff concentration value is found out

Total concentration value minus dyestuff concentration value = % of Retarder.

Starting temperature

Total effective concentration value and fibre type we get temperature value (fibre manufacturers)

From dyestuff type we get temperature correction (dyestuff manufacturers)

Temperature value minus temperature correction = Starting temperature (°C).

In dyestuff mixtures consisting of dyestuffs with different combination indices, the lowest combination index (K index) is used to determine the retarder addition andstarting temperature.

Start dyeing at the determined starting temperature, i.e. enter the material at this temperature or heat the bath straight away to the starting temperature. After a brief period, bring the dye liquor to the boil at a rate of 1 °C in 3 min, and continue to dye at the boil for 30~60 min.

9.2.6.1.7.3.4   Acrylic dyeing by continuous methods

Demands for dyeing large quantities in the same shade has caused the development of continuous and semi continuous dyeing processes which offer technical and economicaladvantages over the conventional dyeing methods.
Generally, the continuous dyeing processes the dye-auxiliary solution is applied to the fibre, e.g., by padding and the dye fixed in another operation. Processes which follow this principle are the

Pad/steam process,

Vigoureux printing process,

Pad/roll process and

Space dyeing.

9.2.6.1.7.3.4.1   Pad-steam methods

In pad-steam method, it is known that the dye and auxiliary is applied on the material by padding and fixed by steaming under certain conditions. Steam fixation can be considered to be a dyeing operation at an extremely short liquor ratio. At a liquor pick-up of 100%, the steaming process is comparable to dyeing from an aqueous bath at a liquor ratio of 1:1.

Since the dyeing process takes place in a short time the selection of dyes, chemicals, auxiliaries and parameters of dyeing, etc. has to be strictly followed as per requirement.

Components of the pad liquor

1. Dyes

In pad steam method of dyeing, the rate at which the dye goes onto the fibre is not determined not by its affinity for the fibre but by its diffusion coefficient alone. This means that dyes with high diffusion coefficients permit short steaming times and thus high running speeds of the goods through the steamer. Economy in pad/steam dyeing is ensured only if dyes are used which meet these requirements.It has been found that dyes with mesomerically distributed charge exhibit a higher diffusion coefficient in acrylic fibres than those which have a local charge.

2. Dyeing accelerator

By adding this auxiliary to the padding liquor, it is possible to increase the speed at which acrylic dyes are fixed in the acrylic fibre. Given below a graph showing the accelerating effect of a dyeing accelerator (Primasol AN –BASF) on an Acrylic dye (Bascryl Blue GL –BASF).

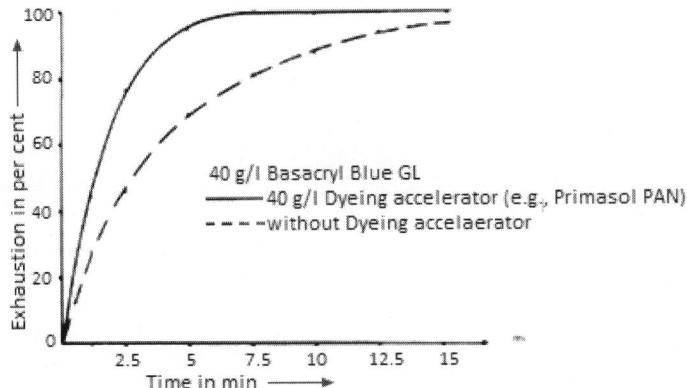

40 g/l Basacryl Blue GL
———40 g/l Dyeing accelerator (e.g., Primasol PAN)
– – – –without Dyeing accelaerator

Effect of dyeing accelerator in dyev exhaustion against time of dyeing

3. Thickening agents

The addition of thickening agents helps in a level dyeing and prevents migration of the dye. But it is important that a non-ionic and an easily washable thickening agent have to be employed. Anionic thickening agent may cause precipitation of acrylic dyes. The most suited are thickening agents derived from locust bean gum and galactomannan compounds, e.g., Meyprogum PAM, Meyprogum PAT and Solvitose OFA.5 – 10 g/l of any of these thickening agents are found sufficient for pad-steam process.

4. Antifoam

If excessive foaming occurs during padding, antifoaming agents may be added to the liquor. It must, however, be ensured that only non-ionic compounds are used. Silicone antifoams are liable to cause precipitation and therefore are not recommended.

5. Retarding agent

Tailing of the dyes can occur due to many reasons like, higher volume of the pad bath, adsorption of the dyes on the surface of the fibre, substantivity of the dyes and to small extent the concentration of the dye in the dye bath. Tailing can be reduced by adding a suitable retarding agent.

Steam

Saturated steam is necessary for the production of perfect shades on acrylic fibres. It is possible to determine the presence of saturated steam or super heated steam by placing two calibrated thermometers in the steaming chamber. The bulb of one thermometer is wrapped in a moist cloth. In a saturated steam atmosphere both thermometers indicate the same temperature but when the atmosphere contains superheated steam, the thermometer with the moist cloth indicatesa lower temperature than the other.

Saturated steam is converted to water by removing heat. It must therefore be ensured that parts of the steamer do not cool by loss of heat. Difficulties are particularly liable to occur when, for instance, the ceiling or other metal parts located above the textile goods do not heat up to the temperature required because the steam will tend to condense at these areas to form water droplets which fall onto the goods. Considerable difficulties of this type can be avoided bysuitable modification of the ceiling construction or by heating these areas with auxiliary heaters.

Super heated steam is unsuitable for dyeing acrylic fibres. This type of steam is liable to dry the padding liquor applied to the goods to be dyed. Serious difficulties are thus encountered in dyeing with cationic dyes.

Dyeing procedure

1) Prepare the pad liquor

Quantity	Unit	Additions
x	g/l	Acrylic dye
y	g/l	Padding and solubilising agent
15–40	g/l	Dyeing accelerator
5–10	g/l	Thickening agent (Locust bean gum)
y	g/l	Retarding agent (if necessary)
z	ml/l	Antifoam (if necessary)

The dye is pasted with a dispersing agent and boiling water added. Thickening is introduced as a 3% stock thickening into the cooling dyestuff solution between 80 and 50 °C. After the thickening agent has been added dilution to final volume is made with water of the same temperature.

Pad temperature 50–80 °C, pick-up not over ٧٠–100%.

2) Steaming 15–30 min at 102–104 °C and subsequently washed in a continuous scouring range at about 60 °C.

3) Rinsing and cross-dyeing of cellulosic component on the winch or jet.

9.2.6.1.7.3.5 *Vigourex printing on acrylic slubbing*

Print paste is prepared as follows:

Quantity	Unit	Additions
x	g/l	Acrylic dye
15–40	g/l	Dyeing accelerator
15–20	g/l	Thickening agent
y	g/l	Antifoam (if necessary)

The material is printed with cold pad liquor and steamed immediately. After printing, the slubbing is usually placed in wire cages or in boxes of perforated sheet metal and fixed with saturated steam. The total steaming time depends on the time required to heat up also the inner layers of the slubbing to the temperatures required. For adequate fixation of dye, all areas of the material in the cage must be subjected to a temperature of 100 °C for 20 min. Further the material is soaped using suitable detergent to remove the thickening agent and the unfixed dyes.

Notes:

1. In comparison to printing on wool slubbing, printing of acrylic slubbing requires no special measures. It must be ensured that the fibrous material has been made adequately antistatic as otherwise it may give rise to difficulties in needling and is liable to adhere to the rollers.

2. It is also advisable to steam the slubbing immediately after printing to ensure that the liquor does not dry up.

3. In steam fixation, strong formation of condensate is liable to cause the printing liquor to run or to wash off. Difficulties in this respect occur particularly if the steam already contains a lot of condensate and when the batched-off slubbing has not been heated rapidly enough. For this reason, the steam should be fed through well insulated pipes of adequate dimensions provided with a condensation water trap. The steamer should be of the type that can be evacuated and it must also be ensured that the layers of slubbing in the steamer are not too thick.

4. For a print with good definition, it is necessary to use a thick print paste. The amount of thickening agent necessary in this case is

15–20 g/1.Only non-ionic products which can be washed out readily should be used. Galactomannans and locust bean gums are very suitable.

9.2.6.1.7.3.6   *Special space dyeing processes*

The space dyeing process is a dyeing or printing process for yarns. The multi-colour effect is obtained by dyeing the yarns at irregular intervals in different shades. This is followed by knitting, tufting or weaving the yarns. The finished piece has a relatively regular colourpattern.

The fibres used for floor coverings which are to be coloured by the differential dyeing process must have different dyeing behaviours. The types of fibres which fulfill this requirement have been developed from polyamide, polyester and acrylic fibres.

In all space dyeing methods the dye/auxiliaries are applied on the yarn and the dye is fixed by steaming with saturated steam. There are differences in applying the dye on the material:

1. Knit/deknitprocess: In this process, the yarn is knitted into a hose which is then printed on both sides with vigoureux or roller printing machines which are installed at different angles. After fixing, washing and drying, the pieces are deknit and the yarn wound up.

2. Yarn printing with adjustable pair of rollers: This machine consists of several pairs of rollers. The lower roller dips into the liquor and is used as the furnishing roller. The upper roller is located at a suitable distance above the furnishing roller. The yarns of approximately 400 packages are united into a group and passed continuously, without contact, through the space between the pair of rollers which are installed a certain distance apart and provided with individual troughs holding the different dye liquors. An electronic or mechanical control lowers the upper rollers at certain intervals for a short time so that the yarns travelling through the space are squeezed and printed.

3. Yarn printing with engraved pair of rollers: Also in this process, a group of yarns passes through pairs of rollers. The upper roller, however, is immovable and the distance between this and the lower liquor furnishing roller is not adjustable. The upper and lower rollers are provided with engravings of different depths. When the group of yarns passes through the pairs of rollers, it is squeezed only when the raised parts of the upper and lower engravings meet.

4. Injection process on wound packages: This method consists of injecting dye solutions into the wound packages with hypodermic needles. The dispersion of dye in the cross-wound package can be improved by subsequent centrifuging. In new methods the injection of the dye solution in "rocket" packages in which the yarn layers are not parallel but lie either perpendicular or at an oblique angle to the axis of the package to avoid the above problems. This type of package is that cross-wound by the Hacoba system with which a uniform dispersion of the dyed zones is obtained along the entire length of the package.

5. Dip centrifuge process for cross wound packages: In this process, the cross-wound packages are placed into the dyebath. The amount of dyeing liquor to be applied can be adjusted by raising or lowering the level of the bath. The package is then centrifuged and a part of the liquor removed. The packages can there upon be turned over and the top corner placed in a second bath. After centrifuging again, a package which has three different colour zones, one of which is white, is obtained.

6. Frauchiger process for wound packages: In this method the dye solution is placed in the inner tube of the package and forced through the interior layers of the wound package by centrifuging. The perforations in the tube can be closed as desired so that certain areas of the wound package are coloured while the rest remains white. It is possible to obtain multi-colour effects by repeating the process with different dye liquors and altering the areas at which the perforations are covered.

Process

In all above processes the dye solution/pad liquor is made in the same manner with small changes such as quantity of thickening agents, etc. The dye solution is made as follows:

Quantity	Unit	Additions
x	g/l	Acrylic dye
15–40	g/l	Dyeing accelerator
0–20	g/l	Thickening agent
y	g/l	Antifoam (if necessary)

The dyes, dyeing accelerator, thickening agents, antifoam, etc. which are suitable for pad-steam are suitable for these processes also.

The dye solution is applied on the fibrous material, steamed with saturated steam, soaped and dried.

9.2.6.1.7.3.7   *Pad-roll process*

This is a semi continuous process. The process involves padding and batching of the material, heating the material to 100 °C to avoid the condensed water falling on the material in the next process of steaming in the steaming chamber.

The set up needs following units:

- a padder,
- an infra-red zone,
- a batching chamber.

The material is padded in following pad liquor:

Quantity	Unit	Additions
x	g/l	Acrylic dye
40	g/l	Dyeing accelerator
1–3	g/l	Thickening agent (Locust bean gum)
y	g/l	Antifoam (if necessary)

The dye is pasted with a dispersing agent and boiling water added. Thickening is introduced as a 3% stock thickening into the cooling dyestuff solution between 80 and 50 °C. After the thickening agent has been added dilution to final volume is made with water of the same temperature.

Pad temperature 50–80 °C, pick-up not over 70%.

The padded cloth is heated to about 100 °C in the infra-red zone.

It is rolled up and batched with continual rotation for 2–4 h. The temperature of the batching chamber is maintained at 102–104 °C by the continuous introduction of fresh steam. Relative humidity to be maintained at100%.

Rinsing at 40–50 °C, then drying.

Notes:

1. Polyacrylonitrile blends with cellulose are cross-dyed at 80–85 °C. This can be done on the jet by reactive dyestuffs or selected direct dyes.

2. Padding can be done in batch sizes of 3000–5000 m a batch.

3. To avoid the condensation of steam in the reaction chambers the material is heated to 100 °C during batching itself. This is done in two stages. The padded material is passed over a shaft ahead of the reaction chamber through which the goods are passed and heated up first in a saturated steam atmosphere and then in an infrared zone in

the descending section of the shaft. Further the batches are rotated inside the reaction chambers filled with saturated steam till the fixation is completed. Soaping and drying is followed.

9.2.6.1.7.3.8   One-bath dyeing and finishing

Acrylic loose stock, yarn and piece goods can be dyed with cationic dyestuffs together with softeners and antistatic agents in a single operation. This one-bath dyeing and finishing process is particularly interesting for finishing acrylic high-bulk yarns. Depending on the required handle, the dyebath containing the usual additions is also set with for e.g.,

2–4% weakly cationic softener or

1–2% cationic softening agent.

Some softening agent can work as levelling agents also, because it is weakly cationic. It goes well with cationic retarders hence it can be combined with them in the dyebath. It goes without saying that the quantities of retarding agents and dispersing agents will then have to be reduced accordingly.

9.2.6.1.7.3.9 Salt treatment for relustering

If the luster of PAN is reduced during dyeing, the silky lustre of acrylic continuous filaments can be retained by treatment with 50 g/l Sodium sulphate (Glauber's a salt) calc. (ora double amount of Glauber's salt cryst.) for 20 minutes at 95 °Cimmediately after dyeing and rinsing. Then cool the bath to about 75 °C) and rinse twice with warm water.

9.2.6.1.7.3.10 Brightening

Depending on the intended use of the continuous filaments, brightening treatment can be done with the suitable brightening agents or softening cum brightening agents.

Chapter 10
Processing of acetate materials

By the term "cellulose acetate", or simply "acetate", we mean fibres consisting of an acetic ester of cellulose, the degree of esterification of which is about 2.4–3. Cellulose acetate exists as bright or dull fibres. As it is sensitive to alkalis and to heat it should not be treated in alkaline baths, or above 100 °C. Even prolonged treatment above 85–90 °C should be avoided.

Acetate filament is a soft, luxurious fibre made from natural resources and widely used in apparel, home furnishings and decorative fabrics. Acetate fibre is used in apparel fabrics for linings, dresses and blouses, active sportswear, bridal gowns and children's party dresses. Woven or knit fabrics for dresses and blouses use 100% acetate or blends with rayon, spandex, nylon, linen, wool, silk, cotton and in some applications, polyester. In woven home furnishings, the fibre is commonly used in the warp with a cotton or rayon filling. Decorative fabrics are satins for ribbons, manufactured flowers and scarves. Another application is linings for luggage and coffins. There are special acetate fibres which has antimicrobial and such special properties. Available as both filament yarn and staple, it is used in bedding, health care textiles, wiping cloths, athletic socks and shoe linings. These special acetate fibres require special handling in bleaching, and this bulletin details a recommended procedure. Otherwise, dyeing and finishing of these acetates are identical to standard acetate. Acetate can be dyed or printed in a wide range of shades with disperse dyes. In the processing of acetate fibres an important factor to be born in mind is that the highest pH that can be used is only 9–9.5. Caustic soda is not recommended because the resultant alkalinity is too high, and at pH 10 and above, acetate starts to saponify (changes chemically to rayon). Saponification is the process of removing part or all of the acetyl groups of acetate, leaving regenerated cellulose. Regenerated cellulose is dyeable with direct and fibre-reactive dyes, and if saponified only partially, the fabric will contain dye streaks. So the alkali used in either scouring or other process is only ammonia, trisodium phosphate (TSP) or tetrasodium pyrophosphate (TSPP) or at the most soda ash. It also should be noted that when acetate comes in contact with hot metal in a jet or on a frame, the fibre glazes as a result of cross-section deformation by the hot metal and it can cause tiny shine spots on dyed acetate fabrics.

10.1   Scouring.

Scouring bath should contain an anionic or non-ionic surfactant in amounts of 0.5–1.5 % on the weight of the fabric and an alkali like ammonia, soda ash, trisodium phosphate (TSP) or tetrasodium pyrophosphate (TSPP). The choice of surface-active agent and alkali for scouring acetate fabrics may vary for different sizes used on the warp. The pH of scouring is adjusted to 9–9.5 with the quantity of the alkali used. The bath also usually contains EDTA or other suitable sequestrant. The scouring is done at temperature of the range 60–80 °C (140–180 °F) for about at least 30 min (the number of ends depends on the load size). After scouring, the fabrics need to be thoroughly rinsed before dyeing.

Scouring can be done on jigs or simple continuous ranges as scouring temperature range is around 60–80 °C. However, the latter is preferred due to its higher productivity and allows better evenness of process.

10.2   Dyeing

It is important to select disperse dyes that are recommended for acetate by their manufacturers. Home furnishings and outerwear require good light, gas and wet fastness properties, while linings and apparel require good perspiration, wet crocking and gas fumefastness. To produce the heavy shades that linings and apparel often require choose dyes that have good colour yield.

Some of the disperse dyes which can be used for acetate dyeings are given below by CI Index numbers. However, it is better to consult the manufacturers also.

Yellow 34	Orange 138	Red 90	Brown 22	
Yellow 34	Orange 31	Red 106	Blue 27	Black 9*
Yellow 42	Orange 73	Red	Blue 60	Violet 33
Yellow 67	Orange 80	Red 135	Blue 102	Violet 27
Yellow 86	Orange 37	Red 35	Blue 200	Black 33*
Yellow 119	Orange 30	Red 30		
Yellow 126		Red 50		
		Red 117		
		Red 72		
		Red 136		
		Red 88		

*It is better to use 55% Blue 102 and 45% Brown 22, may be an easier method of dyeing blacks than using Black 9 or Black 33, which require diazotization.

10.2.1   Dyeing with disperse dyes on jig

Usually, linings and decorative fabrics, satin, taffeta, twill or plain weave constructions are dyed on jig. This helps to avoid cracking and creasing of the fabric. Jigs should have covers to avoid temperature variations across the width of the fabric Modern computer controlled jigs, dye dosing and circulation pumps are preferred. These types of machines can process with temperature controls, tension adjustments, running without stoppages, controls side to side and end to end uniformity and produce creaseless even dyeing compared to the normal jiggers.

Fabrics dyed on jigs usually have some degree of "endiness," meaning that the last 20–30 yards on each dye roll are lighter in colour than the remaining fabric. This yardage shouldbe removed before shipping.

Process

The dye bath is prepared with a scouring/wetting agent, alkali, and EDTA or other suitable sequestrant. The disperse dyes should be pre-dispersed in water at 50 °C, adding a dispersing agent while mixing.

1. Load the fabric on the jig and fill with water at 30–40 °C.

2. Prepare scouring bath containing:

Quantity	Unit	Bath additions
1–2	%	Emulsifier (non-ionic or anionic)
0.1	%	Soda ash
0.25	%	Dispersing and sequestering agent

3. Load the fabric and heat to 60–80 °C.

4. Run for at least 30 min, rinse thoroughly and drain.

5. Prepare fresh dye bath with 0.25% sequestrant and 1.0% dispersing agent at room temperature.

6. Run for two ends and add pre-dispersed dyes.

7. Heat the dye bath at 50 °C and add acetic acid, monosodium phosphate or ammonium sulphate to adjust the pH to 5.5–6.5.

8. Heat gradually to 80–95 °C depending on fabric style and depth of shade.

9. Run as necessary to achieve required shade, cool bath and drain.

10. Rinse with fresh bath containing a detergent in two ends at 80 °C to remove excess dye.

11. Dark shades can be reduction cleared with:

Quantity	Unit	Bath additions
2	g/l	Sodium carbonate
2	g/l	Sodium hydrosulphite
1	g/l	Detergent

12. Run for two ends at 65 °C, rinse, drain and unload.

Notes:

1. Some fabric styles are prone to stretching and should be dyed at the lower end of the dye temperature range.

2. Dyeing at the boil can result in significant stretching and distortion of the fabric. Time at dyeing temperature varies by shade. For a 2750 m batch, light shades can be completed in four ends while dark shades may require additional ends.

10.3   Dyeing of polyvinyl alcohol fibres

Polyvinyl alcohol fibres can be dyed by following dyes.

10.3.1   Dyeing with metal complex dyes

The dye bath is set at 50 °C with:

Quantity	Unit	Bath additions
0.5	%	Acetic acid
20	g/l	Sodium sulphate

and the dye, previously pasted and diluted with hot water is added.

Raise the temperature to 95 °C and dye at that temperature for 2 hours. Allow to cool slowly to 60 °C before rinsing.

10.3.2   Dyeing with vat dyes

The polyvinyl alcohol material can be dyed with vat dyes the same way as we are dyeing a cotton material. Preparation of the dyebaths is identical, and dyeing can be done either in a reduced vat or by pigment impregnation. However, to obtain the best yield, the temperature of dyeing should be 10 °C higher than that recommended for dyeing cotton. The goods are generally dyed on the jig, either in a reduced vat or by pigment pad-jig.

10.3.3   Dyeing with pigment colours

The padding bath is set with:

Quantity	Unit	Bath additions
x	g/l	Pigment paste
100	g/l	Dyeing binder
10	g/l	Ammonium nitrate 25%
y	g/l	Synthetic thickener

After padding, the fabric is dried and then thermo fixed at 140 °C for 5 min.

Part IV

Dyeing of Blended Materials

Chapter 11
Blending and Classification of blends

Blending of fibres are done based on many factors. It can be mostly for covering up some disadvantages of a particular fibre with another fibre – this can be strength, shade brightness, cost, dimensional stability, durability, heat resistance. Principally most of the fibres can bemixed with another fibre by one way or other.

Other than to achieve certain, characteristics, appearance and effect on fabrics there are other reasons to make blended materials. One of the major reason is to avoid or reduce paying high import duty on garments to be exported to certain countries often different fibres are mixed to make fabrics. Instead of using only one kind of fibre, two or three different kinds are used. Even in such cases, it has to be considered that each kind of fibre has its own characteristics, when mixing them, one should consider their characteristics to create a hybrid with the characteristics needed.

11.1 Important characteristics of common fibres considered in blending:

Cotton: Soft, very absorbent, shrinks when washed, easy to crease.

Ramie or Linen: Compared to cotton not so soft, less absorbent, shrinks when washed, but to a less extent, easy to crease.

Rayon: Compared to cotton softer than cotton, equally absorbent, shrinks more than cotton, very easy to crease, has a cluster of silk.

Polyester: Compared to cotton not soft, not absorbent at all, does not shrink in the normal wash, not easy to crease

Wool: The characteristics vary depending upon the kind of wool it is but has the following general characteristics. Compared to cotton not very soft, not absorbent. The staple is slightly crispy which can lead to a bulky effect when finally made into a fabric. It shrinks seriously when washed in hot water. Not easy to crease.

Nylon: Can be soft or firm dependent upon the finish applied, not absorbent, does not shrink when washed, not easy to crease, very strong.

Silk: Finer and softer than cotton equally absorbent, shrinks more than cotton, easy to crease, has a smooth surface with lustre when it is made of fabric

Acrylic: Not soft, not absorbent, does not shrink in the normal wash. The staple is slightly crimpy which can make the final product, the fabric, bulky or beefy not easy to crease

11.2   Major objectives of blending

1. Improvements of functional properties

A pure fibre yarn or fabric cannot impart of all the desired properties. Taking the case of 100% viscose rayon fabric, they suffer from low tensile strength, poor crease resistance and low abrasion resistance. In case of 100 % polyester fabric are not desireable as they are prone to static accumulation, whole melting, pilling and low moisture regain. These negative attributes of polyester, rayon can be reasonably neutralised by addition of certain percentage of each yarn. In a blend of cotton and polyester cotton provides the absorbency and polyester provides the strength. Blending also can provide the fibres light weight with all desireable characteristics, improving spinning, weaving and finishing efficiency and uniformity of product.

2. Improved process performance:

Some fibres like polyester at times are quite troublesome to work in 100% form especially at cards.Addition of fibres like cotton or viscose rayon in the previous process has been seen to facilitate the smooth carding of such fibres.The blending of manmade which are longer and finer to cotton which is shorter influences the spin ability as well as productivity.

3. Economy

In many countries the price of manmade is much more stable than that of natural fibres like cotton. Price stability canenable the mills to pursue optimization of their fibre purchase programme.Blending could also be used for reducing the mixing cost. For example, a fibre like viscose can beblended with cotton for producing specific yarns with reduced raw material costs.

4. Fancy effect

Fibres with a variety of colour mixture or shades can be produced by blending different dyed fibres at theblow room, draw frame or roving stage.

5. Aesthetics

The aesthetics of a fabric can be developed by selecting specific blend components and their properties.

The selection of components for blending, the blending process, methods of blending etc. are not in the scope of this book. However, one has to look into the blending in the point of view of the processor. Hence, we deal with the above topics in short for a better understanding.

11.3   Selection of constituents for the blending

The selection of blend constituents depends upon the following factors:

1. Type of Fibre

Depending upon the end use of the fabric, blend constituents are chosen. For example, it is well known hat a polyester-cotton yarn looks fuller as compared to the lean look ofpolyester-viscose yarn. Therefore for light constructions like shirting, polyester-cotton blend is used. However, polyester-viscose blend is preferred for medium and heavy constituents such as suiting.

2. Compatibility of blend fibres

When selection the constituents for blending it should be compatible in terms of the following properties: (a) Length and Denier of Fibres - The two most important properties which need to be matched for satisfactory processing of blends are the staple length and the fineness (denier) of fibres. For example, for blending of polyester with combed cotton which may have staple length in the region of 34 mm, the polyester staple normally employed is 38 mm.The denier of cotton component is generally in the range of 1.3-1.5 (3.6-4.2 µg/inch) and so is the denier of the polyester which generally 1.5 denier.Similarly in a viscose rayon cotton blend, the rayon staple of 1.5 denier and 29-32 mm length isgenerally used since the cotton component used has a denier of around 1.5 and a length of 28 mm. Excessive difference in the deniers of the blend fibres, say 1.5 denier viscose blended with 3 denier polyester fibres will produce a very weak yarn. (b) Extensibility - The elongation 'balance' of the constituent fibres is equally important especially in regard to yarn quality. A large difference in breaking elongation of fibres in a blend adversely affects the yarn tenacity; the lowering of yarn tenacity being attributed to the unequal sharing of the tensile load at the rupture point. During shed formation in weaving, the breaking extension could significantly influences the weave ability of the warp yarn. It is sometimes preferable to consider the work of rupture in this connection. The change in the breaking extension of a blend yarn with change in blend ratio very much dependent onthe fibres extensibility of the blend components and the structure of yarn. In case of two blend fibres have very different breaking extension like cotton-polyester or viscose/polyesterthen there is an abrupt transition from a lower extension level to a higher extension level. Normalcy breaking extension of two different fibres in a yarn lies between mean of two.A large difference in the breaking elongation of the fibres in a blend adversely affects the yarn tenacity. (c) Density - The blend fibres should preferably have the same density. Any large differences on this account will lead to selective separation while conveying the blended stock through ducts under the influence of airsuction in the blow rooms. The heavier fibres tend to fall in the lower section of the duct and move at a lower speed. Any large differential in fibre densities of the blend components works against the objective of achievinga uniform blend (d) Dispersion properties This property describes the ability of an individual fibre to separate from its group and disperse thoroughly within the fibre matrix of the blend to produce an intimate and homogeneous blend. Poor dispersion may be caused due to factors such as too much or too little crimp, large fibre aggregates with coterminous ends, excessive static accumulation. Deficiency in proper dispersion of individual blend components is particularly undesirable for blends of fibres having different dyeing properties. (e) Drafting properties - Some fibres like viscose are outstanding it terms of draft ability. These fibres, when blended with other fibres act as good carriers to obviate the trouble relating to drafting. (f) Dyeing properties - In case the blend yarn or fabric is to be dyed subsequently, due consideration should be given to the dyeing properties of individual fibre components.

11.4   More common blended fabrics and their properties

Polyester/Cotton or Polyester/Rayon– The tough crease-resistance of polyester combines with the cool comfort of cotton. It is easily laundered, dries quickly and is ironed with lower temperature than pure cotton.

Nylon/Wool – The blending of nylon with wool makes the fabric more absorbent and softer. It becomes more strong and durable.

Nylon/Acetate – This combination makes the fabric more absorbent than nylon alone.

Ramie/Polyester or Ramie/Acrylic – These two blends help the fabric to be easily taken care of and it is less stiff than pure ramie fabrics.

Wool/Cotton – These two fabrics benefit from the inherent qualities of each other after blending. Itgives better comfort, better aesthetics and better performance.

Linen/Silk or Linen/Rayon – This blend helps the fabric to retain the characteristics of linen and makes the fabric drape better and wrinkleless.

Silk/Wool – The blending of silk with wool provides subtle texture to the fabric. It is generally used forties.

Rayon/Cotton – This fabric of rayon and cotton blend wears well and is washable. It is soft and has fuzzy surface. Dresses, suits, sportswear, men's shirts, etc. are made out of this fabric.

Wool/Synthetics or Rayon/Synthetics – This blend has a very clear finish and it drapes better and tailors easily. It has exceptional wearing qualities. The fabric is used for men's and women's suits and coats. Ski slacks are also made out of these blends.

11.5   Methods of Blending

Blending can be done at the opening stage, drawing etc.

1. Blending at blow room - Man-made fibres are used extensively in blends, where the length of the man-made fibre has to be matched to that of the natural fibre. There are three methods of blending at blow room:

   a. Feeder blending - In this method the blends are prepared by feeding different fibres to different hopper feeders with feeds adjusted to the required blend ratio. This method is generally employed when more than two components are required to be blended.

   b. Stack Blending - In this method the blend components from the bale or bale breakers (pre opened) are weighed and laid downin alternate layers, each about 15cm thick and covering about 40 sq. Meter. The stack is normally laid up to a height of 3 m. This stack which laid horizontally is then withdrawn vertically for feeding.

Advantages of Feeder and Stack Blending

   a. More intimate and homogeneous blend achieved.

   b. Only one opening line needed.

   c. Provides simplest control on the use of recovered fibre waste.

   d. Requires minimum man hours for blending

Disadvantages

   a. Difficult to attain uniform blend ratio

   b. Demands greater skill on the part of the operator.

   c. Labour intensive and some what slow.

  d. Lap blending - This type of blending cannot be affected in a single process blow room as it involves the use of a breaker and a finisher scutcher. In this the laps of the components fibers made at the breaker scutcher generally 3 to 4 are blended by feeding them together to the finisher scutcher in the desired ratio.

Advantages of Lap Blending

  a. Ensures good blend homogeneity

  b. Easy to work.

  c. Good control on the use of recovered fibre waste.

  d. Uniform blend ratio is achieved

Disadvantages

  a. Opening line has to be modified to provide for both breaker and finisher scutcher

  b. Need for proper control on lap weights..

2. Card blending - Blending by doubling with laps at the card is sometimes used in high production cards. The procedure involves use of a double lap roll attachment whose feed rolls are positively driven by a link chain by the regular lap roll gear.

Advantages of Card Blending

  a. Most intimate blend is obtained.

  b. This method is used to randomly mix two laps differing in color.

Disadvantages of Card Blending

  a. It can't be obtained from wide range.

  b. Controlled metered blending can't be carried out.

  c. Only make a previously proceeded blend more intimate

3. Blending modification at Draw Frame - This method is normally used for binary blends only. The required blend proportion is adjusted by the number of slivers of each component and the hank of respective slivers. The fleece blending is done on the blending Draw frames specifically designed for this purpose. They are fed with 16-20 slivers at the back and therefore provide a much greater flexibility as regards the blend ratios.

Advantages

  a. Easier to obtain uniform blend ratio.

b. During opening and carding, optimum settings for each blend component can be used for better quality of output with less damage to the fibres.

c. Easy working.

Disadvantages

a. Difficult to attain random arrangement of fibres in the yarn cross section.

b. Additional drawing capacity needed.

c. Separate opening lines needed for each component.

One of the serious problem of blending on a normal Draw frame is the difficulty of achieving random distribution of fibres in the yarn cross-section i.e. the index of blend irregularity is higher for this form of blending than the Blow room blending. This is due to the fact that during drafting there is little lateral movement of fibres. Further, when the web is condensed into sliver there is very little improvement in fibre intermingling as each sliver in the drafted web tends to retain its entity. It is only repeated doubling and drafting that blending really improves and reach near the random state.

4. Blending modification at roving frame - The roller weighting considerations in speed frames are similar to those operating at the Draw frame. The weighting is much higher than that used for cotton and should range between 10-15 kg per spindle.

5. Blending of Combed Cotton Sliver and Polyester - Many Indian mills resort to this practice when the humidity control or conditions of machines is very poor.

Advantages

a. Produces very intimate blend

b. Trouble free running and high productivity at card.

c. Less yarn imperfections due to better fibre individualisation because of reprocessing of the cotton component.

d. Reduced number of d/f passages.

e. Lower end breaks due to fewer slubs.

f. Better uniformity of dyeing due to more intimate blend.

Disadvantages

a. Poor tenacity and evenness in blend yarn.

b. High cotton nep content in blend due to reprocessing

    c. Need of additional b/r and card capacity

    d. Slightly higher waste in b/r and carding.

Optimum Blending Method of various Blends

1. For blends like P/V, blow-room blending is effective as they need similar b/r sequence.
2. For blending of manmade stack blending method is generally used.
3. The polyester /cotton or acrylic/cotton are generally blended at d/f because cotton component needs a severe opening and cleaning action
4. Where there is a problem of running 100% polyester on card, stack blending of polyester stock and combed cotton may be resorted to.
5. In case of v/c blend, they should be blended at the draw/frame as they need quite a different opening sequence.

The blends can be broadly classified into intimate or homogeneous mixtures or staple fibre yarn blends and mixed ply yarns, the former being more common. There are other type blends which do not fit in the broad classification like core spun yarns used in weaving ant to some extent in knitting for special purposes. Even though the blends is usually binary (two components) ever ternary blends are also produced these days.

11.6   Classification of Blends

There are many fibre blends available in the market. Blending can be classified into different categories based on the textile composition or way of blending is used or on their dyeing characteristics. Given below some common combinations and their classifications on different basis :

Classification according fibre mixing

	Blending nature	Type	Examples
A	Natural - Natural	Cellulosic - cellulosic	Cotton/Ramie
		Protein - Protein	Wool/Cashmere
			Silk/Wool
B	Natural/Regenerated	Cellulosic - Regenerated	Cotton/Rayon
		Protein - Regenerated	Woo/Rayon
C	Natural - Synthetic	Cellulosic - Synthetic	Cotton/Polyester
			Cotton/Spandex

		Protein - Synthetic	Wool/Polyester
			Wool/Acrylic
D	Synthetic - Regenerated	Synthetic - Regenerated	Polyester/Rayon
E	Synthetic - Synthetic	Synthetic - Synthetic	Nylon/Spandes
			Nylon/Acrylic

Classification according to the way fibres are blended

	Blending nature	Type	Examples
A	Staple - Staple	Yarn	Cotton/Ramie
B	Core spun (Staple/Filament)	Yarn	Cotton/polyester core yarn
C	Bi constituent or Tri constituent	Synthetic - Filament	Acrylic/Acrylic filament
D	Blended Fabric	Fabric	Polyester warp/Cotton weft

Classification according to dyeing nature

For a processor it is mainly dealing with two fibres in the same fabric. The main problem for the dyer is the properties of both the fibres, and how to achieves the preparation, dyeing finishing of one fibre with in the required norms without affecting the other. In this point of view we can divide the fibres into four classes.

Acid dyeable fibres	(Basic)Cationic dyes	Cellulosic dyes dyeable fibres	Disperse dyes
Class A fibres	Class B fibres	Class C fibres	Class D fibres
Wool and other animal fibres	Acrylic and mod acrylic fibres	Cellulosic dyes dyeable fibres	Cellulose acetate
Silk	Cationic dyeable polyester	Viscose	Cellulose triacetate
Nylon	Cationic dyeable nylon	Lyocell, modal, etc.	Polyester
Polyurethane fibres		Linen and other bastfibres	Polyvinyl chloride fibres
Acid dyeable synthetic fibres			

One method of classifying the fibres is based on their dyeing characteristics. Thus fibres are classified in the following manner based on the most common dyes used to get the fastest dyeing on a fibre.

When you mix these groups within themselves or with other groups following binary mixtures are produced and they can be grouped as follows and the dye selection for each type blends. When naming the blends of polyester with natural fibres it is customary to name the synthetic fibre first, i.e. polyester/cotton (DC) or even 'poly/cotton' is heard far more than cotton/polyester. Also when fibre in a blend is mentioned in the adjective form (e.g., cellulosic, acrylic, etc.) it is mentioned in as second in the blend (e.g., wool/cellulosic,nylon/acrylic, etc.)

11.7   Common binary blends

AA blends	CC blends
Wool/silk	Cotton/viscose
Wool/mohair	Cotton/modal fibre
Wool/cashmere	Cotton/polynosic fibre
Wool/angora	Cotton/linen
Nylon/wool	Linen/viscose
Nylon/silk	Linen/modal fibre
Wool/polyurethane	**DA blends**
Nylon/polyurethane	Cellulose acetate/wool
Wool/acid-dyeable polypropylene	Cellulose acetate/silk
Nylon/acid-dyeable polypropylene	Cellulose acetate/nylon
Normal/deep-dye nylon	Cellulose triacetate/wool
AB blends	Cellulose triacetate/silk
Wool/acrylic fibre	Cellulose triacetate/nylon
Silk/acrylic fibre	Polyester/wool
Nylon/acrylic fibre	Polyester/silk
Polyurethane/acrylic fibre	Polyester/nylon
Acid-dyeable polypropylene/acrylic fibre	Polyester/acid-dyeable polypropylene
Wool/modacrylic fibre	Poly(vinyl chloride)/wool
Mohair/modacrylic fibre	Poly(vinyl chloride)/nylon
Nylon/modacrylic fibre	**DB blends**
Acid-dyeable/basic-dyeable acrylic fibre	Cellulose acetate/acrylic fibre
Modacrylic fibre/acrylic fibre	Cellulose triacetate/acrylic fibre
Deep-dye/basic-dyeable nylon	Polyester/acrylic fibre
Wool/basic-dyeable polyester	Cellulose acetate/modacrylic fibre

Nylon/basic-dyeable polyester	Cellulose triacetate/modacrylic fibre
AC blends	Polyester/modacrylic fibre
Wool/cotton	Normal/basic-dyeable polyester
Silk/cotton	**DC blends**
Nylon/cotton	Cellulose acetate/cotton
Polyurethane/cotton	Cellulose triacetate/cotton
Acid-dyeable polypropylene/cotton	Polyester/cotton
Wool/viscose	Poly(vinyl chloride)/cotton
Silk/viscose	Cellulose acetate/viscose
Nylon/viscose	Cellulose triacetate/viscose
Wool/modal fibre	Polyester/viscose
Nylon/modal fibre	Poly(vinyl chloride)/viscose
Nylon/linen	Cellulose triacetate/modal fibre
CB blends	Polyester/modal fibre
Cotton/acrylic fibre	Polyester/polynosic fibre
Viscose/acrylic fibre	Polyester/linen
Modal fibre/acrylic fibre	**DD blends**
Polynosic fibre/acrylic fibre	Cellulose acetate/triacetate
Cotton/modacrylic fibre	Cellulose acetate/polyester
Viscose/modacrylic fibre	Cellulose triacetate/polyester
	Normal/deep-dye polyester

All these blends may not be important, and hence we will deal with the most common blends. Wherever details of dyeing method is not given one may have to refer the dyeing of the pure fibre materials given in the book and Dyeing of Textile Substrates I – Cotton by the same author.

11.8   Types of shades dyed on blended fabric

Solid shades: Most common requirement is solid shades. Both the components are dyed in the same shade – depth, hue and tone. The shade should look like a shade dyed on a single component substrate. Fibres with different affinity for the same class of dye it may not be difficult but when each component of the blend is dyed with different class of dyes they have to be selected in such a way as that in both blends the dyeing appears the same. According to tinctorial value of the dyes on that fibre the % of dye requirement may be different.

Reserve shades: These are shades where only one component is dyed and the other component is kept reserved or left undyed. This type of dyeing is possible where one class of dyes used for one component does not have affinity for the other fibre (component). The result is a mélange effect with one component appears white. Usually the component is dyed where the used dyes will have the least staining on the other fibrein most of the dyeing there will be some sort of tinting takes place which is reduced by using a suitable reserving agent, or giving and after scouring or a reduction clear to remove the tinted colour on the reserved component. The resultant white effect of the shade will largely depend on the % of that component and also the extent of tinting.

Cross shade: When each component of the blend is dyed different shades, the resultant shade is also a mélange with mixed shade effect. Here the dyeing has to be done (definitely with different class of dye) with highly efficient reserving agent so that the other component is not tinted while dyeing second component. This is necessary because the removal of tinted colour may not be possible. In difficult cases the first component is dyed and given a reduction clear and next dyed the second component which does not or least tints the first component. Normally contrasting colours are selected (orange-black, orange-green, navy blue-yellow, etc.) for either components. One should note that a little tinting on class of dye on the other component will appear as a large shade change in this case.

Shadow effect: in this case both the components of the blend are dyed but the depth of shade of one component shade will be a percentage of the other. Thus the component which is dyed a lighter shade will look like a shadow of the other shade (component). Both the component can be dyed with the same class or different class but the shades should be dyed with same hue but at varying depth. E.g., dyeing of polyester–wool blend with disperse–acid dye system for say, 3% and 1% shades with same blue hue. Shadow shades can be easily produced in blends where the one class of dye having different affinity for the component fibres in the blend. They can be dyed for short time and shades at different depths are produced on these fibres due to their differential receptivity. However, prolonged dyeing may develops a solid shade due to entering of more dye in both the components with increase in time of dyeing thus making practically no difference in depth on component fibres.

11.9   Dyeing methods

The blends are dyed by different methods

- Single bath single step,
- Single bath two stage,
- Single bath two step,
- Two bath two stage,
- Continuous and semi continuous.

Single bath single step will be the most economical process where the water consumption, utility and process time, etc. are reduced drastically compared to two baths two stage processes. But the process is selected as per the individual components of the blends, their dyeing and tinting of each other while dyeing the other component, the dyeing conditions and how it affects the other component, machine availability, compatibility of the dyeing chemicals, fixation pH of each class of dye applied, etc. This will be understood when different dyeing method for each blends are explained.

While following exhaust methods it should be noted that the percentage of shade is usually expressed on the total weight of the blend. But the MLR of exhaustion happens as per the weight of that component which is being dyed and not as per the total weight of the fabric or blend. Recipes are expressed on the total weight of the blend. This is more relevant where the percentage of that component is low. In low percentage component the percentage of dye goes very high as the MLR will be very high when calculated on the basis of the weight of that component.

Chapter 12
Dyeing of AA blends

12.1 Dyeing of nylon/wool

Polyamide fibres are often blended into wool to strengthen its mechanical properties without losing the typical wool character. The main end uses of this blend are woollen fabrics for coats, knitting yarns for socks, carpet yarns. The main blend ratio WO/PA is 80/20.

12.1.1 Pretreatment

The nature of the scouring and processing of wool material containing nylon is determined by the type of goods under treatment and the final effect desired. Due consideration should be paid to the special physical and chemical properties of the polyamide fibre in the individual processes. Singeing, for example, should be avoided if possible since, on account of the relatively low melting point of nylon (250 °C), this process gives rise to the formation of melted fibre ends on the surface of the fabric. When the goods are dyed, these melted ends take on a deeper colour, especially in pale shades. Special care should be taken during scouring. Owing to the factthat nylon has the power to dissolve fats, and since these fats are extremely difficult to remove, the usual scouring is insufficient in many cases if the goods are badly soiled. In such cases it is recommended scouring for 1 h at 50–60 °C with 2% sodium pyrophosphate and 1 g/l.

12.1.1.1 *Bleaching of nylon/wool*

Bleaching this blend is little risky as the bleaching process used for nylon may risk wool and vice versa. The best way is to use peroxide bleach controlling the process where the nylon content is low (25% and below), even though it may affect the nylon but it may not be appreciable in the total blend. Acid bleach is proposed where the percentage of nylon is higher.

Quantity	Unit	Additions
12–15	ml/l	Hydrogen peroxide 35%
2	g/l	Tetra sodium pyrophosphate
1	g/l	EDTA (30%)
0.25	g/l	Protective agent

Treat for 45–60 min at 60–65 °C and wash well.

12.1.2   The behaviour of nylon in dyeing

Two main factors govern the dyeing of wool/polyamide fibre mixtures with water-soluble dyes:

1. The rate and degree of absorption of the polyamide fibre compared with that of wool.
2. The behaviour of individual acid dyeing dyes towards the polyamide fibre.

In practical dyeing, general observation is:

In pale to medium shade depth dyeing polyamide is dyed generally deeper than wool. Hence the dye uptake on polyamide has to be retarded by a blocking/retarding agent (e.g., Erionol RF –Ciba).

In medium to deep shade depth dyeing often tone-in-tone dyeing takes place.

And in case of deep shade depth dyeing wool is often dyed darker than polyamide. In such dyeings polyamide part has to be covered up with dyes having high affinity for polyamide.

The 1:1 metal-complex dyes have low substantivity for nylon and the critical depth on nylon/wool is relatively low compared with monosulphonated acid dyes and 1:2 metal-complexes. Dyes of the 1:1 metal-complex type and the levelling acid disulphonates are dyed at pH 2–3 with formic or sulphuric acid and salt at the boil. Monosulphonated acid dyes and pre metallised 1:2 types dye nylon more readily, so they are applied to nylon/wool at pH 5–6 with ammonium acetate and acetic acid. A near-neutral pH is necessary to control levelness in pale depths, but more acetic acid can be used for full depths.

Besides the dye selection based on end use requirements like wet and light fastness properties.

Each individual dye has a specific affinity to wool and polyamide. This means that some dyes dye wool darker and some dye polyamide darker.

For dye combinations it is very important that the individual dyes distribute similar between wool and polyamide and they react similar to the same amount of blocking agent to give good solidity. Hence before bulk dyeing, lab trials have to be done for fixing the recipe.

PA 6 has a higher dye affinity compared to PA 66 and needs therefore more blocking agent to achieve solid shades in wool/polyamide blends. Besides this differences based on chemical composition, the fibre fineness (detox), fibre surface/cross section (round, trilobal, etc.), degree of delustering, texturing and pretreatment (heat setting) influences the colour depth on polyamide. Wool quality like provenience, fibre diameter, pretreatment, influences the uptake and shade depth on wool. If the wool has been treated by a shrink-resist process the dye affinity is increased appreciably. In most of these cases wool is then dyed darker compared to polyamide.

Blocking agents consist of colourless anionic compounds having high affinity to polyamide. They saturate the polyamide fibre and reduce therefore dye affinity to the fibre and the dye is pushed more to the wool. The amount of blocking agent has to be adjusted so that an even distribution can be achieved between the fibres. The necessary amount of blocking agent depends on the used dyes, dye concentration, fibre type and blend ratio and has to be determined in preliminary lab trials simulating as close as possible bulk conditions.

It has been observed that, at the boil, most dyes exhaust more rapidly on to the polyamide fibre than on to the wool, although the saturation value sets a limit to the depth of shade obtainable. In most cases the dye shows very little tendency to migrate once it is on the polyamide fibre. Thus, it is to be expected that, in pale to medium shades, a marked difference in depth will be observed between the polyamide fibre and the wool.

In full shades, however, the polyamide fibre is, almost without exception, paler than the wool. It thus follows that knowledge of the saturation value is of great assistance when dyeing wool/polyamide fibre mixtures. When one considers these two main factors, plus the fact that the various qualities of polyamide fibre show widely differing rates of absorption, it is evident that the method of dyeing to be adopted depends to a great extent on the type of mixture to be dyed, and also that the range of suitable dyestuffs is limited.

12.1.3   Dyeing with 1:1 metal complex dyes

Metal complex dyes are the best bet for dyeing wool/polyamide, to achieve shades of all round fastness. The dyeing properties of all such dyes are very similar, both on wool and nylon, and for this reason they are to a great extent miscible one with another. As we have already emphasised, where mixtures of wool and nylon are concerned, the saturation figure has to be taken into consideration, especially if the percentage of nylon is high. In other words, there is a fixed limit for deep shades. If this limit is exceeded, discrepancies in shade between the two fibres will occur, or at the very least, the nylon will be dyed paler. In pale shade there is no problem.

There are modified 1:1 metal complex dyes with excellent levelling properties. Dyeing at the wool protecting pH of 3.5 with special levelling agent. In pale to medium shades the PA component is dyed darker than the wool component. Adding a retarding agent (Blocking agent) can solve this problem (e.g., Erional RF –Ciba). In dark shades the wool component is dyed darker than the PA component. The PA component can be covered with selected metal complex dyes.

Dyeing method

The dye bath is set with the previously dissolved dyestuff, 5–6% sulphuric acid 168 ° Tw (60 °Be) and 2–4% levelling agent. The goods are entered at 50 °C., the bath is brought to the boil within ½ h, and dyeing continued at this temperature for 1½ h. The metal complex

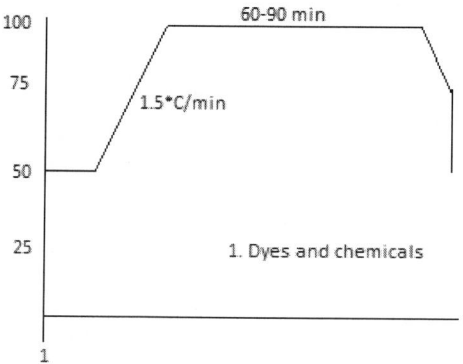

(1:1) dyes show the greatest affinity for nylon when applied at the boil so that loss of heat during dyeing should be prevented. For this reason dyeing is best carried out in closed vessels, thus giving the added advantages of a minimum boiling period and a considerable saving of steam. It has been found that quantities of acid used in this process have hardly any detrimental effect on the polyamide fibre, provided that boiling is not continued for too long; 2½–3 h may be considered as the limit. However, it is advisable to add a little sodium acetate to the last rinsing bath in order to neutralise the mineral acid fixed in the material.

12.1.4   Dyeing method with metal-free levelling acid dyes

Metal-free and brilliant acid dyes for dyeing can be used for wool and WO/ PA in hank and piece form dyeing at wool protecting pH value of 5.5 with a levelling agent. Good partition between WO and PA because all dyes need the same amount of blocking agent.

Dyeing diagram is given below:

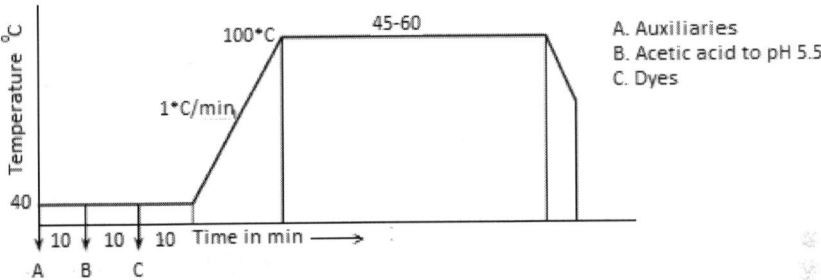

12.1.5   Dyeing with Lanaset (Ciba) dyes

Lanaset dyes are selected tinctorially strong, modified 1:2 metal complex, acid and reactive dyes with a high fastness level, dyeing at the wool protecting pH of 4.5 with a levelling agent.

The partition between WO and PA is given to a careful selection of dyes which needs the same amount of blocking agent.

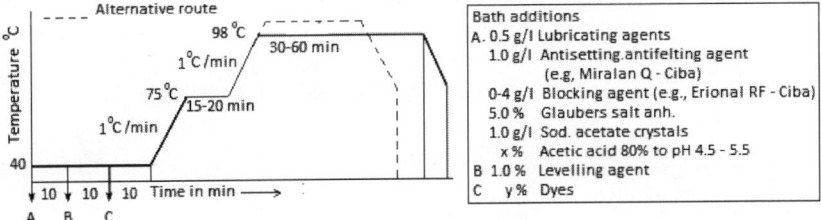

12.1.6   Dyeing with acid dyes

Acid dyes can be used for dyeing nylon and nylon 66, but the dye uptake on nylon is much lower when compared with wool or silk. This is due to the limited number of sites available for dye adsorption; nylon 66 contains only 0.036 equiv./kg of primary amino groups, compared with values for wool and silk of 0.82 and 0.15 equiv./kg, respectively. Thus it is difficult to achieve dark shades on nylon with acid dyes under normal dyeing conditions. But it has been found that the dye uptake increases below pH 4. Dye uptake probably goes higher than proportional to the available amino groups.

However, dyeing wool/nylon blends with acid milling dyes, in pale to medium shades there is preferential dye uptake by nylon compared to wool. Better solid shades can be achieved by using mixtures of monosulphonated and disulphonated dyes of similar hue, whereby due to the preferential uptake of monosulphonated dyes by nylon compared to polysulphonateddyes, exert a blocking effect. Mainly used monosulphated acid milling dyes used for this purpose are C.I. Acid Yellows 25, 29, 49, 172, 196 and 219; C.I. Acid Orange 145; C.I. Acid Reds 42, 57, 266, 361 and 396; C.I. Acid Blues 40, 62, 72, 78, 258 and 277; C.I. Acid Brown 248. One can use an appropriate amount of blocking agent, must also be added to the dyebath to control partitioning between the nylon and wool. When an acid levelling dyes are used, levelling agents are effective in preventing dye uptake by the nylon fibre in the early stages of dyeing, but as boiling continues there is a constant migration of dye from the wool to the nylon; after prolonged boiling times, a distribution is obtained which is little different from that obtained in the absence of blocking agent. Disulphonated acid levelling dyes can be used only for pale and medium shades on wool/nylon 66.

12.1.7   Dyeing with after chrome dyes

After chrome dyes also can be used for dyeing nylon/wool blends. However, there is some problem in achieving solid shades because wool is dyed more readily than the nylon with these dyes. Selected after chrome dyes can give acceptable solidity on wool/nylon. They are mainly monosulphonated monoazo type dyes and majority of them are orange, brown, red and black shades and blocking agents are not required for dyeing. Another way of obtaining balanced shade is to incorporate 1:2 metal complexes to get additional depth necessary for nylon. Other solutions are addition of a reducing agent (such as sodium thiosulphate) to the dyebath in the later stages of dyeing, the use of increased amounts of formic acid and the extension of the chroming time are techniques.

12.2   Dyeing of wool/silk blends

Wool/silk blends in woven or knitted form is employed to give the apparel a luxury character. Silk in the blend may vary from 5% to 50%. Wool/silk blends are dyed in yarn in packages or hank for and piece form. Silk is chemically related to wool since amino groups are an integral part of both the fibres, hence wool and silk can be dyed with the same dyes.

The main parameters affecting the dyeings are – type of silk fibres, degumming of silk, type of wool, fibre diameter, wool preparation, amount of electrolyte (Glauber's salt) used, pH of dyeing and dyeing temperature, etc. Dyeing at low temperature favours a deeper dyeing on silk. However, too low ofdyeing temperature has a negative influence on the wet fastness of wool. Best dyeing temperature for acid dyes is 90 °C.

12.2.1   Solid shades

Dyeing of wool/silk blends is normally carried out on yarn, and to a lesser extent on woven fabric, with solid shades being the most important. Degumming of the silk should be carried out prior to blending in order to avoid alkaline damage to the wool. Like wool, silk can be dyed using milling and 1:2 metal-complex dyes; both portions of a blend can therefore be dyed simultaneously in a single dyebath.

12.2.1.1   Dyeing with acid dyes

Wool/silk blends can be dyed in solid shades with acid dyes. Silk can be dyed by acid from a bath containing high quantities of acid at a low temperature, where as intense boiling andlow quantities of acid promote a deeper dyeing of the wool. Hence, the dyeing is done from a single bath set at room temperature and sufficiently acidified and the temperature is slowly increased to avoid any rushing of the dyes on to silk, to a temperature of 70–80 °C, to complete the exhaustion on silk and the further temperature is raised to 90 °C to complete the dyeing on wool. It is better to use a buffer like sodium acetate. The temperature is held at boil till the wool has picked up all the dyes.

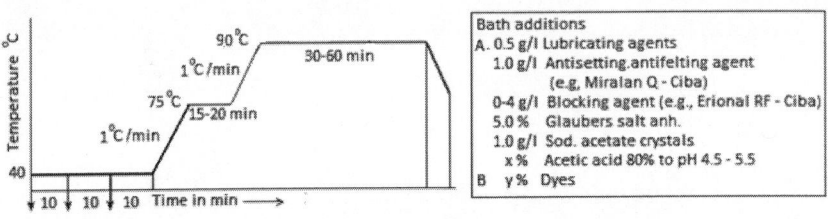

The pH of the dyebath is slightly different brands of metal complex dyes (Lanaset dyes –Ciba –pH =5 and temperature 95 °C; Sandolan MF – Clarient

– pH 4.5–5.5 temperature 70–80 °C; Sandolan Milling N-Clarient –pH 4.5–6.5 and temperature 80–85 °C; Lanasyn pH 4.5–6.5 and temperature 85–90 °C.

Another method of dyeing wool/silk blend using the acid dyestuff is to dye with about 3% acetic acid and 10% Glauber's salt, with the dyeing being carried out at 70 °C and further raising the temperature to boil. Since it is found that the lower temperature favours silkdyeing, after reducing the temperature, a slight amount of dye can be added for shading purpose.

In case of 1:2 metal complex dyes, the dyeing can be started at 40 °C with 15% Glauber's salt and 1% EDTA. The temperature is then raised to 95 °C (rate of heating 0.5 °C per minute) and dyeing is carried out for about an hour.

12.2.1.2   Dyeing with direct dyes

Both wool and silk gets dyed with direct dyes but wool gets dyed deeper shade at a boil. Hence dyeing is done at boil and then the temperature is reduced to 60 °C and extra dyes are added so that the silk fibre picks up the dye equivalent to that of wool.

12.2.2   Dyeing with reactive (MCT)/wool reactive dyes

Reactive/metal complex dyes can achieve better wet fastness and brighter shades. Wool/silk blend can be dyed by a two bath system where by reactive (Monofluorotrazine, MFT) dyeing is done in the first bath for dyeing silk and in the second bath wool is dyed with metal complex dyes. The MFT dyes are dyed a pH of 8.5 and 60 °C temperature in the presence of 20–80 g/l Glauber's salt. After dyeing the bath is neutralised and rinsed. A soaping is done to remove the unfixed and hydrolysed dyes on wool and then the wool portion is dyed with Lanasol dyes (metal complex dyes) as per the process given below (or refer under Wool dyeing–1.1.6.6).

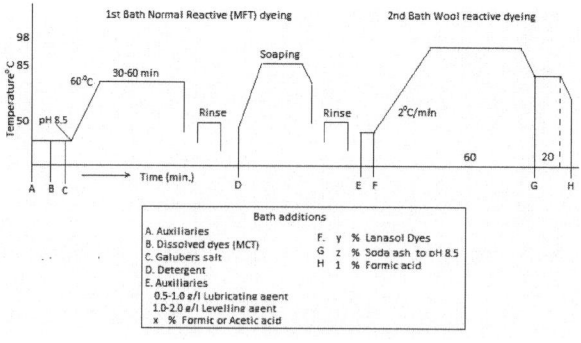

When an FCP dyes (Drimalan – Sandoz) is used, a slightly acidic or neutral dyebath is used. For example (1) It can be dyed at pH 7 with 20–60 g/l Glauber's salt for exhaustion and dyeing temperature of 90 °C, (2) or at pH 4.5–5.5 with 5 g/l Glauber's salt at 70 °C. The dyeing of cotton reactive dyes should be done at a milder alkaline and temperature conditions since there are chances chafe marks in piece dyeing (caused by mechanical damage to the silk fibres at higher temperatures.

12.2.2.1   Dyeing with solubilised vat dyes

Solubilised vat dyes, which are not very common these days due to its exorbitant price, can be used principally for dyeing wool/silk blend in single dyeing as it dyes both the fibre. Since it can be developed in the acid or neutral medium, contrary to vat dyes, it can be used safely for dyeing wool/ silk. They give attractive bright shades and mostly dyed in pastel to light shades and to some extent medium shades only.

Dyeing is carried out in acidic dye bath (for protein fibres); exhaustion is controlled by manipulating temperature and salt (if necessary). Conversion to parent vat dye is done in a strong oxidation bath, usually by nitrite system ($NaNO_2 + H_2SO_4$) but in case of protein fibres it is usual to carry out by chlorate system ($Na_2Cr_2O_7 + NaClO_3$). Development takes place very fast (in minutes) in case of continuous dyeing, but in exhaust dyeing (e.g. Jigs) it can be done in one pass. (See dyeing profile below)

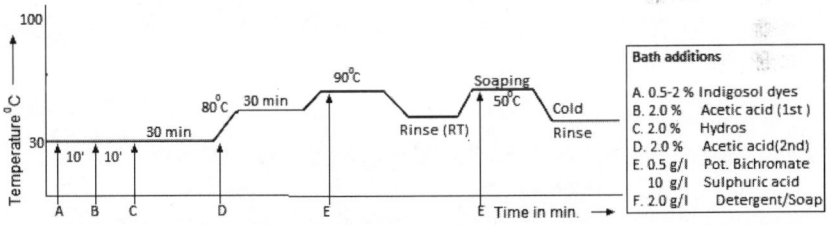

Dyeing profile for Solubilised vat dyes on Wool/Silk Blends

In continuous dyeing padded material (drying – optional) is developed in chlorate system($Na_2Cr_2O_7 + NaClO_3 + H_2SO_4$) oxidation bathsystems.

12.2.3   Contrast shades

First wool is dyed with acid or chrome (mordant) dyes and the silk is cross-dyed with suitable cationic (low rub fastness) acid or direct dyes.

Chapter 13
Dyeing of AB Blends

13.1   Dyeing of wool/acrylic, silk/acrylic blends

Many properties of wool and acrylic are similar and in fact the acrylic fibre was designed to replace wool in many requirements. But a wool/acrylic blend where the percentage of acrylic is above 50% has the dimensional stability superior to those of all-wool fabrics. It is more suited for garments like woven or knitted sweaters, skirts and outerwear where the thermal insulation is very important.

When dyeing acrylic fibre/wool blends, the acrylic component can be dyed with cationic dyes which largely, or entirely, reserve the wool fibres. On the other hand, the 1:1 and 1:2 metal complex dyes and also acid dyes used for the wool component leave the acrylic fibres almost white. Among the wool dyes mentioned, the 1:1 metal complex dyes, occupy a special position. The dyes of this group contain not only one or two anionic groups, but also a positively charged metal atom which depending on the pH of the dye bath, can render the dye cationic. Because of their amphoteric nature, the 1:1 metal complex dyes are more likely than either the acid or 1:2 metal complex dyes to be combinable with acrylic dyes in one bath without producing precipitation.

13.1.1   Scouring

For blends pure acrylic fibre and mass dyed fibres are used. In pretreatment of the blends which usually adopts the wool route as the acrylic fibres does not need much preparation except the removal of any spinning oil. Here we have to take care of the dyed acrylic. During the carbonisation with sulphuric acid process to remove vegetable matter from wool only some of the acrylic colours may change, but in almost all cases a neutralisation treatment with ammonia at 40 °C is able to get back the original colour without any problem.

Wool/acrylic blend can be scoured using a non-ionic detergent in acidic medium at pH 5 and a temperature of 60 °C.

Quantity	Unit	Bath additions
0.5–1	g/l	Non-ionic detergent
0.5–1	g/l	Acetic acid 80%

20–30 min at 60 °C. Rinse thoroughly.

Notes:

1. Blended yarns containing un relaxed high bulk acrylic fibre should be completely relaxed before being dyed, either by steaming in an autoclave at 107 °C for 10 min after preliminary evacuation of the steamer, or by immersion in boiling water for 5 min.

2. Acrylic/wool blended fibre fabrics may be prepared according to established practice for wool, except that precautions are taken to allow for the thermoplastic properties for the acrylic fibres. Relaxation is done to remove the inherent strain.

3. Worsted fabrics are crabbed and then scoured in either a dolly at temperature not exceeding 40 °C or in winch.

13.1.2   Bleaching

For medium colour or dark colour, it may not be necessary to bleach, but, when high whiteness is required in the mixed-spun and mixed-woven products, it is better to adopt 2-bath method in which the fluorescent bleaching is to be applied to acrylic and, after that, then to wool.

Quantity	Unit	Bath additions
0.5	%	Suitable OWA for PAN
		pH 3, MLR 1:10/40 92 °C, 1 h

13.1.3   Full bleach

Quantity	Unit	Bath additions
0.25–0.5	%	Suitable OWA for wool
1	%	Hydrogen peroxide
2	g/l	$Na_2P_2O_7 \cdot 10H_2O$
		NH_4OH to pH 8.8–9

For a better whiteness a third bath of reductive bleach can be proposed. Treat with sodium hydrosulfite 2 g/l solution 50 °C for 60 min.

13.1.4   Dyeing

13.1.4.1 Dyebath additives

13.1.4.1.1   Retarding agent for cationic dye

Cationic dye sticks fast to wool at low temperature, and then moves to acrylic as the temperature rises high. Wool acts as a retarding agent for cationic dye. Cationic dye is also retarded by the mutual action of acid dye and cationic dye, and therefore, it is not necessary to especially use any retarding agent in acrylic/wool mixed spun product. But when the percentage of acrylic is higher, both the volume of wool which acts as a retarding agent and that of acid dye become less,and, therefore it may sometimes require the retarding agent for cationic dye. As the retarding agent of cationic family has a defect which increases acrylic stains of acid dye, it is suggested touse the retarding agent of anionic family after duly selecting the dye.

13.1.4.1.2   Suspending agents

When cationic dye and acid dye are used at the same time in the same bath, there will occur the precipitations through their action against each other as the electric loads of both dyes are in opposition, and, as a result the dyeing will become impossible. There are suspending agents which shows a proper retarding effect on the acid dyes and, in addition, it has a preventing effect for acrylic fibre stains. These suspending agents are the best in one bath dyeing involving cationic and acid dyes.

13.1.4.1.3   Retarding agents for acid dyes

1:2 Metal-containing dye and milling dye are to dye with weak acidity, but when performing one bath dyeing method by using cationic dye, it should be dyed at pH 4–5. As a result, the dyeing speed becomes fast and the retarding agent becomes necessary.

Acid dye and cationic dye retard each other, and therefore they may dye uniformly even without adding any special retarding agent.

13.1.4.2 One bath two stage dyeing (Acid and cationic dyes combination)

Medium and to dark shades are dyed by a two sage one bath dyeing the cationic dyes are applied at first at boil. The bath is cooled to 60 °C the acid dyes are added and again raised to boil to complete wool dyeing.

13.1.4.3   The one-bath one-step dyeing of acrylic/wool (1:1 Metal complex dye and acrylic dye)

This process is based on the use of 1:1 metal complex dyes for dyeing the wool component. It can be used for pale to deep shades, though the one-bath two-step process is to be preferred for deep shades. Apart from 1:1 metal complex dyes, selected acid dyes can also be used for the one-bath method. These latter dyes are readily soluble at pH 2.4–2.8, they dye the wool uniformly, and do not precipitate with cationic dyes.

When using 1:1 type metal-containing dye in the one-bath dyeing method, the colour will be changed if pH becomes high, so cationic dye, too, should be used at pH 2. Accordingly, it is important to use a cationic dye which is not discoloured at pH 2.

The dye bath is set at 60 °C (140 °F) with 4% sulphuric acid, referred to the wool component, together with

Quantity	Unit	Bath additions
0.4	g/l	Sulphuric acid to pH 2.4–2.8
1–2	g/l	Levelling agent

After a short preliminary run, the well-dissolved 1:1 metal complex dyes or acid dye is added, and the dye bath is heated at a rate of 1–2 °C/min to T60, where upon the machine is run for about 10 min at this temperature. A concentrated solution of the acrylic dye is now rapidly introduced, and is allowed to raise to boil exhaust in 60 min at T60–90–95 °C.

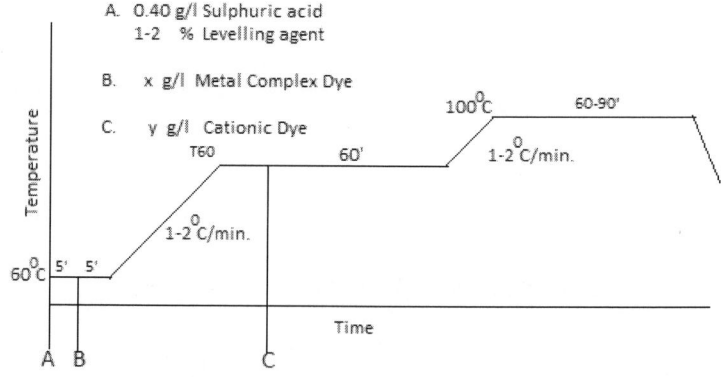

The one-bath one-step dyeing of acrylic/wool yarn with acrylic and 1:1 metal complex dyes

A, Sulphuric acid;B, 1:1 Metal complex dyes;C, Acrylic dyes

The liquor is now heated at a rate of 1–2 °C/min to the boil, where it is maintained for 1–1½ h. The remainder of the treatment proceeds in the conventional manner.

Variant

If dyeing is carried out on package dyeing machines where, for mechanical reasons, the dissolved acrylic dye cannot be added at near the boil, a variant of the foregoing procedure is necessary. The bath, set with sulphuric acid and acrylic dye at 60 °C, is first of all rapidly heated to 70 °C, the dissolved 1:1 metal complex dye is added, the machine is allowed to run for a short time, and the bath then heated at 1–2 °C/min to 80–85 °C the temperature is raised slowly at 1/3–1/2 °C/min to 90 °C. The acrylic dye is allowed

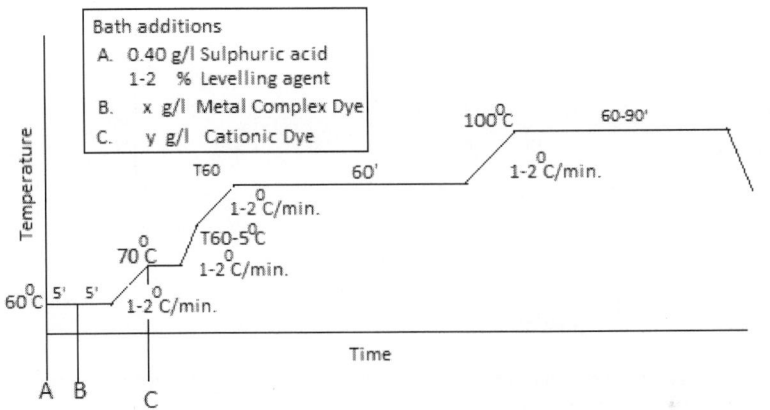

A variant of the one-bath one-step dyeing of acrylic/wool yarn with acrylic and 1:1 Metal Complex dyes

to exhaust at this temperature and dyeing then continues as described earlier.

Notes:

1. When using mixtures of anionic dyes (milling, 1:2 metal-complex or reactive) and cationic dyes, the degree of cross staining of acrylic fibres (of the types which are generally used in blends with wool) by wool dyes is very slight, and is not a problem in blend dyeing. Since wool contains carboxylic acid residues which are ionised under the mildly acidic conditions used when dyeing wool/acrylic blends, sites are available for the adsorption of cationic dyes.

2. Resisting of the wool by cationic dyes depends on the dye, dyeing time and temperature, concentration of electrolyte, pH and auxiliaries used.

3. Selection of cationic dyes for one bath dyeing should be based on the following factors –good solubility, minimal staining of the wool, medium affinity for the acrylic fibre, good diffusibility, stability over a wide pH range, compatibility with other cationic dyes in mixture shades, good fastness to light and water, and adequate thermal stability (for goods which are to be decatised).

4. Most suitable cationic dyes in three colour combinations for the dyeing of wool/acrylics blends suggested are – (1) Dyes with K values of 3: C.I. Basic Yellow 28; C.I. Basic Red 46; C.I. Basic Blue 41. (2) Dyes with K values of 5: C.I. Basic Oranges 29 and 43; C.I. Basic Red 22; C.I. Basic Blue 22.

13.1.4.4   The one-bath two-step dyeing of acrylic/wool

This process should be employed for acrylic dyeing deep shades with acrylic dyes and 1:1 metal complex or selected acid dyes. The two-step procedure is also suitable for dyeing all shades with acrylic dyes and acid or 1:2 metal complex dyes that are applied from a weakly acid bath.

On dyeing deep shades with acrylic dyes and 1:1 metal complex, or selected acid, dyes, the bath is set with sulphuric acid (to pH 2.4–2.8) and rapidly heated to around 90 °C. After running for about 10 min to eliminate any possible temperature differences, the concentrated acrylic dye solution is quickly fed in and dyeing proceeds for 60 min at 90 °C, whereupon the acrylic dye is almost completely taken up by the acrylic fibre. The dye bath is now cooled to 70 °C (158 °F), 1:1 metal complex dyes or selected acid dyes are added to the liquor. After a short preliminary run, the wool is dyed in the usual manner.

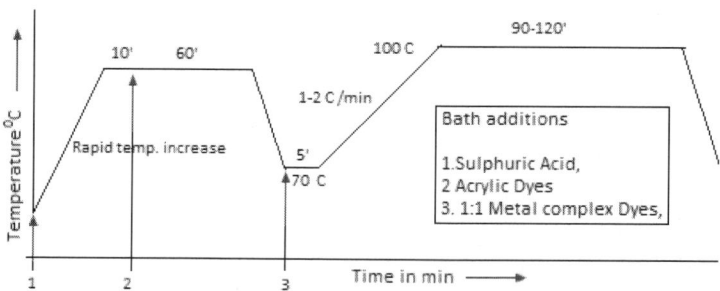

One-bath Two-step dyeing of acrylic/wool yarn with acrylic and 1:1 Metal Complex dyes

When 1:2 metal complex or acid dyes that are dyed from weakly acid liquor are used for dyeing the wool component, the following procedure is to be recommended: The dye bath is set with acetic acid to pH 5–5.5, heated to around 90 °C, and run at this temperature for10 min to eliminate any possible temperature differences in the material. The well-prepared, concentrated solution of acrylic dye is then rapidly introduced into the bath, and dyeing proceeds at 90 °C to exhaustion after 60 min. The dye bath is cooled down to 50 °C (122 °F).

At this temperature, the pH value of the liquor is checked and adjusted to 6–6.5, if necessary. The bath is now set with

1–2% Levelling and penetrating agent,

10% Glauber's salt anhydr., and

well dissolved acid dye. If 1:2 metal complex dyes are used, it is not necessary to have Glauber's salt in the liquor. The temperature of the bath is raised at 1–2 °C/min (1.8–3.6 °F) to

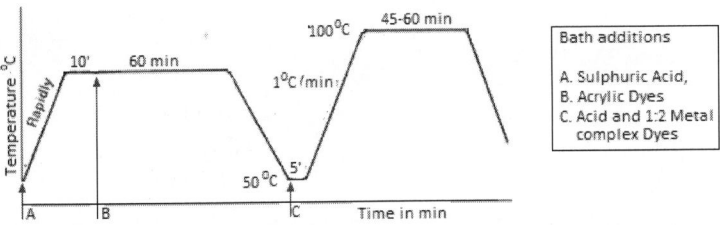

One-bath Two-step dyeing of acrylic/wool yarn with acrylic and 1:2 Metal Complex dyes

the boil, at which temperature the yarn is dyed for 45–60 min, after which the bath is cooled down and the yarn rinsed.

13.1.4.5   One bath method with cationic and acid dyes

Pale shades

For pale shades the dye bath is set at 50 °C and pH 4–5 with acetic acid, salt, an anionic retarder and an alkanolpolyoxyethylene anti-precipitant. The anionic and basic dyes are added separately and then the temperature is raised and held at 90 °C to slow down the rate of uptake of the basic dyes. Finally, the wool component is dyed to shade with the an ionic dyes at the boil.

Medium shades

Better compatibility is obtained in medium depths by starting the dyeing at 50 °C and pH 6–7 with salt, non-ionic anti-precipitant and neutral-dyeing anionic dyes. After rising slowly to 80 °C to promote level dyeing of the wool, formic/acetic acid (to give pH 3.5–4), an anionic retarder and the basic dyes are added and dyeing of the acrylic component completed at the boil.

Bath addition A

Quantity	Unit	Description
0.5	g/l	Penetrating accelerant and antifoam
1.5	%	Wool levelling agent

Bath addition B

Quantity	Unit	Description
10	%	Glauber's salt

Bath addition C

Quantity	Unit	Description
x	%	Acid dye

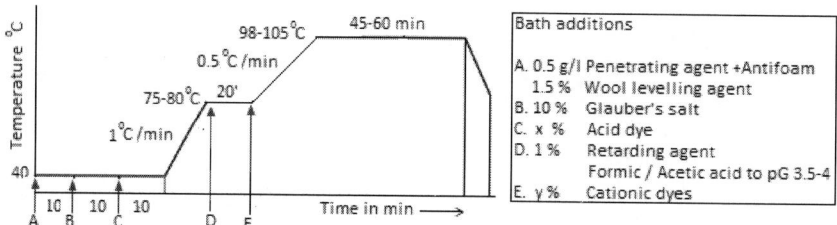

One Bath method with Cationic and Acid dyes

Bath addition D

Quantity	Unit	Description
1	%	Retarding agent
		Formic/acetic acid to pH 3.5–4

Bath addition E

Quantity	Unit	Description
y	%	Cationic dye

Rinses:
1. 10 min warm.
2. 10 min at 60 °C.
3. 10 min at 60 °C.
4. 10 min at 60 °C.
5. 10 min cold.

It is also possible to set the initial dyebath at 60 °C and pH 2–3 (sulphuric acid) with the basic dyes and complexing agent, and to raise the temperature to 80 °C before adding the pre metallised dyes. The pale shades can be dyed with reactive dyes in one bath method.

13.1.4.6 Variant – one bath method with cationic and chrome dyes

When using chrome dye, cool the dyebath down to 70 °C after dyeing and perform the chroming by adding $K_2Cr_2O_7$ up to 1/2 volume of the dye (less than 2.5% (owf) as shown in the dyeing method shown below.

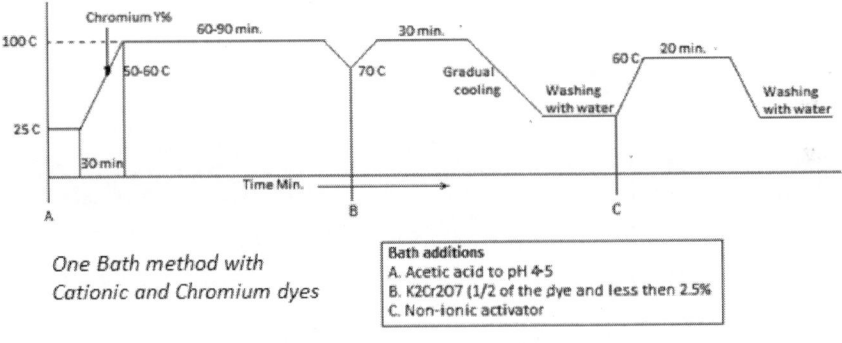

One Bath method with
Cationic and Chromium dyes

Bath additions
A. Acetic acid to pH 4-5
B. K2Cr2O7 (1/2 of the dye and less then 2.5%
C. Non-ionic activator

x% Cationic dye.

1–3% Retarding agent.

1% Sodium acetate.

13.1.4.7 One bath – two step method with cationic and acid dyes

This dyeing method is to be used for dyeing medium to dark shades and in which the dyeing is performed in the same bath by adding retarding and suspending agent and matching dyeafter dyeing of either of cationic dye or wool dye. According to this dyeing method, it is possible to dye even with the dyes which have so far been considered not suitable in-the-same bath and at-the-same-time dyeing as they produced precipitation during one-bath dyeing process due to high concentration of dyes in the bath. There are 2 ways in the one-bath 2-stage dyeing method, namely one is to dye PAN first (when using acid dye and 1: 1 type metal containing dye) and the other is to dye wool first (when using 1:2 metal complex dye and milling dye).

13.1.4.7.1   Method I

In this method dyeing is done similar to the dyeing 100% PAN. PAN is to be dyed with cationic dye for 30 min, and after the cationic dye is absorbed, the bath is cooled to 70 °C and added 2–3% of the suspending agent and the acid dye (x%). The dyeing of acid dye is completed by raising the temperature to boil and dyed for further 60 min.

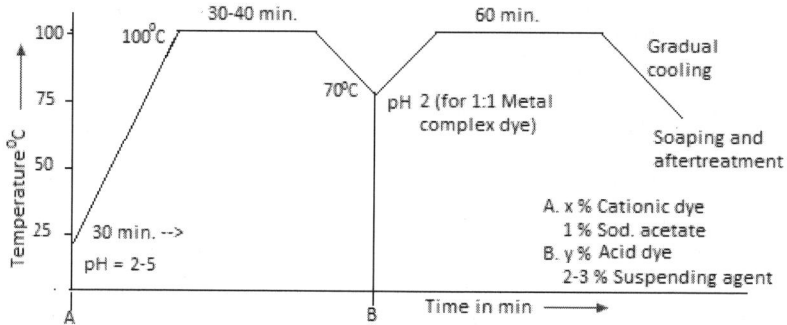

One Bath – Two Step method with Cationic and Acid dyes Method I

Method I

13.1.4.7.2   Method II

In Method II wool is dyed with 1:2 metal complex dye or milling dye for 30 min in neutral or weak acid in accordance with the ordinary method, and further the bath is cooled to 70 °C and pH is adjusted to 5 using acetic acid. Then, add 2–3% suspending agent and cationic dye (x%).

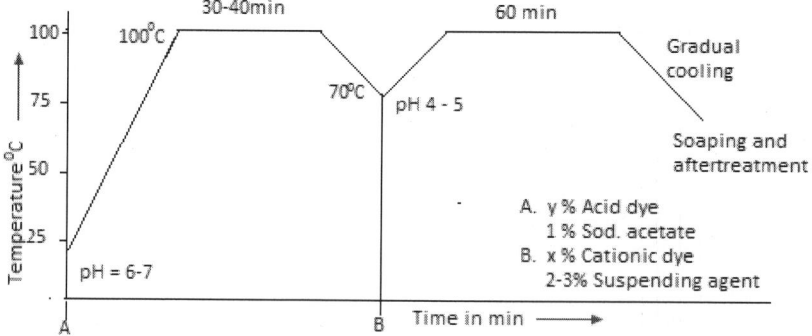

One Bath – Two Step method with Cationic and Acid dyes Method II

The temperature is raised to boil and continued the dyeing for another 60 min. According to this method, it is possible to safely perform neutral dyeing and acid dyeing without causing unevenness.

13.1.4.7.3   Method III

When chrome dye is used, Method III is used. The cationic dye is done as usual and chroming should be applied after the cationic dyeing. The bath is cooled at 70 °C and the suspending agent

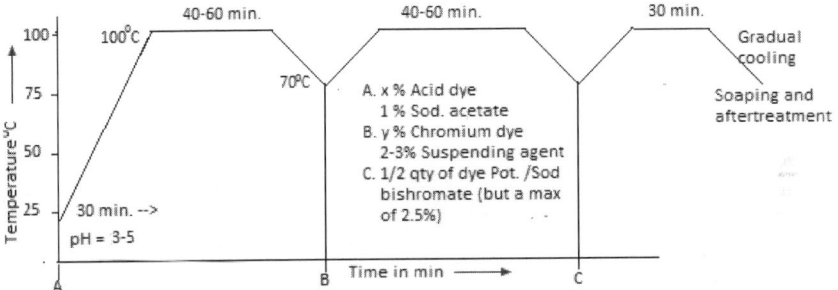

One Bath – Two Step method with Cationic and Acid dyes Method III

and the chromium dye is added. Raise the temperature to boil and dye for 40–60 min and again the bath is cooled to 70 °C. Add the sodium bichromate and run the material in this bath for 5 min and raise the temperature to boil. Chroming is continued for another 30 min cool the bath slowly and drop the bath. Soaping and after treatment is continued as usual.

13.1.4.7.4   Two-bath dyeing method

The two-bath dyeing method is used when the colour is extremely dark and the bath ratio is small.

In this dyeing method, PAN is to be dyed first (if circumstances require, after removing the stains by performing the reducing decolourisation in another bath) and, then, wool is to be dyed in the new dyeing bath. If wool is dyed first, the acid dye for wool will drop into the acrylic dyeing colour, as the former is poor in potting, and so this should be avoided. Therefore, according to the 2-bath dyeing method, all the dyes to be used in dyeing 100% pure material may be used in the process. Thus, this dyeing method is effective in cases where inter colour staining are apt to occur or precipitations are easily produced.

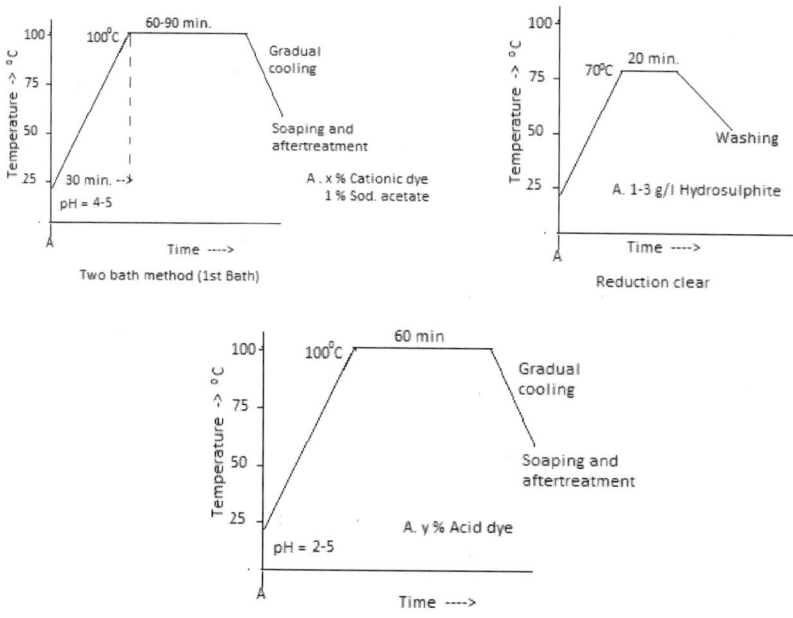

Two bath method (1st Bath)

Reduction clear

Two bath method (2nd Bath)

Guideline recipes
First bath

Quantity	Unit	Dyebath additions
x	%	Cationic dye
1	%	Sodium acetate
		MLR 1:20
		pH 4
		Dye at 100 °C, 60–90 min

Reduction bath (in case of heavy staining on wool)

Quantity	Unit	Dyebath additions
3	g/l	Sodiumhydrosulphite
		70 °C, 20 min

Second bath (dyeing of wool)

Quantity	Unit	Dye bath additions
x	%	Dye
		100 °C, 60 min

13.2   Dyeing of nylon/acrylic blends

Main garments made with polyamide/acrylic wind jackets, ski and bathing pants, polyamide fibres. Acrylic fibres are suitable as component fibre in this blend by virtue of their woolly textile nature and also their tensile strength and their abrasion resistance which are better than those of the natural fibres.

The blends made with 20–30% nylon are used mainly for half-hose, knitted sweaters, sportswear and swim wear, blankets, furnishing fabrics and floor coverings.

13.2.1   Pretreatment

To remove all spin finishes and contamination, the fibrous material is treated with

Quantity	Unit	Additions
0.2–0.5	g/l	Detergent
0.25	ml/l	Ammonia 25%

for 30 min at 50–60 °C. The amounts of detergent required depend on both the degree of contamination and also the liquor ratio.

13.2.2   Full white

Since the bleaching agent used for acrylic (sodium chlorite) bleaches polyamide fibres also the bleaching process for polyamide/acrylic becomes very simple. The goods are entered into the cold bleaching bath which contains

1–2	g/l	Sodium chlorite 80%
1–1.5	g/l	Chlorite stabiliser
0.5–1	ml/l	Formic acid 85% (pH 3–3.5)

at a liquor ratio of 30:1to 50:1 or less. The liquor is heated within approx. 45 min to the boil and the goods bleached for about 1 h at the boil. In many cases, however, it is not necessary to optically brighten acrylic fibres because these fibres are supplied generally with a high degree of white.

Hence, polyamide only to be whitened one can use the optical whitener for polyamide in a long bath along with the bleaching bath or in case both component has to be whitened both components the optical whitener for both has to be added in the bleaching bath. Treatment at 70–80 °C.

13.2.3   Dyeing

13.2.3.1   Dyeing with cationic and acid dyes combination

13.2.3.1.1   One bath two step process

The polyamide portion is dyed first. Dye bath is prepared as follows:

Recipe

Quantity	Unit	Additions
x	%	Acid dye
1	%	Levelling agent for acid dye
0–20	%	Glauber's Salt anhydr.

at the pH prescribed for the acid dyes used. After heating the bath and exhausting the latter as recommended, the liquor is brought to a temperature of 75–80 °C and add

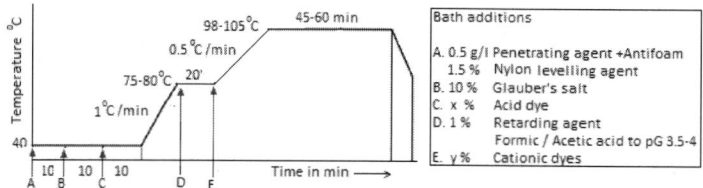

One Bath two-step method with Cationic and Acid dyes

Quantity	Unit	Additions
y	%	Cationic dye
z	%	Levelling agent for acrylic dye

13.2.3.1.2   Alternate method (1)

The acrylic fibre contained in the blend is then dyed by heating the dye bath slowly to the boil and maintaining this temperature until the bath is exhausted. The latter is then cooled slowly to about 60 °C and the goods rinsed.

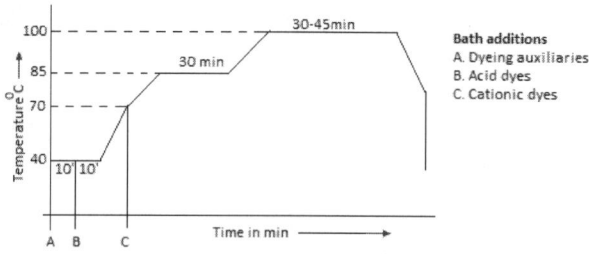

One Bath two-step method with Cationic and Acid dyes -Varient I

Dye bath is set at 40 °C with all the necessary auxiliaries Glauber's salt, acetic acid to a pH of 4–5. After the bath is homogenised and material is wetted out the acid dye is added. After 10 min running along with all auxiliaries and acid dye the temperature is raised 70 °C and the cationic dye is added to the bath. The temperature is again raised to 85 °C and held at this temperature about 30 min to complete the acid dyes exhaustion and again the temperature is raised to 100 °C, (boil) and held at this temperature for the exhaustion of the basic dyes for about 30–45 min. The bath is cooled to 70 °C and dropped. Any staining of nylon by basic dye may be cleared by a light hydros treatment at 70 °C at a pH of 4–5.

13.2.3.2 Dyeing with cationic and 1:1 metal complex dyes
13.2.3.2.1 One bath two-step process

Dye bath is set at 40 °C with auxiliaries and the pH is adjusted at pH 2–3 with sulphuric acid and then the cationic dye is added along with alkanolpolyoxyethylene sulphate complexing agent and an alkanolpolyoxyethylene anti-precipitant. The temperature is raised to 80 °C and the acid 1:1 metal complex dyes are added. The temperature is further raised to boil and the exhaustion is completed.

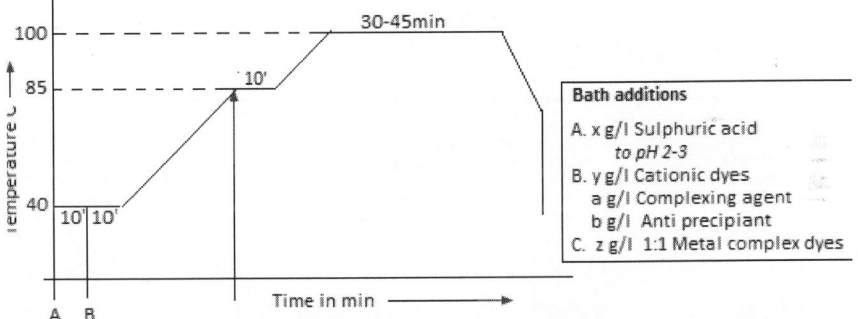

One Bath two-step method with Cationic and 1:1 Metal complex dyes

The one bath dyeing methods are mainly used for light to medium shades.

13.2.3.2.2 Two stage process

The acrylic fibre is first dyed with the basic dyes at the boil and pH 4–5. The dyebath is then cooled slowly to 70 °C and adjusted to pH 2–3 with sulphuric acid. The 1:1 metal-complex dyes and an alkylaminepolyoxyethylene complexing agent are added and the nylon component dyed to the required shade at the boil.

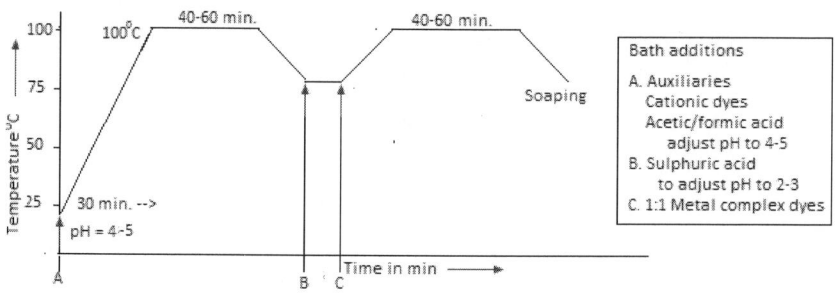

One Bath two-step method with Cationic and 1:1 Metal complex dyes

This combination can be dyed by one bath two step method and two bath method. If there is no chance of precipitation the first method can be applied otherwise two step process has to be followed.

13.2.3.3   One bath one step dyeing with 1:1 metal complex and cationic dyes – reverse method

In case of some cationic dyes cross dyeing may happen, but metal complex reserves the acrylic fibres completely in dyeing. Hence the selection of dyes is important. If dye for one component stains the other component controlling the final shades becomes more difficult.

Set the bath at about 60 °C with

Quantity	Unit	Additions
x	%	1:1 Metal complex dyes
0.5–1	ml/l	Formic acid 85% pH 3–4
1–2	%	Dispersing and levelling agent

Raise the temperature within 20–30 min to about 80 °C and add

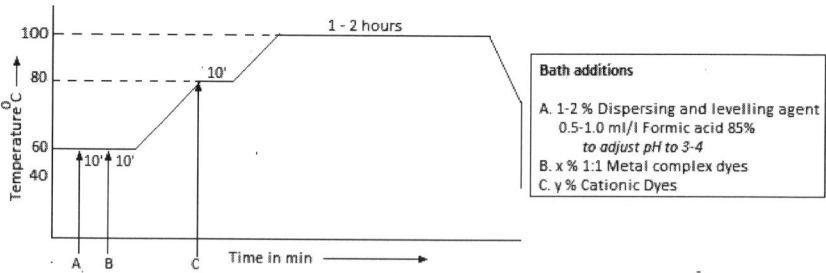

One bath one step dyeing with 1:1 Metal complex and cationic dyes – Reverse Method

Quantity	Unit	Additions
y	%	Cationic dye

After the temperature has been raised slowly in the range starting from 80 to 100 °C, the goods are dyed for 1–2 h at the boil, after which the bath is cooled slowly and the goods are rinsed.

13.2.3.4 One bath two step dyeing with 1:1 metal complex and cationic dyes – reverse method

In dyeing deep shades, it is advisable to pre dye the polyamide fibre with

Quantity	Unit	Additions
x	%	Metal complex dyes
0.5–1	ml/l	Formic acid 85% pH 3–4
1–2	%	Dispersing and levelling agent

Heat the dye bath slowly and uniformly to the boil and maintain this temperature until the dyes have been almost exhausted. After the bath has been cooled to approx. 80 °C, add

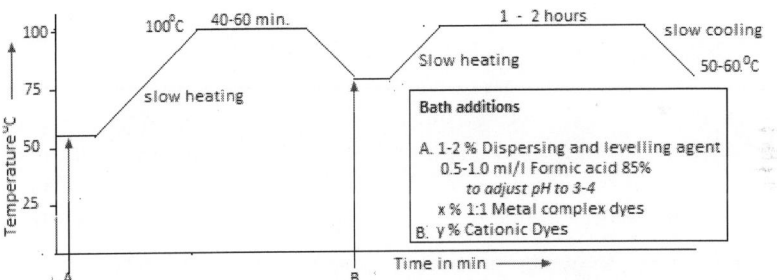

One Bath Two step Dyeing with 1:1 Metal complex and Cationic dyes - Reverse Method

Quantity	Unit	Additions
y	%	Cationic dye

heat the liquor slowly as described, to the boil and dye for 1–2 h at this temperature. Then cool the bath slowly to 50–60 °C and rinse the goods.

Notes:

1. 1:1 Metal complex dyes are amphoteric and therefore can be used only up to medium concentrations together with cationic dyes in the same bath without danger of precipitation. This results in the

interesting possibility of dyeing acrylic/polyamide fibre blends by a one-step method. Only in dyeing deep shades at short liquor ratios and also in dyeing blacks, it is advisable to use the one-bath two-step dyeing method in order to avoid the danger of precipitation.

2. Metal complex dyes have poorer wet fastness properties on polyamide fibres than on wool. To obtain good fastness properties it is therefore necessary only to use selected dyes for the polyamide fibres.

3. To obtain level shades on acrylic fibres, it is advisable to use levelling auxiliaries. The amounts required depend on the composition of the fibrous blend, the dyeing conditions and the depth of shade desired.

4. Polyamide fibres are destroyed hydrolytically by mineral acids. For this reason, it is advisable not to allow the pH of the liquor to drop below 3 and not to use sulphuric acid but formic acid to adjust the pH.

13.2.3.5 *Dyeing with cationic and 1:2 metal complex dyes (1)*

The disadvantages of acid dyes or 1:1 metal complex dyes for polyamide is that the dyeing will have only poor wet fastness and light fastness. If better fastness properties are required one should use 1:2 metal complex dyes. But these metal complex dyes are anionic and if it is used along with the cationic dyes the dyes will be precipitated. Hence there is no possibility of one bath dyeing in this combination of dyes.

In the two step dyeing process the polyamide may be dyed at pH 7–9 to get level dyeing and in the end the full exhaustion is effected by adding acetic acid to the bath. Further the acrylic dyes are dyed.

Procedure

Set the dye bath with

Quantity	Unit	Additions
x	%	1:2 Metal complex dyes
0.5–1	ml/l	Dispersing levelling agent
0.5–1	ml/l	Ammonia 25% to pH 7–9 or
0.5	g/l	Trisodium phosphate

commence dyeing at 30–40 °C, raise the temperature within 45–60 min to the boil and dye at this temperature for approximately 1 h. The exhaustion of the bath, which can be incomplete particularly in dyeing deep shades, can be improved by acidifying the dye bathslowly with acetic acid. The acetic acid should be added in small portions to the dyebath.

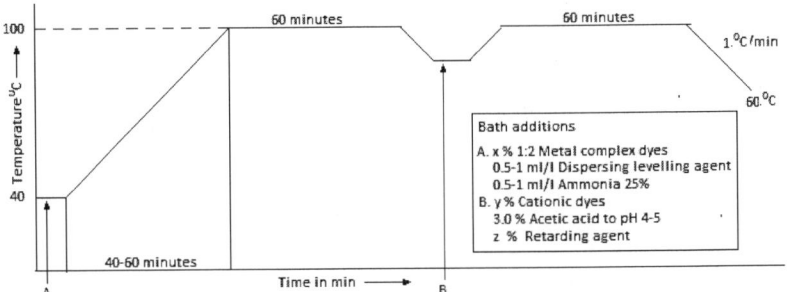

Dyeing with 1:2 Metal Complex and Cationic Dyes - One bath two step process

After cooling the almost completely exhausted dye bath to approx. 80 °C, add

Quantity	Unit	Additions
y	%	Cationic dye
3	%	Acetic acid 30 to pH 4–5
z	%	Retarding agent

heat slowly and dye for about 1 h at the boil. Then cool slowly by about 1 °C/min to approx. 60 °C and rinse the goods.

13.2.3.6   Dyeing with cationic and 1:2 metal complex dyes (2)

When dyeing dark shades the acrylic component may be dyed first. In this method once the cationic dyes are exhausted to acrylic fibres there is no longer any danger of precipitation. In further dyeing of polyamide has to be done in acid medium as the acrylic dye is susceptible to alkali.

Prepare the dye bath with

Quantity	Unit	Additions
x	%	Cationic dye
3	%	Acetic acid 30 to pH 4–5
z	%	Levelling agent

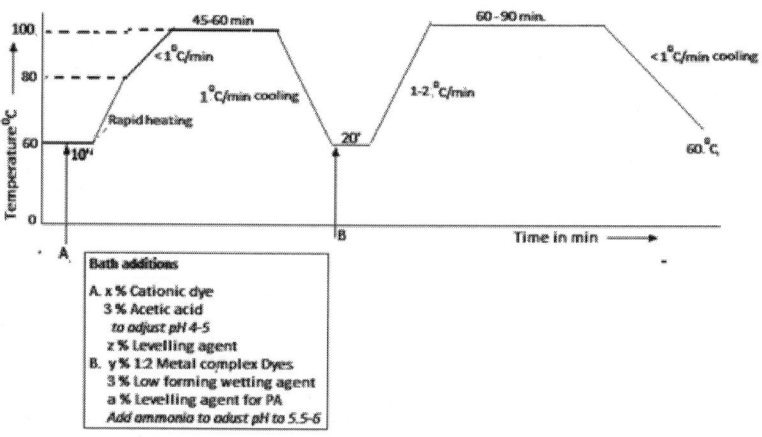

Dyeing with Cationic and 1:2 Metal Complex Dyes - One bath two step method (2)

heat rapidly to about 80 °C and then slowly, by not more than 1 °C/min, to 98 °C and dye until the dye bath is almost exhausted. Then cool to about 60 °C, set the bath to pH 5.5–6 by neutralising a part of the acetic acid with ammonia 25% or tri sodium phosphate and add

Quantity	Unit	Additions
y	%	1:2 Metal complex dye
3	%	Low foaming wetting agent
a	%	Levelling agent for polyamide

Heat the bath slowly and dye for 60–90 min at the boil. Then cool the bath slowly, by not more than 1 °C/min, to approx. 60 °C and rinse.

Notes:

1. When selecting of levelling agents for acrylic dyes care should be taken to avoid any reaction with anionic products in the metal complex colours as additives. Glauber's salt also should be avoided as it can cause precipitation.

2. In dyeing polyamide fibres in ammoniacal baths, it must be considered that ammonia not only improves the levelling properties and slows down the rate at which the dye goes onto the fibre, but also reduces the exhaustion of the bath. It is therefore often advisable to improve the exhaustion of the bath towards the end of the dyeing process by adding acetic acid.

3. It is often difficult to dye level shades on polyamide continuous filament yarns because differences in stretching or in the content of amino end groups, which are present in the fibre, are marked during dyeing. This marking of differences in the structure of polyamide fibres can be prevented neither by careful temperature control nor by adjusting the pH of the bath. Unevenness caused by differences in the substrate can be minimised only by dyeing in an acetic acid bath with suitable wetting and levelling agents.

13.2.3.7 Two bath dyeing methods

Since there is no danger of precipitation, two bathe dyeing are the most suitable and safest method of dyeing acrylic polyamide blends. One can use any cationic or metal complex dyes and there won't be any restriction in the use of levelling agents or other chemicals.

Acrylic dyes can be dyed either at 96–100 °C or at 100–106 °C.

1. Dyeing at 96–100 °C

Dye bath is prepared as

Quantity	Unit	Additions
x	%	Cationic dye
1	ml/l	Acetic acid 30% to pH 4–5 or
0.2–0.5	ml/l	Sulphuric acid 96% to pH 2.5 approx.
0.05–0.1	g/l	Potassiumbichromate (if necessary)
0–5	g/l	Glauber's salt
0–5	ml/l	Levelling agent

Heat the bath, containing the fibrous material, to approximately 60 °C and add the dyes and the auxiliaries or, alternatively, enter the goods at this temperature into the prepared dye bath and treat approximately for 5 min at this temperature. Then heat the bath relatively rapidly to 75–85 °C, depending on the fibre to be dyed. Raise the temperature from 75 or 85 °C to the boil uniformly and adequately slowly. To avoid temperature differences in the dye bath, it is advisable to heat the bath by no more than 1 °C every 2–4 min. The temperature before which the rate of heating is slowed down depends on the glass transition point of the fibre. This temperature is 75 °C with fibres which exhibit a high rate of dye uptake and 85 ° C with those of low rate of dye uptake. The heating rate should be adjusted to suit primarily the circulation conditions.

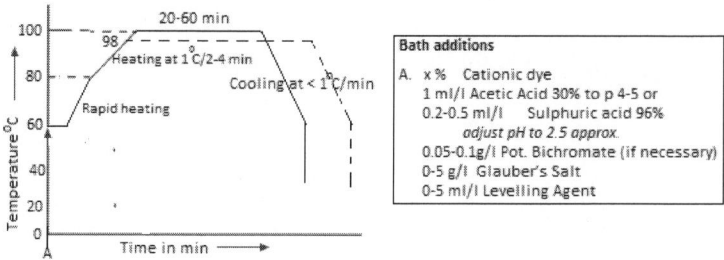

Dyeing of Acrylic fibre at boil in Acrylic/Polyamide blend

When a dyeing temperature of 100 °C is obtained, 20–60 min at the boil is adequate depending on the depth of shade to be dyed and the dyes selected. When a bath temperature of 100 °C is not obtained, it is necessary to prolong the dyeing time. After dyeing, the bath should be cooled by not more than 1 °C/min to 60 °C to ensure that the handle of the goods is not influenced adversely. Care should be taken that the fibrous material is notcooled suddenly by cold water fed into the dyeing vessel. After the liquor has cooled to 50–60 °C, it can be dropped and the material rinsed.

2. Dyeing at 100–106 °C

If the acrylic fibre is to be dyed at temperatures between 100 and 106 °C, it is necessary to prolong the heating phase in order to allow enough time to heat the bath to the higher dyeing temperature. On the other hand, the boiling phase can be cut down because uniform penetration of the acrylic fibre is obtained more rapidly at the higher temperature. Dyeing at temperatures above 100 °C, therefore, usually offers a saving in time which is particularly noticeable when dark shades or when fibres with a high denier are being dyed. It must be considered, however, that dyeing acrylic fibres under high temperature conditions is liable to result in deterioration of the handle and the nature of the goods. This applies in particular to dyeing at temperatures above 106 °C.

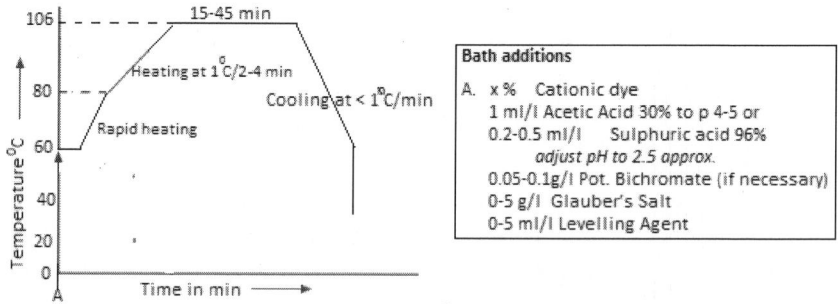

Dyeing of Acrylic fibre at 106°C in Acrylic/Polyamide blend

The second component, polyamide can be dyed by any of the usual method in a second bath described in this book.

13.2.3.8   Dyeing with cationic and disperse dyes

Disperse dyes may be used for dyeing polyamide fibres in blends with acrylic fibresonly if the dyed goods are not subjected to treatments which require good wet fastness properties. When used for dyeing polyamide fibres, disperse dyes not only produce far deeper shades but also a shade which differs from that on acrylic fibres. It is therefore impossible to produce solid shades on blends of acrylic and polyamide fibres with disperse dyes alone.

Acrylic fibres are dyeable only at temperatures above approx. 80 °C. Generally speaking, these fibres do not take up dyes below these temperatures. It is therefore possible to reserve the acrylic fibre almost completely by dyeing the polyamide fibre with disperse dyes at temperatures between 70 and 80 °C. This offers the possibility of producing solid shades and also two-colour shades on blends of acrylic fibres and polyamide fibres with cationic dyes and dispersedyes. The acrylic fibre is dyed at the boil first and subsequently the polyamide contained in the blend at 70–80 °C.If the preliminary trials have shown that precipitation occurs under the conditions of dyeing, the two-bath method must be used.

One bath two step method

Acrylic portion is dyed first. Set the dye bath with

Quantity	Unit	Additions
x	%	Cationic dye
3	%	Acetic acid 30% to pH 4–5
a	%	Levelling agent for cationic dyes

Dyeing is followed as per procedure given in two bath dyeing method above.

The dye bath is then cooled by not more than 1 °C/min to 60 °C and

Quantity	Unit	Additions
x	%	Dispersing dye
1	%	Levelling agent

added. The temperature is then raised to 70–80 °C and the goods dyed for about 1 h at this temperature. The bath should not be heated to a temperature above 800 °C to ensure that the disperse dyes do not go onto the acrylic fibre. The bath is then cooled slowly to about 60 °C and the goods rinsed.

chapter 14
Dyeing of AC blends

14.1 Dyeing of wool/cotton (wool/viscose, silk/ cotton, silk/viscose blends)

Union materials include blended slubbing and yarns, and mixture fabrics made from wool and natural or regenerated cellulose fibres. Apart from cotton, the group of cellulosic fibres includes filament and staple fibres of the viscose types. The yarn may be either spun from an intimate mixture of wool and vegetable fibres or it may be a twisted yarn, one thread being of wool and the other of cotton or cellulosic rayon. The character and the affinity of spun rayons for dyestuffs can vary according to their origin and the method by which they are produced. The dyeing and finishing processes for various fabrics made from these mixtures are very similar. Even though the wool/cotton (viscose) loses some desirable characteristics to some extent, such as handle, drape, pleat retention and crease recovery, but this blend is dearer due to its durability and dimensional stability. Typical examples of traditional union fabrics with a cotton warp and a woollen weft include blazer cloths, gabardine rainwear, shirtings and pyjamas. Blended worsted yarns containing approximately equal proportions of wool and cotton have been long-established in knitwear, dresswear, underwear, children's clothing, lightweight shirtings, pyjama cloths and blankets. Blends with wool percentage ranging from 20% to 70% are very common.

Intimate mixture blends are used for solid or sometimes double shades to give mélange effect. Metal complex (1:1 metal complex dyes are avoided as the sulphuric acid requirement for this dye may affect the strength of cellulosic fibres) and acid milling dyes and direct dyes are used frequently for this blend. High exhaust direst dyes are used to avoid the cross staining of wool even though it may not be a major problem in solid shades. Instead of direct dyes reactive dyes can be used for getting shades with higher fastness but here a two bath process has to be adopted in most cases other than some MCT-VS bifunctional dyes which exhibit a high degree of fixation on wool under acidic conditions and are particularly suitable for wool/cellulosic blends in two stage one bath dyeing techniques.

14.1.1   Pretreatment

Since protein is a weaker fibre in the blend, one has to follow the scouring process of wool/silk has to be followed. Since mercerising cannot be done for cotton, a better quality cotton has to be used for blending.

14.1.1.1   Scouring wool/cotton

Mixture yarns containing reclaimed wool generally have a high spinning oil content and must, therefore, be scoured before dyeing. Wool/cotton yarns, carpet yarns, etc. are scoured with:

1–2 g/l Soda ash and
1–2 g/l Wetting and scouring agent
at 40–50 °C.

Fine yarns made from wool/spun viscose should be scoured with scoring agent and ammonia, provided, of course, that scouring is sufficient.

The normal alkaline scouring conditions required to break down the primary wall of the cotton fibre prior to dyeing are too severe to be used for a blended yarn or fabric, because of the problem of damage to the wool fibre. The need to use relatively mild pretreatments means that the dyeability of the cotton fibre will be less than optimum. In addition, when reactive dyes are used for the cotton, the optimum dyeing conditions with regard to both the amount of alkali required and the dyeing temperature cannot always be employed, because of wool damage. For a yarn-blended fabric, it would be possible to alkaline-scour or mercerise the cotton yarn prior to weaving and subsequent piece dyeing.

14.1.1.2   Scouring of wool/viscose blend fabrics

The pretreatment of fabrics made from wool/cotton or wool/spun viscose is very similar to that of all-wool fabrics, wool being the weaker fibre in the blend. In order to obtain satisfactory dyeings of good fastness to rubbing it is essential that the fabric be thoroughly desized and scoured before dyeing. According to the type of fabric, scouring is carried out either in the open-width washer or in rope form, using scouring agent and soda or ammonia. If necessary, the cloth is then milled and again thoroughly scoured.

Recipe

Quantity	Unit	Additions
3	%	Non-ionic detergent
1	ml/l	Ammonia (S.G. 0.88)

Treat at 15–20 min at 40–45 °C.

14.1.1.3   Bleaching

If mixture yarns or fabrics require bleaching they are treated in a similar manner to all-wool goods, viz. with reducing or oxidising bleaching agents. Hydrosulphite can be used at temperatures up to 45 °C, or hydrogen peroxide can be applied in the usual manner. A very good bleach is obtained by combining the two processes.

Recipe

Quantity	Unit	Additions
2–4	ml/l	Hydrogen peroxide 50%
0.25–0.5	g/l	Sequestering agent
5	g/l	Sodium silicate or sodium pyrophosphate

Adjust pH to 8 and continue bleaching at 30 °C overnight.

Bleaching at 40–50 °C for 4 h.

If bleaching has to be completed in shorter time increase peroxide quantity to 4–5 g/l.

The bleach effect can be improved considerably with the assistance of an OBA which is specially intended for the treatment of mixed yarns and fabrics containing wool and cotton. The goods are treated in a bath containing

0.25–1 % Optical brightening agent (or as per manufacturers recommendations).

At room temperature or around 40 °C. Rinsing is not necessary.

14.1.2   Dyeing

14.1.2.1   Exhaust dyeing methods

When dyeing goods of this nature, the dyer is mainly concerned with the production of solid dyeings; it is seldom that two-tone effects are required. Dyeing may be carried out by the single-bath or two-bath method. The best fastness above all best fastness to washing is achieved either by dyeing the two types of fibre separately, before blending or by first dyeing the vegetable fibre with a cotton dyestuff which is fast to acid cross-dyeing and then dyeing the wool in the woven fabric with metal complex or chrome dyes. Unions of undyed wool and spun dyed viscose staple are dyed in yarn or piece form by conventional wool dyeing methods.

For best fastness properties it is always better to dye in a two bath process, where the two different fixing parameters of two different dyes can be used in dyeing, if they are not affecting the other component in the blend. But this type of a dyeing is time consuming and not always economical. The dyer is always forced to select a cost effective process and hence has to adopt a one bath process or one bath one stage processes if it can achieve the required fastness standards.

14.1.2.1.1.   *Direct/metal complex or acid milling dyes*

As mentioned in the earlier method sometimes it is difficult to get s solid dyeing using direct dyes only as direct dyes exhausts faster on cotton. Hence wool may be dyed lighter hence the solidity of the dyeing gets affected. Hence a solid dyeing is possible by a two bath dyeing wool being further dyed to enhance the depth of wool portion and thereby achieving the solidity.

In single bath dyeing method it is possible to mix the direct dyes and neutral dyeing wool dyes and dyeing in one stroke. Also direct dyes yield dyeings of only moderate fastness to light and wet treatments, whereas along with the neutral dyeing wool dyes can yield dyeings of good fastness to light. There are ready-made mixes of direct dyes and neutral dyeing wool dyes available which can be used directly as one colour or one can mix by lab trials to get the required solidity and fastness requirements. Direct dyes should be selected with good build-up on viscose to minimise cross-staining of the wool. The absorption of direct dyes by wool can be reduced using anionic retarding agents of the syntan type.

14.1.2.1.1.1.   *One bath dyeing*

It is almost the same way as described in the previous method. Dyeing is carried out in the manner usual for union materials.

1–2% Levelling agent

is added to the dyebath plus

10–40% Glauber's salt cryst.,

according to the depth of shade. The temperature of the dyebath is slowly raised to 70–80 °C and dyeing carried out at this temperature for 1–1½ h. The dye bath is then brought to the boil

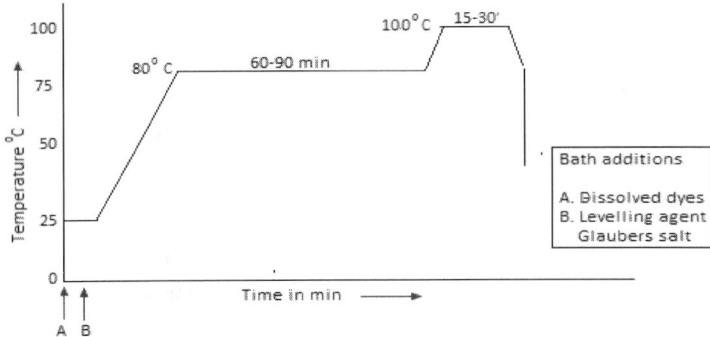

and maintained at a gentle boil for 15–30 min. The steam is then shut off and the goods are allowed to fill up until both fibres are dyed to the same depth. The goods are given a thorough rinse after dyeing. Wet fastness is improved by an after treatment with fixing agents.

14.1.2.1.2.   One bath two stage dyeing

In two stage dyeing one component is dyed with the corresponding dye and the next component dyeing is done from the same bath adding the respective chemicals for the second dye. This method gives better control over dyeing than one bath one stage dyeing but probably takes

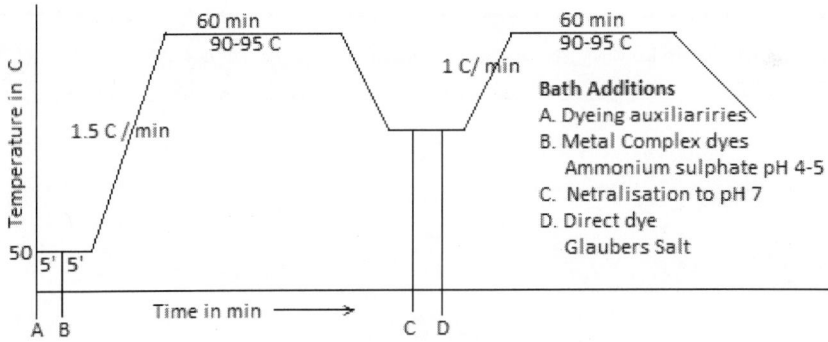

longer time. Rubbing and fastness results are found to be better. The wool component is dyes at pH 4–5 with acid dyes at the boil. Direct dyeing is done at pH 7 and 95 °C.

14.1.2.1.3.   Two bath methods

Ideally it is better to dye the union material with respective dyes of each component. But the problem here is, most of the cotton dyes will dye wool also. Hence the wool dyes are only functioning as a supporting dye to achieve the solidity as all the cotton colours dye wool lighter. Thus in two bath method we can use

Direct/acid	Reactive/acid,
Direct/chrome	Reactive/metal complex,
Direct/metal complex	Reactive only

and many other combinations can be used.

Dyeing may be carried out as per procedure for each dye and each component.

14.1.2.1.4.   Reactive/acid or metal complex dyes combination

Reactive dyeing can be done by either exhaust or pad batch method.
Process
1. Reactive dyeing by exhaust method → Wash off → Wool dyeing with 1:2 metal complex dyes by exhaust method.
2. Reactive dyeing by pad batch method (bicomponent dyes) → Wash off → Exhaust dyeing of wool with 1:2 metal complex dyes.

Reactive dyes on cotton can achieve high fastness compared to direct dyes, even in darker shades. There can be staining of wool with reactive dyes but the dyeing is fast hence pose no problem. But care has to be taken to adjust the wool dyeing (acid/metal complex dyes) has to be adjusted taking the staining into consideration. Normally warm dyeing reactive dyes are used.

14.1.2.1.5.   14.1.2.1.2.1   Cotton dyeing by reactive dyes – exhaust method

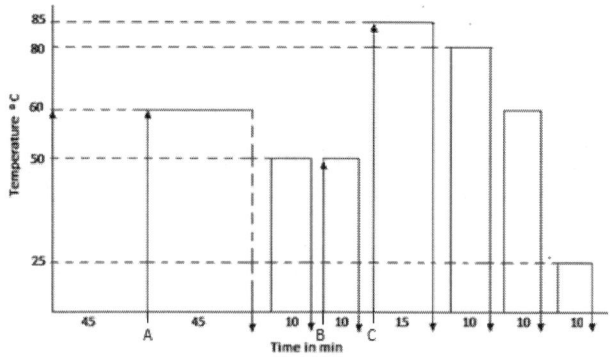

Reactive dyeing by exhaust method

Bath additions

	Levelling agent	g/l	0.5	0.5	0.5	0.5	0.5	0.5	0.5
	Mild oxidising agent	g/l	1	1	1	1	1	1	1
	Warm dyeing dyes	%*	0–0.5	0.5–1	1–2	2–3	3–4	4–5	>5
A	Glauber's salt	g/l	20–30	30–40	40–50	50–60	60–70	70–80	90
B	Soda ash	g/l	2	3	4	5	5	6	7
C	Acetic acid to pH 7	g/l	x	x	x	x	x	x	x
C	Soaping agent	g/l	1	1	1	1	1	1	1

*Based on the cellulosic component in the blend.

Dyeing is proceeded as per the dyeing graph given. Soaping has to be done at neutral pH. Too high pH has a negative influence on the wool quality and at a low pH hydrolysed dyes may exhaust on to wool.

14.1.2.1.6.   Cotton dyeing by reactive (bicomponent dyes) dyes – pad batch method

Pad batch dyeing procedure for the CEL component of wool/cellulosic blends.

For x g/l bicomponent reactive dyes any of the following alkali method can be used.

14.1.2.1.7.   Alkali requirements for warm dyeing dyes (e.g., Cibacron F)

Alkali method	A	B	C
Metering device requirement	Yes	No	Yes
Caustic soda 36 °Be	15–33 ml/l	4–16 ml/l	4–14 ml/l
Sodium silicate 38 °Be	70 ml/l	70 ml	
Soda ash	–	–	10–20 g/l
Padding assistant	1–2 ml/l	1–2 ml/l	1–2 ml/l
Batching time (room temp 25 °C)	6–8 h	12–24 h	6–12 h

Notes:
1. The batching time depends on kind of fabric, alkali method used and batching temperature. The lower the temperature, the longer the fixation time.
2. Method B can be used as tropical option (padding at 35–40 °C), a metering pump is recommended.

3. For dyeing blends with Lyocell (i.e. Tencel) or viscose method B is recommended. The amount of caustic soda should be increased by 50%. The fixation time is longer due to slow diffusion/penetration of the dye into the fibre.

4. For dyeing terry towels method C is recommended. The amount of caustic soda should be increased by 50%.

Washing-off procedure for bicomponent dyes on CEL/PA blends:

Continuous washing machine process

> 1st wash box – Rinse hot, overflow,
> 2nd and 3rd wash box – Rinse at the boil (98 °C),
> 4th wash box soap at the boil with 1–2 g/l detergent,
> 5th wash box – Rinse at the boil,
> 6th wash box – Rinse warm and neutralise with 1 ml/l acetic acid 80%.

14.1.2.1.8.   Alkali requirements for Remazol dyes

Alkali requirement	Up to 30 g/l dyes	Above 30 g/l dyes	Batching time
Metering device requirement	Yes	No	
Caustic soda 38 °Be	19.5 ml/l	27 ml/l	24 h
Sodium silicate 38 °Be	145 ml/l	145 ml	

14.1.2.1.9.   Alkali requirements of other brand dyes

Dye	Temp (∘C)	Alkali	Electrolyte	Time (min.)
Remazol	20	5 g/l soda ash + 4.5ml/l caustic soda (38∘ Be)	80 g/l sodium sulphate	90
Procion MX	30	10-12 g/l soda ash	55 g/l sodium chloride	60
Drimarene K	40	5 g/l soda ash 60 g l-1 sodium	60 g/l sodium sulphate	90
Levafix E-A	40	9 g/l soda ash	60 g/l sodium sulphate	90
Cibacron F	60	2 g/l soda ash	60 g/l sodium sulphate	45

14.1.2.1.10.   Dyeing of woollen component

Generally dyed by exhaust method.

Wool part of the blend is dyed in a new bath with acid or metal complex dyes.

Metal complex dyes

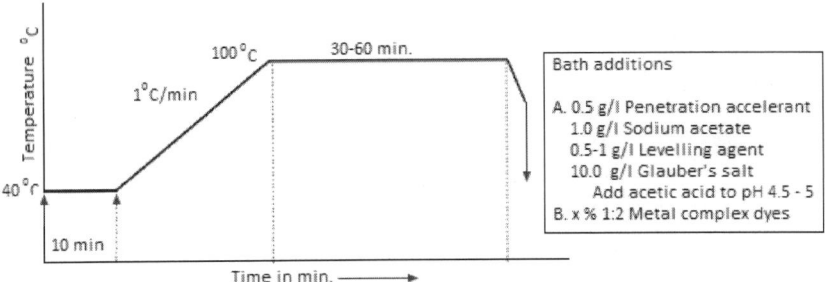

Dyeing of wool component with 1:2 Metal complex dyes

14.1.2.1.11.   14.1.2.1.3   Dyeing with vat dyes

Wool/cotton blends (Mainly 50/50 blends) can be dyed with vat dyes without damaging wool under optimum conditions. The dyes which can dye under these conditions (4 g/l caustic soda and 4 g/l sodium hydrosulphite at 40 °C for 45 min) may be useful for dyeing these blends.

Oxidation can be done with warm (40 °C) 3 g/l sodium perborate or persulphate. This process calls for selection of vat dyes and other severe limitations since the commonly on the choice of vat dyes, since many of the commonly used dyes require larger quantities of caustic soda to ensure sufficient stability in the leuco form, and higher dyeing temperatures to promote an adequate rate of sorption of dye by the fibres.

14.1.2.1.12.   Dyeing wool/cotton with only reactive dyes

When dyeing cotton with reactive dyes, alkaline conditions are usually required to bring about covalent reaction between dye and fibre; these reactions occur preferentially with the primary C6 hydroxy group in the cellulose fibre. In contrast, reactive dyes are applied to wool under neutral or slightly acidic conditions, and reaction occurs mainly with cystine thiolate anions ($-S-$) and amino groups ($-NH_2$).

Keeping these conditions in mind a dyeing process can be designed in such a way that with selected reactive dyes both components can be dyed. In an exhaust dyeing process is done at low temperature like 20-60°C the dyes are exhausted using an electrolyte in the same manner as a direct dye application. As far as the cotton component is concerned no reaction takes place as the bath is neural and the exhaustion is level. But in the case of wool reaction with reactive dye can start and hence an auxiliary has to be added to avoid unlevel dyeing. In the next stage alkali is added (pH 8–12) to promote chemical reaction of adsorbed dye with the fibre and further dye uptake. For those dyes suitable for dyeing of the cotton portion of wool/cotton blends, fixation temperatures normally range from 20 to 60 °C. Next rinse and soaping is followed, for removing the electrolyte, alkali and unfixed dye. The selection of dyes should be in such a way that the removal of hydrolysed dyes from wool is easier. (Even though most of the reactive dyes in hydrolysed dyes have low substantivity but in case of wool cotton dyeing there removal is difficult as there are chances of some hydrolysed dyes act as acid dyes on wool).

The reactive dyes which are suitable for dyeing wool/cotton blends by the above process are β-sulphatoethyl sulphone (Remazol – Dyestar) , Dichlorotriazine (Procion MX)

Monochlorodifluoropyrimidine   (Levafix   E-A   –   Dyestar), Monochlorodifluoropyrimidine Drimarene K (Clarient), Monofluorotriazine (Cibacrone F – Ciba).

Some selected dyes can be applied on wool/cotton by a single step process, avoiding a separate alkali fixation step for reactive dyes on cotton. this depends on using temperature to ionise the cellulose. The main classes of dyes used are fluoro-triazine, VS derivatives and monochloro-triazine derivatives. Dyes are selected which can fix at neutral pH on cotton. Total fixation efficiencies can be obtained by increasing the electrolyte concentration, by working at the lowest possible MLR and by dyeing at 100 °C for Cibacron F and Remazol type dyes and at 140 °C for the less reactive Procion H and HE type dyes. This process have many advantages like it gives dyeings with excellent reproducibility, levelness and penetration, low utility consumption and chemical usage due to low liquor ration employed, hydrolysis of dyes are minimised due to neutral pH in dyeing, the dyeing process is most simplified due to all the chemicals, such as dyes, sodium sulphate and sodium acetate, can be put into the dyeing solution at the same time etc.

Padding method can be employed for dyeing reactive dyes on cotton wool/blends. For example cold pad batch process. The process involves in impregnating of the fabric with a cold solution of dye and alkali, padding, batch and store for 2 to 48 hours, washing off and drying.

14.13   Dyeing of nylon/cotton, nylon/viscose, nylon/ linen blends

Nylon cotton blends are available as intimate blends as well as blends made of nylon/cotton warp and cotton weft. Nylon/cellulosic staple blends containing 10–30% nylon with cotton or viscose are used in lightweight suiting and dress wear, leisure shirts and half-hose. Many of these blends, as well as workwear fabrics with a 25:75 nylon/cotton warp and a cotton weft, or 20:80 nylon/ viscose carpet yarns, pile fabrics which may have nylon pile and woven cotton backing. They contains relatively minor proportions of nylon and acceptable solid effects are not difficult to achieve. Cotton-pile terry to welling with a weft knit nylon backing for beachwear, children's clothing or leisure shirts are also common. In warp-knitted construction velvet fabrics with a viscose pile and a nylon backing, are used as furnishing, outerwear, trimmings and lining fabrics.

14.13.1   Pretreatment

Pretreatment can be done as per pure polyamide fabric (explained in this book under polyamide section) since the cellulosic blends can withstand the preparation chemicals for nylon without any difficulty.

14.13.1.1   Scouring

Recipe

Quantity	Unit	Additions
2–5	%	Synthetic detergent or soap
2–5	%	Trisodium phosphate

Treat for 30 min at 40–70 °C.

But the bleaching chemicals (hydrogen peroxide) used in preparation is reduced correspondingly, where nylon percentage is lower and also organic stabilisers are used.

14.13.1.1   Bleaching of nylon/wool, nylon/viscose (nylon/cellulosics)

Peroxide can degrade nylon, because yellowing rather than bleaching and cellulose can withstand most of the bleaching chemicals. Peroxide bleaching is adopted where the nylon content is less than 30% where the peroxide is preferentially absorbed and hence the nylon is not attacked.

Bleaching bath is prepared with the following recipe at 40 °C.

Quantity	Unit	Additions
3–4	ml/l	Hydrogen peroxide 50%
1	g/l	Sodium hydroxide solid
0.2	g/l	Peroxide stabiliser
0.002–0.5	g/l	Free radical suppressor

The material is entered and the temperature is increased to 85 °C and the bleaching is continued for 1 h. Temperature is cooled to 60 °C, drained and washed. The optical brightener can be added in recipe wherever necessary.

Sodium chlorite bleaching is a better bet for bleaching nylon/cotton since it bleaches both components without any problem.

Recipe for batch wise process

Quantity	Unit	Additions
2–5	g/l	Sodium chlorite
1	g/l	Sodium chlorite stabiliser
x	ml/l	Formic or acetic acid to pH 3–4

Treat for 90–120min at 90°C. After bleaching bath, a fresh bath is taken with 2g/l sodium carbonate to neutralise the alkali and neutralised.

Continuous process also can be adopted for nylon/cotton blends. Please refer 'Pretreatment of Textile Substrates' by the same author.

14.13.1   Dyeing

There is no common class of dye which can dye both component of this blend. But where the percentage of cellulosic component is high (nylon is minor component) it plays a significant part in the surface appearance of the blend fabric, can be dyed with cotton dyes as per fastness requirements, brightness requirements (direct, reactive, even vat, etc.).

14.13.1.1   Exhaust dyeing

Reserve shades can be dyed by dyeing only the nylon component with acid/metal complex dye in 1-bath dyeing method. Cellulosic component is not dyed by acid dyes. Any tinting on cellulose can be cleared with a good soaping, scouring or mild reduction clear without affecting the dyed shade.

Pale shades and tone in tone dyeings can be achieved using vat dyes dyed by the semi-pigmentation process or other appropriate dyeing methods.

Solid shades pale or medium depths dyes can be dyed with selected direct dyes (mainly of yellow to red disazo dyes with two solubilising groups and a disulphonated phthalocyanine blue etc.) with good level ling characteristics. Solidity can be achieved by dyeing by applying these dyes at pH 5–6 with sodium dihydrogen phosphate as buffer and with limited salt addition at the boil.

They are economical, both the different fibres are dyed simultaneously in one bath with short dyeing times. It is specially suited for ternary combination dyeings. High reproducibility when reserving agent is used. Both fibre components can be shaded during the dyeing process with high flexibility compared with the pad batch processes. The polyamide portion is dyed with acid colours and the cotton portion is dyed with direct colours from the same bath. The acid dyes can be selected which are having the property of high cotton reserve and can be dyed without using a cotton reserving agent, otherwise reserving agents has to be added in the bath.

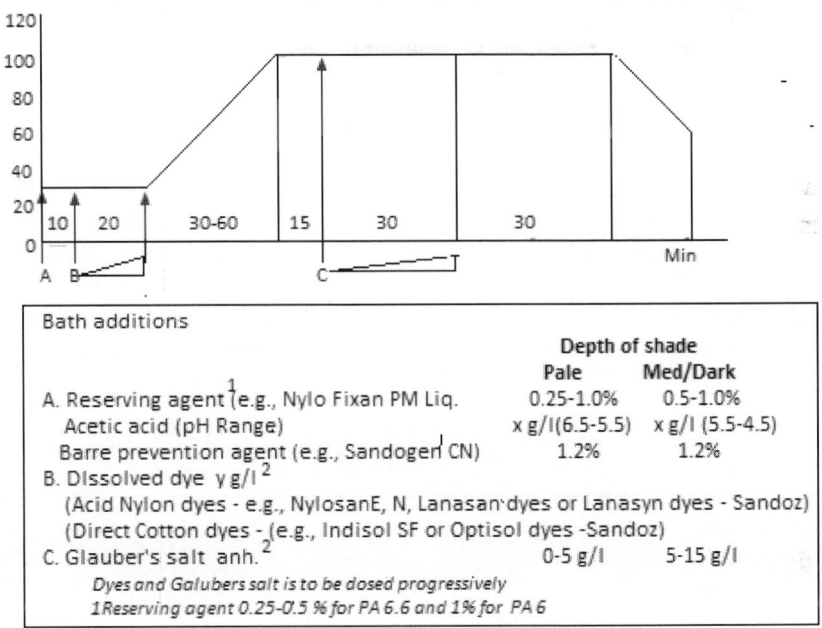

One bath exhaustion dyeing preocesswith Acid/Direct dyes combination

The dyeing is done as per the dyeing graph given above.

14.13.1.1 Disperse/direct dyes combination

Disperse/direct dyes combination is one method used for solid shades of widest range. They can be conveniently applied by a one-bath method at pH 8 and 70 °C. An alkanol polyoxyethylene is recommended as dispersing and levelling agent, together with a syntan to control uptake of direct dyes by the nylon. Multisulphonated direct dyes of the salt-controllable type are used and the preferred disperse dyes are low-energy types with good levelling properties on nylon. The wet fastness properties of nylon/cellulosic blends dyed in this way are severely limited and the method is restricted to low-quality fabrics.

Another commonly practiced combination for nylon/cellulosic blends where both component fibres make a major contribution to the appearance of the material, is the metal complex and acid milling dyes for nylon and direct dyes for cotton. Solid dyeings with good overall fastness can be achieved in one bath dyeing. The dyebath is prepared with a retarding agent for acid milling dyes (if necessary) and set the pH 5–6 for acid dyes and 7–8 for metal complex dyes. The acid dyes are

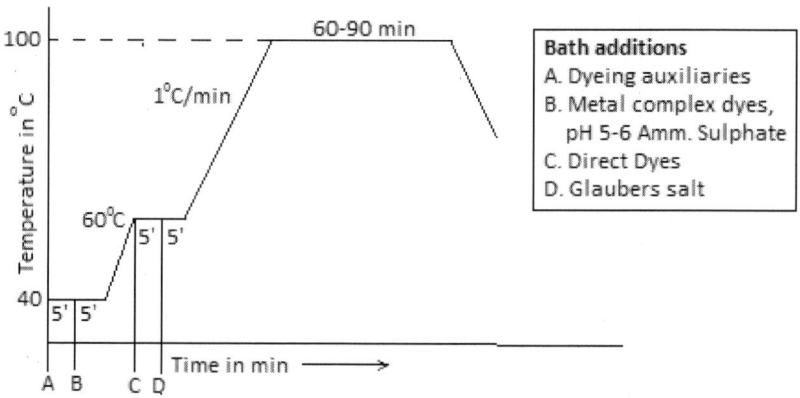

added and the dyeing is started with temperature raised to 60 °C and the direct dye is added. The Glauber's salt is added and further raised the temperature to boil when full exhaustion takes place. The bath is cooled to 70 °C and dropped, aftertreat direct dyes if necessary after rinsing.

In the above method dull shades are only possible because most of the metal complex dyes and direct dyes are not so bright. Bright shades of good fastness are possible with reactive dyes. Dyeing of the nylon by reactive dyes depends on the type of nylon, filament or staple, etc. A blend with very low percentage of nylon and the fabric where nylon is hidden from the face of the fabric can be dyed directly by reactive dyes as we dye a normal cotton fabric by any of the suitable methods. Here even though the nylon gets dyed partly it does not show on the face of the fabric.

14.13.1.1   Single bath two stage method

A single bath two-stage method may also be used with selected reactive dyes. In this case, either the cellulose is dyed first under alkaline conditions followed, after acidifying the bath, by the polyamide or, the polyamide is dyed first at pH 5 and followed, after neutralisation, by the cellulose.

14.13.1.1   Use of reactive dyes

Where the nylon and cotton has to be dyed equally and bright shades of overall good fastness is required, we can dye the nylon part with acid milling or metal complex dyes and then it can be further dyed with reactive dyes. The depth of the shade to be dyed by acid milling dyes should take into consideration of part dyeing of nylon by reactive dyes.

Selected reactive dyes can be used for getting solid shade in a three stage dyeing process. First the reactive dyes are exhausted on nylon at pH of about 4–5 for 30 min at 80–90 °C in the absence of any electrolyte. In the second stage the electrolyte is added and the reactive dye remained in the bath is exhausted on to cellulosic material in the neutral medium. Once exhaustion is

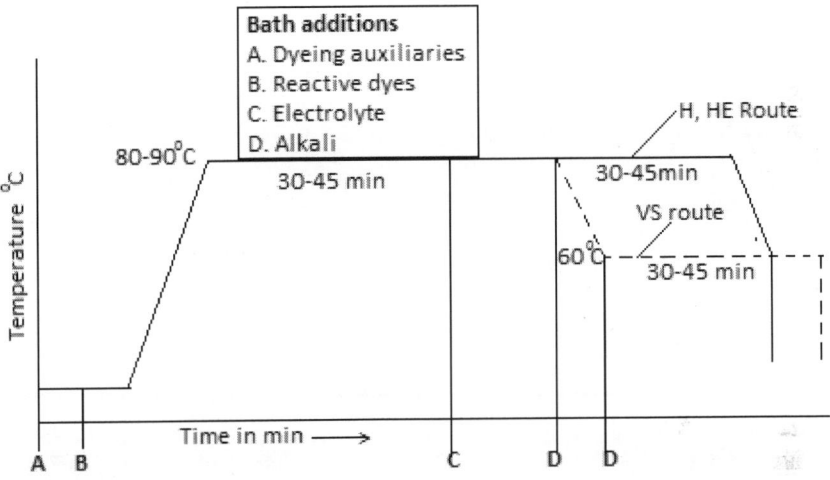

completed, in the third stage the required alkali and at the temperature as per the reactive dye used helping the fixation of the reactive dye on the cellulosic material. Here the main difficulty is in the control of the quantity of the dyes exhausted on nylon and cellulose. This is done by the pH, electrolyte and temperature control in the dye bath. Nylon is favoured at low applied depths but the distribution shifts in favour of the cellulosic fibre as the saturation level of nylon is approached.

Another two stage process is one in which the reactive dyeing of cotton is done in alkaline medium and the bath is neutralised and the acid milling and 1:2 metal complex dyes are allowed to dye the nylon portion. The acid dyes selected should have the stability in the alkaline

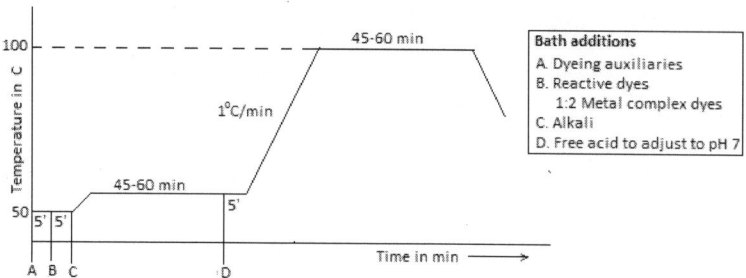

dyeing bath. Dye bath is prepared with the necessary dyeing auxiliaries and the reactive and alkaline stable acid dyes. The alkali necessary for the fixation of the reactive dye is added and the temperature of fixation is maintained. After the fixation is complete the bath is neutralised pH 7 and temperature is raised to boil for the exhaustion of acid dyes on nylon. After the dyeing is completed the bath is dropped and soaping, rinsing, etc. is followed.

When one is using a one bath two stage method the selection of reactive dyes are limited. For a safer method is to use a two bath method where the above two stage process is adopted into a two bath dyeing separately. No limitation of dyes used, wider gamut of colours and no neutralisation are required. The process can be taken from the dyeing of pure nylon and cotton given in this book (as the dyeing is done independently of other component in the blend).

14.13.1.1   Disperse reactive/reactive dye combination

Nylon is dyed with disperse reactive in acidic pH and normal reactive dyeing procedure to dye cotton. Set the bath at 50 °C, with all the dyeing auxiliaries and after 5 min both the dyes are added. Add dispersing agent (1 g/l) for dispersion of disperse reactive dye and electrolyte (100 g/l) is added for exhaustion of conventional reactive dye simultaneously on cellulose for 15 min at neutral pH; Na2CO3 (10–20 g/l) is added for fixation at 85–95 °C for 1 h on cotton.

14.13.1.1   Continuous processes

The blends where the cotton component is higher (nylon 10–20%) and where the cotton is seen on the face of the fabric can be dyed continuous method using direct dyes.

In a two component continuous dyeing system reactive and disperse dyes are padded urea and an anionic migration inhibitor, followed by thermofixation at 180–200 °C. Batching for 2 h after padding may improve fixation of the reactive dyes when viscose is the cellulosic component.

Another possibility is to use metal complex and reactive dyes combination in continuous method. In the first stage both dyes are padded from a neutral bath along with padding auxiliaries, dried and thermofixed the metal complex dyes on nylon. Next, the alkali is padded from a saturated common salt bath to avoid the reactive dye leeching into the pad bath. The dye is fixed bay batching at room temperature for 8–16 h. (Padding recipe and process can be seen under cotton dyeing in the Dyeing Textile Substrates I – Cotton by the same author.)

Other continuous dyeing possibilities are given below:

Blend	Colour effect	Dyeing method	Dyes combination
Nylon/ cellulosic	Solid shades	Pad–dry–thermofix– acid shock	Selected premetallised or milling acid shock acid dyes and reactive dyes
		Pad–dry–thermofix– chemical pad – steam	Selected premetallised dyes, then pre-reduced sulphur dyes
		Pad–dry–steam	Premetallised or milling acid dyes and salt-controllable direct dyes
		Pad–dry–steam, pad–dry–chemical pad–steam	Selected 1:2 metal-complex dyes, pad–dry–chemical then vat dyes

14.13.1.1   Amination of cellulosic fibres and blending with wool and dyeing

By amination the cotton/viscose can be made dyeable with wool dyes and the blend with wool can be dyed with wool dyes in acid bath. Treatment of cotton with a reactive cationic polymer under alkaline conditions and dyeing with reactive dyes under neutral or mildly acidic conditions was suggested by researchers but it was not been a commercial success. The main reason for this was the shade was becoming duller to normal dyeing of reactive dyes in alkaline conditions and reduction in light fastness by 1-2 grades. Lenzing has introduced viscose dyeable with wool dyes by adding adding a cationic polymer to viscose dope and spinning (Rainbow fibre). This fibre was blended with wool fibre and could be dyed with wool dyes in one bath.

The effort of amination of cotton was continued and many methods were propsed by researchers. One method was modification of cotton or viscose by an alkaline pad–bake (200 °C) or by an alkaline pad–batch procedure with glycidyl-trimethyl-ammonium chloride (Glytac A from Protex) or its precursor 3-chloro-2-hydroxy-N,N,N-trimethyl propanaminium chloride. Due to the problem in in handling the epoxide form of the reactant the more common method is to undertake the modification by a pad–batch procedure, using a

mixture of the chlorohydrin analogue 3-chloro-2-hydroxy-N,N,N-trimethyl propanaminium

chloride and sodium hydroxide. The high alkalinity converts the agent, during the batching

procedure, to the corresponding epoxide, which then fixes covalently to the fibre. This product could be dyed with reactive dyes at neutral pH without the addition of

$$\text{Cell - O - CH}_2\text{- CH - CH}_2 \quad \xrightarrow{-HCl} \quad \text{Cell - O - CH}_2\text{- CH - CH}_2$$

$$\underset{\overset{|}{OH^+}}{\underset{N(CH_3)_3Cl^-}{|}} \qquad\qquad \underset{\overset{|}{O^-}}{\underset{{}^+N(CH_3)_3}{|}}$$

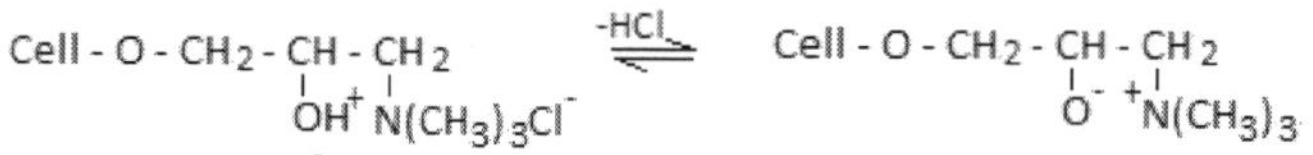

A   **B**

Amination products (Different forms) by glycidyl-trimethyl-ammonium chloride (Glytac A)

salt. The dye is added to bath and the temperature is raised to boil over 30 minutes and dyeing continued at the boil for a further 60 minutes. Exhaustion can be of the level of 99-100% with almost fully (99%) fixed with hardly any dye to be washed off.

Another method suggested is to pretreat the cotton fabric with N-methylol acrylamide (NMA) by pad-bake method where by different aliphatic amino groups are introduced into into cotton

The reaction is shown below where NMA reacts covalently with cellulose by a Lewis acid-catalysed baking reaction where appropriate amines gave a series of amino-substituted cellulosic substrates (Structures A – E)

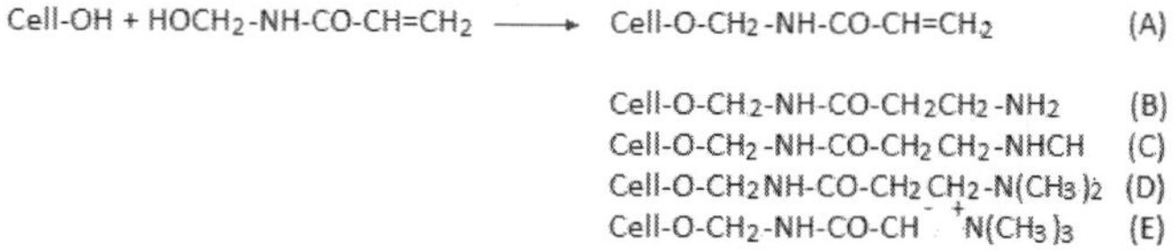

$$\text{Cell-OH + HOCH}_2\text{-NH-CO-CH=CH}_2 \longrightarrow \text{Cell-O-CH}_2\text{-NH-CO-CH=CH}_2 \qquad \text{(A)}$$

$$\text{Cell-O-CH}_2\text{-NH-CO-CH}_2\text{CH}_2\text{-NH}_2 \qquad \text{(B)}$$
$$\text{Cell-O-CH}_2\text{-NH-CO-CH}_2\text{CH}_2\text{-NHCH} \qquad \text{(C)}$$
$$\text{Cell-O-CH}_2\text{NH-CO-CH}_2\text{CH}_2\text{-N(CH}_3\text{)}_2 \qquad \text{(D)}$$
$$\text{Cell-O-CH}_2\text{-NH-CO-CH}^-\ {}^+\text{N(CH}_3\text{)}_3 \qquad \text{(E)}$$

Reaction of NMA with cotton and Amino derivatives of NMA-treated cotton.

These substrates can be dyed from weakly acidic bath (pH 5) without the addition of salt, process being the same as above with running at boil for 1 hour. Except the structure D, substrates B,C and E gives good fixation.

Chapter 15
Dyeing of BC Blends

15.1 Dyeing cotton/acrylic, viscose/acrylic fibre blends

Generally when PAN is dyed with cationic dye there are chances of cellulose fibre staining, especially in case of medium and dark shades. This depends on the dye used, dyeing conditions (temperature, time, etc.). Thus on dyeing acrylic/cellulosic fibre blends, it is necessary to employ cationic dyes that largely reserve the cellulosic fibre component. Most of the stains on cellulose fibre are usually eliminated by exhausting cationic dye fully into PAN. The cellulosic component can be dyed with any appropriate class of dye provided that this does not stain the acrylic fibre or adversely affect the shade of the acrylic fibre. When selecting direct dyes, it should be noted that only those dyes that can be applied from a weakly acid medium are suitable for the one-bath procedure.

When the cellulose fibre is considerably stained, such stains may be eliminated mostly by soaping or reducing decolourisation with a chemical such as Rongalite C. But, when dyeing both fibres in various colours including same colour and different colours, it is necessary to use the dye with less stains by making careful selection in respect to the colour tone and fastness.

PAN also can be dyed with disperse in light and medium. The absorbing capacity of disperse dye for PAN is limited may up to 2%. It is found that the disperse dye can give level dyeing and good light fastness and also it can be dyed along with direct dye (for cellulose) in one-bath dyeing method.

In the dyeing of cellulose fibre, the direct dyes are generally used. However, when more fast dyeings or special colours are demanded, reactive dyes, naphthol dyes, vat colours are also employed. If proper dye selection is done we can do cotton dyeing without staining PAN.

Suspending agent

If cationic dye and direct dye are used in the same bath and at the same time, the dyeing will become impossible due to the inter-dye precipitation caused by the opposite electric loads in the solution. This assistant prevents precipitation without changing the dyeing characteristics of cationic dye and direct dye, and may be dyed by one bath in the similar colour tone,colour depth, fastness as in the cases dyed in the separate baths.

Retarding agent

When cationic retarding agent, which is generally used as a retarding agent for cationic dye, may sometimes reduce the solubility of the direct dye and produce precipitation by forming a complex with the direct dye. Even when precipitation is not produced, it can have retarding effect on the direct dye. Hence, this type of retarding agent should not be used in the dyeing of acrylics.

15.1.1   Pretreatment

15.1.1.1   Scouring

Scouring same way, as in the case of 100% PAN, the scouring of yarns will be performed by using a suitable non-ionic scouring agent (2–3% owf) and 0.5 g/l acetic acid 80% at 60–65 °C for 20 min.

Fabrics should be desized and scoured to remove sizes, spinning assistants and other impurities. Scouring can be carried out with 1 g/l non-ionic detergent, 1 g/l trisodium phosphate at   60 °C for 30 min followed by rinsing well with sufficient water.

15.1.1.2   Scouring of viscose/acrylic

Quantity	Unit	Additions
0.25	%	Synthetic detergent or soap
0.5	%	Sodiumacetate or acetic acid to pH 5.5–6

Treat for 30 min at 90 °C.

15.1.1.3   Bleaching and full white

Acrylic supplied enough white and may not need a bleaching. However, if cotton is the second component it will need bleaching. Peroxide or chlorite bleaching can be employed. Usually chlorite bleach is used where acrylic also needs bleaching.

Guideline recipes

15.1.1.3.1   One bath method with sodium chlorite

Set the bleaching bath at 35 °C.

Quantity	Unit	Additions
1.5	g/l	Sodium chlorite
2	g/l	Oxalic acid
1	g/l	Tetrasodium phosphate
1	g/l	Corrosion inhibitor
		pH is adjusted 3.5–4

The temperature is raised to 9095 °C in 30 min and held at this temperature for 30–60 min. The bath is cooled slowly to 50 °0C and then rinsed thoroughly.

Dechlorination can be done with 2 g/l sodium bisulphite (or 1.5 g/l sodium thiosulphate) and 1.5 g/l tetrasodiumphosphate. At 600 °C for 20–30 min.

Along with OWA for cotton at 60 °C, 20 min.

15.1.1.3.2   Two-bath method with sodium chlorite

First bath fluorescent bleaching for PAN:

Quantity	Unit	Bath additions
1	g/l	Sodium chlorite 80%
0.7–1	%	Optical whitener for PAN (owf)
		at pH 3–4 (MLR 1:10–40)

at 90–95 °C for 60 min.

Dechlorination can be done with 2 g/l sodium bisulpite (or 2 g/l sodium thio sulphate) at 60 °C 20 min.

Second bath fluorescent whitening of the cellulose side:

OWA for cellulose 0.3–0.6% (owf) (Only in case of viscose rayon, add 10% Glauber's salt).

pH neutral bath ratio 1:10–40, 20–30 °C (in case of viscose rayon 40–60°) for 30 min.

(Note: % is the volume for the total fibre).

15.1.1.3.3   Bleaching with hydrogen peroxide

Quantity	Unit	Additions
5–7	g/l	Hydrogen peroxide 50%
3	g/l	Sodium silicate
1	g/l	Sodium carbonate

Treat at 90 °C, for 45–60 min. After bleaching the bath is cooled slowly to 50 °C rinsed and neutralise.

15.1.2   Dyeing

15.1.2.1   One-bath two-step dyeing (in circulating liquor machines) with acrylic dyes and direct dyes

The dye bath is adjusted to pH 4.5–5 with acetic acid, and the temperature raised rapidly to around 90 °C at which circulation proceeds for 10 min to eliminate temperature differences in the material. A well-prepared, concentrated solution of the acrylic dye is rapidly introduced and is allowed to exhaust in 60 min at this temperature.

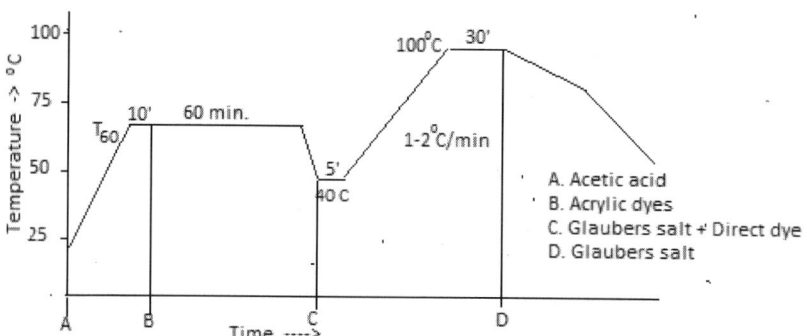

One-bath two-step dyeing of acrylic/cellulosic fibre blends acrylic dyes and direct dyes

The dye bath is now rapidly cooled down to about 40 °C (104 °F) and set with 5% Glauber's salt anhydr., 1–2% Levelling agent, and well-dissolved direct dye. After a short preliminary run, the bath is heated at a rate of 1–2 °C/min (1.8–3.6 °F) to the boil. After 30 min at this temperature, 5% Glauber's salt anhydr. is added, and the direct dye is allowed to exhaust as the temperature falls over about 30 min. The bath is now cooled to 60 °C (140 °F) and the dyeing finished off in the usual manner.

Note: After treatment for improving wet fastness and light fastness of direct dye can be done.

15.1.2.2   One-bath two-step dyeing in circulating liquor machines with acrylic and direct dyes without cooling

Here again, only those direct dyes can be used that dye from weakly acid baths. This process differs from the one described previously only in that the bath is not cooled down to 40 °C (104 °F) after dyeing at around 90 °C, but raised to the boil at a rate of 1 °C/min (1.8 °F).

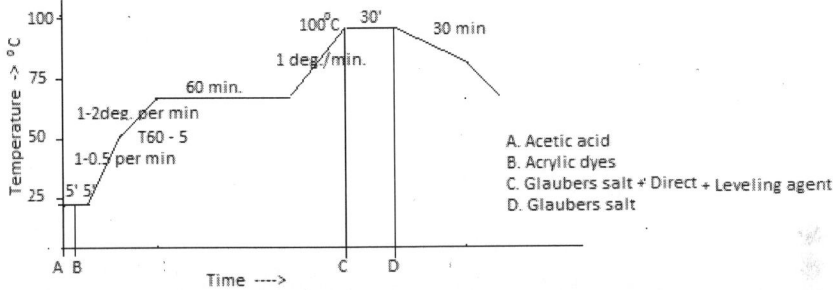

One-bath two-step dyeing in circulating liquor machines with acrylic and direct dyes

The addition of the auxiliary (1–2% levelling agent) and the direct dye is made at the boil. After 15 min, the required amount of Glauber's salt is added in lots, and dyeing proceeds at the boil for 30 min, and then for a further 30 min in a bath that is cooled off.

Note: In all above dyeing methods where the retarders are not used one should use more levelling acrylic dyes.

15.1.2.3   Dyeing with cationic/vat dyes combination

Two essentials are required in this process, namely,

    a) the acrylic fibre should suffer no damage under the processing conditions, and

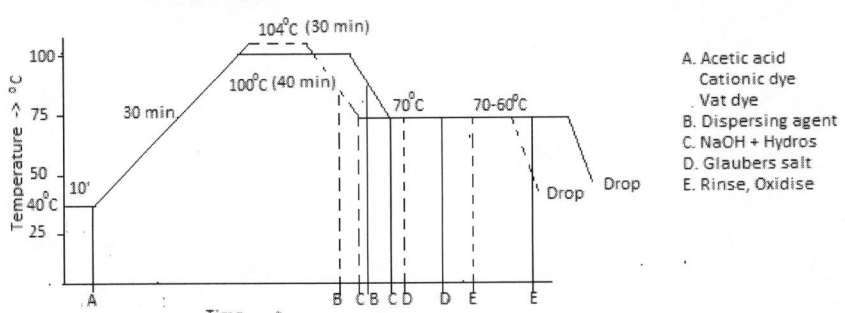

b) the acrylic fibre dyeings should be sufficiently resistant to cross dyeing with vat dyes.

15.1.2.3.1   Dyeing with cationic/vat dyes combination – Method I

The first and simplest possible method would be, as shown above, the joint introduction of the two classes of dye into the bath from the very beginning. At this stage, where the bath contains dissolved cationic dye and the vat dye as pigment, the acrylic fibre is dyed by heating up the bath, and the goods are simultaneously pigmented with vat dye, so that this latter can, after cooling, be fixed by vatting and development.

15.1.2.3.2   Dyeing with cationic/vat dyes combination – Method II

The second possible method, shown schematically below, would be to pigment the goods by heating up with the vat dye, and then to add acetic acid and the cationic dye at about 70–75 °C, i.e., at a temperature below the glass transition point of the acrylic fibre, which is then dyed by

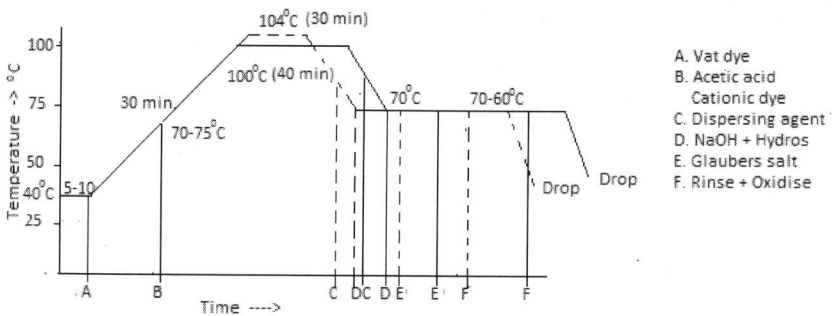

heating the bath up to 100 °C or 104 °C. The bath is cooled down after dyeing for 30 min, whereupon the dye for the cellulosic fibre is applied by vatting and is then developed.

15.1.2.3.3   Dyeing of acrylic/cellulosic blends with cationic/ vat dyes combination – Method III

A third possible method is in which the acrylic fibre component is first dyed with the cationic dye in the conventional manner, whereupon the bath is cooled down and the cellulosic fibre is

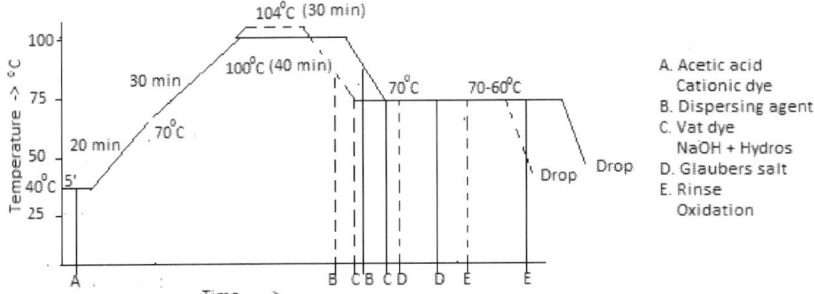

treated with the vat dye in the same bath, again in the conventional manner. The method is schematically shown above.

Notes:

1. Most of the dyes in the market contain additives to make possible the supply of a dye of constant strength. In the case of the vat dyes, these additives simultaneously comprise dispersing agents which, as with the disperse dyes, have the object of maintaining the fineness of the dye during manufacture and the dyeing process. These dispersing agents are anionic, e.g., based on condensation products of naphthalene sulphonic acid and formaldehyde, or on lingo sulphonic acid, and so on. With cationic dyes, these products may form salt-like compounds that are sparingly, or totally, insoluble and are liable to Precipitate. In order to prevent such precipitation and resultant filtration in circulating liquor dyeing machines with some degree of certainty, vat dye preparations containing almost no dispersing agent, e. g., Colloisol Liquid brand dyes, can be used, or dispersing agent can be added to excess. The latter choice is based on the fact that small amounts of this type of anionic dispersing agent act as precipitating agents, but as soon as a certain ratio of auxiliary to dye is exceeded, they exert a dispersing effect.

2. The second method mentioned offers no marked advantages over the one just discussed.

3. The third method proved to be the best, and based on this following procedure was developed.

15.1.2.4 The one-bath two-step dyeing process (cationic/ vat dyes combination)

The bath is heated to 45 °C with acetic acid and run for 5 min. Add cationic dye and retarding agent and check the pH around 4–5. Cationic dyes are exhausted using retarders at around in presence of 45 °C for about 30 min.

Proceed exhaustion at this temperature 30 min then raise the temperature to boil, or at 104 °C (219 °F), in order to improve penetration of the dye (15 min).

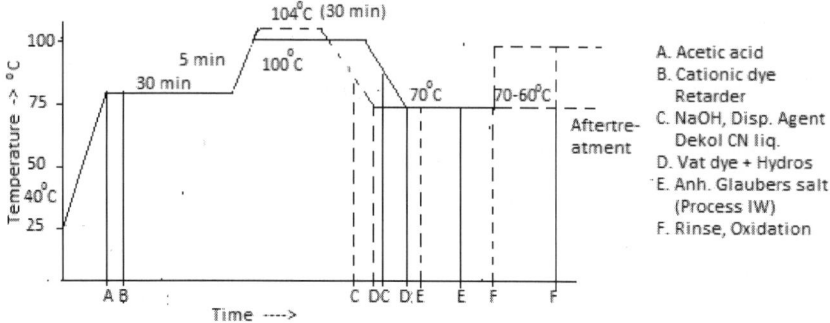

The one-bath two-step dyeing process

The bath is cooled down prior to dyeing the cellulosic component. Dispersing agent is added as the temperature falls below 80 °C (176 °F), as is also the calculated amount of caustic soda. The dispersed vat dye is introduced at 75–70 °C (167–158 °F), and finally hydrosulphite together with levelling agent, where necessary. It is essential to follow this sequence. The temperature should be maintained at 75 °C (167 °F) for process IN dyes, whereas the Indanthren dyes applied by the IW dyes should be dyed with decreasing temperature, and Glauber's salt can be added after 20 min, where necessary.

The usual oxidation and soaping for vat dyes follow the rinse. Best soaping conditions are: 80 °C (176 °F), pH 6–7, 20 min.

Guideline recipes

Step 1

Quantity	Unit	Description
3	ml/l	Acetic acid 30%, pH 4–5
x	%	Cationic dye
y	%	Retarding agent, wherever necessary

Step 2

Quantity	Unit	Description
1	g/l	Dispersing agent
10–15	ml/l	Caustic soda 36 °Be
z	%	Vat dye

5–6	g/l	Hydrosulphite
0.5–0.75	g/l	Levelling agent
10–20	g/l	Glauber's salt anhydr.

15.1.2.5 Two bath dyeing with acrylic/reactive dyes

Reactive dyes are recommended for medium and heavy shades.

Bath 1

The acrylic component is dyed first using the following profile.

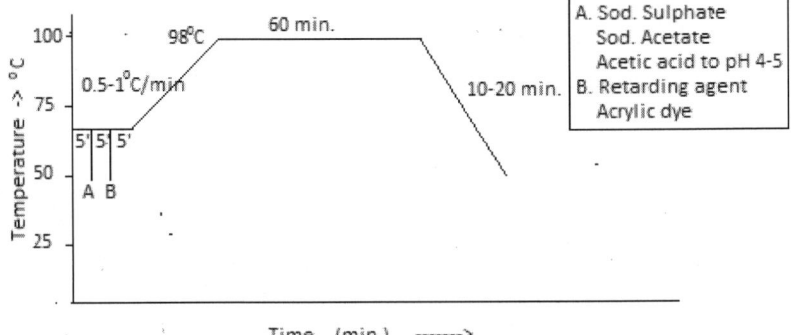

Two Bath Dyeing Acrylic/Cotton with Acrylic/Reactive Dyes – Bath I

A

Quantity	Unit	Description
0–5	%	Glauber's salt anhydr.
0–3	%	Sodium acetate
		Acetic acid to pH 4–5

B

Quantity	Unit	Description
1–0	%	Retarding agent
x	%	Acrylic dye

Rinsing

10 min warm,

10 min at 60 °C,

10–20 min at 70 °C with suitable detergent,

10 min at 80 °C,

10 min at 60 °C,

10 min cold.

Bath 2

Dyeing of the cellulosic component with reactive dye. This method follows a warm dyeing

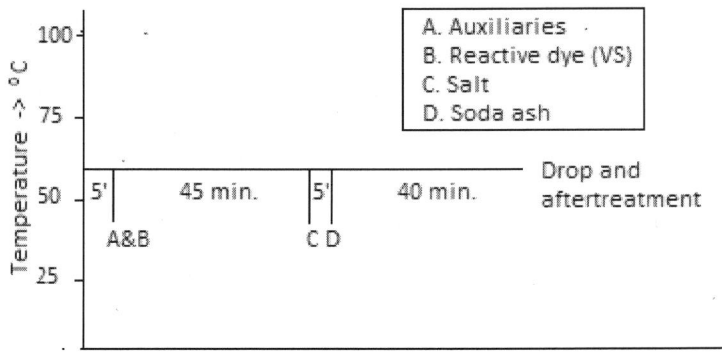

Two Bath Dyeing Acrylic/Cotton with Acrylic/Reactive Dyes - Bath II

dye (Remazols). For the addition of Glauber's salt, soda ash,etc. please follow dye manufacturer's instructions.

15.1.2.6   Continuous methods

Only pale thermosol dyeings are obtained with disperse dyes on acrylic fibres. On the other hand, there are fundamental difficulties on thermosol dyeing with cationic dyes, owing to the salt-like nature of these dyes, the result being that only dull and weak dyeings are produced when padding liquors containing cationic dyes in true solution are employed.

However, there is some special process with which cationic dyes can be applied by the thermosol process to some acrylic fibres to produce deep and brilliant dyeings with excellent fastness properties, and comparable with dyeings made from a long liquor. This process is based on creating a dispersion of cationic dyes by means of a special system of auxiliaries. A definite amount of special dispersing agent (e.g., Basopal NA Conc. Liquid) is added slowly with stirring to a solution of basic dyes containing a like wise definite amount of special levelling agent for thermosol dyeing (e.g., Peregal ON Highly Conc.). The anionic dispersing agent itself acts as a precipitating agent on the cationic Basic 1 dyes, but the second non-ionic dispersing agent, has the effect of dispersing the insoluble cationic dye/first dispersing agent product as it is formed. The Basacryl dyes are largely protected from precipitation in a dispersion prepared in this manner.

15.1.2.6.1 One-stage process (for pale and medium shades)

Procedure

Padding liquor:

Quantity	Unit	Bath additions
x	%	Basic dyes
$x/10–x/5$	g/l	Levelling cum dispersing agent (e.g., Peregal ON highly conc. (BASF)
$x/4–x/2$	g/l	Dispersing agent (e.g., Basopal NA conc. liquid –BASF)
0–5	g/l	Thickening agent
y	g/l	Disperse dyes

The liquid basic dye and dispersing cum levelling agent are dissolved in 50–70% of the water required for the padding liquor, whereupon dispersing agent, which has previously been mixed with double its amount of water, is added slowly with stirring. The thickening agent is then added if necessary, and, finally, the liquid or powder dispersed disperse dye is poured in slowly with stirring, this having first been mixed with the amount of water required to bring the padding liquor up to volume.

Pad at 20–30 °C to a liquor pick-up of 30–50%.

Dry in hot air at about 100 °C.

Apply thermosol treatment at 200 °C for about 60 s, using hot air.

After clearing:

The rubbing fastness is improved by treating on the winch or continuously on an open-width scouring machine. Good clearing is obtained on the winch by using the following recipe:

Quantity	Unit	Bath additions
3–5	%	Caustic soda 38 °Be (32.5%)
1–3	g/l	Sodium hydrosulphite
0.5–1	g/l	Dispersing agent

Treat for 20–30 min at 50–70 °C, rinse and acidify with acetic acid.

It has been found that a treatment on the winch also imparts a pleasing handle to the material.

15.1.2.6.2   Two-stage process (mainly for dark shades)

First of all, the material is padded with the liquor described for the one-stage process but not containing disperse dye. The drying and thermosol operations are also the same.

After fixing the basic dye, a fresh padding is made with

Quantity	Unit	Bath additions
x	g/l	Disperse dye
2	g/l	Dispersing agent cum fixing accelerator

After a second drying and thermosol treatment (60 s at 200 °C), the material is given a reduction clear in the manner described.

15.1.2.6.3   Continuous method where only one component is dyed

In a blend where acrylic is the major component we can probably dye only acrylic portion using disperse or cationic dye in a continuous manner. Disperse dye along with padding assistant and acetic acid to adjust pH to 4–5 is padded dried and thermo fixed. This method can achieve only pale to light shades only. But pale to medium shades and to some extent dark shades can be dyed continuously applying acrylic only by cationic dyes. The padding bath is set at pH 5 using citric acid with all the auxiliaries at 70 °C. After padding the material is steamed at 100–103 °C like any other pad steam method. In some cases the steaming time may be increased to get the full colour value. If staining on cotton has to be removed it may be removed by a soaping with an anionic detergent at 90 °Cor in case of darker shades a reduction clearing can be given with hydros at 70 °C. Another possibility is dyeing with vat dyes, only for pale –light shades, which will dye the both components. Hence for this method blends of any percentage can be used.

15.1.2.6.4   Continuous method when both components are dyed

Any shades can be dyed by cationic/vat dyes combination. But the dyes are not compatible. But there are selected basic dyes which is being marketed in complex form as liquids stabilised with anionic dispersing agents which is compatible with vat dyes. But for applying this, the anionic groups in the acrylic fibre must be converted to the ammonium salt form by pretreatment with an ammonium salt at 80 °C. Once the fabric is treated with ammonia, it is padded with the complexed basic dyes together with selected vat dyes at 50 °C and pH 8–9 (phosphate buffer). During thermo fixation at 200 °C the complexes are transferred from the cellulosic to the acrylic fibre and also dissociate to give the parent basic dyes. Vat dyes are further developed by chemical pad–pad steam method, where by any unfixed cationic dyes are removed improving fastness properties of the final dyeing.

Since the cationic and reactive dye are non-compatible and there is no special method to keep them not precipitating till dyeing as in the previous method, if required it has to be dyed each dye separately, for example, two bath pad-steam method. This combination gives bright shades of all round fastness on all type of blends percentages.

Chapter 16
Dyeing of CC blends

16.1 Cotton/viscose Blends

It is general doubt in the mind of a processor, why cotton/viscose blend since both are cellulosic fibres. Cotton is a natural fibre (also Linen, Ramie, etc.) and viscose a regenerated fibre. Cotton is considered to be a perfect fibre when its overall characteristics but slightly lacking lustre, drape, absorbency and firmness of handle, etc. Viscose is more absorbent, having good drape but lacking in strength especially wet strength which gives some limitation in certain end uses say, sports wear. The elongation property of pure cotton is also improved by using viscose, especially Lyocell and modal, even though the importance of this has been reduced after the introduction of Lycra. Given below table is comparing the strength, elongation and moisture regain properties of viscose fibres and cotton:

	Fibre tenacity (cN/Tex)		Elongation	Moisture regain	Water imbibition	Crystallinity
Fibre	Dry	Wet	%	%	%	%
Cotton	20–24	26–30	7–9	8	50	70
Viscose	22–26	10–15	20–25	13	90	40
Modal	34–36	19–21	13–15	12.5	75	50
Lyocell	40–42	34–38	13–15	11.5	65	70

It is a natural thinking to blend these two fibres to get more acceptable perfect fibre. This is a very important blend to compensate shortage of good quality cotton and lustre on cotton.

In addition to lustre regenerated fibre mixed with cotton gives additional absorbency than pure cotton, which is useful in to welling constructions. Terry to welling may be made from 50:50 or 65:35 cotton/viscose blends for greater absorbency than all-cotton cloths. Apparel uses for cotton/viscose blends include poplin shirts, blouses, dress wear, knitwear, leisure garments, T-shirts, underwear and children's clothing. These blends offer comfort appeal with good wear and laundering properties. The optimum blend composition for wear resistance is approximately 70:30 cotton/viscose.

Another area where the mixing of viscose with cotton is advantageous is to welling. Here the viscose percentage is increases to 35–50% to achieve greater absorbency than all-cotton to welling. Taking the advantage of the lustre of viscose especially filament yarn, traditionally cotton/viscose unions is used in brocade material for curtains and furnishings in which the viscose appears on the surface in the form of floral designs. These constructions often contain both filament and staple yarns. Pile fabrics are sometimes made with a viscose pile in a cotton backing fabric. These are an economical alternative to wool pile/cotton backing fabrics. Thus viscose/ cotton (or linen/viscose or ramie/viscose) blends are made as intimate mixture or yarn filament mixtures.

The greatest challenge of the processor in dyeing these blends is to get a solid shade even though both being cellulosic fibres. The fibres are having different substantivity, affinity and dyeability due to difference in surface area, accessibility, surface charge, etc. The introduction of viscose microfibres has made the processors life easier. These micro fibres have colour yields and reflectance values to be obtained with direct and reactive dyes that are close to those on cotton. But in the case of brocades and other furnishing fabrics woven into designs formed by raised viscose wefts on a cotton warp ground are not as critical as intimate blends where lack of solidity gives an objectionable skittery appearance. The proportions of crystalline material in viscose fibres and cotton given above prove that polynosic is the closest to cotton and hence water ambition and disability which makes a polynosic/ cotton blend easier to achieve a solid dyeing.

16.1.1   Pre treatment of woven fabrics

Pre treatment of viscose/cotton blend is usually done according to the weaker component in blend – viscose. Generally, warp sizes consist mainly of water-soluble polymers but they are often blended with partly insoluble starch components and fatty or oily size additives. Removal of spinning oils and preparations may happen during desizing and washing processes with effective surfactants, which should be compatible with the desiring recipe. Other impurities: Besides the washing/emulsifying surfactants the use of multipurpose complexing agents is helpful to remove inorganic or metallic impurities.

Desizing

After singeing the fabric is passed through the quench box of a singeing machine and pad with following desizing bath.

Quantity	Unit	Chemical
2–4	ml/l	Amylase enzyme
1.0–2.0	ml/l	Wetting agent

Impregnate at 60°C (as per enzyme requirement); pick-up 80–100%; batch minimum 4–24°h – normally overnight; hot wash-off.

16.1.1.1   Demineralising

Demineralising is not a must for viscose fabric because as such it is clean fibre. However, for better white demineralising can be done at room temperature.

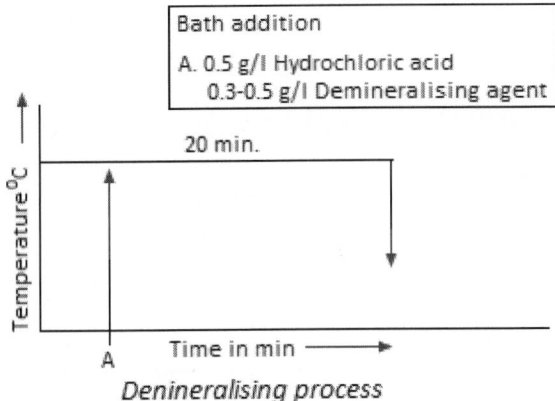

Demineralising process

The process can be carried out at room temperature for 20°min with hydrochloric acid (If the machine metallurgy allows the same) and a demineralising agent.

16.1.1.2   Discontinuous scouring/desizing by surfactants

Woven viscose fabric with water soluble starches and for knit goods.

Quantity	Unit	Chemical
1–3	ml/l	Wetting and detergent
0.5–2.0	ml/l	Chelating agent for heavy metal ions
0–4.0	g/l	Soda ash

Treat for 20–40 °min at 60–80 °C at liquor ratio 5:1–20:1 and hot rinse. pH of the desizing bath should be maintained depending on the sizing material used.

16.1.1.3   Relax/scouring

Viscose rayon fibres, unlike natural cellulosic fibres, are free from natural fats and waxes, motes and seeds, and the scouring process, therefore, need not be as severe as for cotton, and can be based on soda ash or tetra sodium pyrophosphate recipes rather than caustic soda. A typical scouring recipe on a fivebox continuous open width washing range would be:

Box 1

Quantity	Unit	Chemicals
3	g/l	Scouring agent
2	g/l	Soda ash
1	g/l	Wetting agent or detergent

Box 2

Quantity	Unit	Chemicals
1	g/l	Wetting agent or detergent at 95 °C

Box 3

Quantity	Unit	Chemicals
1	g/l	Wetting agent or detergent at 95 °C

Box 4

Water only at 70 °C.

Box 5

Water only – Hot.

16.1.2   Pretreatment of Knitted fabric

16.1.2.1   Scouring on jet or overflow machine

Quantity	Unit	Additions
2–4	g/l	Low foaming jet scouring agent
0.5–1.0	g/l	Wetting agent
0.5–2.0	g/l	Defoamer (if necessary)
1–3	g/l	Soda ash (pH about 9)

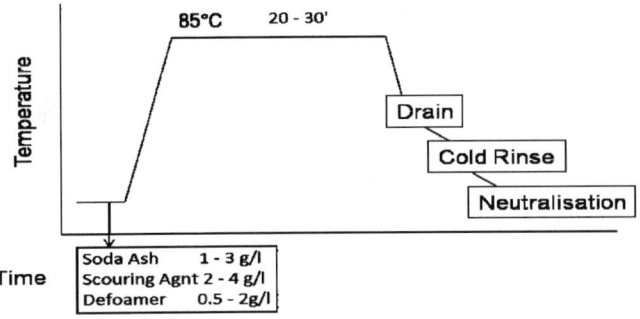

Run 20–30 °min at 60–90 °C.

Cold rinse, neutralise.

16.1.2.2   Peroxide bleach on jet or overflow machine

For full white

Quantity	Unit	Additions
1–3	ml/l	Peroxide stabiliser (as recommended)
2–4	g/l	Wetting cum scouring agent
0.5–2.0	g/l	Defoamer (if necessary)
2–6	ml/l	Hydrogen peroxide 35%
0.3–0.7	%	Optical whitener

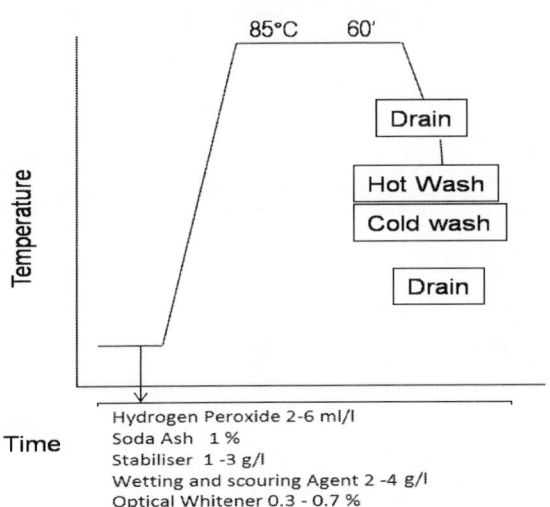

Load the fabric, run for 5 min, add chemicals, auxiliaries and optical whitener, run at room temperature at 5 min and raise the temperature to 85°C, and hold at this temperature for 60 min, cool and drain. Neutralise with core neutralising agent, hot wash and cold wash.

Notes on soft-flow processing

   (a) Remove oils and waxes in preparation.

   (b) No twists in loading – load machine below capacity.

   (c) Sewing – sew more than once – at an angle.

   (d) Abrasion – avoid slippage – always run at 50 °C or higher.

   (e) Cooling creases – cool at 1 °C/min.

   (f) Danger of sulphur residues and reduction.

   (g) Tension in soft flow – problems with shrinkage.

   (h) Migration of unfixed dye while waiting for drying.

   (i) Fabric drying out in patches–watermarks (physical/optical effect) Keep fabric wet right up to the dryer.

16.1.2.3   Pad steam peroxide bleach (PS PB)

Quantity	Unit	Chemical
4.0–6.0	ml/kg	Suitable peroxide stabiliser
4	g/kg	NaOH 100%
6.0–15	ml/kg	H_2O_2 50%
1.0–2.0	ml/kg	Low foaming washing and rewetting agent

Process: Impregnate cold

   3–12 °min steaming (saturated conditions),

   Hot wash off.

16.1.2.4   Discontinuous sodium chlorite bleach

For viscose knit goods and woven goods

Quantity	Unit	Chemical
1.0–2.0	ml/l	Wetting and detergent
1.0–2.0	ml/l	Sodium chlorite 80%

1.0–2.0	g/l	Buffering salt (e.g. PK2 from Degussa-Hulis)
2.0–3.0	g/l	Sodium nitrate (corrosion inhibitor)
1.0–2.0	ml/l	Formic acid 80% pH about 4.5

Process: Treat 30–60 °min at 80–85 °C.

Liquor ratio 5:1–20:1.

Rinse, antichlor.

Hot rinse.

16.1.2.5   Discontinuous reductive bleach

For viscose knit goods and woven goods

Quantity	Unit	Chemical
3.0–5.0	g/l	Stabilized sodium dithionate
0.5–1.0	m/l	Wetting and detergent

Process: T reat for 30–60 °min at 75–85 °C.

Liquor ratio 5:1–20:1.

Wash off at 40 °C with addition of 0.5–1.0 °ml/l H2O2 35%.

Warm rinse.

16.1.2.6   Mercerising (dry)

Mercerising of cotton/viscose fabrics can be carried out on grey goods. Since mercerising in grey is not as effective as on scoured fabrics for cotton it is advantageous in case of cotton viscose to get better even effect in dyeing. Another advantage in the dry mercerising of this blend is that the swelling and dissolving range of 12–13 °Be caustic soda (Viscose fibres have a solubility optimum in caustic soda liquor of 13 °Be at 20 °C) where viscose fibres is particularly at risk is not undergone twice, as happens in the mercerising of wet goods. The mercerising of grey yarns and fabrics is what is usually understood by dry mercerising. (If moist yarns or fabric are mercerised from cotton/viscose, these run through twice the dangerous concentration range of 13 °Be for viscose – one during impregnation and one during washing. That is why it is better to dry it beforehand. The solubility is less in hotter liquors, greater in colder ones. The rinsing process is altered compared with the treatment of pure cotton in that it is washed immediately with very hot water (50–80 °C) in order to achieve a throughput of the danger zone as quickly as possible. Mercerisation with caustic potash solution is less critical, as viscose fibres swell significantly less in that than in caustic soda liquor. However, as the effect on cotton is less, one works with mixtures of both liquors. Protection of the viscose fibres can be achieved by adding 40–50 g/l of sodium chloride to the caustic soda liquor. Mixed fabric made from cotton and polynosic fibres can be mercerised without special measures. Certain differences should though exist between polynosic and HWM fibres in their reaction to mercerising liquors.

16.1.2.7   Causticising

Since there is danger of weakening of viscose in normal mercerising usually a causticisation process is more advisable, during which a partial mercerisation and increase in dye uptake of cotton also takes place. Typical conditions are treatment with 6–8 Becaustic soda at 25–30 °C for at least 2 min followed by low tension washing with boiling water to assist the rapid

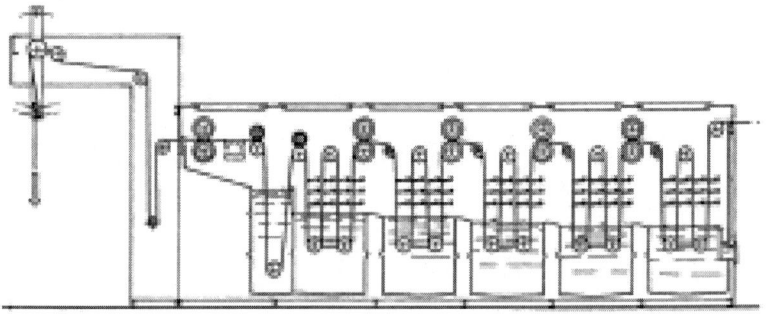

A typical causticising machine

removal of alkali. Liquor pick-up of about 120% should be achieved. The best after-washing device is probably a sieve drum continuous open-width range. Low uniform processing tensions are essential for consistent results and good quality. Rinsing should be done as hot as possible to minimise swelling and ensure rapid and complete removal of caustic soda. An addition of 2–4 g/l soda ash in the initial wash boxes will promote removal of alkali, and help maintain a good fabric handle. Neutralisation with suitable acidis also advisable.

Pad-batch processing is also popular although dedicated continuous plant, employing a scray or convey or for tensionless swelling and reaction, is preferable.

This step is preferably carried out by:

(j) Continuous pad dwell methods on conveyors or scrays,

(k) Pad-batch semi continuous methods on A-frames,

(l)   Jig,

(m) Discontinuous methods in rope form only to obtain special surface effects.

To get higher colour yield the causticising should be applied with 6–8°Be NaOH.

To reduce excessive shrinkage and harsh handle, the treatment with pure caustic soda may be modified: (any one)

1.   KOH up to 7 °Be,

2.   Mixture of both,

3.   NaOH in mixture with common salt (50–150 °g/l NaCl).

Process: Causticising bath: 40–50 °g/l NaOH 100% containing 2–4 ml/l detergent.

-Padding

Impregnate cold; liquor pick-up 80–120%,

Dwell for 2–4 h at room temperature,

Hot wash off.

-Or continuous/relaxing

Treat 1–15 °min at 60–80 °C,

Hot wash off

16.1.3   Dyeing

Since both are cellulosic components of similar dye ability reserve and cross shades cannot be produced. Shadow and solid shades are produced with any cotton dye in one bath one step method. The shadow effect is developed on short time dyeing whereas solid shade on prolonged dyeing.

Blends of mercerised cotton and viscose will often give good solidity with direct dyes even in full depths by dyeing at the boil with little or no salt because of the inherently higher dye ability of mercerised cotton. The development of viscose microfibres has enabled colour yields and reflectance values to be obtained with direct or reactive dyes that are close to those on mercerised cotton. As a result, the attainment of solidity on fabrics containing these two fibre types is now easier than on conventional cotton/viscose blends. The proportions of crystalline material in regenerated cellulosic fibres are about 40% in regular viscose, 50% in modal fibres and 65% in polynosics, compared with 70% in cotton. As crystallinity increases the water imbibition and dye ability decrease accordingly. Thus direct dye uptake under a given set of conditions generally increases in the order: cotton<polynosics<modal fibres<regular viscose.

Given below the better dye selections for various cotton/viscose blends

Blend	Solidity	Dye selection
Mercerised cotton/ viscose	Solid	Solid direct dyes at low salt concentration at the boil
Unmercerised cotton/viscose	Solid	Disazo tetrasulphonated direct dyes at 60 °C
Cotton/viscose	Solid or shadow	Selected vat dyes at 20–30 °C
Cotton/viscose microfibres	Solid or shadow	Direct or reactive dyes
Cotton/modal	Solid or shadow	Direct or reactive dyes
Cotton/lyocell	Solid	1. Selected self-levelling and temperature-controllable direct dyes 2. Selected vat dyes at 50 °C 3. Selected high-reactivity dyes by pad-batch
Viscose/polynosic Viscose/lyocell	Solid or shadow	Direct, reactive or vat dyes

16.1.3.1   Dyeing with direct dyes

Dyeing methods for dyeing cotton/viscose blends can be same as for the cotton fabrics. Solidity of shade with direct dyes on cotton/viscose blends varies considerably with dye structure but it can be controlled by adjusting salt concentration and dyeing temperature. The substantivity of direct dyes for cellulose is approximately inversely related to their degree of sulphonation. Direct dyes giving good solidity on cotton/viscose blends tend to be mainly disazo tetra sulphonates, including some copper-complex types.

Pales shades can be dyed without any problems by dyeing in the absence of salt, which can be referred under cotton dyeing (Dyeing of textile substrates I – Cotton). Solidity in deeper shades is achieved more readily with little or no salt present at the boil. Where salt must be used for medium and full depths in order to attain economical exhaustion, sometimes solidity can be ensured only by dyeing at a temperature as low at 60°C. The optimum conditions vary from one dye to another and result in reduced penetration and lower wet fastness. Blends of mercerised cotton and viscose, however, will often give good solidity in full depths by dyeing at the boil with only low concentrations of salt (0–5 °g/l) because of the higher dye ability of mercerised cotton. If necessary, dyeing is started at the boil with a portion of the dye, and the rest is added to the bath after cooling down.

The development of viscose microfibres has enabled colour yields and reflectance values to be obtained with direct and reactive dyes that are close to those on cotton. As a result, the attainment of solid effects on fabrics containing viscose microfibres and cotton is now easier than on conventional cotton/viscose blends.

16.1.3.2   Dyeing with vat and sulphur dyes

With the use of vat and sulphur dyes more solidity can be achieved compared to direct and reactive dyes. But even in this case the selection of vat dyes (acridones, carbazoles, indanthrone blues and violanthrone blues and greens) or cold dyeing dyes (coder dyeing prefers cotton dyeing as far as vat dyes are concerned) helps in better solidity. Other than these, better solidity can be got by dyeing at lower temperatures, pad dyeing methods (pad jig at 30–50 °C gives a better solid dyeing than direct exhaust dyeing on jig).

16.1.3.3   Dyeing with reactive dyes

Exhaustion and fixation of reactive dyes on viscose or modal fibres are normally higher than on cotton or linen, but this is not generally true for polynosic or high-tenacity fibres. Many reactive dyes show significantly higher fastness to light on viscose than on cotton at equivalent applied depths. Lightweight viscose materials can be dyed at relatively low temperatures with high-reactivity dyes using processes that are not much different from corresponding methods for cotton. Vat or sulphur dyes are often used because it is generally more difficult to achieve solidity with direct or reactive dyes.

Notes:

1. Salt and alkali requirements are generally lower for reactive dyes on viscose than for the corresponding dyeings on cotton but most green or turquoise hues based on phthalocyanine dyes are applied according to special recommendations.

2. Yarns and fabrics made from coarser denier viscose or heavy weight fabric constructions may show poor penetration and surface frostiness in reactive dyeings.

3. Dyes of low to moderate reactivity requiring fixation temperatures of 60°C or higher give better results on materials that show inferior levelling or inadequate penetration at lower temperatures.

4. Under alkaline conditions in enclosed machines, viscose tends to give problems of reduction with certain sensitive azo reactive dyes. Addition of a reduction inhibitor from the start of dyeing is essential.

5. Semi continuous dyeing process for cotton and cotton/viscose tubular knit goods for crease-free dyeing with reactive dye stuffs according to the pad-roll dyeing process.

6. The dye liquor application in the pad trough, there follows compressed air injection and inflation of the fabric tube before squeezing and winding onto the perforated cylinder where the fabric will dwell (2–6h). Washing then follows on the same perforated beam.

7. Padding methods are more suitable than exhaust methods for vat, leuco-vat ester and naphthol dyes. Causticising before dyeing improves the affinity of the cotton.

Chapter 17
Dyeing of DA blends

17.1 Dyeing of polyester/wool, polyester/silk blends

Dyeing of polyester/wool materials are little complicated because the wool may undergo thermal decomposition at the dyeing conditions of polyester, which is usually at high temperature (130 $\pm$ 2°C) and high pressure. Thus wool may not with stand a dyeing temperature above 105–110 °C, and hence we have to design the processes at a maximum temperature of this temperature or below but still there can be some sort of degradation on woollen part. Due to this, wool and polyester fibres are invariably dyed in top form with desired shade and blended to produce fancy yarn and fabric.

17.1.1 Pretreatment

Spotting: The material is inspected for manufacturing faults and spots. Wet spotting is suitable for dealing with pronounced but localised soiling, and non-ionic wetting agent in an aqueous solution (1:5–1:10) is excellent for this purpose. Dirt and grease spots, especially mineral oil stains, can be relatively easily removed by this method. Persistent spots are removed more easily if the material is allowed to lie for, say, ½–1 h after treating the spots. However, the treated areas should not be allowed to dry out before scouring, otherwise, rings may remain. If a scour is not to be given soon after spotting, the material should be wrapped in a damp cloth or plastic sheet. Solvents, or solvent-containing detergents, should only be used with great care for localised spotting on account of rapid drying out and ring formation.

17.1.1.1 Desizing/scouring

Scouring does not only remove spinning oil, sizing agents and impurities of all kinds, but also serves to eliminate stresses in the polyester/wool fabric, and to close up the material. Scouring is usually carried out in open-width machines or rope scouring machines, of the type usually employed in wool factories. The scouring effect is more intensive in rope scouring machines, but there is a danger of the formation of folds and breaks which produce streakiness on subsequent dyeing. Scouring is also frequently carried out on the winch, but the effect is less than with a scouring machine. If a light milling effect is required, the goods can be treated on an open-width machine with a stuffing device or on a jet scouring machine. The use of a milling machine is also possible, no top roller pressure being applied. Folds and creases which have arisen during scouring or dyeing can also be eliminated by mild milling.

Recipe

Quantity	Unit	Bath additions
0.25–0.5	%	Non-ionic detergent
0.25–0.5	ml/l	Ammonia 25% (or 0.2–0.4 g/l soda ash)
		Adjust pH 8–9

Scouring is done at 30–40°C for 30–45 min. The material is afterwards rinsed cold and acidified with acetic or formic acid. (Where soiling is heavy, a further scour is given for 15–20 min after a short intermediate rinse. In this case, detergent can be increased to 1.5 times than in normal scouring – recipe above). Normal water soluble sizes are removed by the above scouring process. Any starch size is used (usually not used) an enzyme desizing can be given.

17.1.1.2   Crabbing

To obtain optimum setting of polyester fibre/wool fabrics, the two components must be dealt with in different ways. The wool component is subjected to a hot water treatment (crabbing), while the polyester component is usually heat-set by a hot-air treatment. Heat-setting hardly has any effect in setting the wool, while the polyester fibre is little affected by crabbing. The hot water setting of the wool component is carried out in the crabbing machine. Normally, piece goods are crabbed at the boil after scouring. The water and goods should be neutral or weakly acid.

Procedure:

The material is entered in open width and under low tension into the water at 50–60°C under a pressure roller; the water is then heated up to the boil. After running for a few minutes, the pressure roller is removed and treatment continues for a further 15–20 min. With heavy materials, it is better to crab at the boil over two passages, each involving 10 min running time, so that uniform setting over the whole length of the material is ensured. This also eliminates the danger on "ending" on dyeing. Finally, the material, depending on its type and desired effect, is either quenched by passing through cold water, or it is allowed to cool down slowly in roll form over 4–6 h (possibly overnight).

On crabbing coloured woven pieces, the fastness of the dyeings must be considered. It is not advisable for the water temperature to exceed about 60 °C, and a little acetic or formic acid can be added to the treatment liquor. If there is a danger of running folds with sensitive materials being scoured on a rope-scouring machine or winch, crabbing is carried out before scouring. Often, a short treatment of 10–15 min at 50–60°C suffices. In special cases, and depending on the type of material, the temperature is raised up to about 90 °C. After scouring, another crabbing treatment can be given if required. Crabbing does not have such a permanent effect on the set of the wool as does heat-setting on the polyester component. It is therefore often useful to repeat the crabbing treatment during after finishing, in order to smooth out folds or to improve the handle of the mixture fabric as far as this is affected by the wool component. If a crabbing or smoothing treatment is given after dyeing, the conditions observed for crabbing coloured woven materials, i.e., a temperature of 50–60 °C, should be observed.

17.1.1.3   Drying

The goods, which have first been subjected to suction or squeezing, are dried at about 100°C with as little warp wise or weft wise tension as possible. Stenters (either flat or in multi-stage units on top of one another), and short loop driers are suitable for this. In order to obtain a fold-free and well relaxed material on a stenter, the width to which the frame is adjusted should only be slightly greater, say 1–1.5 cm, than the wet width of the material. In addition, should there be any lengthwise shrinkage in the material, it should be overfed into the machine by, say, 4–5%.

A temperature of over 100°C can be employed over a correspondingly shorter heating time. Excessive drying must be avoided at all costs.

17.1.1.4   Heat-setting

Heat-setting polyester fibre/wool fabrics improves the handle, resilience, crease resistance, dimensional stability, shrink- and pilling-resistance of the polyester fibre component. Similar effects are obtained on the wool component by crabbing; polyester fibre/wool fabrics to be dyed in the piece should be given a prior heat-setting treatment in order to avoid trouble during dyeing. The quality is improved if the material is steamed, sprayed with a finely dispersed mist of water or allowed to lie in a moist chamber for a sufficient length of time between drying and heat-setting, so that the wool component can regain its natural moisture, i.e., about 15%.

A hot-air treatment on the stenter for

30 s at 180–195°C,

is generally sufficient. The material should be given the opportunity to shrink freely during the hot-air treatment by a suitable adjustment of the width and an adequate over feed. Depending on the type of material, the shrinkage in warp and weft is usually of the order of 3–8%. Tightly woven fabrics shrink less. The handle of the material is impaired by raising the temperature or the treatment time, whereas dimensional stability, crease-resistance and pilling resistance become worse if temperature and time are reduced. Heat-setting under a fairly high tension, that is, with little opportunity for shrinking, also leads to poorer crease recovery and a greater tendency to pilling; furthermore, a higher residual shrinkage may well become apparent at a later stage. An optimum setting effect is obtained when the hot goods are cooled extremely rapidly in a cooling zone after leaving the last heat-setting unit. Hardening of the handle (setting stiffness) arising during heat-setting disappears again during subsequent dyeing so that it is not normally necessary to give the piece an intermediate wash before dyeing. Very densely woven materials provide an exception here, because they are made very stiff by heat-setting. It is best to wash these once more before dyeing, otherwise the goods may prove very difficult to handle. The same wash recipes are employed as are given in the next section for coloured woven materials, though the temperature and time can be made higher for the undyed goods where necessary.

With coloured woven fabrics, setting stiffness must be removed by an after wash in order to improve the handle, crease recovery and resilience. In this case the material should be washed on an open-width, rope or jet washing machine, or on a winch with

Quantity	Unit	Bath additions
0.5	%	Non-ionic detergent
0.5–1.0	ml/l	Acetic acid 30%
		Adjust pH 5–6

for 20–30 min at 45–60°C. The material is then rinsed and the excess water removed by suction or squeezing, where upon drying proceeds without tension. It is advantageous to give an after wash with an anionic agent to obtain an even softer handle.

17.1.1.5   Polyester/silk

Polyester/silk may be washed in a minimum tension in open soaper containing 1 g/1 non-ionic detergent and 1 ml/1 NH$_3$ (25%) at 50 °C.

17.1.1.6   Bleaching of polyester/wool and polyester/silk

Wool has an inherent creamy colour. The polyester/wool blend can be bleached with peroxide in either alkaline or acid medium. There is a tendency for the original cream colour to return, which is prevented by reductive bleach after peroxide bleach.

17.1.1.6.1   **Acid bleach**

Quantity	Unit	Additions
30–40	ml/l	Hydrogen peroxide 35%
2–4	g/l	Organic stabiliser
0.25	g/l	Wetting agent
0.25	g/l	Detergent
x	ml/l	Acetic acid to pH 5.5–6

Treat for 40–60 min at 80 °C or 2–2.5 h at 65 °C. The treated fabrics are then given warm and cold rinse.

17.1.1.6.2   **Alkaline bleach**

Since the wool cannot withstand alkaline bath, care has to be taken regarding the alkali and pH.

Mild alkali like ammonia is used to maintain the pH of about 8.5.

Quantity	Unit	Additions
30–40	ml/l	Hydrogen peroxide 35%
2–4	g/l	Sodium pyrophosphate
0.25	g/l	Wetting agent
0.25	g/l	Detergent
x	ml/l	Ammonia to adjust the pH to 8.5–9

Treat at 40 °C for 2–4 h and wash well and neutralise, if necessary.

17.1.1.6.3   *Reductive bleach*

Quantity	Unit	Additions
3–4	g/l	Stabilised sodium hydrosulphite
1	g/l	Synthetic detergent

Run at 50 °C for 30–40 min and rinse well.

Polyester/silk also can be bleached following the same process as above.

17.1.2   Dyeing polyester/wool and polyester/silk

Since the polyester/wool blends are made of two fibres of two different natures, it can be dyed only by the two bath two step method. This method of dyeing is more popular mainly because dyeing of components in two separate baths eliminates complications arising from interaction of dyes and chemicals if one bath is used.

Most commonly followed dye combinations are disperse–acid or metal–complex dye. Disperse dye heavily strains wool during dyeing of polyester which reduces the wash fastness and light fastness of blend; the problem becomes severe if carriers are added. Even though the wool is stained by disperse dyes, at higher temperatures these dyes migrate to polyester and thereby ultimately less staining. Since there both the dyes are dyed in acidic pH there are possibilities of dyeing by a one bath one step method by dyeing polyester at 130 °C, the bath is cooled down to 80°C followed by dyeing of wool at boil. However, wool is degraded up to some extent. The safest method is to take process in between – dye at 110°C and reduction clear and dye wool as a second step.

17.1.2.1   Dyebath additives

17.1.2.1.1   *Dispersing agents*

Dispersing agents are necessary in the dyeing baths of disperse dyes since it is dyed from dispersion, rather than a solution in the case of most other dyes. The dye molecules of disperse dyes are in the size range 0.5–1.0 μm, but during dyeing there are chances of agglomeration to form bigger sizes which can result in dye precipitation which can further cause dye spots, slow exhaustion, poor colour yield, unlevel dyeings, poor reproducibility and deposition on dyeing machines. In such cases dispersing agents helps in dyeing by avoiding such problems.

Commonly used dispersing agents are mainly sulphonic acid salts of naphthalene/formaldehyde condensates (see figure below) and lignosulphonates are used widely as dispersing agents. The

hydrophobic group of the dispersant is adsorbed on to the surfaces of the dispersed dye particles, and this effectively encapsulates the dye. The sulphonate groups are solvated by the water (ensuring water solubility of the agent), with the formation of an electrical double layer. Mutual repulsion of the negatively charged dye particles inhibits breakdown of the dispersion.

17.1.2.1.2   Levelling agent

Polyester dyeing rate being slow the necessity of a levelling agent may not be of that importance. However, if a levelling agent is required mostly non-ionic surface active agents like poly(ethylene glycol) derivatives such as $C_{16}H_{33}(CH_2CH_2O)_nCH_2CH_2OH$ (n = about 16) have been found to be useful. These agents are supposed to increase the solubility of disperse dyes in water there by reducing dye uptake and thus the levelling action. It has to be noted that if the cloud point of the levelling agent is below the dyebath temperature there are chances of precipitation of the dyes and can cause uneven dyeing. The cloud point of the surfactant is affected by other dyebath additions like electrolytes, anionic agents, etc. – the former decreases the cloud point while the latter increases the cloud point.

17.1.2.1.3   Carrier

After the introduction of HT/HP machines the requirements of carrier for dyeing of polyester has been necessary. It was a blessing in disguise as there was lot of environmental restriction in the use of carriers due to its toxicity obnoxious smell, etc. There are many theories regarding the action of the carriers. They said to act by increasing the rate of dyeing of the polyester fibre by swelling the polyester and increasing the pore size and thereby increasing the diffusion of the disperse dye molecules. There are other theories like water-insoluble carriers form a film on the fibre surface having high solubility of disperse dyes which supports dye uptake by the fibre. This is used to be helpful to dye the polyester at lower temperature when high temperature machines were not popular. However, by the introduction of blends of polyester with other fibres which cannot withstand the temperature of dyeing of polyester like polyester/wool. Polyester/silk, etc., the importance of carriers has again increased. Most of the carriers are not soluble in water and normally used in the emulsion form. Carriers like di- and trichlorobenzene are highly volatile and hence mainly used in enclosed machines. They have little or no effect on the light fastness on the fibre and they are easily removed from the material since they are volatile. Other type carriers like o-phenyl phenol, benzyl phenol, etc. are less volatile and can be used in machines like winches, jigs, etc. But they have the problems of lowering the light fastness due the difficulty in removing them after dyeing. Whichever is the case the carrier has to be removed from the fabric by stentering at 160–170 °C.

Di- and trichlorobenzene are particularly effective carriers at low concentrations, giving good colour yields on polyester and minimal disperse dye staining on the wool, along with good economy in usage. Their use is restricted to closed machinery, however, because of their volatility and toxicity; furthermore, they are not readily biodegradable. o-Phenylphenol (OPP) has been used widely for the dyeing of wool/polyester blends at temperatures close to the boil. It is relatively non-volatile, has a weak odour and can be used in open machines. Compare to the former carriers OPP type carriers has to be used relatively high concentrations due to its slight solubility in water and its absorption by wool. OPP is not readily biodegradable, has no influence on levelling and increases staining of the wool by disperse dyes.

17.1.2.1.4   *One-bath dyeing process with disperse and acid dyes with carrier*

Acid dyes have very good fastness properties on wool and do not dye polyester fibres, or at least, only slightly. They dye on wool from a weakly acid liquor and are therefore highly suitable for the one-bath dyeing of polyester fibre/wool mixtures in company with selected low energy disperse dyes which stain the wool very little, or which can be readily removed from the wool by washing. (This process is not followed much presently due to the restrictions on the usage of carriers).

Recipe

Quantity	Unit	Bath additions
x	%	Disperse dyes
y	%	Acid dyes
z	g/l	Carrier
2–3	%	Acetic acid 30% to pH 5–6
1–2	%	Dispersing agent
5–10	%	Glauber's salt anhydr.

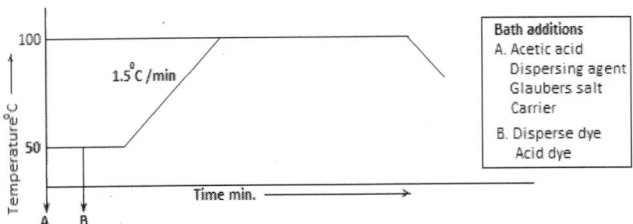

The dye liquor is first set with acetic acid and dispersing agent and Glauber's salt. The initial temperature depends on the dyeing accelerator employed. The material is allowed to run for 10–15 minutes whereupon the dispersed disperse dyes and dissolved acid are added. After running for a further 5 minutes, the temperature is raised within 30–45 min either to the boil. Depending on the depth of shade, dyeing proceeds at the boil for 90–120 min. After dyeing and slow cooling, the material is first rinsed hot to remove the carrier, and then cold. Finally it is given an after scour as follows.

The same dyeing can be done at 103–108°C, for 60–90 min provided the carrier withstands this temperature.

After scour

On dyeing polyester fibre/wool mixtures, staining (non-fast) of the wool by disperse dyes cannot be completely avoided even when using most suitable dyes and favourable dyeing procedure. To obtain better fastness properties, therefore, it is necessary to remove the disperse dye staining the wool, as well as that which is only superficially held by the polyester fibre, by clearing. Dyeings which have been prepared by the one-bath process cannot normally be subjected to reduction clearing, since wool dyes cannot stand up to this treatment. An after scour at not too high temperature is given.

Quantity	Unit	Bath additions
1–2	g/l	Dispersing and clearing agent
0.5–1.0	g/l	Acetic acid to pH 5–6

Treat for 20–30 min at 50–60 °C. The material is then well rinsed. This treatment is repeated with deep shades wherever necessary.

17.1.2.1.5 *One-bath dyeing process at 110–115 °C (in the presence of wool protective agent)*

The HT dyeing of polyester/wool mixtures normally proceeds over the temperature range 103–108°C. Sometimes, the temperature of the liquor is raised briefly up to about 110 °C towards the end of dyeing in order to obtain better exhaustion of the liquor and to reduce staining of the wool component by the disperse dye. The resulting slight degradation of has to be tolerated, but it is not advisable to dye polyester fibre/wool mixtures at a higher temperature, say 110–115°C without the addition of a suitable wool protective agent, since, at this temperature over the usual dyeing time of approximately 60–90min, a considerable change in chemical and mechanical properties of the wool can take place. Even though of a number of known wool protective agents which were tested for this purpose, formaldehyde (3–5% formaldehyde 30%) proved to be the best but there are restrictions in using the same because of reasons obvious. It is better to use other wool protective agents available. Usually, pale to medium shades are dyed by this method from one bath and without the addition of carrier.

Recipe

Quantity	Unit	Bath additions
x	%	Disperse dyes
y	%	Acid dyes
2–3	%	Acetic acid to adjust pH 5–6
3–5	%	Wool protective agent
0.5–1.5	%	Dispersing agent

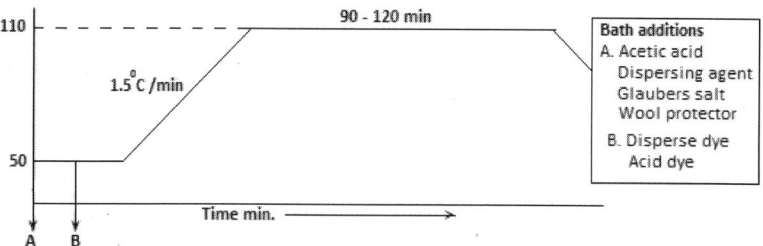

The dye liquor, at approximately 50 °C, is first set with acetic acid, wool protector, dispersing agent. If acid dyes are to be included, Glauber's salt is also added. After running the material in the liquor for a short time, the dispersed disperse dyes and the dissolved acid dyes are added one after another. After a further 5–10 min running, the bath is raised to 110–115 °C within about 45 min. Dyeing proceeds at this temperature for a total dyeing time (including shading) of about 90–120 min should not be exceeded on account of possible damage to the wool. After dyeing, the liquor is cooled slowly and the material rinsed.

After scouring is done as given in the previous method.

Notes:

1. 120 °C gives better dyeing but can cause significant wool damage and calls for the use of a wool protective agent. However, dyeing in the range 110–115 °C is far more safe. Studies have shown that the use of 3–5% omf of a 30% solution of formaldehyde for a blend such as 45:55 wool/polyester has shown an effective wool protective agent.

2. Later, to avoid the use of free formaldehyde, auxiliaries such as Irgasol HTW (Huntsman) and Lanasan PW Liquid (Clariant), based on formaldehyde/urea condensates, were promoted.

3. Polyester/silk with 15–30% of silk can be dyed using this method as silk can withstand the temperature of about 115 °C. In the first stage, polyester is dyed with disperse colour in the temperature range of 105–115 °C for about an hour in order to reduce staining of the disperse dye on silk. Subsequently, after reduction clearing, the silk is dyed with acid and/or metal complex dye.

17.1.2.1.6   One-bath dyeing process with reduction clearing with disperse/metal complex dyes combination

As a rule, polyester fibre/wool mixtures dyed from one bath cannot be given a reduction clearing, since the wool dyes are frequently unstable under the reducing conditions. Nevertheless, an intensive reduction clearing treatment is often advantageous after dyeing deep shades, especially navy blues and blacks. In this event, selected 1:2 metal complex dyes can be used, since these are stable to reduction clearing under certain conditions.

Quantity	Unit	Bath additions
x	%	Disperse dyes
y	%	Selected 1:2 metal complex dyes
z	g/l	Carrier
2–3	%	Acetic acid 30% to pH 5–6
1–2	%	Dispersing agent

Dyeing is carried out the same way as described under 'Sect. 17.1.2.1.4' 3% acetic acid 30% or 2% formic acid 85% are added after dyeing at the boil for 60 min in order to ensure exhaustion of the dye liquor.

Reduction clearing

After rinsing, the material is treated with:

Quantity	Unit	Bath additions
3	g/l	Sodium hydrosulphite
0.5–1.0	ml/l	Ammonia 25%
1–2	g/l	Dispersing and wool clearing agent

for 30 min at 45–50°C. This is followed by a thorough rinse and acetic acid is added to the last rinsing bath.

17.1.2.1.7   Two bath dyeing with disperse and acid dyes

When dyeing deep shades, and where the fastness requirements are very high, a two-bath process is often resorted to, and with this method disperse dyes can be used which, although they strongly stain the wool component, can largely be removed from the wool subsequently by reduction clearing or thorough scouring.

However, the efficiency of the two-bath process as against the one-bath process should not be over-estimated for the following reasons:

(a) The effect obtained by the intermediate clearing is not completely maintained, because some of the disperse dye migrates to the wool during the second dyeing process, and further clearing, as after the one-bath process, is advisable upon dyeing the wool component.

(b) Dyeings made with anthraquinonoid disperse dyes, included in almost every dye recipe, usually become much paler after reduction clearing, but are not decomposed to colourless products. A change in shade of the disperse dye on the wool when given reduction clearing can be very troublesome. An intensive intermediate washing without reducing agent might prove preferable in such cases. The intermediate clearing of polyester fibre/wool blend yarns in the form of cheeses can prove particularly difficult, since the removal and rinsing out of dye residues can become very tedious on account of the filtering action of the material.

Procedure

Dyeing of polyester component

Recipe

Quantity	Unit	Additions
x	%	Disperse dyes
y	g/l	Carrier
2–3	%	Acetic acid to pH 5–6
1–2	%	Dispersing agent cum wool reserve

Starting bath temperature (50–70 °C) is adjusted as per the carrier used. Add acetic acid and dispersing agent. The carrier emulsion is then added and the material is allowed to run for 10–15 min. The disperse dye after dispersing is then poured in, and after running for a further 5 min, the liquor is raised to the boil, or to 103–108 °C, within 30—45 min. Depending on the depth of shade, dyeing proceeds for 1–3 h at the prescribed temperature. After cooling the liquor down slowly, the material is given a hot and cold rinse.

Intermediate clearing

As a rule, reduction clearing is the most effective method, but a milder intermediate scour is often preferred on account of the wool. Since there is no need to be concerned about any wool dye, this scour can be made more vigorous than the after scour. The efficiency can be improved by raising the scouring temperature and the amount of detergent. Prolonging the time over 30 min does not usually produce better results, but, on the other hand, a repetition of the treatment in a fresh bath is very effective, particularly with dark shades.

Recipe for reduction clearing

Quantity	Unit	Additions
3	g/l	Sodium hydrosulphite
0.5–1.0	ml/l	Ammonia 25%
1–2	%	Dispersing agent

Treat for 30 min at 45–50 °C. The material is then rinsed very thoroughly, and acetic acid or formic acid is used to acidify the last rinsing bath.

Recipe for clearing by scouring

Quantity	Unit	Additions
1–2	g/l	Dispersing agent

| 0.5–1.0 | ml/l | Acetic acid 30% pH 5–6 |

Dyeing wool component

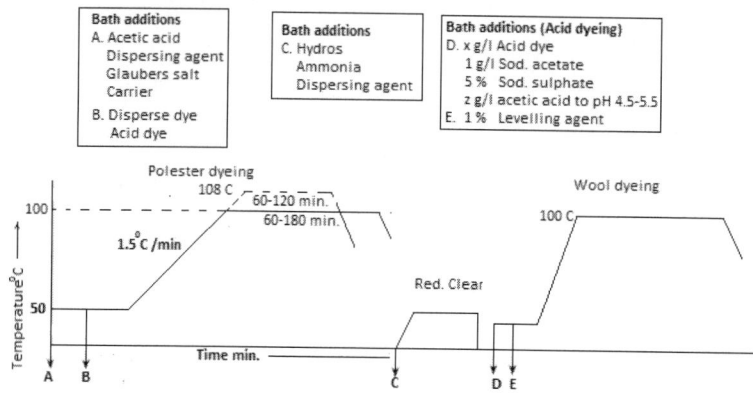

Bath additions
A. Acetic acid
   Dispersing agent
   Glauber's salt
   Carrier
B. Disperse dye
   Acid dye

Bath additions
C. Hydros
   Ammonia
   Dispersing agent

Bath additions (Acid dyeing)
D. x g/l Acid dye
   1 g/l Sod. acetate
   5 % Sod. sulphate
   z g/l acetic acid to pH 4.5-5.5
E. 1 % Levelling agent

Normally, acid dyes, do not stain polyester fibres, or only slightly so. They are, without exception, suitable for the subsequent dyeing of the wool component. (The details of process can be seen under dyeing of wool with acid dyes).

Quantity	Unit	Additions
x	%	Acid dyes
2–3	%	Acetic acid 30% pH 5–6*
1–2	%	Dispersing/levelling agent
5–10	%	Glauber's salt calc. (Optional)

*Can be replaced by 2–3% ammonium sulphate or 3–5% ammonium acetate or 2–3% sodium acetate.

The liquor, at about 45 °C is first set with the chemicals and disperse dyes, and the material is allowed to run for a few minutes, a pH check being made. The dissolved dye is then added and the bath brought to the boil or to approximately 105 °C in about 45 min. Dyeing then proceeds for about 45 min at the desired temperature. After dyeing, the liquor is cooled slowly and the material rinsed.

If a beam-dyeing machine used for dyeing, the mixture fabric must be particularly well set. Residual shrinkage should be so low that none occurs during dyeing. Polyester fibre/wool fabrics can be rolled on in either the dry or wet state. The material is rolled on edge to edge onto the fabric-covered beam, the tension being kept uniform taking care of the coverage from the perforations. After batching, the material is covered with a loosely-woven but firm material or netting. The danger of moire effects is reduced if the liquor is circulated in an outward direction only, and the material allowed to cool down slowly after completion of dyeing.

Usually, in the two-bath process, only the polyester component is dyed in the beam-dyeing machine, while intermediate washing or reduction clearing and crossdyeing of the wool is carried out on the winch or jet/soft flow.

17.1.2.1.8   *Continuous methods*

17.1.2.1.8.1   *Semi-continuous processes*

In the case of polyester wool, generally polyester wool can be continuously dyed by thermosol method and wool can be dyed by batch wise process in jig, winch or jet/overflow machines. The polyester dyeing by thermosol method, which can be followed here also, has been discussed under polyester section (Sect. 17.2) in detail. And wool dyeing by exhaust has been described above. Hence we are not going into details of these processes.

17.1.2.1.8.2   *Continuous process*

The polyester/wool can be continuously dyed by special process.

Principle of the process
1. Padding the material with Palanil dyes,
2. Intermediate drying,
3. Thermosol treatment,
4. Intermediate clearing,
5. Cross-dyeing the wool component with metal complex and/or acid dyes.

Procedure:

Padding liquor:

Quantity	Unit	Bath additions
x	%	Selected disperse dyes
10–15	g/l	Padding asst.
2–5	g/l	Dispersing cum levelling agent
1	g/l	Rapid wetting agent

Pad at 20–30 °C.

Dry in hot air at 100–110 °C.

Thermosol in hot air at 190-200 °C for 45–60 s, or at 210 °C for 30 s. The staining of wool by disperse dyes is considerably less in the thermosol process than when dyed in a liquor. Thus, those disperse dyes that would otherwise stain wool considerably in a liquor can also be used, and actually in greater strength. If liquid dyes are available it is preferred because it is fixed more rapidly, it is often possible either to use a somewhat lower temperature, or to heat for a somewhat shorter time in the thermosol process than when using powder dyes. In powder dyes also if final fastness requirements are acceptable we can select medium energy dyes so that the fixation temperature is lower than high energy dyes. This protects the wool, and the handle of the mixture fabric undergoes less change. Drying should be as mild as possible so as not to alter the handle and the character of the material. A favourable influence is exerted on the quality of the material if the wool component is given the opportunity to acquire its normal moisture content (about 15%) after drying prior to the thermosol treatment, for instance, by applying a very fine spray of water.

On thermosol dyeing polyester fibre/wool fabrics, there is an increasing hardening of the handle with prolonged heat treatment, and the wool begins to yellow. However, with a normally conducted thermosol process, the wool exhibits neither a notable loss in mechanical properties (tensile strength, elongation, abrasion resistance), nor any marked change in its chemical properties (cystine content, solubility in alkalis and acids, ability to take up dye, and so on). The harsher handle is usually largely removed during the subsequent cross-dyeing of the wool on the winch. Yellowing is normally slight and does not cause any annoying change in shade.

An intermediate clearing can be given as usual after the thermosol process. In many cases, because of the staining of the wool being only slight, a thorough intermediate rinse, or perhaps a mild scour, may be sufficient (both recipes and procedure is given above).

17.1.2.1.8.3   *Thermosol dyeing wool with acid/metal complex dyes*

The following procedure is used with this class of dyes:
1. Padding with acid/metal complex dyes;
2. Intermediate drying at around 100 °C;
3. Thermosol process at 190–200 °C for about 60 s; and
4. After treatment on a winch with

acetic acid 30% and/or ammonium sulphate pH 5.5–6.5*

Glauber's salt anhydr.

Dispersing agent

(*or 3–5% ammonium acetate or 2–3% sodium acetate)

material at about 50 °C, and heat the liquor to the boil within 45 min. Dye for 1 h at this temperature, then rinse warm and cold.

17.2   Polyester/nylon blends

17.2.1   Pretreatment

Polyester/polyamide fibre mixtures are precleansed in precisely the same way as polyester fibres alone which is given earlier.

17.2.2   Heat-setting

Woven and knitted polyester/polyamide fibre fabrics are heat-set before dyeing or bleaching to impart shape and dimensional stability. On applying heat, the differing softening ranges of the two types of fibre must be taken into account. The conditions for hot-air setting are mainly governed by the type of polyamide fibre.

Usually, the following conditions apply:

(a) Polyester/nylon 6 fibre mixtures, e.g. Diolen/Perlon fabrics,

20 s at 192–195 °C;

(b) Polyester/nylon 66 fibre mixtures,

20 s at about 210 °C.

The affinity for disperse dyes by polyamide fibres is not affected as much by heatsetting as is that of polyester fibres.

On treating polyamide fibres in an aqueous medium above the boil, e.g., for 20 min at 130 °C, the material is hydroset, the effect being similar to setting with hot air under the conditions mentioned above. The polyester fibre, however, is not set as much as it would be with hot air.

17.2.3   Bleaching and Optical Brightening

Chlorite bleaching imparts excellent whiteness to both types of fibre. The procedure is as for polyester fibres alone. Quite a good white effect is also obtained on the polyamide fibres alone by means of reductive bleach.

Procedure:

The bleaching bath is set at about 50 °C with

3 g/l suitable OBA Blankit I N,

whereupon the temperature is raised within about 15 min to 80–90 °C, and treatment proceeds at this temperature for 30–60 min, i.e., until the reducing agent is almost completely used up. The temperature of the bath can be raised to 130 °C, when the treatment lasts for 20 min.

17.2.2   Dyeing

The possibilities of obtaining fast and reproducible dyeings on polyester/ polyamide fibre blends are limited by the process. Changes in the type and nature of the fibre, (e.g., nylon 66 instead of nylon 6, staple fibre instead of filament, standard instead of bulked fibre, count, degree of delustering, and so on), the location of the fibres in the textile material (i.e. whether the fibres are intimately mixed or are separate as warp and weft) and the kind of pretreatment given, can all affect the dyeing behaviour considerably. When dyeing mixtures of polyester and polyamide fibres, it is necessary, therefore, to restrict oneself to a limited selection of recipes which have already been tried out.

The following facts are of fundamental significance in dyeing polyester/ polyamide fibre mixtures:

1. All disperse dyes dye both polyester and polyamide fibres, the latter usually to a deeper shade. On the other hand, with disperse dyes, the two kinds of fibre do not dye to the same shade and, in fact, the resulting shades may differ quite remarkably.

2. The fastness properties of disperse dyes on polyamide fibres are lower than on polyester fibres.

3. Since disperse dyes are not so firmly attached in the polyamide fibre as in the polyester fibre, the former fibre can be stripped, or partially stripped, by a treatment appropriate to the individual case.

4. The polyamide component can be dyed selectively with 1:1 or 1:2 metal complex dyes, or with acid dyes. Selected dyes of these groups leave the polyester fibres a pure white.

5. So far, no dyes have been found for this mixture of fibres which are selectively taken up by the polyester fibres.

17.2.2.1   One-bath dyeing

When the fastness properties need not be quite so high, the mixture can also be dyed from one bath. Only those disperse dyes should be used for this process which stain the polyamide fibres relatively little. Metal complex or acid dyes are used for shading the polyamide component. Union dyeing by the one-bath process is possible only in a few cases. One-bath dyeing can be carried out at the boil in the presence of a dyeing accelerator, or under HT conditions, along with the toning acid colours along with disperse dyes. The method of dyeing can be seen under polyester/wool, etc.

17.2.2.2   Two-bath dyeing

(a) In two-bath dyeing of polyester polyamide blend polyester is dyed with disperse dyes. Since the polyester dyes polyamide also to a lesser extent one has to select dyes which are taken up relatively little by the polyamide component or which can be relatively easily stripped from the polyamide fibre. The method of dyeing polyester by HT or carrier method which is given under polyester or polyester/acrylic section may be followed. When following carrier method at boil is followed the carrier which is not suitable for polyamide may be avoided.

After rinsing, the goods are given an intermediate clearing by one or more of the procedures cited below:

(i) After scouring

Recipe:

The goods are treated with

Quantity	Unit	Additions
1–2	g/l	Dispersing agent
0.5	ml/l	Acetic acid 30%

for 20 min at 60–70 °C, after which they are rinsed.

(ii) Reduction clearing

Recipe:

The material is treated with

Quantity	Unit	Additions
3	g/l	Sodium hydrosulphite
2	g/l	Soda ash
1–2	%	Dispersing agent

for 15–30 min at 60–70 °C, after which it is thoroughly rinsed, warm and cold, the last rinsing bath being acidified with acetic acid.

(b) After an intermediate clearing and rinsing, the polyamide component is cross-dyed with metal complex or acid dyes (1:1, 1:2 metal complex, milling dyes, etc.). It should be checked these dyes leave the polyester fibres pure white. Since most 1:2 metal complex dyes produce comparatively dull shades, they are often combined with the more brilliant acid dyes. It should also be noted that it is not all that easy to obtain dyeings free of barriness with polyamide fibres when using highly wet-fast dyes.

17.2.2.3   Dyeing polyamide component with acid and 1:2 metal complex dyes

Recipe:

Quantity	Unit	Additions
x	%	Acid or 1:2 metal complex dyes
3	%	Wetting agent
1	%	Levelling agent for PA differences
0.5	ml/l	Acetic acid 30% to pH 6

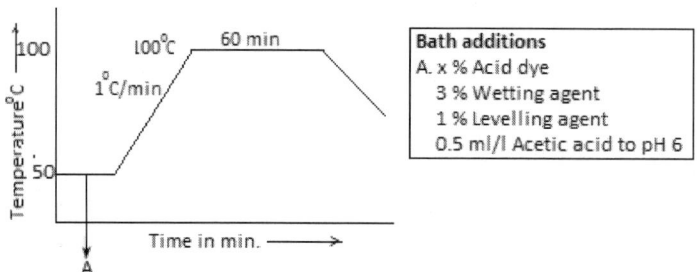

Dyeing commences at about 50 °C. For good levelling, it is important that the temperature of the bath is raised slowly and uniformly (about 1 °C/min) to the boil. Dyeing proceeds at the boil for about 1 h, whereupon the material is rinsed.

17.2.2.4   Dyeing polyamide part with 1:1 metal complex dyes

Recipe:

Quantity	Unit	Additions
x	%	1:1 Metal complex dyes
3	%	Wetting agent
1	%	Levelling agent for PA-fibre
2	% owf PA	Sulphuric acid 96% or
2	% owf PA	Formic acid 85% to pH 2.5–3

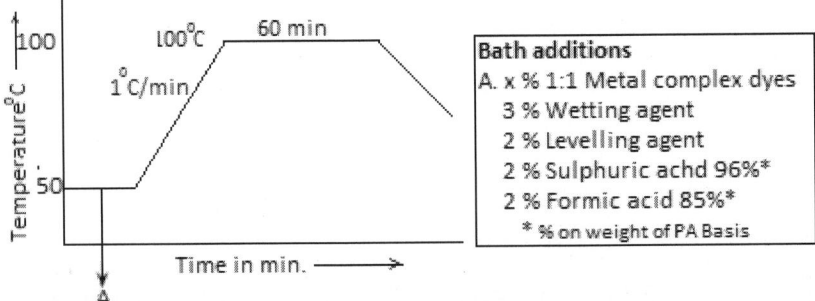

The material is entered into the dyebath at 50 °C, the liquor then being brought to the boil in 30–45 min; dyeing proceeds for 1 h at the boil, whereupon the material is rinsed.

The processing steps after dyeing are largely the same as those for polyester fibres; After-finishing (antistatic finishing) is effected by the same products and by the same recipes as for polyester fibres alone. If the material is to be after set or pleated, the thermal stability of the disperse dyes on the polyamide component must be considered and prior trials are advisable.

17.3   Dyeing of acetate/nylon blends

Dyeing with disperse/acid or metal complex dyes

Blends of acetate and nylon can be suitably dyed with disperse/acid or metal complex dyes combination. Care should be taken to select the disperse dyes which has least affinity for nylon.

Process

1. Fill with water to an MLR of 20:1 to 30:1.
2. Heat to 50 °C and add 1% dyebath conditioner and 0.1% soda ash.
3. Load the fabric run for 10–15 minutes while heating to 60–70 °C and drain.

4. Fill the bath again with fresh water and heat to 40 °C and add the disperse dye and dispersing agent and run for 5 min.

5. Add acid dyes at 80 °C and 2% ammonium acetate. Heat at 1 °C/min to 90 °C.

6. Run for 30 min and check the sample.

7. If shade is acceptable, cool at 1 °C/min to 60 °C.

8. Overflow wash slowly to 40 °C and drain.

9. Soap at 70–80 °C with 1% detergent, cool, drain.

10. Fill and unload.

17.3.1   Disperse/Remazol dye combination at 95 °C

Acetate/nylon blend can be dyed with disperse/remazol dyes combination at 95 °C with the following chemical additions:

Dye % owf	Sodium sulphate (g/l)	Soda ash (%)	Sodium bicarbonate (%)
0.0–0.50	10–20	1	1
0.51–1.00	20–30	1.5	1.5
1.01–2.00	30–40	1.5	1.5
2.01–3.00	40–50	1.5	1.5
3.01 or higher	50–60	1.5	1.5

17.3.2   Disperse/Remazol dye combination can be dyed at 45 °C

Acetate/nylon blends can also be dyed with disperse/remazol dyes combination at 45 °C with following chemical additions:

Dye % owf	Sodium sulphate (g/l)	Soda ash (%)	Sodium bicarbonate (%)
0.0–0.50	10–20	1	1
0.51–1.00	20–30	1.5	1.5
1.01–2.00	30–40	2	2
2.01–3.00	40–50	2	2
3.01 or higher	50–60	2	2

17.3.3   Disperse/direct dye combination

Acetate/nylon blends can also be dyed with disperse/direct dyes combination at boil with following chemical additions:

Dye % owf	Sodium sulphate (g/l)	Alkali
0.0–0.50	3–5	
0.51–1.00	10–15	
1.01–2.00	15–20	
2.01–3.00	20–25	
3.01 or higher	30	
Black shades	20	1% Buffer cum dispersing agent

17.4   Dyeing of acetate/wool, acetate/silk and acetate/linen blends

The acetate/protein fibre blend can be dyed by acetate/nylon process and the acetate/linen can be dyed by acetate/viscose process as linen is a cellulosic fibre.

The major process difference is in the preparation process. In case of acetate/nylon or acetate/viscose both the fibres are manufactured fibre and hence does not need a severe preparation as these fibres are more or less clean except for some spinning oils and handling stains. But in the case of natural fibres like wool, silk and linen needs proper scouring treatments. Since they are blended with acetate fibre which is not much stable to strong alkali and strong acids, care should be taken in using these in the preparation of these fibres. Wool needs carbonising using strong acids to remove cellulose and other impurities. But this is not a major problem as the carbonising is usually done in fibre form and before blending with acetate. In case of silk, the degumming process which is an alkaline process may affect the acetate. Here also in case of an intimate blends the degumming is done before mixing but blends which are made using silk yarn as such needs to be degummed by other methods like enzyme degumming or using synthetic detergents in less alkaline bath. Acetate/linen blends can be prepared as in case of acetate/cotton blends (see above) provided well boiled, scoured linen is used in blending.

Once the preparation is done well with specially designed processes, dyeing wcan be done with acetate/nylon process for wool and silk blends and acetate/viscose process for linen blends. In case of disperse dyeing of woollen blends the disperse dye may stain wool which may be removed by a reduction clear using ammonia, or by an intense soaping by a suitable detergent. A much safer method is to use an efficient wool reserving agent in the disperse dyeing bath. Because linen is extremely prone to creasing and fabric distortion, the fabrics should be dyed on a jig.

Chapter 18
Dyeing of DB blends

18.1 Dyeing of polyester/acrylic blends

Normal polyester and acrylic fibres blends are in the proportions of 50:50. This mixture fabrics are generally used for ladies' and gentlemen's apparel, casual and children's clothing.

Garments made of polyester/acrylic fibre mixtures possess the dimensional stability of the polyester fibre and the wool-like loftiness of the acrylic fibre. They can be laundered in a household washing machine, while spots can be removed relatively easily. No ironing, or hardly any, is needed after laundering. There are no fundamental difficulties associated with the dyeing of this fibre mixture. Both solid and two-colour dyeings are possible.

18.1.1 Pretreatment

Blend fabrics are first inspected for faults and spots. If the goods are strongly soiled locally, it is recommended that the spots be removed with a 10–20% solution of non-ionic detergent before washing; where possible, the material is treated on an open-width washing machine (with a stuffing device), or they can be dealt with on a normal rope or jet washing machine with only slight upper-roller pressure, or even on a winch.

18.1.1.1 *Scouring*

Scouring can be carried out in weakly acidic or weakly alkaline liquor, using one of the following recipes.

Recipe I:

Quantity	Unit	Additions
0.5–1.0	g/l	Non ionic detergent
0.5–1.0	ml/l	Acetic acid 30% pH 5–6*

Treat for 20–30 min at around 60°C.
The goods are rinsed afterwards.

Recipe II

Quantity	Unit	Additions
0.5–1.0	g/l	Non-ionic detergent
0.25–0.5	ml/l	Ammonia 25% or 0.2–0.4 soda ash to pH 8–9

Treat at 40–50 °C for 20–30 min.

The goods are then rinsed and acidified in the final rinsing bath.

Notes:

1. The quantity of detergent may be adjusted as per soiling of the goods.
2. After treatment give a good wash. In case of alkaline treatment rinse till the fabric is free of alkali and sour the last rinsing with a little acetic acid.

18.1.1.2 *Heatsetting*

Polyester/acrylic fibre fabrics are heat-set before dyeing. This operation imparts the necessary shape and dimensional stability for dyeing and subsequent use. Furthermore, the tendency to pilling is diminished. The material is passed through the stenter for 30–60 s at 190–200 °C.

The fastness of coloured goods must be taken into consideration. To obtain an adequate heat setting effect, a temperature of at least 175°C is essential. It is advisable to scour heat-set coloured materials in order to improve the handle and crease resistance. This after-scour is best affected by employing Recipe I as given in Sect .18.1.1.1.

18.1.1.3 *Bleaching and optical whitening*

Polyester/acrylic fibre mixtures can be bleached under the same conditions as described under the bleaching of pure polyester fibres (5.2.3.2, 5.2.3.3), i.e., with

Quantity	Unit	Additions
1–2	g/l	Sodium chlorite 80%
0.8–1.5	g/l	Sodium chlorite stabiliser
1–2	ml/l	Formic acid 85% to pH 3–3.5

for 30–45 min at the boil. After this treatment, the material is slowly cooled at the rate of 1 °C/min down to 60–50 °C. Only at this stage, the liquor run off and the material rinsed.

For optically whitening OBA for polyester and OBA for acrylic (e.g. Ultraphor AL) are used in separate bath (in one bath it may precipitate due to different iconicity). But there is OBA available which can whiten both PE and acrylic (e.g., Ultraphor NA) at boil even without a carrier.

Quantity	Unit	Additions
0.5–2	g/l	Ultraphor NA
1–2	g/l	Formic acid 85% to pH 3–3.5

Treat for 20 min at boil. The liquor cooled slowly at 1 °C/min down to 60–50 °C and drained.

18.1.2   Dyeing methods

Selected disperse dyes and cationic dyes can be used for dyeing polyester acrylic blend for all round shades with high fastness required for mixtures of polyester fibres and anionic acrylic fibres. Most of the disperse dyes are partly taken up by acrylic fibres and produce very much the same shade on the two types of fibre, but at different depths. Further more, they possess similar fastness properties on the acrylic fibres as they do on the polyester fibres. It is also important to note that the acrylic dyes can be selected for this type of mixture have been selected in such a way that it is also taken up at the same rate in almost quantitatively similar amounts on the acrylic fibres, even with dark shades. It is possible to dye blends both at the boil in the presence of a suitable carrier and under HT conditions. The following methods can be employed:

1. At the boil:

    (a) One bath one stage,

    (b) One bath two stage,

    (c) Two bath.

    Under HT conditions:

    (a) One bath two stage,

    (b) Two bath.

18.1.2.1 *Light colours with only disperse dye (dyeing at boil 100 °C)*

This is a method to dye both acrylic and polyester fibre by the disperse dye. It is applicable only to the extremely light colours. In this method, after dyeing the mixed-spun and mixed-woven products of acrylic/polyester fibre at 100 °C without using a carrier, the soaping treatment is given. Accordingly, the dyes usable for the purpose are extremely limited.

18.1.2.2   *Single bath dyeing of polyester/acrylic*

Dye polyester at boil with carrier and disperse dye followed by dyeing acrylic in the same bath with basic dye. Dispersing agents used to disperse the dye are mostly anionic in nature which may interact with basic dye causing precipitation. This in turn creates a two fold problem, i.e. inter action with basic dye will produce light shade on acrylic due to precipitation of a part of basic dye and at the same time a part of disperse dye will be precipitated due to in efficient dispersion through loss in effective concentration of dispersing agent. A non-ionic anti-precipitate may be used to handle this problem. Even cationic retarding agent cannot be used due to this problem. Hence it is better to go for two bath dyeing method which is described below:

18.1.2.3   *Two bath dyeing methods (disperse/cationic dye combination)*

In this method of dyeing acrylic dyes and disperse dyeing in separate baths one can dye from pale shades to dark shades.

In the first bath, the polyester fibre side is dyed by the disperse dye at 100 °C with the help of a carrier and after that a reduction clearing is given. And then, in the second bath, the PAN side will be dyed by the cationic dye and, after that, the soaping is performed for it.

In the second bath, the disperse dye stained into the PAN will be eliminated to some extent by the reduction cleaning and the second bath.

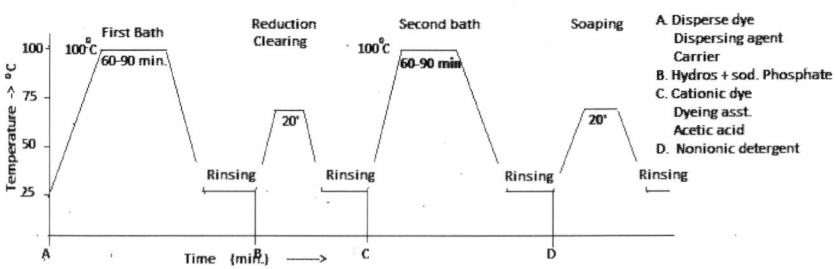

Two bath dyeing method for dyeing PES/PAN blend

Recipe for disperse dyeing:

Quantity	Unit	Bath addition
x	%	Disperse dye
8	g/l	Disperse dye developer
8	g/l	Carrier
0.5	g/l	Anionic dispersing agent
		MLR 1:10–30

Reduction cleaning

Reduction cleaning has an effect to eliminate the disperse dye adhered to the surface of the fibre, and at the same time, to remove part of the disperse dye which stains the acrylic side.

An example of the reduction cleaning

Sodium hydrosulfite	1–2 g/l
Trisodium phosphate (crystal)	1–2 g/l
Bath ratio	1:10–40
Temperature and time	70–75 °C
	20–30 min

Soaping

In case of medium colour, a light soaping will be given in the final process with a view to improving the fastness to friction. An example of the condition is shown in the following.

Detergent 1–2 g/l, 70 °C, 20 min.

Notes:

1. Polyester fibres are dyed at the boil in the presence of a carrier, this should not be done in open vessels, but in vessels that are closed or at least covered. This ensures a higher dyeing temperature and usually a better colour yield, while the un welcome odour of the carrier can be largely suppressed.

2. Since any residual carrier in the fibre could impair the light fastness properties of the dyeing, great care must always be taken to ensure its complete removal, and this is best effected by hot rinsing (possibly using the overflow) by an after wash (preferably in an alkaline

liquor), or by reduction clearing; such residual carrier can also be removed by drying at a high temperature or during after setting. In order to obtain level dyeings even with very pale shades, it may well be advisable to commence dyeing at a some what lower temperature. However, it is better to follow manufacturer's directions for best results.

3. In order to obtain level dyeings, very pale shades can be dyed without a carrier, there being, with some limitations. Use can also be made of the following method to obtain level dyeings in difficult cases – The carrier is applied uniformly to the material which is then dyed in a fresh bath with the usual additives but without further addition of carrier. Thus, the material can be pretreated with carriers for 30–40 min in an acetic acid liquor at 90–95°C. It should be noted, by this method, 50% of more carrier is needed to obtain the same depth of shade as that given by the normal dyeing process with the carrier in the dye liquor itself.

Dyeing of the acrylic

Cationic dyes which causes less stains to polyester fibre will be used. After dyeing the acrylic side at 100 °C for 60–90 min, the soaping is applied.

18.1.2.4 *One-bath dyeing method (disperse/cationic dyes combination)*

The main problem in the one-bath dyeing are the dispersant of the disperse dye and the emulsifying agent of the carrier are anionic and it causes an ionic interaction with the cationic dye. As a result, if the dispersing conditions of the disperse dye and carrier become worse if the quantity of dyes used is high precipitation happens between both dyes and such a precipitation will sometimes be adhere as specks on the surface of the material. At the same time, the dyeing affinity of the cationic dye will become worse because of the adverse influence of the anionic activator mentioned above exerted on the cationic dye.

The carrier to be used in the one-bath dyeing, has to be selected such that it causes no interaction with the cationic dye. In order to achieve a satisfactory uniform dyeing by controlling these actions mentioned above, it is necessary to use a suspending agent.

Notes:

1. In the one-bath dyeing method, the hue on the acrylic side takes up a shade consisting of the cationic dye plus the disperse dye, and

therefore it is necessary to pay attention to the bath ratio and pH when making the colour matching.

2. In some combinations of the cationic dye and disperse dye, there will occur precipitations. In case of blue dyes, this tendency is especially frequent.

3. General tendency for causing the precipitation is the condition of cation > anion in the dye bath (wholly or partially).

4. In order to examine the existence of the precipitation occurrence, the following method should be performed prior to the dyeing – If x, y, z are the concentrations to be used in the dyeing mix three times of the concentration in cold and boil for 10 min and filter and see whether there is precipitation.

5. Some of the disperse dyes may change colours at low pH (below pH 5). In the one-bath dyeing, especially the hue of the disperse dye staining acrylic may sometimes differ or become deeper as pH becomes 4–5. Prior to the dyeing, it is necessary to examine the pH 7 and pH 4–5. Polyester fibre has none of this tendency.

Method 1

This method is applicable to light to deep shades. First, acrylic is dyed by normal method. As most part of the cationic dye is absorbed into acrylic by boiling for 30 min, reduce the temperature of the dye bath to near 70°C for a while in order to pour the suspending agent (dispersant), carrier, disperse dye in the order mentioned, and, then, dye polyester fibre side by raising the temperature again. After the dyeing process, reduction clearing and soaping are done as usual.

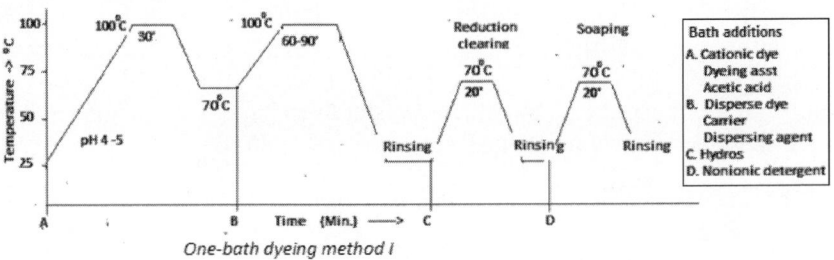

One-bath dyeing method I

Method 2

This method is applicable to the dyeing of light to medium shades and is possible to dye acrylic and polyester fibre at the same time. At first, add the cationic dye and assistant agent to the dye bath and raise the temperature up to 70–75°C. Then, at this temperature, add suspending agent 2–4 g/l, carrier and disperse dye, in the same order, and gradually raise the

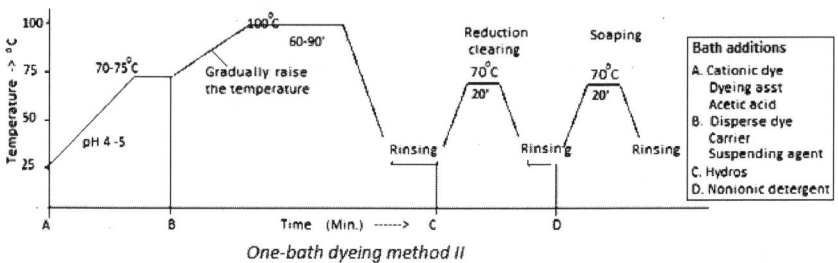

One-bath dyeing method II

temperature up to 100°C. After dyeing for 60–90 min, perform the reduction cleaning and soaping.

18.1.2.5   *One bath two stage HT process*

The polyester component is first dyed with disperse dyes at 120–125°C. The acrylic fibre component is then dyed in the same bath at 100–105°C with Basacryl dyes.

Procedure:

The bath at 60°C is set with

Quantity	Unit	Additions
0.5–1.0	ml/l	Acetic acid 30% to pH 5–6
0.5–1.0	g/l	Dispersing agent
x	%	Disperse dye

and the material run in this liquor for a few minutes. The temperature is now raised to 120–125 °C in 30–45 min, dyeing proceeding at this temperature for about 1 h. After the polyester dyeing has been completed, the liquor is cooled slowly (at about 1°C/min) down to 80 °C.

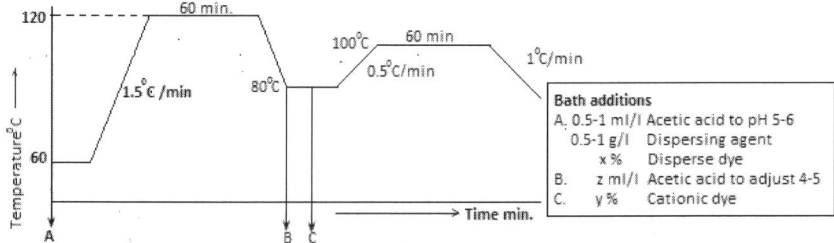

The pH value of the dye liquor is now adjusted with acetic acid to approx. 4–5, and

Quantity	Unit	Additions
x	%	Cationic dye
0.05–0.1	g/l	Potassium bichromate (optional)

are added. The liquor is brought up to 100–105 °C in 30–45 min, and dyeing proceeds for about 1 h at this temperature. As soon as dyeing is completed, the liquor is cooled slowly within 45 min to 60–50 °C, and the material well rinsed.

18.1.2.6   Two bath HT dyeing

The polyester component is first dyed with disperse dyes 120–125 °C, after which the acrylic fibre component is dyed from a fresh bath with acrylic dyes.

Procedure:

The dyebath at approx. 60°C is set with

Quantity	Unit	Additions
0.5–1.0	ml/l	Acetic acid 30% to pH 5–6
0.5–1.0	g/l	Dispersing agent
x	%	Disperse dye

The material is run in this liquor for a few minutes before the temperature is raised within 30–45 min to 120–125 °C, dyeing then proceeding for 1 h at this temperature.

After the bath has been slowly and carefully cooled down to 60–50 °C, the material is well rinsed.

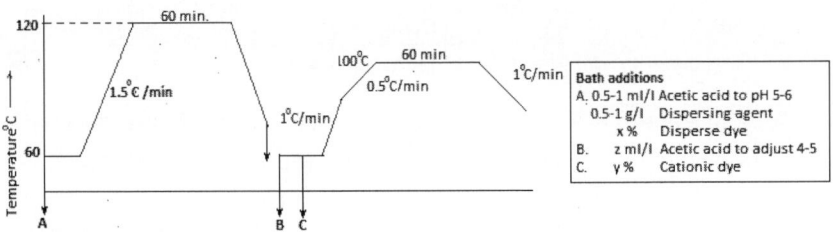

The second fresh bath is taken for acrylic dyeing, also at about 60 °C, is first set with

Quantity	Unit	Additions
1	ml/l	Acetic acid to pH 4–5
0–5	g/l	Glauber's salt calc.
0–2	%	Levelling/retarding agent
0.05–0.1	g/l	Potassium bichromate (Optional)

And then add

y% Basic dye.

The bath temperature is first raised rapidly to 80–85 °C, and then slowly, within 30–45 min, to 100–105 °C. After dyeing at this temperature for about 1 h, the liquor is cooled slowly (1 °C/min) to 60–50 °C, and the material well rinsed.

Notes

1. Under high-temperature conditions, cationic dyes are usually less stable than disperse dyes. Whereas disperse dyes can still be applied at temperatures around 130 °C, acrylic dyes should not be dyed at temperatures above 110 °C.

2. In high-temperature dyeing by the one bath two step method, it is also possible that the dye cations and anionic dispersing agents precipitate like in dyeing at 100°C. Hence, precautions have to be taken in the use of auxiliaries and chemicals. Thus Glauber's salt, levelling and dispersing agents which will form complexes with cationic dyes should not be used. Generally anionic products have to be avoided.Precipitation is more liable to occur in high-temperature dyeing than in carrier dyeing because the liquor ratio in high-temperature machines is usually shorter.

3. Deep shades usually require a reduction treatment with

Quantity	Unit	Additions
3	g/l	Sodium hydrosulphite
5	g/l	Soda ash
1–2	ml/l	Levelling agent

for 15–30 min at 60–70°C. Following this treatment, the goods must be given a warm and cold rinse. It is advisable to add some acetic acid to the last rinse bath.

18.2   Dyeing of polyester/cationic dyeable polyester blends

Cationic dyeable polyester is dyed at 110–115°C with basic dye and dyeing beyond 120°C degrades fibre while dyeing below boil requires addition of carrier in bath. Disperse dyeing gives 10–20% higher dye uptake over that on normal polyester. It is difficult to produce solid shades. Only lower molecular weights disperse dyes which can be applied up to 120 °C are used on both the components in one bath method.

18.3   Acetate/acrylic blends

The importance of blends of acrylic fibre and acetate is that it has the property of the woolly nature of the acrylic fibre and also the silky lustre of acetate. But it has not got popularity that it deserves. The reason for this may lie in the difference in the dyeing behaviour of acrylic fibres and that of acetate which always confronts the dyer with serious problems.

18.3.1   Pretreatment

To remove spin finishes, sizes and contamination from the fibrous material, the goods are treated with 0.2–0.5g/l detergent and approx. 0.25ml/l ammonia 25% for about 30 min at 40–50°C. The goods are then rinsed until they no longer show an alkaline reaction. The amount of detergent depends on the degree of soiling and the liquor ratio. If the material is contaminated too much with fats and oils, it may be treated with stronger oil emulsifying agents to the wash liquor.

18.3.1.1   *Bleaching and optical whitening (full white)*

Most suitable bleaching is oxidation bleaching by sodium chlorite. Generally both the synthetic fibres are enough white without bleaching itself. Hence, the bleaching is aimed at only any stains and discolouration.

Process

Treat the material for 30–45 min at 75 °C with

Quantity	Unit	Additions
1–2	g/l	Sodium chlorite 80%
1–1.5	g/l	Chlorite stabiliser
1	ml/l	Formic acid 85% to pH 3.5–4

30–45 min at the boil. The bath must be heated slowly, particularly from 80 °C to the boil. After the treatment, the bath is cooled slowly,at not more than 1°C/min to 70–80°C, the liquor dropped and the material given a good warm and cold rinse.

If full white is required one may add the suitable optical whitening agents along with the bleaching bath.

Reduction bleaching is also applicable to this blend. This can be carried out with

2–3 g/l Zinc sulphoxylate formaldehyde (e.g., Rongolite C, Blankit IN) and

0.5–1 g/l Dispersing cum levelling agent for synthetics.

for about 60 min at 700 °C. Optical whitening agent can be added along with the bath if required. The process given above produces a very good white on acetate. The white shade of the acrylic fibre, however, usually remains the same which is generally sufficient, because of the low treatment temperatures.

Notes:

Temperatures of 95–100 °C which are necessary for bleaching the acrylic fibre should not be used because the acetate is liable to be damaged in a boiling bath in the presence of formic acid and sodium chlorite.

18.3.2   Dyeing

18.3.2.1 *Acrylic/disperse dye combination*

There are a lot of limitations in this combination. The main problem is that the acetate looses its luster and strength at near boil temperature. At the same time, acrylic material cannot be dyed at temperature lower than boil. Hence it is difficult to dye this blend without affecting one component. Generally, the acetate is bound to undergo some degree of damage during dyeing with this combination.

Acrylic fibre remains reserved under certain conditions in dyeing acetate with disperse dyes. It may be noted that disperse dyes dye/tint only certain acrylic fibres and this advantage can be taken into consideration in selecting this combination. It must be ensured only that the dyeing temperature remains below 80°C. Also acrylic fibres do not take up dyes until the glass transition point of the fibre concerned, which usually lies between 75 and 85 °C, is exceeded. At temperatures below this point, the acrylic fibre remains practically undyeable and the disperse dyes used go onto the acetate almost exclusively. This offers a possibility of dyeing acrylic fibre acetate blends in solid shades and also in two-colour shades. The acrylic fibre is dyed first at a temperature of 95–98°C. The acetate is dyed in the second step with disperse dyes at the dyeing temperature of 75°C.

Prepare the bath at about 60 °C with

Quantity	Unit	Additions
x	%	Cationic dye
3	%	Acetic acid 30% to pH 4–5
0–5	%	Levelling agent for cationic dyes

raise the temperature rapidly to about 80 °C and then slowly by not more than 1°C per 2–3 min to 95–98 °C and dye until the bath is almost completely exhausted. Then cool slowly by not more than 1 °C/min to 50–60 °C, add

y g/l Selected disperse dye
b g/l Dispersing agent

heat to 75 °C and dye for 1–2 h at this temperature. Then cool to about 60 °C and rinse.

Blends of acrylic fibres and acetate can be dried rapidly at temperatures as low as 60 °C. Higher temperatures should not be used.

18.4   Dyeing of triacetate/acrylic blends

18.4.1   Pretreatment

Acrylic fibre/triacetate blends can be pre treated in acid or in weakly alkaline baths. The acid scouring is carried out with
approx. 0.5 g/l detergent
approx. 0.5 ml/l acetic acid 30% (pH 4–5)
    and the goods are treated for about 30 min at approx. 60 °C.

A weakly alkaline treatment is more suitable for heavily contaminated goods. This is carried out with

0.25–0.5 g/l Suitable detergent
0.5 ml/l ammonia 25%

for 30 min at 60 °C. The goods are subsequently given a good rinse. It is advisable to add some acetic acid to the last rinse bath to ensure that the goods do not transport alkali into the dye bath.

There are some disadvantage for triacetate fibres like hydrphobicity, static charge, less moisture content which are arise problems in wear ability. Some manufacturers recommend an alkaline pretreatment known as the S-finish. This treatment causes a superficial saponification of triacetate to produce regenerated cellulose. The layer of regenerated cellulose formed on the surface of the fibre, which covers the hydrophobic triacetate fibre completely, has the ability to absorb moisture. This reduces the tendency of triacetate to build up electrostatic charges, i.e., the fibre is given a permanent antistatic finish. This in turn reduces the soiling properties of the triacetate fibre. The S-finish also improves the thermal stability and thus the ironing properties and the tear strength of the goods. The treatment also imparts a soft handle.

An S-finish is mainly given to articles of 100% triacetate. It is occasionally carried out on blends of triacetate and acrylic fibres. In this case, the blended fabric is prewashed in the usual manner and subsequently treated in a bath, e.g., with

6 g/l caustic soda

for 2 h at 50 °C at a long liquor ratio. The temperature should not be allowed to go over beyond 50 °C as otherwise some commercially available acrylic fibres are liable to yellow and be damaged.

18.4.2   Bleaching and full white

Bleaching can be done the same way as explained above in acrylic/acetate blends. For full white an optical whitener suitable for both triacetate and acrylic may be added in the bleaching bath.

Since the triacetate fibre is supplied with a high degree of whiteness and you need only optical whitening of acrylic fibre, then acrylic specific OWA may be used. And if the triacetate only to be whitened a OWA specific to triacetate may be used.

18.4.3   Single bath dyeing

Blends of triacetate and acrylic fibres can usually be dyed by the one-step method using the following recipe:

Quantity	Unit	Additions
x	%	Acrylic dye
y	%	Disperse dye
z	%	Levelling agent for acrylic dye
0.05–0.1	g/l	Dispersing agent
1	ml/l	Acetic acid 30% to pH 4–5

Adjust the pH of the bath to 4–5, add the requisite amounts of dispersing agent and, if necessary, the emulsified carrier and treat the goods for a few minutes at 60 °C. Then add the disperse dyes which have been dispersed in warm water and heat the bath to 80 °C. At this point add the acrylic dye, which has been dissolved in acetic acid, and levelling agent for acrylic dye. Then heat within 30–40 min to the boil and dye for 1–2 h depending on the depth of shade. When the desired shade has been dyed, cool slowly to 80 °C but do not add cold water to the bath directly. Complete the dyeing by giving the goods a warm rinse. The one-bath dyeing method involves the danger of the cationic dye precipitating with the anionic dispersing agents in the disperse dyes. This can be reduced by suitable precautionary measures such as not using any anionic auxiliaries and chemicals like Glauber's salt, etc. (see Sect. 18.1 for dyeing acrylic fibre/polyester fibre blends).

Notes:

1. When the one-step method does not ensure the safety required, it is possible to dye this blend either by the one bath two step or the two bath method.

2. The carriers used for dyeing polyester fibres are not suitable for dyeing triacetate with disperse dyes because these products usually cause the triacetate fibre to swell excessively. Products based on phthalates, salicylates or benzoates and diphenyl have been found to give far better results. A butyl benzoate carrier is suitable.

3. If adding carrier the amount of the carrier will depend on the depth of shade, dyeing temperature, dyeing equipment and liquor ratio.

4. The depth of shade on triacetate and acrylic fibres depends also on the dyeing conditions. For instance, increasing or reducing the dyeing temperature or changing the liquor ratio is liable to cause changes in the depth of shade produced on both fibres contained in the blend. In order to reproduce a dyed shade, it is therefore necessary to use exactly the same dyeing conditions.

5. Disperse dyes are particularly suitable for shading triacetate/acrylic fibre blends as these dyes can be added to the boiling bath. Cationic dyes, in contrast, cannot be added to the bath until the latter has been cooled slowly to 800°C. After the shading addition, the bath must be heated slowly to the boil again. Dye for about 20 min after each replenishing addition and match off again.

Soaping and after treatment

When high requirements are placed on the rub and wet fastness properties of the dyed shade, it is advisable to after-cleanse the goods, particularly when they have been dyed in deep shades, in order to remove unfixed dye on the surface of the fibre. An after-wash with

Quantity	Unit	Additions
1–2	g/l	Dispersing agent
0.5	ml/l	Acetic acid 30% (pH 5–6)

for 20 min at 60–70 °C is usually adequate. Only occasionally is it necessary to carry out a reduction after-cleansing treatment with

Quantity	Unit	Additions
2–3	g/l	Zinc sulphoxylate formaldehyde
1–2	g/l	Dispersing agent

for 20 min at 60°C. The goods are then given a good warm and cold rinse.

Chapter 19
Dyeing of DC blends

19.1 Dyeing of acetate/cotton, acetate/viscose blends

Acetate/rayon and acetate/cotton blends can be dyed at neutral pH with disperse/direct dyes mixture in the same bath by exhausting with a suitable electrolyte (10–30% NaCl or Na_2SO_4 according to depth of shade). After the dyes exhaustion, rinse the fabric with water at 20–25 °C containing a small amount of salt added to hold the direct dyes on the rayon or cotton. In a new bath, fix the direct dyes with 2% owf of a fixing agent.

19.1.1 Pretreatment

19.1.1.1 Scouring of acetate/viscose

Recipe

Quantity	Unit	Additions
0.2	%	Nonionic detergent
0.2	ml/l	Ammonium hydroxide (S.G 0.88%)

Treat at 90 °C for 30 min.

19.1.1.2 Full white

When the acetate is blended with cotton, the fabric may need a bleaching to remove the neps, etc. compared to acetate or acetate/rayon blends which needs only a simple scouring to remove the handling stains spinning oil, etc. only.

The task here is that we cannot give highly alkaline bleach as the higher affects acetate as mentioned earlier. Hence the process has to be designed at a pH of maximum 10 or lower.

Procedure

Prepare bath at 50 °C with

Quantity	Unit	Bath additions
5	g/l	Hydrogen peroxide 50%
1	g/l	Peroxide stabiliser (Sodiummetasilicate)
1	g/l	Sequestering agent
1	g/l	Emulsifying/scouring agent

For optical brightening add: 0.25–0.5% Leucophor EFR (for acetate) and 0.25–0.5% Leucophor BMBI (for cotton) (Clariant) or equivalent optical whitening agents.

Start the process at 35 °C(room temperature) and heat at 1 °C/min to 90 °C.

Run for 60 min at this temperature.

Cool at 1 °C/min to 60 °C.

Overflow rinse slowly and drop the bath.

Acidify with 0.5 g/lacetic acid for two ends at 80 °C, drain and unload.

19.1.2   Dyeing acetate fabric in black shades

Since a major use of acetate fabric is used as linings, the black shades are of higher demands with special requirements like better rubbing fastness, softness, etc. Disperse dyes for pure acetate fabric and disperse/direct or disperse/ reactive dye mixture can be used for acetate/cotton and acetate/ viscose blends as per procedure below on soft flow.

19.1.3   Jet dyeing acetate and blended fabrics

Acetates and acetate blends can be dyed on soft flow dyeing machines safely. However, normal jet dyeing is not generally recommended for acetate fabrics due to their tendency to form crack and creases. In addition, the pressure dyeing capability of jets is not required. Soft flow machines with higher MLR are more suitable.

1. Soft flow machine is filled with water at 50 °C and add

Quantity	Unit	Bath additions
1	%	Emulsifier/wetting agent
0.5	%	Buffer and chelating agent (e.g. Diaserver brands of Dystar)

2. Load the fabric.

3. Check pH and adjust to 9–9.5

4. Add disperse dyes premixed with 0.25% dispersing agent at 50 °C and run for 5 min.

5. Add direct dye and run for 5 min.

6. Heat 1 °C/min to 75–80 °C and run for 30 min.

7. Add 50% of the suggested electrolyte (sodium sulphate or sodium chloride) and run for 10 min.

8. Add the remaining electrolyte and run for 45 min and check the shade.

9. Cool to 60 °C overflow rinse slowly till the temperature is cooled down to 35 °C and drain.

10. Fill with fresh water and add 3–5% electrolyte (sodium chloride) and run for 10 min and drain.

11. Fill again with fresh water and overflow wash till the water is clear and drain.

12. After treat with suitable fixing agent for the direct dye (e.g. 1–2% Sandofix WEI (Clariant) at 20–25 °C for 15 min and drain.

13. Fill with fresh water at room temperature for 5 min and unload.

19.1.4   Exhaust dyeing of acetate/cotton and acetate/viscose blends with disperse/remazol combination

1. Set the dye bath with the following recipe and heat the bath to 50 °C and load the material

Quantity	Unit	Bath additions
1	%	Emulsifier/wetting agent
0.1	%	Soda ash

2. While running heat the bath to 60–70 °C. Run for 15 min and drain.

3. Fill the vessel again and add x g/l electrolyte (as per reactive dye percentage) run for 10 min and check the salt level and if necessary add more salt.

4. Add the well dissolved reactive dye and run for 5 min.

5. Next add the predispersed reactive dye along with 0.25% dispersing agent and run for 10 min.

6. Add x g/l soda ash and 1 g/l sodium bicarbonate (as per dye manufacturers recipe) and run for 5 min check and adjust the pH for 8.5–9.

7. Heat the bath to 60 °C at 1 °C/min and run for 30 min.

8. Raise the temperature to 85 °C and run for 90 min or 60 min at 95 °C.

9. Cool the bath to 60 °C at 1 °C/min. Over flow slowly to reach the temperature to 40 °C and drain.

10. Fill the bath again with fresh water and add 0.5–1% detergent and heat to 85 °C and soap at this temperature for15 min cool to 60 °C at 1 °C/min.

11. Overflow rinse slowly till the temperature reaches 40 °C and drain.

12. Refill with fresh water and neutralise with 0.25% acetic acid at 50 °C for 5 min.

13. After treat with fixing agent as per previous procedure if necessary and unload.

19.1.5   Two bath dyeing method for acetate/cellulose blends with disperse/remazol combination

1. Fill the vessel at 40 °C and add dye bath conditioner and x g/l electrolyte (as per dye manufacturer's recommendation prescoured goods are loaded and run for 10 min and check the salt level.

2. Add well dissolved reactive dyes and run for 15 min.

3. Add x g/l sodium bicarbonate and y g/lsodaash (see recommendations from dye manufacturer as per depth of shade).

4. Heat the bath at 1 °C/min to 60 °C and run at this temperature for 30 min.

5. Again heat the bath to 80 °C and run for another 30 min.

6. Check the sample and if the shade of cellulosic component is alright cool to 40 °C at 1 °C/min and drain.

7. Fill with fresh water and heat it to 50 °C run for 5 min and drain.

8. Fill again and heat to 40 °C and add 0.5% sequestering agent and run for 5 min.

9. Add well dispersed disperse dye and 0.5% dispersing agent.

10. Heat the bath to 90–95 °C at 1 °C/min.

11. Run for 60–90 min. Overflow slowly to cool to 60 °C and check the shade.

12. If shade is acceptable, overflow rinse slowly till the bath temperature cools to 40 °C and drain.

13. Fill and heat to 50 °C and add 0.5–1.0% suitable detergent.

14. Heat the bath at 1–2 °C/min to 80 °C and run for 10–15 min.

15. Overflow wash slowly to 40 °C and drain.

16. Fill and heat to 50 °C, run for 5 min and drain. Unload.

Polyester/cellulose blends
19.2.1   Pretreatment

Blends of polyester/cotton is one of the most widely used blend in both woven and knitted goods. The majority of the PE/C blends are used in woven presently but the usage of these blend in the knitted is increasing.

Generally used blends are 67/33, 80/20 and 50/50 but many other blends are being produced as per customers' requirements and government tax regulations. When we talk about polyester cellulosic blends, the main fabric comes under preview are polyester/cotton, polyester/spun viscose and polyester/filament viscose.

19.2.2   Pretreatment of polyester/regenerated cellulose blends

Recipe

0.5–1.0	g/l	Suitable wetting and scouring agent
0.5–1.0	g/l	Soda ash

Treat for 30 min at 70–80 °C.

Since spun rayon swell more in cold alkaline bath, the treatment should start at 70 °C itself.

In case of blend where large amount of spinning oils and paraffin wax (e.g., knitted goods) are present, then it has to be washed with higher amount of detergents.

1.0–3.0	g/l	Suitable wetting and scouring agent
0–3.0	g/l	Soda ash

Treat for 30 min at 70–80 °C, warm and cold rinse and acidify.

Pad roll pretreatment

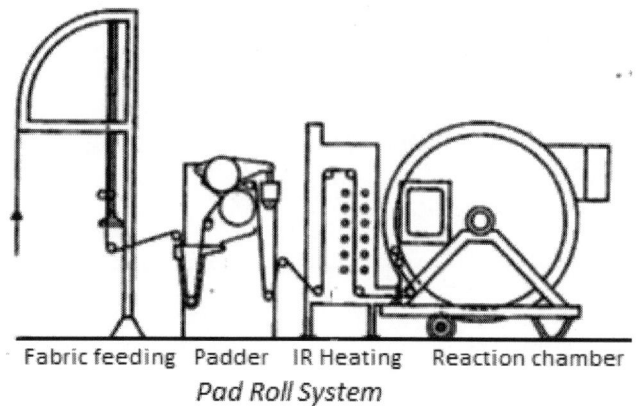

Fabric feeding   Padder   IR Heating   Reaction chamber

Pad Roll System

Pad Roll System

Recipe for pad liquor

5–15	g/l	Suitable wetting and scouring agent
1–5	g/l	Sequestering agent
0–5	g/l	Soda ash

Pad at 40–60 °C at an expression of 60–100%.

Reaction time 20–60 min.

Then wash in an open soaper as follows:

- First washing compartment with water at 60–95 °C.
- Second washing compartment as above.
- Third washing compartment with water at about 40 °C.

You may add 0.3 g/l detergent in the first and second compartment in the beginning.

19.2.2.1   Desizing

Woven polyester/cellulosic fabric may contain sizing materials. In such cases the size has to be removed before the cotton blend has to be prepared for dyeing.

Starch based size has to be degraded and made water soluble and washed off as in the case of cotton materials. The synthetic sizes are generally water soluble which normally swells with the help of wetting agents and can be washed off. Only it has to be noted that enough time has to be given for the size to swell before washing off.

There are three type of desizing which are followed generally:

- Enzymatic desizing (Enzyme)
- Bromite desizing (Oxidative)
- Persulphate desizing (Oxidative)

19.2.2.2   Enzymatic desizing

This is the most widely followed desizing method. It is important due to its mild treatment.

Recipe for pad/batch process

5–10	g/l	Enzymatic desizing agent
5–15	g/l	Common salt
5–10	g/l	Wetting agent

Padding temperature 60 °C.

Batching time minimum 3 h.

19.2.2.3   Bromite desizing

Advantage of this process is in the rapid reaction. But if the pH, time and temperature are not followed accurately it can cause damage to the fibres.

Recipe for pad batch process:

1–3	g/l	Active bromine
20–30	g/l	Caustic soda
5–10	g/l	Wetting agent

Padding temperature 90–95 °C.

Batching time 90–60 min.

19.2.2.4   Persulphate desizing

Recipe for pad batch process

3–6	g/l	Active bromine
8–10	g/l	Caustic soda
5–10	g/l	Wetting agent

Padding temperature 90 °C.

Batching time 60 min.

Persulphate desizing and peroxide bleaching can be combined on one bath.

Recipe

40	ml/l	Hydrogen peroxide 35%
10	g/l	Persulphate
10	ml/l	Sodiumsilicate 38 °Be
10	ml/l	NaOH 50%
5	g/l	Peroxide stabiliser
5	g/l	Wetting agent

Padding temperature 20–40 °C.

Batching time 1–2 h.

In all cases after the reaction time the material is washed in an open soaper.

19.2.2.5   Alkaline scouring (boiling off)

From time immemorial alkaline boiling using caustic soda, Turkey Red Oil, trisodiumphosphate, etc., is used to clean cotton. The conversion of wax, oil, etc. to soluble form with the action of caustic soda is well known.

Now a days synthetic scouring agent is having solvent action on the insoluble contamination in cotton without any deteriorating action on the polyester is plenty available from almost all main auxiliary manufacturers.

Recipe for jigger treatment

10–15	g/l	Caustic soda
4–8	g/l	Synthetic scouring agent
1–2	g/l	Wetting and detergent

Liquor ratio 2:1 to 7:1.

Process; 6–8 ends at boil. Wash off at boil two ends with overflow. Cold rinse and neutralise if necessary.

Jet and such circulating liquor machines

Recipe for boiling off at 100 °C

2–4	g/l	Synthetic scouring agent
0.5–1	g/l	Low forming detergent
5	g/l	Caustic soda

Liquor ratio 10:1.

Time of treatment 60–120 min.

Rinse as hot as possible with overflow. Cold rinse and neutralise if necessary.

Recipe for boiling off at 115 °C

2–4	g/l	Synthetic scouring agent
0.5–1	g/l	Low forming detergent
4	g/l	Caustic soda

Liquor ratio 10:1.

Time of treatment 30–60 min.

Rinse as hot as possible with overflow. Cold rinse and neutralise if necessary.

Recipe for boiling off at 130 °C

2–4	g/l	Synthetic scouring agent
0.5–1	g/l	Low forming detergent
3	g/l	Caustic soda

Liquor ratio 10:1.

Time of treatment 30 min.

Rinse as hot as possible with overflow. Cold rinse and neutralise if necessary.

Notes:

1. In all cases if foaming problem occurs, add silicon free defoaming agents.
2. It is important to drop the bath while it is hot and to rinse as hot as possible to remove all the dissolved and emulsified contaminations.
3. If the whiteness of the fabric is sufficient, you can directly go for dyeing.
4. In case more whiteness is required 3–5 g/l hydrogen peroxide 35% may be added to the above recipe along with peroxide stabiliser or silicate.

19.2.2.6   *Continuous and semi-continuous methods of scouring*

Schematic diagram

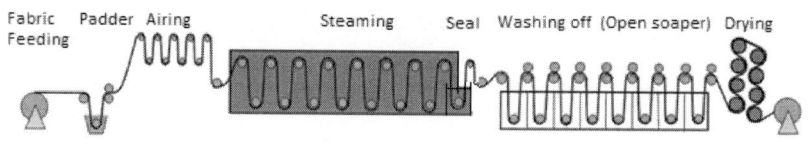

Fabric   Padder   Airing   Steaming   Seal   Washing off (Open soaper)   Drying
Feeding

Process parameters					
Running speed	60 m/m	Steaming temperature	103.^{0}C	Soaping baths	
Moisture at entry	70 %	Heating time approx.	3 s.	First wash box	98^{0}C:
Moisture at exit	100 %	Reaction time	100-180 s.	Second wash box	90^{0}C
Impregnation time	10 s.	water seal	90^{0}C	Third wash box	80^{0}C etc.
Liquor temperature	20.^{0}C			Last wash box (Neutralisation)	40 C

Continuous Methods of Scouring

Recipes for various equipments

Equipment	Synthetic scouring agent (g/l)	Wetting and detergent (g/l)	NaOH (g/l)	Temp. (°C)	Pick up	Treatment temp.	Treatment time
High temperature and pressure steamer	10–15	3–5	20–30	60–95	60–70	130	1–2 min
Normal ager (atmospheric)	10–20	4–8	40–50	60–95	60–70	100–103	1–3 min
Open width J box (a)	10–15	2–5	10–20	40–90	60–70	100	10–15 min
(b)	10–15	2–5	20–25	40–90	60–70	100	5–10 min
J box, conveyer, rebatchingrange	10–15	2–3	10–20	20–60	90–100	90	1–2 h
–do– Open width	10–15	2–3	10–20	40–60	70	90	1–2 h
Pad roll	10–20	2–5	10–15	40–90	70	90	1–2 h
Pad batch (a)	15–20	3–6	25–30	50–90	70	While cooling 3–6 h	
(b)	15–20	3–6	25–30	20	70	20	10–20 h

Notes:

1. In all cases the fabric should be rinsed as hot as possible after the above treatment preferentially on an open width soaper.

2. To avoid moiré effect due to the swelling of cotton in the alkaline bath when done in batch system, depending on the working conditions 30–120 s time has to be before it is batched up. This can be done in various ways.

   • Impregnating in large roller vats or in several compartments of an open soaper.

   • Impregnating in several steps (1) 1st step: in extinguishing trough in the singeing machine and (2) 2nd step: in the actual pretreating bath.

   • An air passage before batching.

   • Running the fabric through a steamer before batching.

3. When dine in rope form fabric is liable to crease permanently.

4. A good alkaline boiling gives a good absorbent fabric and these can be bleached perfectly with less peroxide.

19.2.2.7   Bleaching

19.2.2.7.1   *Neutral peroxide bleach*

Recipe for long liquor processes

	Jig	Vat	Circulating liquor machines	Winch beck
Liquor ratio	7:1–3:1	8:1–5:1	15:1–8:1	50:1–20:1
Hydrogen peroxide 35% (ml/l)	4–10	4–8	2–6	2–6
Peroxide stabiliser (g/l)	2–5	2–4	1–3	1–3
Soda ash(g/l) pH 7–7.5	0.1–0.5	0.2–0.4	0.1–0.3	0.1–0.3
Wetting/detergent	10–15	2–3	10–20	20–60
Temperature (°C)	85–90	90–98	a. 90–98	90–98
Time (h)	1–3	2–4	1–2	1–2
Time (h)			¾–1	

Give hot and cold rinses after bleaching.

19.2.2.7.2   *Wet on wet pad steam process*

Recipe

15–20	m/l	Hydrogen peroxide 35%
7–10	g/l	Stabiliser
1.4–2	g/l	Soda ash to pH 7–7.5
0.2–0.5	ml/l	Detergent/wetting agent

Procedure on different equipments as follows:

	Pad roll	Rebatch
Padding temperature	20–60 °C	20–60 °C
Liquor pick up	70%	70%
Heating temperature	90–95 °C	95–100 °C
Reaction temperature	80–85 °C	90–95 °C
Reaction time	3–4 h	1–2½ h
Temperature (°C)	85–90	90–98
Time (h)	1–3	2–4

Strength of the replenishing 3–5 times in the initial strength. Follow up with hot and cold wash.

19.2.2.7.3   *Alkaline peroxide bleach*

Pad hot batch recipe

		Pad roll and rebatch
10–20	m/l	Hydrogen peroxide 35%
5–10	ml/l	Sodium silicate 38 °Be
3–5	g/l	Stabiliser
3–5	ml/l	Caustic soda 50%
1–3	g/l	Scouring and stain removing agent

Padding temperature 20–40 °C.
Liquor pick up 70%.
Reaction temperature 80–85 °C.
Reaction time 3–4 h.
Strength of replenishing addition 3–5 times in the original strength.

Pad cold batch recipe

40–50	ml/l	Hydrogen peroxide 35%
10–15	ml/l	Sodium silicate 38 °Be
5–10	g/l	Stabiliser
10–15	ml/l	Caustic soda 50%
2–5	g/l	Scouring and stain removing agent

Padding temperature 20–40 °C.
Liquor pick up 70%.
Reaction temperature 20–40 °C.
Reaction time 6–20 h.
Pad steam process

Recipe

	Pressure steamer		Ager	Open width J box		
	a	b		a	b	c
Hydrogen Peroxide 35% (ml/l)	20–30	20–30	20–40	30–40	20–30	20
Sodium silicate 38 °Be (ml/l)	8–10	nil	0–8	10–15	10–15	10
Peroxide stabiliser (g/l)	8–10	4–6	8–10	5–10	5–10	5–7
Caustic soda (g/l)	2–4	2–4	3–6	2–4	2–3	2–3
Wetting/detergent (g/l)	0.5–1					
Liquor pick up (%)	70	70	70	70	70	70
Padding temp (°C)	20–35	20–35	20–35	20–35	20–35	20–35
Dwell time (min)	1–3	1–3	1–3	5–10	10–20	20–30
Temperature (°C)	130	130	100–103	100	100	100

Hot soap in an open soaper. Temperature soaping should be as hot as possible (90–95 °C).

Recipes for pad batch processes

	Pad batch	J box rebatching	Open width J box			
Active chlorine (g/l)	2–4	2–5	15	10	8	6
Wetting/detergent (g/l)	0.5–1.0	0.5–1.0	0.5–1.0	0.5–1.0	0.5–1.0	0.5–1.0
Soda ash (g/l)	3	3	3	3	3	3
Padding temp (°C)	20	20	20	20	20	20
Liquor pick up (%)	70	70–100	70	70	70	70
Reaction temp (°C)	20	20	20	20	20	20
Reaction time (min)	120	60–120	10	15	20	30
Temperature (°C)	130	130	100–103	100	100	100

Follow up with washing and anti chlorination treatments in an open width soaper.

19.2.2.7.4   Sodium chlorite bleaching

Sodium chlorite bleaches polyester and cellulose. A high degree of whiteness is achieved. Bleaching can be done in long liquor and pad steam processes.

Pad steam process

10–20	g/l	Sodium chlorite 80%
2	ml/l	Ammonia 25%
2–3	g/l	Wetting agent
		Adjust pH 5.5–6 with formic acid
20–40	°C	Padding temperature
70	%	Liquor pick up
85–90	°C	Reaction temperature
2–4	h	Reaction time

If necessary activator can be added in the padding bath.

After the process follow up with hot and cold rinse and dechlorinate with sodium bisulphite.

Long liquor machines

Recipe

	Jig	Vat	Jet	Winch
Liquor ratio	7:1–3:1	8:1–5:1	15:1–8:1	50:1-20:1
Sodium chlorite 80% (g/l)	5–7	4–6	1–3	1–2
Stabiliser for chlorite (g/l)	2–4	2–3	1–1.5	0.5–1.0
Sodium nitrite (g/l)	2–3	2–3	1–2	1–2
Formic acid to pH 3.5–4				
Temperature(°C)	85–95	80–95	80–95	80–95
Reaction time (h)	1–3	2–4	1	1

19.2.2.7.5   Hypochlorite bleach

Long liquor machines

Recipes

	Jig	Vat	Jet	Winch
Liquor ratio	7:1–3:1	8:1–5:1	15:1-8:1	50:1-20:1
Active chlorine (g/l)	2–4	2–4	1–2	1–2
Wetting agent (g/l)	0.2–0.5	0.1–0.2	0.1–0.2	0.1–0.2
Soda ash (g/l)	3	3	3	3
Adjust to pH 11–11.5 with NaOH				
Temperature (°C)	20–25	20	20	20
Reaction time (h)	1–2	3–4	1	1–2

Any hypochlorite treatment has to be followed by anti-chlorination treatment, to remove the hypochlorite that has not been consumed and chloramines. This is done by rinsing the material with 3–5 g/l Sodium bisulphite.

Add enough formic or acetic acid so that the goods have a neutral or weakly acid reaction. If peroxide bleach is followed after hypochlorite bleach antichlor treatment is not necessary.

19.2.3   Dyeing of polyester/cellulosic blends

Many classes of dyes and dyeing processes are used for polyester/cellulosic blends. Polyester is almost always dyed with disperse dyes, while cellulosic components are dyed with one of the following dyes:

- Direct dyes,
- Vat dyes,
- Reactive dyes,
- Sulphur dyes,
- Naphthol dyes.

Some dyes manufacturers have introduced blends of two classes of dyes i.e., disperse and another class of cellulosic dyes so that both components can be dyed together. But this method has not become very popular except in case of light shades and some medium shades. However, efforts are continued by manufacturers to introduce dyes which can dye any shades in one go.

There are many dyeing methods to dye polyester/cellulosic blends which are described below. Selection of a dyeing method will depend on many factors:

- form of substrate (yarn/piece, etc.),
- type of goods (type of cellulosic component),
- dyeing equipment available,
- shade, brilliancy, depth of shade required, etc.
- fastness properties requirement of the shade,
- quantity (batch size),
- blend ratio.

19.2.3.1   Batch-wise exhaustion method

19.2.3.1.1   With carrier

This method is of academic interest as it is not widely practiced now, due to the environmental problems of carriers.

1. Disperse/direct dye combination

Machines: Circulating liquor machines, winches, jigs, beam dyeing, etc.

Generally bright shades are not possible with this combination. Even though wet fastness of the direct dyes are poor, it can be increased considerably by after treatments. The dyeing procedure is as per the dyeing graphs given below:

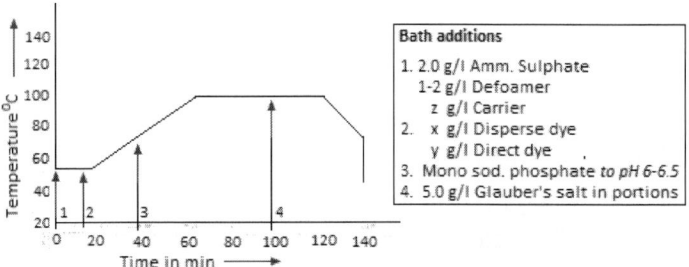

Disperse/Direct Dye Combination - Carrier dyeing

Bath additions

(1)

Quantity	Unit	Bath additions
2	g/l	Ammonium sulphate
1–2	g/l	Defoamer
z	g/l	Carrier

(2)

Quantity	Unit	Bath additions
x	g/l	Disperse dye
y	g/l	Direct dye

pH 6–6.5 by adding mono sodium phosphate.

5 g/l Glauber's salt in portions.

Alternatively, the dye bath is set at about 50 °C with dispersing agent and the disperse and direct dyes. The liquor is brought up to 120–125 °C within **30**–40 min. Dyeing proceeds at this temperature for 1–2 h. After the disperse dye has been exhausted, the dye bath is cooled to 90–80 °C, Glauber's salt is added and the direct dye is allowed to be taken up by the cellulosic fibre at this temperature.

Notes:
1. The holding time after addition of Glauber's salt or common salt can be adjusted according to the depth of shades.
2. Please follow the dye manufacturer's instruction for the addition of Glauber's salt.
3. In case of jet dyeing machines the foaming has to be controlled by the usage of defoamer.
4. After treatments: Rinse cold till water is clean in the overflow machines. For the improvement of wet fastness the goods can be treated for 15 min at 40–60 °C with fixing agent and copper sulphate.
5. Diphenyl carriers are added at 80 °C only.

19.2.3.2   Exhaust two bath methods

19.2.3.2.1   *Disperse/direct dyes combination*

Machines: Circulating liquor units, winches and jigs, jet dyeing machines, beam dyeing machines, etc.

Even though it takes longer time the twobath process is particularly advantageous for dark shades, since several of the limitations applying to the selection of dyes no longer exist and better fastness properties are usually obtained.

a. Carrier method

This method is almost the same as the earlier method, but the two dyes are dyed in separate baths. This method has the advantage that we can do the reduction clear of the disperse dyeing wherever necessary to remove the unfixed dye from cotton as well as polyester in between the two stages which will increase the fastness properties of the dyeing especially light fastness.

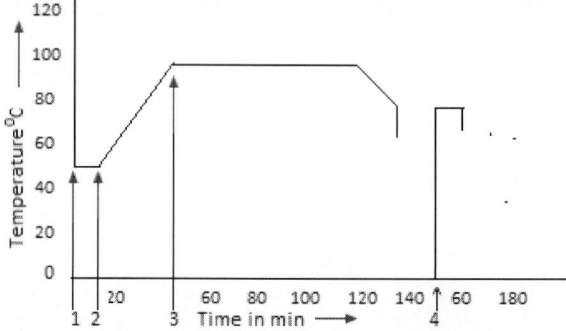

Disperse / Direct Dye Combination - Two bath carrier Method

Dye bath additions

1.

Quantity	Unit	Bath additions
2	g/l	Ammonium sulphate
1–0.5	g/l	Defoamer
y	g/l	Carrier

2. *x*% Disperse dye

3. Adjust pH 6–6.5 with mono sodium phosphate.

4. Reduction clear if necessary.

5. Dye bath 2 – Normal direct dyes dyeing method for cotton.

19.2.3.2.2   *Disperse/reactive (Cold brand) combination*

In this method, the cellulosic component is dyed with reactive dyes Cold brand dyes and then washed. Then the disperse dyes are dyed in a separate bath and then soaped to remove both reactive and disperse unfixed dye.

Machines: Circulating liquor units, high temperature winch and jigs, jet dyeing machines, beam dyeing machines, etc.

Stage A – Reactive (Cold brand) dyeing

Machines: Circulating liquor units, high temperature winch and jigs, jet dyeing machines, beam dyeing machines, etc.

Dye bath additions

Quantity	Unit	Bath additions
30–50	g/l	Glauber's salt
10–15	g/l	Soda ash

Dyeing diagram for stage A

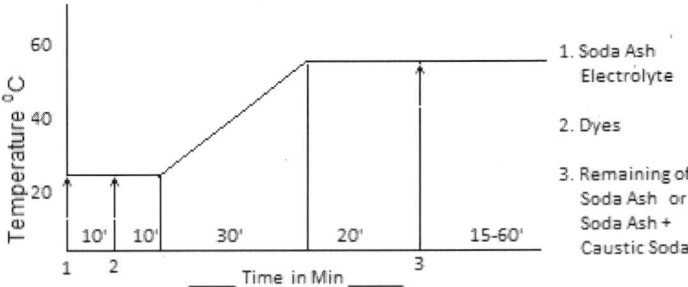

Disperse / Reactive (Cold Brand) - Two bath method -Reactive dyeing

Proceed as per the diagram given, drain and soap.

Notes:

1. Soda ash addition as per depth of shade.

2. Instead of soda ash a combination of soda ash and caustic soda also can be employed. In this case 2 g/l soda ash and 0.5–1 ml/l NaOH (38 °Be) for pale shades and 1–1.5 ml/l for medium and heavy shades may be added. Please follow dye manufacturer's recommendations as in some cases you may need higher amount of caustic soda.

Stage B – Disperse dyeing

Dyeing diagram

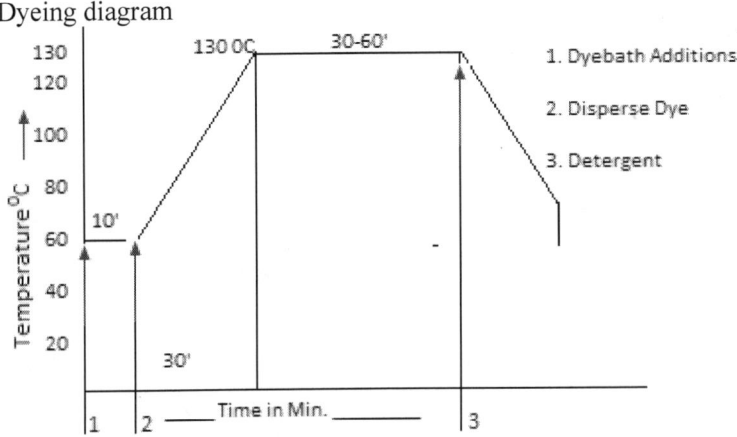

Disperse / Reactive (Cold Brand) - Two bath method- Disperse dyeing

Dye bath additions

Quantity	Unit	Bath additions
0.5–1.0	g/l	Defoamer
1–2	g/l	Mild oxidising agent (Resist salt)
x	ml/l	Acetic acid to adjust pH to 5

Notes:

1. Anti foaming agent and lubricating agent (for crease prone fabrics) may be added wherever it is must as it is not a necessary component for actual dyeing.

2. Mild oxidising agent may be added in the range of 1 g/l for polyester/cotton and 2 g/l for polyester/viscose blends – as chances of higher reducing atmosphere inside the machine in case of viscose material.

3. In the cooling stage add 1 g/l detergent and 1 g/l tripolyphosphate to clear the unfixed dyes from the goods.

4. Holding time at 130 °C can be varied from 30 to 60 min as per the depth of shade.

5. Cooling and soaping can be done for 20–45 min as per the fabric, cooling capacity of the machine, depth of the shade and the dyes involved.

19.2.3.3 *High temperature batch wise exhaustion*

19.2.3.3.1 *One bath exhaust process*

a. Disperse/direct dye combination

Machines: Circulating liquor units, high temperature winches and jigs, jet dyeing machines, beam dyeing machines, etc.

Dyeing program

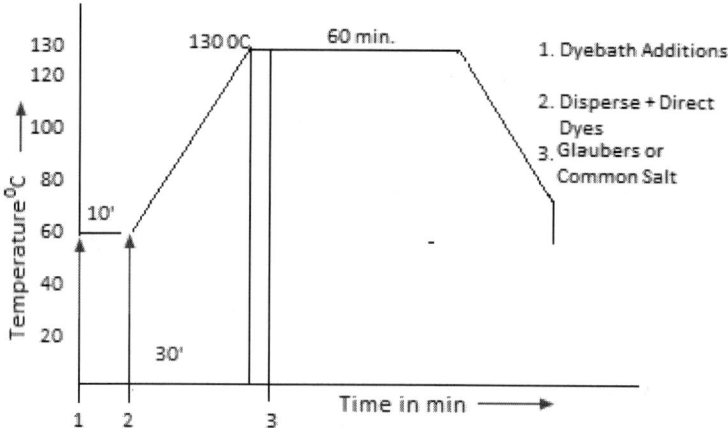

One Bath Exhaust Process Disperse / Direct Dye Combination

Recipe

Quantity	Unit	Bath additions
1–2	g/l	Ammonium sulphate (Buffer)
0.5–1	g/l	Dispersing agent
3–15	g/l	Glauber's salt or common salt
x	ml/l	Formic acid 85% (to adjust pH to 6–6.5)

Notes:

1. The holding time after the addition of Glauber's salt can be added as per depth of shade.

2. The cooling time from 130 °C can be adjusted as per requirement of the fabric dyed and cooling capacity of the machine.

3. Where the high temperature liquor draining provision is available also it is better to cool to 75–80 °C. Before draining this type of dyeing.

4. After treatment: Rinse cold till the bath is clear. After treatment with fixing agent and copper sulphate can be done at around 40–60 °C wherever necessary.

5. In case of jet dyeing machine – anti foaming agent may be added to avoid foam formation while dyeing and entanglement and other problems.

19.2.3.3.2   One bath exhaust – disperse reactive combination (hot brand)

Machines: Circulating liquor units, high temperature winches and jigs, jet dyeing machines, beam dyeing machines, etc.

Disperse dye and reactive dyes are added together and the cellulosic component is dyed first in the alkaline medium. Hence the disperse dye selected should be alkaline stable. Large amount of electrolytes for the exhaustion of reactive dye usually can cause problem in the subsequent dyeing of disperse dyes. Hence this method is recommended mostly for pale to medium shades only.

Dissolve the dye in boiling water and disperse the disperse dye in water at 30–40 °C. Add both to the cold bath, then the Glauber's salt and the amount of alkali used is given in the below chart (refer manufacturer's recommendation also).

Percentage shade	Dyeing temp (°C)	Caustic soda(38 °Be) (ml/l)	Trisodium phosphate (g/l)
Upton 2% dye	40	1	10
	60	1	5
Over 2% shade	40	2	15
	60	1	10

Dyeing graph

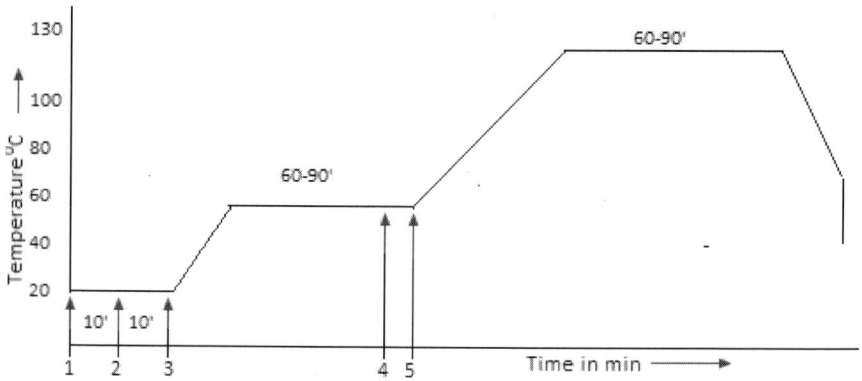

One Bath Exhaust – Disperse Reactive Combination (Hot Brand)

Dye bath additions

	Disperse dye
	Reactive dye
1	Sequestering agent
2	Glauber's salt
	5 g/l soda ash
3	TSP and soda as per table
4	Acetic acid to pH 5–6
5	Dispersing agent

Start dyeing at room temperature suitable for the reactive dyes used (40, 60 or 80 °C). Dye for 60 min at 60–80 °C or for 90 min at 40 °C when the shades dyed on cellulosic fibre matches the standard the bath is set to pH 5–6 with an addition of 10 ml/l or so acetic acid 60%.

Further add

0.5 g/l Dispersing agent,

x g/l Carrier if necessary.

Heat the dye bath within 30–45 min to 130 °C (or to boil, if carrier dyeing). Run about 60 min, drain, rinse hot. This is followed by soaping at boil in neutral or weakly acid medium. Finally the material is given another warm and cold rinse.

19.2.3.3.3   *One bath exhaust methods – disperse/reactive Cold brand*

Machines: Circulating liquor units, high temperature winches and jigs, jet dyeing machines, beam dyeing machines, etc.

Stage A – The Cold brand reactive dye is fixed with 0.2–1.0 g/l soda ash only at 60–70 °C

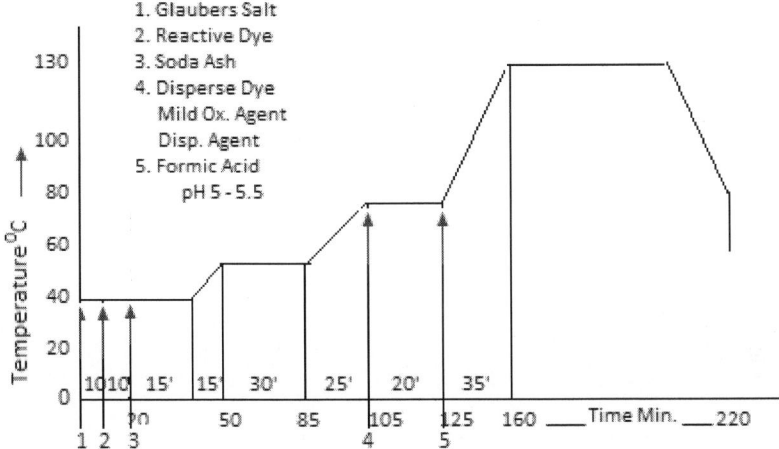

One Bath Exhaust Methods - Disperse /Reactive Cold Brand

Dye bath additions
1. Glauber's salt 10–40 g/l (for 0–1% shade)
                   40–50 g/l (for 1–3% shade)
                   60–70 g/l (for above 3%)
2. Soda ash   0.2–1.5 g/l according to the shade and liquor ratio
3. Mild oxidising agent 3 g/l
4. Dispersing agent   2 g/l
5. Anti foam   1 %

Stage B – The dye bath pH is adjusted to 5–5.5 with formic or acetic acid. Since only a small amount of soda ash is added in Stage A only a little foam forms on adding acetic acid (Neutralisation). Then the disperse dye is added and the temperature is raised to 130 °C and the disperse dye is fixed at this temperature.

After treatment

Rinse 1–2 times at 60 °C.

Soap 1–3 times at 60–70 for 20min.

19.2.3.4   Exhaust one bath two stage methods
19.2.3.4.1   Disperse/reactive hot brand method

Machines: Circulating liquor units, high temperature winches and jigs, jet dyeing machines, beam dyeing machines, etc.

This method is suited for polyester/cotton, polyester/viscose materials. Suitable for bright shades, which is not possible with disperse/direct combination.

Dyeing diagram

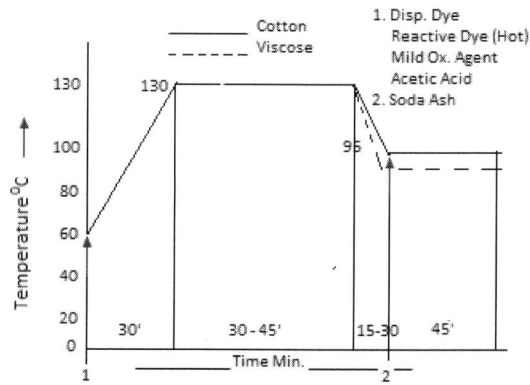

Exhaust One Bath Two Stage Methods - Disperse / Reactive Hot Brand

Dye bath addition (At the start – Stage 1)

Quantity	Unit	Additions
40–60	g/l	Glauber's salt
1–2	g/l	Mild oxidising agent
x	g/l	Acetic acid to pH 5–5.5
y	g/l	Disperse dye
z	g/l	Reactive hot brand dye

Notes:

1. For exhaustion and high yield it is better to add Glauber's salt as electrolyte.

2. Mild oxidising agent is added to avoid the reducing action of cellulose at high temperature on reactive dye to avoid dullening of the shade.

3. 0–0.3 g/l anti foaming agent can be added to the dye bath to avoid any foam generation inside the machine.

4. To protect from crease formation suitable lubricating agent can be added.

Dye bath addition (At Stage 2)

5–15 g/l Soda ash (5–10 g/l for pale and medium shades and 15 g/l for dark shades)

Notes:

1. It is always better to add half the amount of Glauber's salt in the first stage and the remaining half in the second stage.

2. Soap at the longest practical liquor ration possible with 1 ml/l Detergent (low foaming in liquor circulating machines) 1 g/l Sodium tripolyphosphate

   0.5 g/l Soda ash.

19.2.3.4.2   *Disperse/vat combination (1)*

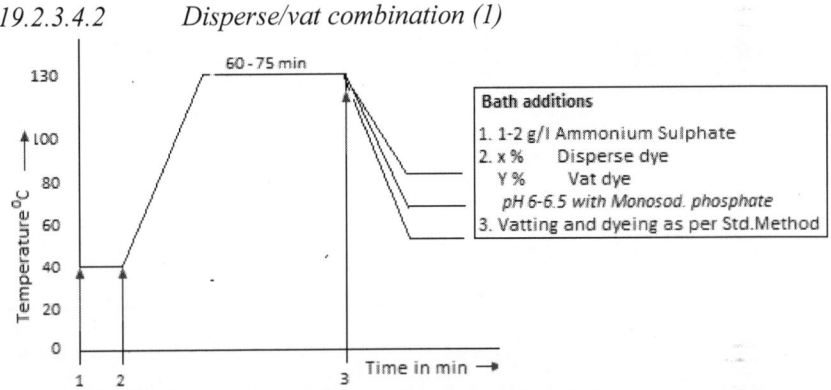

Exhaust One Bath Two Stage Methods - Disperse / Vat dyes I

Dye bath additions

1	1–2	g/l	Ammonium sulphate
	x	%	Disperse dye
	y	%	Vat dye
2			pH 6–6.5 with mono sodium phosphate
3			Vatting as per standard method

The disperse and vat dyes are added at 40 °C after the addition of auxiliaries and adjust the pH to 6–6.5 with mono sodium phosphate. The temperature is raised to 130 °C and the disperse dye is exhausted (60–75 min). Then the bath is cooled to the dyeing temperature of the vat dye as per the class of dye used and vatting, exhaustion, oxidation rinsing and soaping, etc. is done as per standard vat dyeing procedure.

19.2.3.4.3   *Disperse/vat combination (2)*

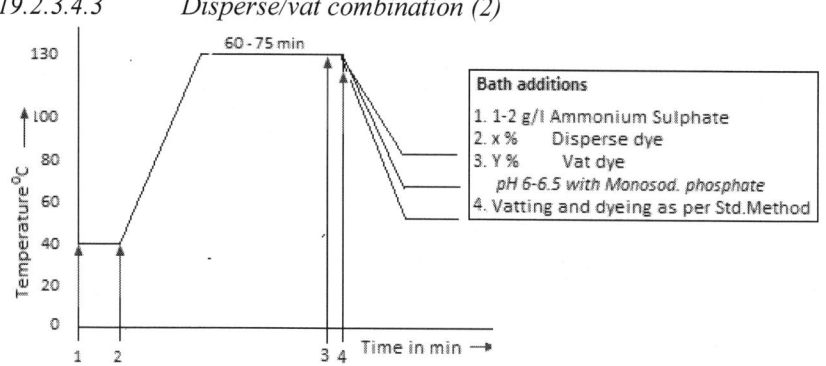

Exhaust One Bath Two Stage Methods - Disperse / Vat dyes II

Dye bath additions

1	2	g/l	Ammonium sulphate
	x	%	Disperse dye
2			pH 6–6.5 with mono sodium phosphate
3	y	%	Vat dye
4			Vatting as per standard method

In this method the procedure is almost the same as in method (1) but the vat dye is not dosed along with the disperse dye but after exhaustion of the disperse dye and cooling the bath. After cooling to the required temperature as per the vat dyes used, the dye is dosed, vatted and dyed as usual.

19.2.3.4.4   *Disperse/reactive combination (conventional method)*

In Stage 1: Disperse dyeing is done as usual. Afterwards the bath is drained. Reduction clear is done and washed.

In Stage 2: Reactive dyeing is done.

Stage 1 – Add the dyeing auxiliaries and the buffer and then the dye dispersion at 50–60 °C run the goods for few minutes and adjust the pH with acetic acid/mono sodium phosphate/formic acid to 5–6.5. Run the goods at 50–60 °C for few minutes until the dye is evenly dispersed. Then raise the temperature to 120–135 °C (normally 130 °C) and dye at this temperature for 60–90 min, depending on the depth of shade. Cool to 80 °C and drain. Reduction clear is done as per recipes given below.

Chemical	Jet	Jig	Winch
Caustic soda 38 °Be (ml/l)	3–5	8–10	3–5
Hydros (g/l)	2–3	3–4	2–3
Dispersing agent (g/l)	0.5–1.0	1–2	0.3–1.0
Levelling agent (g/l)	1–2	2–3	0.5–1

Afterwards reactive dyeing is done by any one of the method described in reactive dyeing section or elsewhere.

19.2.3.4.5   Disperse/reactive combination (non-conventional method)

Conventionally, disperse dyes are dyed first and reduction clear is done. In this unconventional method the cellulosic part is dyed first reactive dyes and then disperse dyes using cotton reserving agent in the disperse dye bath and by careful selection of the dyes.

Bath I –Dyeing of cellulosic part

Cellulosic portion is dyed first with reactive dyes using soda ash/ sodium hydroxide as alkali.

Quantity	Unit	Additions
x	g/l	Reactive dye
5	g/l	Soda ash
1	ml/l	Caustic soda (32.5% or 38 °Be)
50	g/l	Glauber's salt

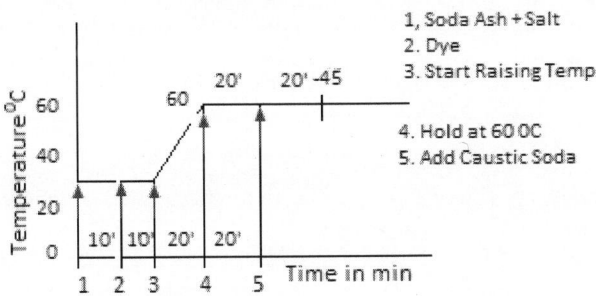

Disperse/ Reactive Non-conventional Method - Reactive

Notes:

1. 50 g/l Glauber's salt has been recommended as a standard requirement. For light shades the quantity may be reduced to 30 g/l.

After the dyeing of cellulosic portion, the bath is dropped and a short rinsing may be carried out.

Bath II –Dyeing of polyester portion.

Dyeing graph

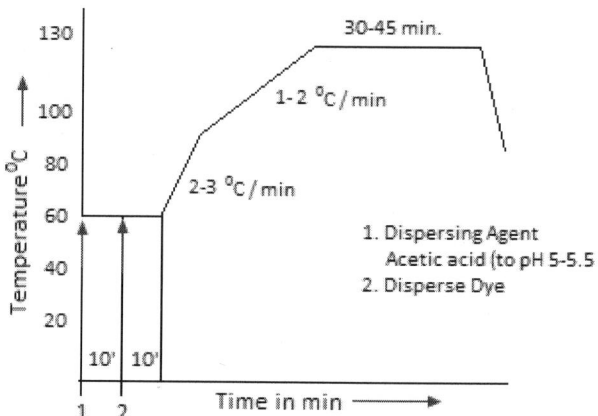

Disperse/ Reactive Non-conventional Method - Disperse

Notes:

1. In this process reduction clear is not possible. The tinting of cotton may be minimised by selecting the dyes (cotton reserve dyes) or adding a cotton reserving agent along with the dye bath.

2. Since the reactive dyed fabric runs throughout the disperse dyeing time we can expect a better fastness properties of reactive portion

19.2.3.5   Exhaustion –two bath methods

19.2.3.5.1   Disperse/vat combination

Machines: Circulating liquor units, high temperature winches, high temperature jigs, jet dyeing machines, etc.

Dye bath additions

1. Ammonium sulphate (2 g/l), levelling agents, etc.

2. Disperse dye, mono sodium phosphate to pH 6–6.5 (or sodium acetate and acetic acid can be used).

3. Caustic soda + hydros (reduction clear optional).

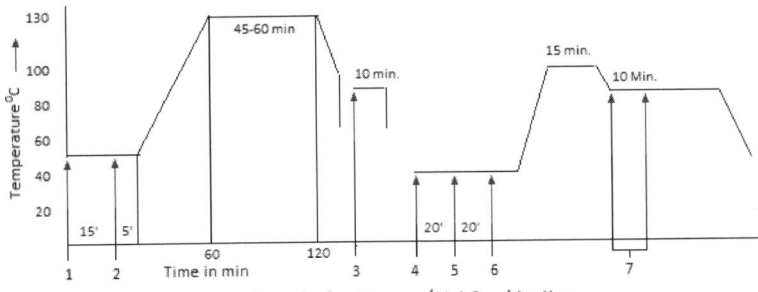

Exhaustion - Two Bath Methods - Disperse / Vat Combination

1. Wetting and levelling agent.
2. Caustic soda.
3. Vat dye x g/l.
4. Sodium hydro sulphite.

Process details

1. Set the dye bath I: set the machine with water at 60 °C and the fabric then add 0.25–1.0 g/l dispersing agent.
2. Add the disperse dye.
3. Adjust the pH to 5–6 with sodium acetate/acetic acid or any other alternatives.
4. Raise the temperature to 130 °C.
5. Continue dyeing for 30–45 min.
6. Rinse warm at 40-50⁰C for 10 min.
7. Reduction clear if necessary and drop.
8. Set the dye bath II at 40⁰C with soft water containing 2 g/l wetting agent and run the fabric for 5 min.
9. Add the required amount of caustic soda.
10. Add the vat dyes dispersed with warm water. Run the cloth for 10 min.

Guideline for dye bath additions

Chemical		Percentage dye on total weight of goods				
		0.3	1	2	3	Black only
Caustic soda 38 °Be (ml/l)						
Method I		14	18	24	28	45
Method II		8	11	14	17	
Sodium hydro sulphite (g/l)		3.5	4.5	6.5	8	10
Glauber's salt or common salt for Method II		nil	nil	25	45	nil

Caustic soda added should be previously diluted.

11. Raise the temperature to 80 °C and run the fabric for 15 min.

12. Cool the bath to 70 °C (60 °C for bright blues) and add the hydros dissolved in a portion of the dye bath over 10 min.

13. Continue dyeing for 30–40 min at 70 °C.

14. Rinse with cold water and oxidise, running speed of the fabric should be maintained and the bath should not be drained completely. In the overflow and jet dyeing machines an overflow rinse (maximum 3 min) is advised. Drop 2/3rd of the dye bath and refill with soft water as quickly as possible. If tests with vat yellow paper fail to show reduction, add 0.5 g/l hydros to avoid premature oxidation of the dye. Repeat the procedure 1–2 times and finally oxidise with

Quantity	Unit	Bath additions
2–3	ml/l	Hydrogen peroxide 130 vol.
1.0	ml/l	Dispersing agent

Or   2–3 g/l Sodium perborate or percarbonate.

19.2.3.6   *Semi continuous methods*

In semi continuous method we include here where one component dyed by continuous method and the other component dyed by batch wise or semi continuous method and actual semi continuous method.

19.2.3.6.1   *Polyester continuous/cotton batch wise*

In most semi continuous methods polyester portion can be done continuous and cotton dyeing can be done batch wise.

Process: Disperse dye padding → Thermosoling → Reduction clear → Batch wise dyeing of cellulosic component.

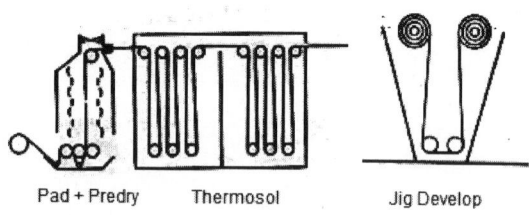

Pad + Predry   Thermosol   Jig Develop

Polyester dyeing

Padding

Quantity	Unit	Additions
z	g/l	Disperse dye
20–200	g/l	Thickening
x	g/l	Mono sodium phosphate for pH 6–6.5

Drying

100–102 °C

Thermosoling

60 s at 200–210 °C

Dyeing of cellulosic component

If necessary the goods are given an intermediate reduction clear, then the cellulosic component is dyed by suitable dyeing method (batch wise) or pad batch (reactive dyes). These dyeing methods are explained in details elsewhere in this book.

Dye combinations which can be dyed by this method

1. Disperse/direct dyes combination

2. Disperse/vat dye combination

3. Disperse/reactive dye combination

19.2.3.6.2   Disperse/Vat Dye Combination

Thermosol/pad-batch process

The process offers the advantage of developing a 3–5000 m thermosol batch in one go as pad batch rather than 3–5 jig batches. The danger of tailing may be avoided by this method. These methods are of academic interest only since it is not widely practiced now. The vat dyes have to be selected carefully to confirm cold pad batch method.

Thermosoling is done after padding the disperse dye and vat dye along with the auxiliaries which is explained earlier. Mixture of selected disperse dye + vat dye for a particular shades were also available in the market earlier. Any finer adjustment for a shade is adjusted in the vat dyeing (e.g. Cottestron dyes of BASF).

The reduction treatment is carried out by padding the thermosoled goods by padding the same with cold caustic alkaline reduction liquor that contain

Quantity	Unit	Additions
80–120	ml/l	Caustic soda 38 °Be
40–60	g/l	Hydros
10	g/l	Antimigrating agent
1–2	g/l	Alkali resistant wetting agent

Liquor temperature 25 °C.

Liquor pick up wherever possible 100% (dripping wet)

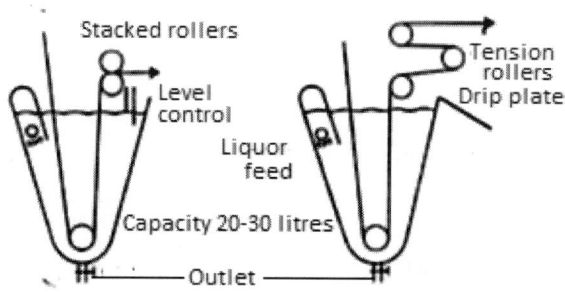

After the dip in the chemical trough fabric passes over adjustable tension rollers or in between two stacked rollers to remove excess chemical liquor. The selvedge must have an adequate amount of moisture i.e., they must remain dripping wet.

The excess liquor has been removed is allowed to flow off a dripping plate. The goods are batched with the highest possible pick up. Although the hydro sulphite is consumed by oxidation at the selvedges during the batching time of ½–1 h, the increasing batch pressure passes the fresh reducing agent continuously from the centre of the fabric towards the selvedges so that the selvedges won't get premature oxidation.

If oxidation stains or even dry areas appear at the selvedges during batching they must be supplied with additional fresh chemical liquor. The fabric batch must be wrapped airtight with plastic sheets to prevent atmospheric oxygen from entering the fabric batch. To get an air tight seal wrap strips of fabric saturated with chemical liquor on the beam alongside the left and right walls of the fabric batch. The ends of the plastic sheets are then tied to these sealing strips. It must be ensured that no air gets to the fabric batch not even through the beam itself. It may be necessary to cover the entire beam with a plastic sheet before batching the fabric on to it. The reaction within the fabric forms a vacuum that is liable to suck in air if the seal is not adequately airtight.

The batch is then kept rotating for 1½–3 h, in case of some shades of dark blue, navy, etc. for 5–6 h.

The fabric is then oxidised and soaped in an open soaper as follows:

Compartment 1 – Cold rinse.

Compartment 2 – Cold rinse

Compartment 3 – Oxidising bath 2–3 ml/l Hydrogen peroxide 35%

Adjust pH of 8–9

Temperature 50 °C

Or 3–5 g/l Sodium perborate

Temperature 65 °C

Or 2 g/l Potassium bichromate

Adjust the liquor pH to 5–6 with acetic acid

Temperature 50 °C.

Compartment 4 – Same as above.

Compartment 5 – Soaping bath.

Quantity	Unit	Additions
1–2	g/l	Ammonium sulphate
1	g/l	Sequestering agent
1	g/l	Soda ash

Temperature – Boil.

Compartment 6 – Same as above.

Compartment 7 – Hot rinse.

Compartment 8 – Cold rinse at pH 6 with acetic acid.

It is advisable to check and maintain the concentration of the oxidising agent and pH continuously (Or can have a continuous dosing system).

Notes on cold pad-batch:

This method is explained in the cotton dyeing section of Dyeing of Textile substrates Volume I – Cotton by the same author

1. Refer manufacturer's pamphlet to find out the fixation time for the dyes involved.

2. Additions of chemicals in the cold batch process (General addition).

Liquor pick up	Hydros (g/l)	Caustic soda 38 °Be (ml/l)
20%	150–160	170–180
25%	120–160	140–180
30%	100–160	120–180
35%	85–140	110–160
40%	75–120	100–140
45%	65–110	90–130
50%	60–100	85–120
60%	50–85	75–115
75%	40–65	60–90
100%	30–50	45–75
150%	20–35	30–50

3. Addition of thickening agent to the padding liquor.

Addition of thickening agent to pad liquor has following advantages:

Increases the liquor pick up.

Prevents dye migration.

Reduces the pigment bleeding in the chemical liquor.

In dyeing with intermediate drying and in dyeing fabrics with high liquor pick up by wet-in-wet process an anti migrating agent is added to both dye liquor and chemical liquor because excessive accumulation of pigment in the chemical liquor is prevented. The addition of thickening agent to both liquors also ensures adequate viscosity of the liquor on the fabric, so that the dye pigment is not liable to migrate to the selvedges during the batching time.

For dyeing other qualities of fabrics (e.g., Twill) by the wet-on-wet process, better results are obtained by using a highly viscous alginate as addition to the dye padding liquor into the chemical liquor. Quantity – around 5 g/l alginate.

19.2.3.6.3 *Disperse/vat combination – pad-dry–thermosol–jig development*

This method involves padding of the disperse and vat dyes and drying followed by fixing the disperse dye continuously by thermosoling. The vat colour is further developed on Jiggers.

Padding recipe

Quantity	Unit	Additions
x	g/l	Disperse dye
y	g/l	Vat dye (Colloisol/UD/MF, etc. brands)
0.3–0.5	g/l	Low foaming alkaline stable wetting agent
0.1–0.3	g/l	Anti migrating agent
5	g/l	Levelling agent
0.5–1	g/l	Sequestering agent
1–3	g/l	Alginate thickening (dry basis) for fine and loose structure fabrics

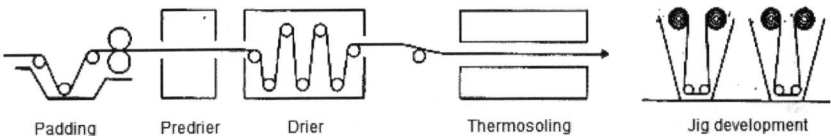

Padding   Predrier   Drier   Thermosoling   Jig development

Padded goods are passed through an infra-red predrier and further dried in a hot flue or cylinder drier. After this the material is thermosoled. The thermosoling time, temperature are given below:

Notes:

1. An infra-red pre drier is absolutely necessary when drying cylinders are used in drying. For hot flue drying it may not be a must.

2. Pre drying prevents migration. It also reduces the soiling of the guide rolls in the next drying unit (e.g., hot flue).

3. Fabric containing yarn of different thickness or polyester cellulosic blend which does not have homogeneous blends tends to give skitter drying. This is prevented by the addition of a levelling agent.

Machine type	Medium diffusing disperse dyes (°C)	Time (s)	Low diffusing disperse dyes (°C)	Time (s)
Hot air e.g., stenter, hot flue	200–215	60–30	215–225	60–30
Hot air contact heat e.g., perforated drum driers	200–215	45–20	215–225	45–20
Contact heat e.g., cylinder driers	215	30–15	220–225	30–15

4. Padding liquor should be neutral to slightly acidic. The liquor pick up should be as low as possible because it reduces the danger of migration during drying.

5. Some of the auxiliaries may have opposite side effect which may create problem in another front. Thus auxiliaries have to be carefully chosen.

Development of vat dyes

The development method is same as for pad/jig development. We give below the approximate recipes for jig development of vat dyes. However, it is better to refer vat dyes in cotton section where these are dealt with in detail.(Dyeing of Textile Substrates Vol. I – Cotton by the same author)

Recipes for reduction treatment in jiggers

Quantity	Unit	Additions
...	ml/l	Caustic soda 38 °Be
...	g/l	Hydros
2	g/l	Dispersing agent
1–2	g/l	Sequestering agent (EDTA)
...	g/l	Glauber's salt

The requisite amount of caustic soda, hydros and Glauber's salt is given below:

Shade	Dye (g/kg)	NaOH 38 °Be (ml/l)		Hydros (g/l)		Glauber's salt calc.
		MLR 5:1	MLR 2.5:1	MLR 5:1	MLR 2.5:1	
Pale shade	Upto 20	18–20	30–35	6	10–12	
Medium shade	20–50	20–25	35–40	6–8	12–14	20
Dark shades	50–100	25–33	40–50	8–12	14–18	30
Very dark shades	Over 100	33–38	50–80	12–16	18–24	35–40

Normally the developing liquor is made to the volume after the first end when the batches are very large (300–400 kg) and the liquor is short, it may be advisable to maintain the level of the jig liquor during the first end. This however, often involves serious problems in practice. The chemical additions are calculated on the total volume. Usually, two thirds of this amount is added during the first end and one third at the beginning of the second end. It must be ensured that the temperature of the liquor remains constant while dyeing is completed in the cooling bath (down to about 60 °C). Generally speaking, large batches (300–400 kg, 15–20 min per head) are dyed by giving four ends and small batches 6–8 ends.

Other precautions to be taken are given in the cotton dyeing section.

19.2.3.6.4   Disperse/sulphur or naphthol combination

Sequence of operation

Pad-dry–Thermosol–Sulphur/Naphthol dye development (batch wise)

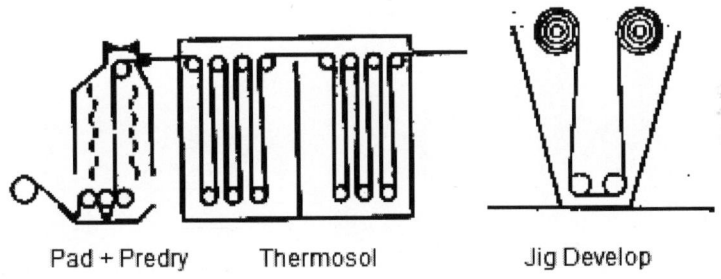

Pad + Predry   Thermosol   Jig Develop

Padding

x g/l Disperse dye

20–200 g/l Thickening

pH adjusted to 6–6.5 with mono sodium phosphate.

Drying 100–120 °C.

Thermosol 60 s at 200–210 °C.

Development of naphthol or sulphur dye on jigs as usual.

19.2.3.6.5   First Component by batch wise process and second component by continuous

19.2.3.6.6   Disperse/vat combination

When we have large quantities for dyeing same shade but do not have continuous machine for polyester dyeing it can be done on HTHP machine in batch wise process and the cotton dyeing can be done continuously –by pad-steam or pad dry pad steam.

The process sequence

Dye disperse dye in HTHP – reduction clear – rinse – unload – dry – join many batches – pad steam (pad–dry–Pad Steam) – Chemical oxidation – After treat – Dry.

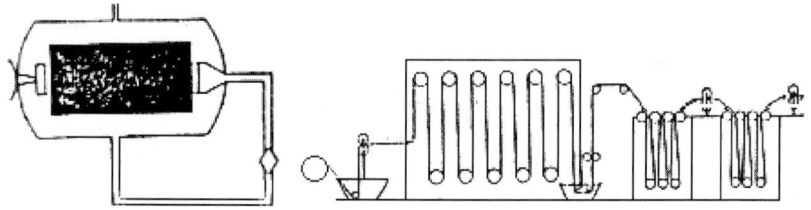

Beam Dyeing (Polyester dyeing) Padder   teamer (Cotton dyeing)   Wash-off

Semi- Continuous Dyeing - Disperse / Vat Combination

Dyeing in HTHP beam dyeing/jet dyeing (or any suitable machine available)can be done as explained in earlier processes.

Vat dyes can be developed after reduction clearing and drying by straight pad steam (we on wet) or pad dry pad – steam method. These dyeing methods have been explained in cotton dyeing section.

Other combinations which can be dyed by this method

19.2.3.6.7   *Disperse/direct*

Disperse dyes can be dyed by any of the batch wise process and the direct dyeing by normal dyeing process. Reduction clearing is possible in between the process.

19.2.3.6.8   *Disperse/reactive*

Disperse dyes can be dyes by any of the batch wise process explained above and reduction cleared. Reactive dyes can be dyed by any process, according the reactivity of dyes used. Since the dyeings are done in separate baths and the components are resistant to most of the dyeing chemicals there are no restrictions to any method of dyeing. This method can achieve the best fastness possible.

19.2.3.6.9   *First component continuous and second component batch wise*

19.2.3.6.10   *Disperse/reactive combination*

19.2.3.6.10.1 *Pad thermosol cold pad batch process*

Disperse dyes and reactive dyes are padded simultaneously. After drying and thermosoling leaves the reactive dye unfixed. The goods are then padded with alkali and batched. After allowing the dwelling time it is washed and after treated.

Process sequence

Pad–Dry–Thermosol–Chemical pad – Cold pad batch – Rotate – Wash off – After treat – Dry

Polyester dyeing - Pad-Thermosol   Cotton dyeing - Reactive cold pad batch   wash off   Drying

Pad Thermosol Cold Pad Batch Process

PE padding–Guideline recipes

Quantity	Unit	Additions
x	g/l	Disperse dye
y	g/l	Reactive dye
2–5	g/l	Defoamer
20–200	g/l	Thickening (alginate)
50	g/l	Urea
		pH adjusted to 6–6.5 by monosdiumphosphate

Pad →Predry→Dry → Thermosol

Pre dry   70% IR

Drying   130 °C,1 min

Thermosol   215 °C, 1 min

CEL part dyeing –Cold pad batch –guideline recipes

Quantity	Quantity	Quantity	Unit	Chemical
<10	10–50	>50	g/l	Special disperse dyes
50	50	50	ml/l	Sodium silicate 38 °Be
5–12	12–24	24–36	ml/l	Caustic soda 36 °Be
5–10	5–10	5–10	ml/l	Pick up increaser

1–2	1–2	1–2	ml/l	Low foaming wetting agent
0.5–1	0.5–1	0.5–1	ml/l	Non foaming sequestering agent

Consult the manufacturers shade card and calculation program for all information about:

- Alkali recipe
- Bath stability
- Batching time

Batching time 8–24 h

Notes:

1. Reactive dyes selected should be thermally stable.
2. Reactive chemical pad should be adjusted as per the dyes selected and the equipment (dosing) availability.
3. Electrolyte addition in chemical pad can avoid bleeding into the pad liquor

200 g/l   Glauber's salt calc. PE/cotton mercerised Or

100–200 g/l   Glauber's salt calc. PE/unmercerised cotton Or

10–50 g/l   Glauber's salt calc.PE/viscose.

19.2.3.6.10.2   *Indigosol semi continuous method*

Generally used for bright pastel shades to pale shades

Sequence of dyeing

Pad–Batch–Develop–Thermosol.

Procedure – Method 1

1. Pad the goods with

Quantity	Unit	Bath additions
x	g/l	Indigosol dye
6	g/l	Sodium nitrite

At 20–25 °C.

2. Batch for 4 h.
3. Continuously develop at 70 °C with

Quantity	Unit	Bath additions
20	ml/l	Sulphuric acid 168 °Tw
1–2	g/l	Dispersing agent
0.5–1	g/l	Thiourea

Sky for 20–30 s

4. Rinse with cold water.
5. Neutralise at 30–50 °C, with 2–4 g/l soda ash.
6. Rinse neutral, dry and thermosol at 210 °C.
7. Soap at boil with detergent and soda ash.
8. Rinse with warm water.
9. Rinse with cold water and dry.

Procedure – Method II

1. Pad the goods with

Quantity	Unit	Bath additions
x	g/l	Indigosol dye
6	g/l	Sodium nitrite

At 20–25 °C.

2. Batch for 4 h.
3. Develop in jig (MLR 5:1) for 15–20 min at 20–30 °C with

Quantity	Unit	Bath additions
20	ml/l	Sulphuric acid 168 °Tw
1–2	g/l	Dispersing agent
0.5–1	g/l	Thiourea, wherever necessary

4. Rinse with cold water.
5. Neutralise with 1–2 g/l soda ash at 30–50 °C.
6. Rinse neutral and dry.
7. Thermosol at 210 °C for 60 s.
8. Soap at boil with 1 g/l detergent and 2 g/l soda ash.
9. Rinse with warm water.
10. Rinse with cold water and dry.

Notes:

1. Urea increases the solubility of Indigosol dyestuffs, hence it can be used wherever necessary.

2. Care has to be taken to prevent tailing in padding.

3. This can be done by using a short liquor padding mangle and also by feeding the dye bath continuously (refer cotton dyeing by Indigosols).

19.2.3.7   Pad/roll methods

In principle, polyester/cellulosic fibre fabrics can be dyed with the following combinations in the pad-roll process:

1. Disperse and direct dyes;

2. Disperse and reactive dyes; and

3. Disperse and vat dyes (two-stage).

In case of disperse/direct, disperse reactive combinations it is not possible to give a reduction clear. Since a certain agglomeration of the disperse dyes occurs outside the fibre during the prolonged dyeing time in the extremely short liquor, a thorough reduction clear is, however, particularly important in achieving good fastness properties with pad-roll dyeings. Hence, in these two combinations only limited fastness properties only can be achieved. In case of direct dyes where possible, only concentrated, low-salt-content direct dyes which have a sufficiently high solubility in the padding liquor should be used. When dyeing in conjunction with reactive dyes, account must also be taken of the sensitivity of some disperse dyes towards alkalis.

Disperse/vat combination by pad/roll method

Dyeing of polyester component

Quantity	Unit	Bath additions
x	g/l	Disperse dye
y	g/l	Vat dye
8–10	g/l	Suitable thickening agent (easily washable)
2	g/l	Dispersing agent
0.5–1	g/l	Rapid wetting agent
z	ml/l	Acetic acid to pH 5

Pad at room temperature to a liquor pick-up of about 60%.

Rate of entering the material: 20–50 m/min, depending on the type of material.

Reaction chamber temperature:

Wet thermometer: 100 °**C;**

Dry thermometer: 100.5–102 °**C.**

Dwell-time: 4–8 h.

Dyeing the cellulosic fibre component

The vat dyes are subsequently developed, this also effecting a reduction clearing of unfixed disperse dye.

The fixation of the vat dyes can be carried out:

(a) in the jigger, using process IW or IN, or

(b) by the pad-steam process, or

(c) by the pad-roll process.

Afterwards, the vat dyeing is completed as usual by rinsing, oxidising and soaping. The material is subsequently washed and after set if required. The material must first be dried if the vat dyes are to be fixed by either the pad-steam or pad-roll process. Refer above or in Dyeing of textile substrate I – Cotton by the same author.

For pad roll process:

The development of vat dyes by the pad-roll process proceeds as follows:

Chemical pad:

Quantity	Unit	Bath additions
80–100	ml/l	Caustic soda 38 °Be (depending on the depth of shade)
40–50	g/l	Sodium hydro sulphite
2–5	g/l	Sequestering agent* for water hardness

*This amount is for use with softened water. With hard water, the amount of Trilon B depends on the degree of hardness.

Pad at room temperature with a large liquor pick-up.

Dwell-time: 2–4 h.

When using the pad-roll process for vat dyes, the steam atmosphere in the chamber must be absolutely free of air. The other conditions are the same as for the pad-roll dyeings of the polyester component.

It should be noted that the pad-roll development of vat dyes may bring about a certain lightening of the polyester component previously dyed with disperse dyes.

There is a further interesting method for dyeing polyester/cellulosic fibre blends, and this consists in first dyeing the polyester component by the thermosol process, and then the cellulosic fibres by the pad-roll method.

Procedure:

Pad with disperse and vat dyes,

Give intermediate drying,

Apply thermosol treatment,

Impregnate with reducing agent and alkali, and

Pad-roll develop the Indanthren dyes.

19.2.3.8 Continuous methods

19.2.3.8.1   Disperse/direct dyes combination

Sequence

Pad–Pre dry–Dry–Thermosol–Chemical pad–Pad steam–After treat.

1. Pad at 20–25 °C with

Quantity	Unit	Additions
x	g/l	Disperse dye
y	g/l	Direct dye
10	g/l	Anti migrating agent
1–2	g/l	Wetting agent

Predry if provision is available

2. Dry at 100–120 °C.

3. Thermosol at 200–210 °C for 30–60 s.

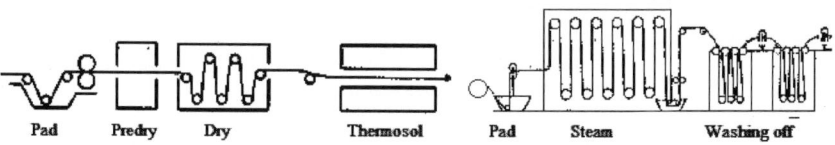

Pad   Predry   Dry   Thermosol   Pad   Steam   Washing off

Continuous Methods - Disperse / Direct Dyes Combination - Pad-Dry-Thermosol-Pad-Steam Method

4. Pad with chemical liquor.

Quantity	Unit	Additions
100–150	g/l	Glauber's salt calc. or common salt
1–2	g/l	Wetting agent
1–2	g/l	Sustilan N (Bayer)

At 20–25 °C

5. Steam at 102–108 °C for 60–90 **s.**

6. Rinse, soap, rinse.

7. After treat and dry.

19.2.3.8.2   Disperse/vat combination–Thermosol–pad steam method

This method is the most versatile and most practiced method and can be used for almost all shades with all round fastness.

Sequence

Pad–Predry–Dry–Thermosol–Chemical pad–Steam–After treat.

Procedure

1. Pad at 20–25 °C with

Quantity	Unit	Additions
x	g/l	Disperse dye
y	g/l	Vat dye (UD, colloisol, MF brands)
10	g/l	Anti migrating agent
1–2	g/l	Wetting agent adjust pH 6–6.5 with mono sodium phosphate

And predry

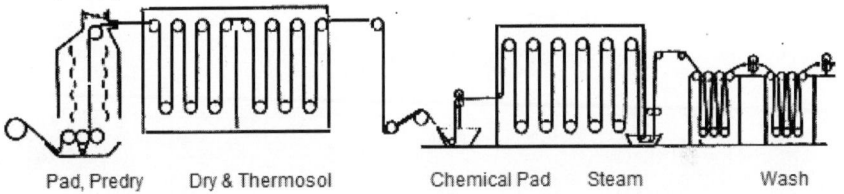

Pad, Predry   Dry & Thermosol   Chemical Pad   Steam   Wash

Continuous Methods - Disperse / Vat Combination – Thermosol – Pad Steam Method

Dry at a temperature of 100–150 °C.

2. Thermosol at 200–210 °C.

3. Pad with chemical liquor.

Quantity	Unit	Bath additions
30–60	g/l	Hydros
60–100	ml/l	Caustic soda 38 °Be

At 18–20 °C

4. Steam at 102–103 °C for 40–60 s.

Notes:

1. Vat dyes which are used for padding should finely disperse like collusion, ultra disperse, micro fine, etc. brands.

2. Oxidation can be done with hydrogen peroxide, sodium perborate or even with sodium bichromate and acetic acid for browns and green shades (dull shades probably – shades to be checked and decided). Sodium bichromate oxidation may be avoided for pale, bright and blue shades.

3. Ready mixed disperse dyes were available where the same process can be followed (e.g., Resithren Dyes – Resolin (Disperse dye) + Indanthrene (Vat Dyes) by Bayer). The dyes are mixed matching shade tone.

19.2.3.8.3   *Disperse/reactive combination – Thermosol–pad steam method*

Sequence

Pad–Pre dry–Dry–Thermosol–Chemical pad–Pad steam–After treat.

Process

1. The goods are padded with

Quantity	Unit	Additions
x	g/l	Disperse dye
y	g/l	Reactive dye
10	g/l	Anti migrating agent
1–2	g/l	Wetting agent

At 20–25 °C.

2. Dry at 100–120 °C.
3. Thermosol at 200–210 °C for 30–60 °C.
4. Chemical pad.

Quantity	Unit	Bath additions
10–30	g/l	Soda ash
200–250	g/l	Glauber's salt/common salt

Or

Quantity	Unit	Bath additions
10	ml/l	NaOH 38 °Be
200–250	g/l	Glauber's salt/common salt

5. Steam at 102–103 °C for 40–60 **s.**

6. Rinse, soap, rinse.

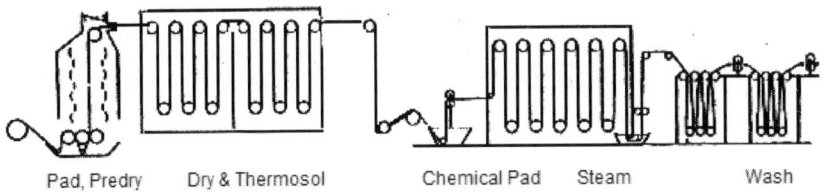

Pad, Predry   Dry & Thermosol   Chemical Pad   Steam   Wash

Continuous Methods - Disperse / Reactive Combination – Thermosol – Pad Steam

Notes:

1. Also we can add 2.5 g/l sodium bicarbonate along with the dye liquor and follow the same procedure.

2. It is better to follow the recommendations of the dye manufacturer for chemical pad as there any many new types of reactive dyes are available in the market.

3. Suitability of the dyes for pad steam method has to be checked.

19.2.3.8.4   Disperse/reactive combination – Thermosol-alkali shock process

Sequence

Pad–Pre dry–Dry–Thermosol–Alkali shock–After treat.

Procedure

1. Pad and predry with

2.

Quantity	Unit	Additions
x	g/l	Disperse dye
y	g/l	Suitable reactive dye
10	g/l	Anti migrating agent
1–2	g/l	Wetting agent

3. Dry at 100–150 °C.

4. Thermosol at 200–210 °C for 30–60 °C.s

5. Alkali shock with

Quantity	Unit	Bath additions
80–50	g/l	Soda ash
200–250	g/l	Glauber's salt/common salt

Or

Quantity	Unit	Bath additions
20–50	ml/l	NaOH 38 °Be
250–300	g/l	Glauber's salt/common salt

At 98 °C for 10–15 s

6. Rinse, soap, rinse and dry.

Note: 3–5 g/l Sodium bicarbonate can be added in the dye pad liquor and the same procedure can be followed.

19.2.3.8.5   Disperse/phthalogen dye combination *(Obsolete)*

Some novel pastel shades, especially turquoise and greens can be developed by this method at a very low cost.

Sequence

Pad–Predry–Dry–Thermosol–Aftertreat

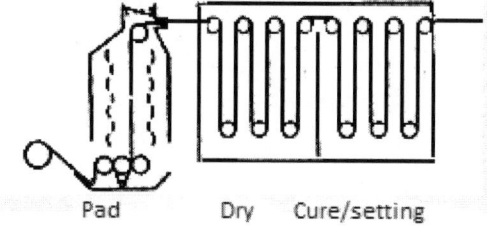

Pad   Dry   Cure/setting

Continuous - Disperse / Phthalogen Dye Combination

Procedure:

1. The goods are padded with

Additions					
Phthalogen dye (g/l)	0.2	0.4	0.6	0.8	1
Auxiliaries required for dissolving phthalogen dyes	As required				
Sodium sulphate (g/l)	20	20	20	20	20
Disperse dyes (g/l)	a	b	c	d	e

At 20–25 °C

1. Dry at 80–120 °C.

2. Thermosol at 200–220 °C for 30–60 s.

3. After treat.

19.2.3.8.6   Dyeing with pigment–pad cure method

Polyester cellulosic material can be dyed to pastel bright shades by padding through pigment dye solution.

Sequence:

Pad–dry–Cure–Wash

Process:

1. The material is padded with

Quantity	Unit	Bath additions
0.2–2.0	g/l	Pigment colours
40–60	g/l	Binder
10	g/l	Formic acid

At 20–25 °C

2. Dry at 100–125 °C.
3. Cure at 140–170 °C for 2–5 min.
4. Wash.

Notes:

1. In padding anti migrating agent and other padding auxiliaries like defoamers can be added as required.
2. Maximum up to medium shades can only be dyed by this method.
3. Since binder has to be used for fixing the pigment colour the handle of the fabric will be slightly affected. Correspondingly finishing recipe has to be adjusted to get the final handle required.
4. There are soft binder combination available in the market which can help in to improve the handle.
5. The last process (washing) can be avoided in certain cases.
6. For fuller shades, especially when polyester percentage in the blend is higher or darker shades are required, disperse dyes can be added in the padding recipe and can be fixed by an additional thermosoling. This way we can achieve even darker shades also.
7. More suitable for blends with lower polyester content.

19.2.3.8.7   Continuous process with Indigosols

Applicable for pastel to pale shades only. Not widely practiced due to non-availability of Indigosols. Some shades will have poor light fatness.

Sequence

Pad – Develop – Thermosol – After treat if necessary.

Process:

1. Pad the blended fabric with following pad liquor:

Quantity	Unit	Additions
x	g/l	Indigosol dyes
2	g/l	Anti migrating agent
6	g/l	Sodium nitrite

At 20–25 °C.

2. Develop continuously with

Quantity	Unit	Additions
20	ml/l	Sulphuric acid 168 °Tw
1–2	g/l	Dispersing agent
0.5–1	g/l	Thiourea

3. Rinse with cold water.
4. Neutralise at 30–50 °C with 2–4 g/l soda ash.
5. Rinse, neutralise and dry.
6. Thermosol at 210 °C for 30-60 s
7. Soap at boil with detergent and soda ash.
8. Rinse with warm water.
9. Rinse with cold water.

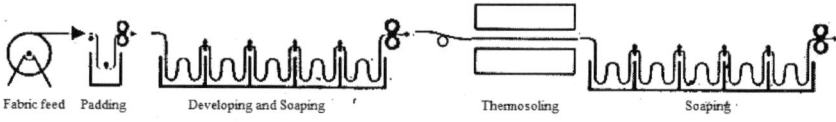

Fabric feed   Padding   Developing and Soaping   Thermosoling   Soaping

Continuous Process with Indigosols - Pad – Develop – Thermosol – Aftertreat

Notes:

1. While padding care should be taken to avoid tailing.
2. In developing thiourea can be added when oxidation sensitive dyestuffs are used (e.g., Indigosol Blue IBC, Indigosol Olive IB, etc.).
3. Soaping is a must after thermosol to remove any browning and also to get brightness of the shades and also to improve light fastness.

4. After developing (step 2 above)soaping for 2–5 min. is necessary to give time for full development of Indigosols.

19.2.3.8.8   *Continuous dyeing without intermediate reduction clear*

There are some disperse dyes which has the property of minimum staining on cotton while dyeing polyester. These dyes allow to dye PE/CEL blends without intermediate reduction clearing. These dyes can be cleared by mere washing (e.g. Foron SWF group). These dyes allows considerable reduction in the process time and cost at the same time giving required fastness properties.

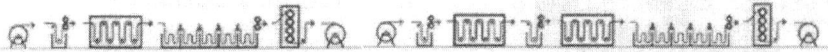

Standard process

Faster process

19.2.3.8.9   *Pad thermosol pad moist (E-control) process*

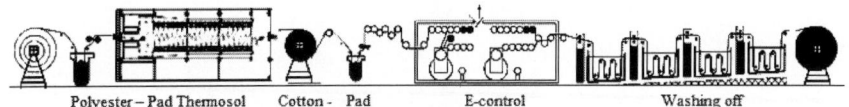

Polyester – Pad Thermosol   Cotton - Pad   E-control   Washing off

Pad Thermosol Pad Moist Process

PES part

Padding –Guideline recipes

Quantity	Unit	Chemical
x	g/l	Special disperse dyes
5–20	g/l	Anti migrating agent
1–2	ml/l	Low foaming wetting agent

Pad →Predry→ Dry → Thermosol

Predry	70% IR.
Drying	130 °C,1 min.
Thermosol	215 °C, 1 min.

CELpart – guideline recipes

Quantity	Quantity	Quantity	Unit	Chemical
<10	10–50	>50	g/l	Drimaren CUHF
10	10	10	ml/l	Soda ash
0–1	1–5	5–10	ml/l	Caustic soda 36 °Be
1–2	1–2	1–2	ml/l	Low foaming wetting agent
10	10	10	ml/l	Mild oxidising agent
0.5–1	0.5–1	0.5–1	ml/l	Sequestering agent

Padding:20–25 °C.

Steaming:2–4 min at 110–140 °C.

Humidity: 25–30% volume.

Wash off

Consult the manufacturers shade card and calculation program "Moist fix/E control, etc."for all information about:

- Alkali recipe.
- Bath stability.
- Fixation time.

19.2.3.8.10   Pad thermosol pad wet steam process

Pad disperse dye → Dry →Thermosol→ Pad reactive dye →Steam → Wash → Soap → Dry

Pad   Thermosol   Pad   Wet Steam   Soap   Dry
Disperse   Reactive

Pad Thermosol Pad Wet Steam Process

PES part

Padding – guideline recipe

Quantity	Unit	Chemical
x	g/l	Special disperse dyes
5–20	g/l	Anti migrating agent
1–2	ml/l	Low foaming wetting agent

Pad → Perry → Dry → Thermosol

Perry   70% IR.

Drying   130 °C,1 min.

Thermosol   215 °C, 1 min.

CEL part

Guideline recipe

Quantity	Quantity	Quantity	Unit	Chemical
>10	10–50	>50	g/l	Suitable reactive dye
5	10	15	g/l	Soda ash
15	10	5	g/l	Sodium bicarbonate
	10	20–40	ml/l	Na_2SO_4 or Nancy
1–2	1–2	1–2	ml/l	Low foaming wetting agent
10	10	10	ml/l	Mild oxidizing agent
0.5–1	0.5–1	0.5–1	ml/l	Sequestering agent

Padding:20–25 °C.

Steaming:60–90 s at 102 °C.

Wash off

Notes:

1. No negative influence of light fastness, no migration, less two-sidedness, less listing problems.

2. Especially recommended for pale to medium shades.

19.2.3.8.11   Pad thermosol chemical pad steam process

Pad Disperse + Reactive   Chemical pad + Steam   Wash   Dry
Dye and thermosol

Pad disperse+Reactive Dye → Dry → Thermosol → Pad Chemical→Steam → Wash → Soap → Dry

PES and CEL part

Padding – guideline recipe

Quantity	Unit	Chemical
x	g/l	Special disperse dyes
y	g/l	Reactive dye
5–20	g/l	Antimigrating agent
1–2	ml/l	Low foaming wetting agent

Pre-drying 70% IR.

Drying 1 min, 130 °C.

Fixation 1 min,215 °C.

Chemical pad – guideline recipes

Pale	Medium	Dark	Unit	Chemical
100	150	200	g/l	Na_2SO_4 or NaCl
20	20	20	g/l	Soda ash
5	10	20	ml/l	Caustic soda 36 °Be
10	10	10	ml/l	Mild oxidising agent

Steaming: 90 s at 102 °C.

Washing off.

Notes: High build-up. Especially recommended for medium to dark shades.

19.2.3.8.12   Pad thermosol pad thermo fix process

Pad Disperse Thermosol   Pad Reactive Thermo fix   wash   Dry

Pad Disperse Dye → Dry → Thermosol → Pad Reactive Dye→Thermo fix→
Wash → Dry

Quantity	Unit	Chemical
x	g/l	Special disperse dyes
y	g/l	Reactive dye
5–20	g/l	Anti migrating agent
1–2	ml/l	Low foaming wetting agent

Pre-drying 70% IR.

Drying 1 min 130 °C.

Fixation 1 min 215 °C.

CEL part

Pale	Medium	Dark	Unit	Chemical
100	Up to 5	>5	g/l	Suitable dye
20	15	15	g/l	Sodium bicarbonate
5	150	150	g/l	Urea
10	10	10	ml/l	Mild oxidising agent
1–2	1–2	1–2	ml/l	Low foaming wetting agent
0.5–1	0.5–1	0.5–1	ml/l	Sequestering agent

Padding: 20–25 °C.

Drying: 100–130 °C.

Thermofixation: 60–90 s at 160 °C.

Washing off.

Notes: Possibility of migration, two-sidedness and listing problems, decreased light fastness on the cellulosic part. No steamer required. Especially recommended for medium to dark shades when no high light fastness is required.

19.3   Dyeing cellulose–cationic dyeable polyester

All shades can be developed as in the case of cellulose–acrylic. However, cationic dyeable polyester is to be dyed first with low molecular weight disperse dye below 120 °C or with basic dye at 110–115 °C followed by dyeing of cotton. Cellulose is to be dyed first if vat and sulphur dyes are used.

Chapter 20
Dyeing of DD blends

20.1   Dyeing of polyester/triacetate blends

As far as dyeing properties are concerned the triacetate fibre falls at an intermediate position between acetate fibre and polyester fibre. As with polyester fibre material, triacetate material can be given permanent shape by heat setting or pleating. Polyester/triacetate fabrics are manufactured primarily on account of the relatively lower priced triacetate fibre as compared with the synthetic fibres.

Even though triacetate has a lower strength but polyester can compensate in a blend. Other than various blends in the blends (35– 5 parts polyester to 65–25 triacetate), knitted goods may contain mixed plies of continuous filament yarns, woven fabrics made of staple fibre yarns are also manufactured. The blend is usually dyed with disperse dyes. But many disperse dyes do not, however, give the same colour and depth of shade on the two fibres. It is therefore necessary to select dyes to obtain solid shades. Also, dyeing accelerators (carriers) applicable may not be applicable to triacetate. Hence to dye at boil the special carriers may be employed. At temperatures of around 120 °C, a carrier is not usually added, though reduced amounts of carriers are sometimes used at 105–110 °C. Polyester/triacetate piece goods can also be dyed by the thermosol process.

20.1.1   Pretreatment

The material is scoured with

Quantity	Unit	Additions
0.5–1.0	g/l	Suitable scouring agent (e.g. Kierlon B conc.)
0.5–1.0	g/l	Soda ash

for 20–30 min at 70–80 °C. It is then rinsed and acidified with acetic acid. Drying is done at about 100 °C.

20.1.2   Heat-Setting

For about 30s at 200°C with hot air.

20.1.3   S-finish

Triacetate woven or knitted materials can have their fibre properties, such as handle, dimensional stability, ironing fastness, gas-fume fastness, and others, improved by an S-finish. The material is treated under defined conditions with dilute caustic soda, whereupon the triacetate is superficially saponified. An S-finish is generally unnecessary with polyester/triacetate mixtures. If the material is to be dyed by the thermosol process, an S-finish must not precede, because of the resulting poor dyestuff uptake and un level dyeing.

20.1.4   Exhaust – dyeing at boil

Machines: Jigs, winch or circulating machines.

Recipe:

Quantity	Unit	Additions
x	%	Disperse dyes
0.5–1.0	g/l	Acetic acid 30% to pH 5–6
0.5–1.0	g/l	Dispersing agent

Dye for 1½ h at the boil, then rinse thoroughly hot and cold. Needs higher dye accelerator quantity is necessary when dyeing on the jigger, the liquor ratio being about 5:1.

20.1.5   Exhaust – HT dyeing

Machine: Beam-dyeing machine

The material should be batched wet onto the beam.

Recipe:

Quantity	Unit	Additions
x	%	Disperse dyes
0.5–1.0	g/l	Acetic acid 30% to pH 5–6
0.5–1.0	g/l	Dispersing agent

Dye for about 60 min at 120 °C, and then rinse.

20.1.6   Continuous – thermosol method

Recipe for pad liquor:

Quantity	Unit	Additions
x	g/l	Disperse dyes
5–8	g/l	Thickener or padding asst
2	g/l	Thermosol fixer

Padding is carried out at room temperature and dried at around 100 °C. The thermosol process proceeds at 210 °C for 60 s.

20.2   Secondary acetate/triacetate blends

Dyeing solid shade in this blend is difficult. To get dark shades the triacetate has to be dyed at boil at least with disperse dyes. Even though secondary acetate is also dyed by disperse dyes, its qualities like lustre deteriorates at temperatures beyond 85 °C which is necessary for dyeing triacetate. If carriers are used to reduce the dyeing temperature, it also affects the secondary acetate which makes the carrier dyeing also impractical to produce a solid shade.

Shadow effect dyeing is possible by dyeing secondary acetate with low energy disperse dyes at 50–60 °C and partly triacetate at 75–85 °C (not beyond 85 °C) with same disperse dye to produce shadow shade.

20.3   Dyeing of acetate/polyester blends

Polyester is normally dyed at temperatures of 120–135 °C, but at these temperatures acetate loses significant tensile strength. In fabric construction in which 100% acetate is used as the warp or filling yarn, the fabrics will not be acceptable in wearing apparel. If the acetate yarns are plied or twisted with polyester, the strength loss of the acetate is compensated by the strength of the polyester, and fabric properties may not be compromised significantly. An alternative method is to use cationic dyeable polyester instead of homo polymer fibres. Cationic-dyeable polyester can be dyed atmospherically as can acetate. Both acetate and cationic-dyeable polyester are dyeable with disperse as well as cationic dyes. However, cationic dyes do not have very good fastness properties on either acetate or polyester. Also, the dye affinity of both cationic and disperse dyes is fibre specific, so dyes need to be screened for their ability to produce union shades. Some cross dyeings are possible, using disperse and cationic dyes, but repeatability of shade and fastness properties is challenges.

20.3.1   Dyeing at boil with carrier

Procedure

1. Set the bath at 40 °C with

Quantity	Unit	Bath additions
1	%	Sequestrant
1	g/l	Monosodium phosphate (0.25% acetic acid)
2–6	%	Biphenyl carrier (for dyeings below 100 °C)

2. Run for 5 min.

3. Add disperse dyes, predispersed with dispersing agent.

4. Run for 10 min. Adjust pH to 6.0–6.5.

5. Heat the bath at 1 °C/min to 80 °C.

6. Run the material for 5 min. and again heat to 100 °C at the rate of 1 °C/min.

7. Run for 30 min for light shades or 60 min for medium and dark shades.

8. Cool to 75 °C at 1 °C/min and check the sample and if shade is acceptable further cool to 60 °C.

9. Overflow wash slowly to 40 °C and drain.

10. Set the bath again at 40 °C and add 0.5–1% detergent as per shade and heat at 2 °C/min to 80 °C. Run for 10 min.

11. Overflow wash slowly to 40 °C and drain.

12. Fill again and heat to 50 °C, run for 10 min, drain and unload in fresh bath.

20.3.2   Dyeing on beam dyeing machines

Woven tricots and Warp-knits that are dyed to light and medium shades can be dyed on beam dyeing machines. But in case beck or jet dyeing can cause streakiness in these shades. But shades like navy blue blacks and such dark shades are dyed preferentially on soft flow due to the moiré effect caused by beam dyeing machines. Care may be taken to batch the material properly by either by dry or wet batching.

Procedure:

1. Load beam into beam dyer and fill. Start liquor flow slowly, circulating inside-out without chemicals for 10–15 min. Run the pump continuously without stopping throughout the procedure.

2. Prescour in a bath containing:

Quantity	Unit	Bath additions
1–2	%	Wetting/scouring agent
0.1	g/l	Soda ash

1. Run for 20 min at 50 °C.
2. Slowly rinse by overflow.
3. Prepare bath for disperse dyeing with dispersing agent and buffer.
4. Run for 15 min at 40 °C and slowly add disperse dyes.
5. Heat 1 °C/min to 90 °C and run for 60–90 min.
6. Check sample and cool to 60 °C and adjust for shade if necessary.
7. Rinse overflowing slowly with water at 20–25°C drops the bath.
8. Unload the fabric.

20.3.3   Alternate methods

Since both the components dye disperse dyes a solid shade dyeing is not that difficult. Dyeing can be done by using carrier at boil or in HTHP dyeing method.

For two stage dyeing (for blends explained earlier) in which polyester is dyed at 130 °C with disperse dyeing and then cooled the bath and add second dye (low energy disperse dyes or cationic dyes) and carry out dyeing at 100 °C for acetate.

In case of blends where the percentage of acrylic less, only disperse dyeing (preferably low energy disperse dyes) may be required. In this dyeing disperse dye will dye polyester and to some extent acrylic fibre which will appear as solid shade.

20.3.4   Drying

Acetate should normally be dried at temperatures of 95–120 °C. An exception is an acetate/spandex or polyester blend in which the spandex/polyester requires heat setting at 190–200 °C. Acetate can tolerate these temperatures, but dyes should be selected with care because of sublimation problems with certain low-energy dyes. Above 205 °C, acetate begins to soften and become sticky.

Notes:

1. Acetate swells in steam. In woven fabrics, the weaving crimp swells along with the yarn, causing fabric contraction. After the steam evaporates, the crimp swelling remains and the fabric contraction is permanent. Subsequent steamings will repeat the process and continuing shrinkage occurs.

Chapter 21
Dyeing of BB blends

21.1 Dyeing of cationic dyeable polyester/acrylic blends

21.1.1 Dyeing with cationic dyes

The advantage of this blend is that both acrylic and polyester can be dyed by cationic dyes. For solid shades the dyeing can be started at 40 °C with auxiliaries and with or without retarding agent pH adjusted to 4–4.5 with sodium acetate and acetic acid. After addition of the well dissolved cationic dyes without retarding agent the temperature is raised to 70 °C at 1.5–2 °C and then up to 85 °C and the normal heating to 90–95 °C after the exhaustion and fixation is over the temperature is further raised further to 105–110 °C for the dyeing of polyester with cationic dyes. (In case, dyeing with retarding agents, temperature gradient can be faster – please

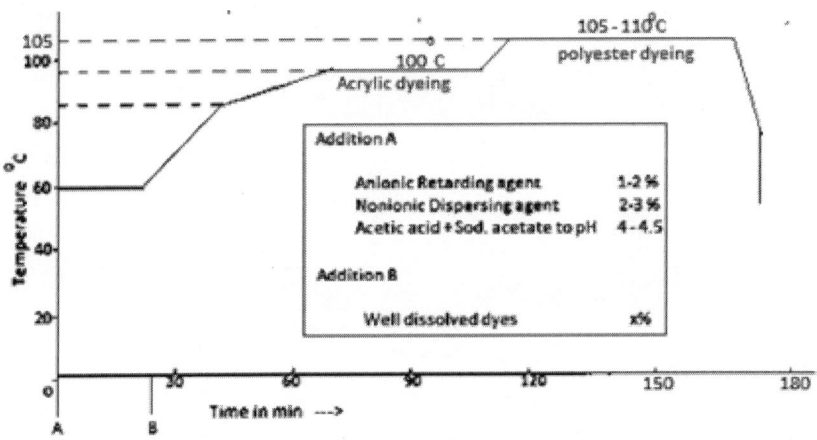

Dyeing of Cationic dyeable polyester/Acrylic Blends - Cationic dyes

refer above, dyeing of pure acrylic with cationic dyes in thisthe book only – Chapter 9.2.6) of beyond 110 °C there are chances of acrylic fibre turning yellowish. Hence, dyeing is restricted below 110 °C.

Cationic dyeable polyester can also be dyed by dispersing dyes but only below 110 °C. Disperse dyeing polyester can be used for dyeing cross shades and reserve shades.

21.2   Elastane blends

In normal terminology elastane a fabric means elastane fibre blend with a major hard fibre. In these blends the percentage of elastane may vary from 2%to 10%. But there are sportswear, corsetry, medical hosiery, ladies hosiery where elastane is used between 10% and35%. In the former case these yarns are virtually invisible and do not show through in the finished fabric, hence they can be left undyed. The choice of dyestuffs can thus best suit the companion hard fibre, if these dyes are compatible with the chemical and physical properties of elastane. But in case of the latter elastane fibre dyeing may necessary. Elastane has affinity for many types of dyes like acid, chrome, premetallised, disperse and vat dyes as well as some reactive dyes and a few direct, sulphur and vat dyes, can all be used for,dyeing elastane to varying degrees. Premetallised, sulphur and vat dyes exhibit good light and wash fastness as well as good colour build-up. Chrome dyes lack colour brilliance and, in some shades, fastness is inadequate for outerwear goods. Disperse dyes give good results in pale shades, but their wet fastness in full shades is limited. However, many colour fastness standards can be met, when the percentage of LYCRA is low.

21.2.1   Dyeability and general fastness properties if different dyes on lycra

Dye type	Dyeability	Light fastness	Wet fastness
Disperse	Good	Fair-good	Poor
Acid	Medium	Fair-good	Fair-good
2:1 Metal complex	Good	Fair-good	Fair-good
1:1 Metal complex	Good	Fair-good	Good

Chrome	Good	Good	Good
Direct	Medium	Fair	Fair
Fibre reactive	Medium	Fair-good	Good
Vat	Medium	Poor-fair	Fair-good
Solubilised vat	Medium	Poor-fair	Good
Sulphur	Medium	Poor-good	Good
Naphthol	Good	Poor	Poor
Basic (cationic)	Fair	Poor	Fair

Many acid dyes give an acceptable all-round fastness on elastane, but the faster dye strike and heavier build-up on its companion yarn can cause the dull elastane can affect the final shade and fastness. Bright and clear elastane yarns have greatly improved the colour wet fastness of stretch fabrics. The table below lists hard fibres used in stretch textiles with elastane and the general dye classes suitable to colour such fabrics.

Dye class	Polyamide/ elastane	Cotton/ elastane	Reg. cellulose/ elastane	Wool/ elastane	Silk/ elastane	Polyester/ elastane	Acrylic/ elastane	Triacetate/ elastane	Acetate/ elastane
Disperse	Y	N	N	N	N	Y	S	Y	Y
Acid	Y	N	N	Y	Y	N	N	N	N
2:1 Metal complex	Y	N	N	Y	Y	N	N	N	N
1:1 Metal complex	Y	N	N	Y	N	N	N	N	N
Chrome	Y	N	N	Y	N	N	N	N	N
Direct	S	Y	Y	N	Y	N	N	N	N
Fibre reactive	Y	Y	Y	Y	Y	N	N	N	N
Vat	S	Y	N	N	N	N	N	N	N
Solubilised vat	N	X	X	S	X	N	N	N	N
Sulphur	S	Y	N	N	N	N	N	N	N
Naphthol	N	Y	N	N	N	N	N	N	N
Basic (cationic)	S	N	N	N	N	N	Y	N	N
Pigments	N	Y	N	N	N	Y	N	N	N

Y = Suitable, N = Not suitable, S – Selected dyes used.

21.2.2   Dyeing of cellulose/spandex blend

a. If only cellulosics has to be dyes– Dye cellulosic with reactive dyes (Elastane will be reserved).

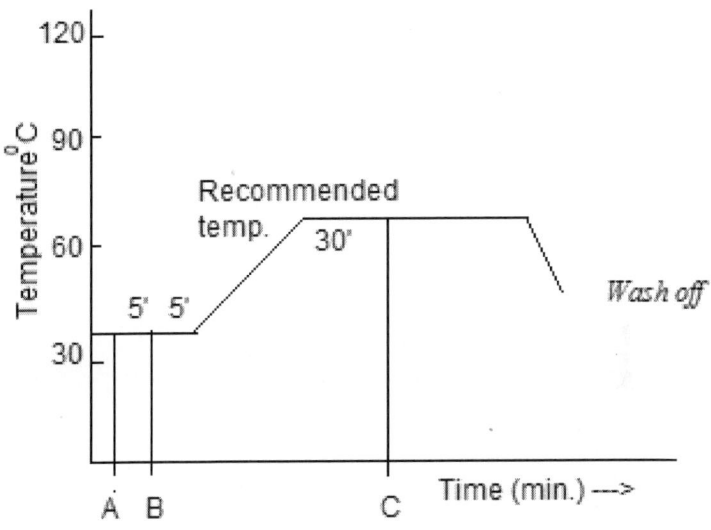

Figure 21.6. Reactive dyeing

Recipes

A

1.0 g/l	Lubricating agent
0.5 g/l	Wetting agent
1.0 g/l	Levelling/dispersing agent
x g/l	Salt

B

x%	Dyestuff

C

y g/l	Soda ash

D

1 g/l	Core neutralising agent

b. If both components has to be dyed

First dye elastanewith disperse dye and the cotton with reactive dyes.

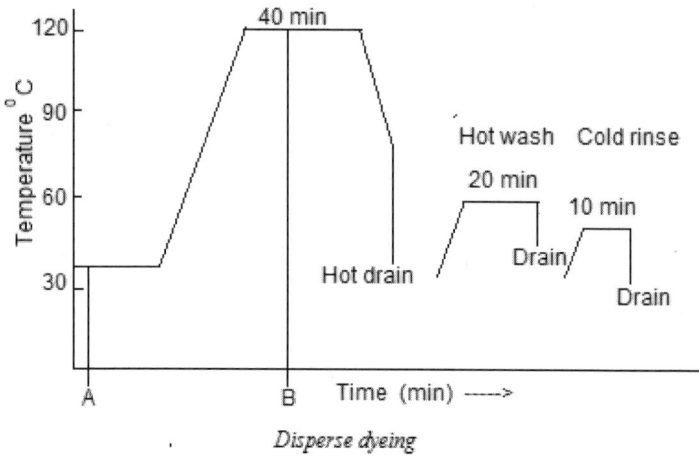

Disperse dyeing

A

1–2 g/l	Disperse levelling agent. Adjust pH 4–5 with acetic or formic acid

B

x%	Disperse dye
1 g/l	Cotton reserve

The cellulosic part is dyed as per procedure given earlier.

Dye elastaneportion with acid dyes and the cellulosic portion are dyed as usual.

Set up the dyebath at 4–6 with formic acid or acetic acid, add the premetallised acid dye to the dyebath. Introduce the fabric into the dyebath. Heat the dye bath to boiling, hold at this temperature for 45 min to 1 h. Cool, drain, rinse, cold. Further, dye the cellulosic portion with any suitable dye and method and soap and dry.

Consolidated table of dyeing methods of various blends

Given below a consolidated statement of all the common blends and their dyeing possibilities and method of dyeing and the dyes involved in the dyeing. Dyeing methods can be taken from this book mainly from pure fibre dyeing methods and cotton dyeing from Vol. III in this series by the same author.

Blend	Shade	Dyeing method	Dye selection	Dyeing conditions
AA type				
Wool/Silk	Solid	Two stage	Reactive dyes and acid milling or 1:2 metal complex	Reactive dyes on silk and 1:2 metal complex on wool after neutralisation at 90 °C
	Silk reserve	Wool dyeing	Monosulphonated acid milling dyes	Selected acid milling on wool at pH 5–6
Nylon/Wool	Solid	Single dyeing	Monosulphonated 1:2 metal-complex and acid dyes with	Anionic retarder to be included in dyeing bath
AB type				
Wool/Acrylic	Acrylic reserve	Single bath dyeing	Reactive, metal complex or acid milling	Only wool dyeing with reactive, metal complex or acid milling dyes
	Solid or cross dyeing	Onebath pale shades	Premetallised or milling acid dyes and basic dyes with anti-precipitant	Acid dyes on wool and cationic dyes on acrylic dyes with anti-precipitant in dye bath
	Solid or cross dyeing	Onebath colour fast dyeing	Reactive dyes and migrating basic dyes	With levelling/stabilising agents in the dyebath
Nylon/Acrylic	Solid/cross dyeing	Onebath (Light shades)	Levelling acid or 1:1 metal-complex dyes and basic dyes	Add anti-precipitant in dyebath
		Two stage dark shades	Basic dyes and metal complex dyes	Basic dyes at pH 4–5 and metal complex dyes as per the dye used
AC blends				
Wool/Cotton	Solid	Onebath	Acid dyes and low-sulphonated direct dyes	Normal neutral dyeing acid dyeing on wool and direct dyes in one bath
	Wool reserve	Onebath	Multisulphonated salt-controllable direct dyes	Only cotton dyeing with selected direct dyes with wool reserve in the dyebath

Substrate	Category	Method	Dyes	Remarks
	Solid fast shades	Reactive dyeing	Monofunctional or bifunctional reactive dyes	Reactive dyeing to dye both component to get solid shades
	Solid fast shades	Twostage	Premetallised or milling and reactant-fixable dyes wool-reactive and then cellulose-reactive dyes	First stage dye wool with wool reactive or acid dyes and second stage dye cellulose with reactive dye
	Solid fast shades	Twobath	Chrome dyes, then reactant-fixable dyes	Wool dyeing with chrome dyes and reactive dyeing of cotton with reactive dyes (wool gets slightly dyed with reactive dyes)
Wool/Viscose	Only wool dyeing	One bath	Disulphonated milling acid dyes or unsulphonated 1:2 metal-complex dyes	Dyeing of wool with acid colours with cotton reserve in the dyeing bath
	Solid or cross dyeing	One bath	1. Wool reactive and direct dyes 2. Acid milling and direct dyes 3. Acid milling and cotton reactive dyes	Wool dyeing with acid dyes and viscose dyeing with reactive or direct dyes in the same bath or selected direct dyes for both components together
Nylon/Cotton	Only cotton dyeing	One bath	Salt controllable direct dyes	Dyeing in presence of syntan so that there is no staining on nylon
	Only nylon dyeing	One bath	Acid milling or 1:2 mono or disulphonated metal complex dyes	Dye both wool and nylon with selected acid milling or 1:2 metal complex from the same bath
	Solid shades	One bath –light shades-single dyeing	Disulphonated disazo and phthalocyanine direct dyes	Light shades can be dyed with selected direct colours for both components
	Solid shades	One bath –single dyeing	Reactive dyes	First dye reactive dye at pH4–5 and the add salt and alkali

Solid shades	One bath two dyes	1. Low energy disperse dye and salt controllable direct dyes. 2. Metal complex or acid milling dyes and salt controllable direct dyes 3. Metal complex or acid milling dyes and reactant fixable dyes 4. Selected acid dyes and neutral fixing dyes	Low energy disperse dyes can be dyed on nylon at boil and cotton to be dyed with direct dyes OR Nylon to dye with metal complex or acid dyes and cotton with direct or reactive dyes
	Two stage one bath	Reactive dyes and acid milling or metal complex dyes	Reactive dyes are fixed on cotton in the presence of salt and alkali and the neutralise the bath and fix neutral dyeing acid dyes
	Continuous dyeing	1. Direct dyes and metal complex and acid milling dyes 2. Salt controllable direct dyes and reactant fixable dyes 3. Salt controllable direct dyes	Pad–dry–pad steam
	Continuous dyeing	Low energy disperse dyes and reactive dyes	Pad–dry–thermofix
	Continuous dyeing	Metal complex or acid milling dyes reactive dyes	Pad–dry–thermofix and acid shock
	Continuous dyeing	Unsulphonated 1:2 metal complex and reactive dyes	Pad–dry–thermofix and pad batch

Blend	Type	Method	Dye class	Technique
			Selected 1:2 premetalised dyes, then vat dyes	Pad–dry–thermofix and chemical pad steam
			Selected 1:2 metal complex the vat dyes	Pad–dry–steam
			Selected 1:2 metal complex the vat dyes	Pad–dry–chemical then vat dyes pad–steam
CB blends				
Cellulosic/ Acrylic	Cellulose reserve	Acrylic dyeing only	Salt controllable direct dyes after syntan treatment	
	Acrylic reserve	Cellulose dyeing only	One bath cationic dyeing only	Cationic dyeing of acrylic only
	Solid or cross dyeing	Single bath	Multisulphonated direct dyes and migrating basic dyes with anti-precipitant	
		One bath two stage	Basic dyes and direct dyes	Basic dyes are dyed at boil and then direct dyes at 80 °C
		Two bath	Basic dyes and aminochlorotriazine reactive dyes	First bath cationic dyes at boil and second bath reactive dyes at 80 °C
Cellulosic/ Modacrylic	Solid	Single bath	Direct and disperse dyes	Basic dyes are dyed at boil and then direct dyes at 80 °C
		Two bath or two stage	Cationic and reactive, vat, or sulphur dyes	Cationic dye at boil and vat, sulphur, or reactive dyeing as usual
CC blends				
Cotton/linen/ ramie	Solid	Single bath	Direct, reactive , vat, sulphur, etc.	Dyed in one bath as normal cotton dyeing

Mercerised cotton/Viscose	Solid	Single bath	Direct dyes	Normal cotton dyeing at boil with low salt concentration
Unmercerised cotton/Viscose	Solid	Single bath	Direct dyes	Selected direct dyes (Disazo tetrasulphonated direct dyes)dyeing process at 60 °C
Cotton/Viscose	Solid	Single bath	All cotton dyes (Direct, reactive, vat, sulphur)	Normal cotton dyeing process at lower temp (e.g. Vat at 20–30 °C) lyocell, polynosic, etc. can be dyed as normal cotton
DA blends				
Polyester/Wool	Polyester reserve	Single bath one class of dye	1:1 Metal complex or acid milling dyes	Normal dyeing of wool with metal complex or acid milling dyes
	Solid shades	Single bath	Medium energy disperse dyes and 1:2 metal complex dyes	Disperse dye on polyester dyed at 105 °C and metal complex also exhaust in the same dyeing process
		Single bath	Selected disperse and 1:2 metal complex (disulphonated)	Both dyes exhausted in acidic bath at 120 °C with wool protective agent in the bath
		Two bath	Selected disperse and acid milling or metal dyes	Disperse dyeing at 120 °C with wool protective agent and in second bath normal acid dyeing with acid milling, 1:1 or 1:2 metal complex dyeing
Acetate/Wool	Acetate reserve	Single bath	Only wool dyeing	Single bath dyeing of wool with acid milling or metal complex dyes
	Solid or reserve dyeing	Single bath	Disperse/neutral dyeing acid dyes	Selected disperse dyes and neutral-dyeing acid dyes at pH 6

Substrate	Shade	Method	Dyes	Remarks
Triacetate/wool	Solid	Two bath	Selected disperse dyes and acid dyes	Disperse dyes at 80 °C, then 1:2 metal-complex or milling acid dyes
	Solid	Single bath	Medium energy disperse dyes and acid milling dyes	Intermediate-energy disperse dyes and wool milling acid dyes at the boil with ester carrier
Polyester/Nylon		Two bath	Mediumenergy disperse dyes and 1:2 metal complex dyes	Disperse dyes dyed on triacetate at 105 °C and second bath 1:2 metal complex at boil
	Polyester reserve	Single bath	Acid dyes for wool	Any acid dyes can be used and normal acid dyeing on wool
	Solid or shadow	Single bath	Selected disperse dyes only	Upto light shades with low energy disperse dyes at 110–120 °C
	Solid and contrast	Single bath two stage	Neutral dyeing acid and low energy disperse dyes	Darker shade with neutral dyeing acid dyes at 70 °C and the mediumenergy disperse dye at 120 °C
	Solid and contrast		Mediumenergy disperse dyes and acid dyes	Disperse dyes at(full depths) 120 °C, anionic dyes at 70 °C with syntanaftertreatment
Acetate/nylon	Acetate reserve	Only nylon dyeing	Metal complex or acid milling dyes	Metal complex or milling acid dyes at 80 °C
	Solid or shadow	Single bath	Disperse dyes	Low energy disperse dyes when dyed at 80 °C, both the component get dyed to get a shadow effect in mediumshades and solid in light shades
	Solid/contrast	Single bath two stage	Disperse and acid milling or metal complex dyes	First stage disperse dyeing at 100–105 °C and then acid milling or metal complex dyes for nylon at 80 °C

DB blends

Blend	Reserve	Method	Dye class	Remarks
Polyester/Acrylic	Acrylic reserve	Single bath	Disperse dyes	Disperse dyes with methyl naphthalene carrier
	Polyester reserve	Single bath	Cationic dyes	Normal dyeing acrylic in the presence of a retarder with cationic dyes
	Solid or contrast	Two bath	Cationic dyes and disperse dyes	Polyester to dye at 120 °C, and in second bath acrylic with cationic dyes at boil with a retarder
Acetate/acrylic	Solid	One bath one class dyeing	Disperse dyes	Disperse dye are dyed at 80 °C for light shades only
		Two bath	Cationic dyes and disperse dyes up to dark shades	First bath acrylic with cationic dyes at boil with a retarder briefly and second bath polyester to dye at 80 °C
Triacetate/Acrylic	Triacetate reserve	Only acrylic dyeing	Cationic dyes only	Cationic dyeing of acrylic along with a non-ionic dispersing agent
	Solid or cross dyeing	Single bath	Disperse and special cationic dyes	Disperse dyes and delocalised charge basic dyes at the boil with anionic retarder and carrier

DC blends

Blend	Reserve	Method	Dye class	Remarks
Polyester/cellulosic	Polyester reserve	Single bath only cotton dyeing	Direct, vat and reactive dyes	Normal cotton dyeing with vat, direct or reactive dyes and after clearing
	Cotton reserve	Single bath polyester dyeing	Disperse dyeing	Normal disperse dyeing by exhaust at HT (125–130 °C) at pH 4–5 with cotton reserve or after clearing or thermo fix
	Solid	Single bath, Single class	Selected vat or Indigosol or Pigment, pale shades only	Normal vat dyeing of benzamidoanthraquinone vat dyes or pigment dyeing and curing or suitable Indigosol dyeing and heatsetting

Solid	Single bath	Disperse and selected direct	First bath dyeing of disperse dyes and second bath disazomultisulphonated direct dyes at pH 6
Solid	Single bath	Disperse and reactant fixable reactive dyes	Disperse dyes and reactant-fixable dyes with copper-specific sequestrant
Solid	Single bath	Disperse dyes and nicotinotriazine reactive dyes	Disperse dyes and nicotinotriazine reactive dyes at 130 °C
Solid all depths	Single bath two stage	Disperse and low reactivity reactive dyes	In first stage dye disperse dye at 130 °C, cool neutralise and H/HE reactive dyes at 80–95 °C
Solid all depths	Single bath two stage	Cold and warm reactive dyes and disperse dyes	First stage cold or VS reactive dyeing and the next stage disperse dyeing at pH 5–6 and 130 °C
Solid all depths	Single bath two stage	Disperse dye and selected vat dyes	First stage disperse dyeing at 130 °C and then vat dyeing at 20–60 °C
Solid all depths	Two bath	Disperse and vat, reactive, sulphur or direct	No restriction first disperse dye on polyester at 130 °C then reduction clear and dye vat, sulphur, reactive or direct in fresh bath
Solid all depths	Continuous dyeing	Disperse dyes and high reactive dyes with bicarbonate and urea	Pad–dry–thermo fix method
Solid all depths	Semicontinuous	Disperse and reactive dyes (Pad–thermofix–Pad-Batch method)	Pad–thermofix– pad batch. Disperse dyeing by pad thermofix and cotton dyeing pad batch with all pad batch suitable reactive dyes

	Solid all depths	Semicontinuous	Disperse and vinyl sulphone reactive dyes (pad batch and HTHP dyeing)	Vinyl sulphone is dyed on cellulose by pad batch and disperse dyeing on batch wise methods (beam or jet dyeing)
	Solid all depths	Semicontinuous	Vat dyes and disperse dyes	Dyeing of vat dyes by pad jig and disperse by thermofix
	Solid all depths	Continuous	Vat and disperse dyes (pad thermofix and pad-steam)	Disperse dyes on polyester by pad thermofix and vat dyes by pad-steam method
	Solid dark shades	Continuous	Disperse and solubilised sulphur dyes	Any disperse dye by thermofix and solubilised sulphur dyes on cellulose by pad-steam
Acetate/Cellulosic	Acetate reserve	Single class dyeing	Multisulphonated direct dyes only	Salt controllable multisulphonated direct dyes applied on cellulose at 80 °C
	Cellulose reserve	Single class dyeing	Low energy disperse dyes only	Low energy disperse dyes on acetate at 80 °C with cellulose reserve or reduction clearing
	Solid	Single bath	Low energy disperse dyes and salt controllable direct dyes	Both low energy disperse dyes and salt controllable direct dyes are dyed from same bath at 80 °C and pH 6–7
	Solid or contrast	Two bath	Low energy disperse dyes and direct dyes	Polyester dyes at 80 °C with low energy disp. dyes and reduction clear and dye with salt controllable direct dyes at 80 °C
Triacetate/Cellulosic	Triacetate reserve	Single class dyeing	Selected multisulphonated direct dyes	Normal dyeing with multisulphonated direct dyeing and aftertreatment

	Cellulose reserve	Single class dyeing	Medium energy disperse dyes	Medium energy disperse dyes at 120 °C for triacetate
	Solid or contrast	Two-stage	Disperse dyes and direct dyes	First stage dye triacetate with med. energy disperse dyes at 120 °C and the normal direct dyeing of cellulosic at 90 °C
	Solid or contrast	Two bath	Vat dyes and disperse dyes	Cellulosic is dyed with selected vat dyes at 45 °C and then in fresh bath disperse dyes at 120 °C
DD blends				
Triacetate/ Polyester	Polyester reserve and shadow	One bath single stage	Low energy disperse dyes	Low-energy monoazo and nitro dyes at 90 °C with ester carrier
	Solid	One bath single stage	Medium and high energy disperse dyes	Only polyester dyeing with selected disperse dyes at 120 °C
	Solid	One bath single stage	Medium and high energy disperse dyes	Medium energy disperse dye on both fibres at boil in presence of a carrier
	Solid	One bath single stage	High energy disperse dyes for polyester and low energy disperse dyes for triacetate	High-energy dyes for polyester and low-energy dyes for triacetate with ester carrier
	Solid	Two bath	Selected disperse dyes	Polyester dyeing at 130 °C with disperse dyes and low energy disperse dyes are dyed on triacetate at boil
Triacetate/ Acetate	Shadow	Single bath single stage	Selected disperse dyes	Low energy disperse dyes are dyed on both the fibres at 60–80 °C which gives shadow effect due to different dye ability of fibres
Polyester/ Acetate	Polyester reserve	Single bath single stage	Selected disperse dyes	High-energy monoazo and acetate reserve low-energy anthraquinone dyes at 60–70 °C

References

1. Ahura Chemical Products India Ltd. (1977), Bombay·400 022 India, Samacryl dyes on acrylic (PAC) Fibres.

2. Asahi Chemical Industry Co. Ltd., Technical Leaflet, Dyeing and Finishing of Cashmilon, General.

3. BASF Manual, Dyeing and finishing of polyester fibres.

4. BASF Manual, Palanil Dyestuffs on Polyester fibres Lanestren Dyestuffs on Polyester/wool Blends, Cottestren Dyestuffs on Polyester/Cellulosic blends, MK 201/1 e, 6.64 WIN.

5. BASF Manual, Textile finishing.

6. BASF (July 1971), Technical Information Textile Industry Dyeing of Polyester/ Cellulosic Fibre Blends with Cottestren Dyes by the Thermosol Process.

7. BASF,Manual for dyeing and finishing of acrylic fibres, alone and in blends with other fibres basacryl dyes, Acryl Dyes on Acrylic Fibres.

8. BasuArindam (Ed.) (2015), *Advances in silk science and technology*, 80 High Street, Sawston, Cambridge, CB22 3HJ, UK, Published by Woodhead Publishing Limited in association with The Textile Institute Woodhead Publishing is an imprint of Elsevier.

9. Bayer Farben Revue, Special series, Resolin Dyestuffs for Polyester Fibres and Blends, Procedures for the Dyeing of Polyester Fibre/Cellulosic Fibre Blends.

10. Bayer Farben Revue, Special series, Resolin Dyestuffs for Polyester Fibres and Blends, The Thermosol Process for the Dyeing of Polyester Fibre Fabrics and Blended Fabrics.

11. Bayer Farben Revue, Special series, Resolin Dyestuffs for Polyester Fibres and Blends, Heat-setting of Polyester Fibre Fabrics and Blended Fabrics.

12. Dr. BerndtFriedrich, The one-bath, fast dyeing of acrylic/cellulosic fibre blends with basacryl, Acryl and Indanthren Dyes.

13. BroadbentArthur D (2001), *Basic principles of textile coloration*, Society of Dyers and Colourists, Perkin House, 82 Grattan Road, Bradford, West Yorkshire, BD1 2LU, United Kingdom

14. Calvin Woodings (Ed.) (2001), *Regenerated cellulose fibres*; Woodhead Publishing Limited in Association with The Textile Institute Woodhead Publishing Limited, 80 High Street, Sawston, Cambridge CB22 3HJ, UK.

15. Cashmilon Technical Information, Cashmilon in General, Asahi Kasei.

16. Cegarra J and Puente P (May 1967), 'Considerations on the kinetics of the dyeing process of polyester fibers with dispersed dyes', *Textile Res J*, 37 (5), 343–350.

17. CesareAndreoli and FabrizioFreti (March 2004), Reference Book of Textile Technologies – Manmade fibres, Fondazione ACIMIT.

18. ChakrabortyJ N, *Fundamentals and practices in colouration of textiles*,Woodhead Publishing India Pvt. Ltd., 303, Vardaan House, 7/28, Ansari Road,Daryaganj, New Delhi - 110002, 2010

19. IndiaChao Y Cand Chen S S (September 1994),'Dyes for polyester microfibres', *Dyes Pigments*, 24 (3), 205–222.

20. CHT Bezema (June 2014), Technical literature – Dyeing, PES-Dyeing – 100% PE Knitwear, 100% PE Yarn, Polyester/Cotton.

21. Ciba (2002), ERIOFAST® Dyes for polyamide – Highly fast, metal-free dyes with outstanding fastness, Ciba Speciality chemicals.

22. Ciba Specialty Chemicals Inc., Ciba® LANASET® PA acid/metal complex dyes for polyamide, Technical Information, November 2002/PA dyeing.

23. Ciba Technical Bulletin PES 202 – Volume II F, Survey of dyeing methods for polyester/cellulosic blends.

24. Ciba Technical Literature, *Neolandyes standard and fashion shades on Wool Piece*, Ciba Ltd.,Basle, Switzerland, Ciba Limited.

25. Ciba, Dorasyn XL dyes from M, Dohmen, Textile effects, GTC 4.51.

26. Clark M,*Handbook of textile and industrial dyeing – Volume 1: Principles, processes and types of dyes*,New Delhi, Woodhead Publishing Limited,2011

27. CookJ Gordon (2001), Handbook of textile fibres – II Manmade fibres, Wood head Publishing Limited,2001.

28. DattaRajat K and NanavathyMahesh (2005), *Global silk industry a complete source book*, USA, Universal Publishers.

29. Dave AbIard (October 2007), *The growth of regenerated cellulosic blends with elastomer – A dyer's blessing or a nightmare?*, International Dyer.

30. Dawson J F (July 1983),'*The structure and properties of disperse dyes in polyester coloration*', *J Soc Dyers Colour*, 99 (7–8), 183–191.

31. DeopuraB L, AlagirusamyR, JoshiM and GuptaB (Ed.) (2008), *Polyesters and polyamides*, Abington Hall, Granta Park, Great Abington Cambridge CB21 6AH, England, Published by Woodhead Publishing Limited in association with The Textile Institute Woodhead Publishing Limited.

32. Du Pont Technical Information (December 1976), Physical and chemical properties of lycra as related to end use requirements, "Bulletin L-86".

33. Du Pont Technical information, Wet processing of fabric containing lycraElastane, Basic Information, Bulletin L-517.

34. Durand &Huguenin SA (1936), *Basle, recipes and processes, Part I*,Ingosols in Textile Printing and Padding, Second edition.

35. Encyclopedia of polymer science and technology, John Wiley & Sons, Inc.

36. Dr. Ernst Schonpflug, Dyeing Polyester/Cellulosic Fibre Blend Yarns and Fabrics with Cottestren Dyes, Symposium on the Dyeing and Finishing of Polyester Fibres and their Blends.

37. Feiz M and Salimpour S, 'Dyeingof pile silk yarn with reactive and basic dyes: Effect of aftertreatments on fastness properties', *Colourage*, June 2008

38. Fleischer Herbert,The"Defitherm" Process and its Application in Dyeing Acrylic Fibre Blends.

39. Francaise Matieres des Colorantes S.A, FRANCOLOR, Technical Literature, Synthetic fibres dye selection and dyeing processes.

40. FranckRobert R (Ed.) (2001), *Silk, mohair, cashmere and other luxury fibres*, 80 High Street, Sawston, Cambridge CB22 3HJ, UK, Wood head Publishing Limited in association with The Textile Institute Wood head Publishing Limited.

41. Frtischjacques and AroraOraveen (2007),*Innovative techniques in manmade cellulosic wet processing*, Colour Trends Seminar Papers, The society of Dyers and Colourists.

42. Gries, Veit, Wulfhorst (2015), *Textile technology – An introduction*, Munich, Carl HanserVerlag.

43. Griffiths J (Ed.) (1984),*Developments in the chemistry and technology of organic dyes*, Oxford, UK, Blackwell Scientific Publications.

44. GulrajaniM L (Ed.) (2013), *Advances in the dyeing and finishing of technical textiles*, 80 High Street, Sawston, Cambridge CB22 3HJ, UK, Wood head Publishing Limited in association with The Textile Institute Wood head Publishing Limited.

45. HauserPeter J (Ed) (2011),*Textile dyeing*, JanezaTrdine 9, 51000 Rijeka, Croatia, InTech.

46. Dr. Heinz-Dietrich Jordan (1979E),*Phthalogen dyestuffs in the exhaust process possibilities and limitations*, Bayer Farben Revue, Nr. 30.

47. Helmut Sieber, Developments in Polyester/Cotton Blends dyeing, International Conference by SDC, "Color Trends 2006 Processing of Polyester and Blends", 10th June 2006.

48. Hirning H and Schluter H (1970), 'Translation fromdyeing of polyester/cellulosic fibre blend fabric by the thermosol process', *Textil-Praxis* 25 (11),682–689.

49. Hoechst Technical information, CELANESE@ACETATE – Dyeing Printing and Finishing Acetate Fabrics and Yarns.

50. HonguTatsuya, PhillipsGlyn O and TakigamiMachiko (2005), *New millennium fibers*, Abington Hall, Abington, Cambridge CB1 6AH, England, Woodhead Publishing Limited in association with The Textile Institute.

51. Horst Charwat,*Dyeing of wool/cotton fibre blends with reactive dyestuffs*, Bayer Farben Revue 32.

52. Hosam El-Din M Saleh (Ed) (2012), *Polyester*, JanezaTrdine 9, 51000 Rijeka, Croatia, Published by InTech.

53. Hunger K (Ed.) (2003), Industrial dyes – Chemistry, properties, applications, Weinheim, Wiley-VCH Verlag GmbH & Co. KGaA.

54. Huntsman (2006), Technical leaflets, Textile Effects Terasil, Cibacron Fast in process Fit to pocket.

55. ICI Technical Information (1954), Dyehouse, no. 176, Dyeing of Acrilan.

56. ICI Technical Information, Acid dyes on Nylon – The extended Nylomine Range, Dyehouse no. 844, New Fibres Dyeing.

57. ICI Technical Information, Continuous thermofix dyeing of polyester-cellulose blends, D967 Textile Blend Dyeing.

58. ICI Technical Information, Dyeing of polyester/cellulosic unions – Continuous methods, Dyehouse no. 733.

59. ICI Technical Information, Dyeing of Polyester/Wool Unions, D1217 Textile Blend Dyeing.

60. ICI Technical Information, Polyester/cellulose blends: One-bath batchwise dyeing with disperse, Procion Hand Procion H-E Dyes, D1220 Textile Blend Dyeing.

61. ICI Technical Information, Polyester/cellulosic blends: batchwise dyeing methods for yarn and piece goods, D1082, Textile Blend Dyeing.

62. ICI Technical Leaflet: Disperse dyestuffs for dyeing acrylic fibres.

63. Imperial Chemical Industries, 'Terylene' Technical Information Manual, Fibres Division, ICI.

64. IngamellsW,*Colour for textiles: A user's handbook*, Bradford, UK, SDC.1993

65. Iyer N D, 'Silk: The queen of textile fibres –X, Continuous processing of 'silk fabrics", *Colourage*.April 2005

66. Iyer N D, 'Silk: The queen of textile fibres – XIV, Dyeing of silk – Part 2',*Colourage*. August 2005

67. Iyer N D, 'Silk – The queen of textile fibres', *Colourage*, June 200451, 77–78.

68. Iyer N D, 'Silk: The queen of textile fibres – XI, Bleaching',*Colourage*.May 2005

69. Johnson A (1989),*The theory of coloration of textiles*, Bradford, UK, SDC.

70. Johnson N A G and Russell I M (2013), *Advances in wool technology*, 80 High Street, Sawston, Cambridge CB22 3HJ, UK, Woodhead Publishing Limited in association with The Textile Institute Woodhead Publishing Limited.

71. Karger-Kocsis J (Ed.) (1999), *Polypropylene – An A–Z reference*, P.O. Box 17, 3300 AA Dordrecht, The Netherlands, Published by Kluwer Academic Publishers.

72. Dr. Karl Thurner, The Computing of Recipes for Dyeing Polyester Fibres and Polyester/Cellulosic Fibre Blends, Symposium on the Dyeing and Finishing of Polyester Fibres and their Blends.

73. Koh J Sand Kim J P (April 1998),'Application of phthalimide based alkali-clearable azo disperse dyes onto polyester and polyester/cotton blends', *J Soc Dyers Colour*, 114 (4), 121–124.

74. KozłowskiRyszard M (Ed.) (2013), *Handbook of natural fibres*, 80 High Street, Sawston, Cambridge CB22 3HJ, UK, Woodhead Publishing Limited in association with The Textile Institute Woodhead Publishing Limited.

75. Larry Bright (August 1996), Troubleshooting continuous thermosol dyeing of polyester fiberand blends, American Dyestuff Reporter.

76. Lenzing Technical Literature, Dyeing and finishing of modal and its blends, Lenzing AG, Austria.

77. Lewin Mand Pearce E M (Ed.) (1985),*Handbook of fiber science and technology*, Volume IV, New York, USA, Marcel Dekker Inc.

78. LewinMenachem (Ed.) (2006), *Handbook of fiber chemistry: Third edition*, CRC Press LLC, 6000 Broken Sound Parkway, NW, Suite 300, Boca Raton, FL 33487, USA

79. MahapatraN N (2016), *Textile dyes*, 303, Vardaan House, 7/28, Ansari Road, Daryaganj, New Delhi – 110002, India, Woodhead Publishing India Pvt Ltd.

80. Mahapatra N N (October 2004), 'Degumming of silk in industries', *Colourage*, 51, 53–54.

81. Dr. MayerUdo and SuflowManfred, Recent advances in the "Defitherm" Process.

82. Dr. MayerUdo, The basic principles of the optimum combinability of cationic dyes on dyeing acrylic fibres.

83. McIntyreJ E (Ed.), *Synthetic fibres: Nylon, polyester, acrylic, polyolefin*, 80 High Street, Sawston, Cambridge CB22 3HJ, UK, Woodhead Publishing Limited in association with The Textile Institute.2005

84. MenezesEdward, 'Wet processing of polyester microfibres', *Colourage*.October, 2007

85. MurugeshBabu K (2013), *Silk – Processing, properties and applications*,Woodhead Publishing Limited in association with The Textile Institute Woodhead Publishing Limited, 80 High Street, Sawston, Cambridge CB22 3HJ, UK.

86. Near Cimica, Technical information, Viscose – Recommendations for Pre-treatment, Dyeing and Finishing.

87. Nearchimica, Technical literature, Problems with oligomer in dyeing polyester yarns and fabrics.

88. NeedlesHoward L (1986), *Textile fibres, dyes, finishing and processes – A concise guide*, Park Ridhge, New Jersey, USA, Noyes Publications.

89. Dr.Padhye RN &Ms.Sakha M, *Degumming silk at low temperature*, Sandoz, Colour Chronicle.January–March 1992

90. PrisacariuCristina (2011), *Polyurethane elastomers – From morphology to mechanical aspects*, Wien, Springer-Verlag.

91. RouetteHans-Karl (2000), *Encyclopedia of textile finishing*, Aachen, Springer.

92. Sadov F, Korchagin M, Matesky A (1973),*Chemical technology of fibrous Materials*, Moscow, MIR publishers.

93. Sandoz Manual, Silk and colour.

94. Scheirs John and LongThimothy E (Ed.) (2003), *Modern polyesters: Chemistry and technology of polyesters and copolyesters*, The Atrium, Southern Gate, Chichester, West Sussex PO19 8SQ, England, John Wiley & Sons Ltd.

95. SchomakersPeter, SeilerGuido, SutcliffeSteve, TindalRichard (2002), *A handbook for the fabric piece dyer*, Germany, Thies GmbH & Co. KG.

96. Shore John (1998), *Blends dyeing*, Bradford, Society of Dyers and Colourists.

97. Siegfried Weigold, *Dyeing of texturized polyester fibre yarns and knitgoods on package machines*, Winch Becks, HT Winch Becks, Beam and Jet Dyeing Machines, Symposium on the Dyeing and Finishing of Polyester Fibres and their Blends.

98. SimpsonW S andCrawshawG H (Ed.) (2002), *Wool: Science and technology*, 80 High Street, Sawston, Cambridge CB22 3HJ, UK, Woodhead Publishing Limited in association with The Textile Institute Woodhead Publishing Limited.

99. Sreenivasa Murthy H V (2016), *Introduction to textile fibres*,Woodhead Publishing India Pvt. Ltd. 303, Vardaan House, 7/28, Ansari Road, Daryaganj, New Delhi – 110002, India.

100. SuflowManfred and Dr. MayerUdo, Practical experiences in dyeing acrylic fibres in the form of loose stock, tops and yarns by the "Defitherm" Process.

101. Technical Information from Hoechst, Dyeing PA 11, The dyeing of polyamide piece goods with selected acid dyestuffs, Textile no. 92.

102. Technical Information from Hoechst, The thermosol dyeing process for polyester blends with cellulose, Dyeing,PE/C 33, Textile 95.

103. TENCEL@A100 – Dyeing& finishing dyeing observations.

104. ThomsonEliza B (1922),*Merchandise manual series – Silk*, New York,The Ronald Press Company.

105. Trotman E R (1984), *Dyeing and chemical technology of textile fibres*, New York, USA, John Wiley & Son.

106. TzanovTzanko, CostaSilgia,GuebitzGeorg M,Cavaco-PauloaArtur (2001), 'Dyeing in catalase-treated bleaching baths', *ColorTechnol*, 117, 1–5.

107. Ulrich Keuser& Paul Richter (1970), 'Pretreatment and dyeing of yarns and piecegoods made of texturized polyester fibres', *Textil-Praxis*, 748–754.

108. VigoTyrone L (1994), *Textile processing properties – Preparation, dyeing, finishing and performance*, The Netherlands, Elsevier Science B.V.

109. Vijayendra M K, HadimuniVV, VenugopalBR, SrinathK, KalavathiN,RadhaHB (1991), *Comparative dyeing studies of reactive dyes and conventional direct and acid dyes on silk*, Special supplement on BTRA, Colourage Seminar on Reactive Dyes, September 28.

110. Waring D RandHallas G (Ed.) (1990),*The chemistry and application of dyes*, New York, USA, Plenum Press.

111. Werner Marschner&Dr. Dietrich Hildebrand,*AT process-a new method for the continuous dyeing of polyester/cellulosic blended fabrics*, Bayer Farben Revue 32.

112. Wet processing of Spandex fiber and its blends, Rossera, March–April 2002.

113. White H J (1960),'Some theoretical considerations of the dyeing of cellulose acetate with disperse dyes', *Textile Res J*, 30 (5), 329–338.

114. Willi Pullen & Fritz lesszmskv,The dyeing and finishing of blended fabrics of polyamide/®DORLASTAN or other elastanes, Bayer Farben Revue 32.

115. WolfensbergerArnold, Theory of silk weaving, Graduate of the Textile Institute of Zurich, Switzerland.

116. Dr.Wolfhard Beckmann (1964E),*The mode of action of auxiliaries for dyeing polyacrylonitrile fibres with cationic dyestuffs*, Bayer Farben Revue, Nr. 9.

117. WurzAlbert, Introduction to the Symposium on the Dyeing of Acrylic Fibres by the "Defitherm" Process.BASF Technical Literature.

118. Xinru Liu (2010), *The silk road in world history*, NewYork, USA, Oxford University Press, Inc..

119. Zollinger H (1991), *Colorchemistry*, ISBN 3-527-828352-8, Weinheim, Germany, VCH.